Simulating Enzyme Reactivity
Computational Methods in Enzyme Catalysis

RSC Theoretical and Computational Chemistry Series

Titles in the Series:

1: Knowledge-based Expert Systems in Chemistry: Not Counting on Computers
2: Non-Covalent Interactions: Theory and Experiment
3: Single-Ion Solvation: Experimental and Theoretical Approaches to Elusive Thermodynamic Quantities
4: Computational Nanoscience
5: Computational Quantum Chemistry: Molecular Structure and Properties *in Silico*
6: Reaction Rate Constant Computations: Theories and Applications
7: Theory of Molecular Collisions
8: *In Silico* Medicinal Chemistry: Computational Methods to Support Drug Design
9: Simulating Enzyme Reactivity: Computational Methods in Enzyme Catalysis

How to obtain future titles on publication:
A standing order plan is available for this series. A standing order will bring delivery of each new volume immediately on publication.

For further information please contact:
Book Sales Department, Royal Society of Chemistry, Thomas Graham House, Science Park, Milton Road, Cambridge, CB4 0WF, UK
Telephone: +44 (0)1223 420066, Fax: +44 (0)1223 420247
Email: booksales@rsc.org
Visit our website at www.rsc.org/books

Simulating Enzyme Reactivity
Computational Methods in Enzyme Catalysis

Edited by

Iñaki Tuñón
Universitat de València, València, Spain
Email: ignacio.tunon@uv.es

Vicent Moliner
Universitat Jaume I, Castellón, Spain
Email: moliner@uji.es

ROYAL SOCIETY OF CHEMISTRY

RSC Theoretical and Computational Chemistry Series No. 9

Print ISBN: 978-1-78262-429-5
PDF eISBN: 978-1-78262-683-1
EPUB eISBN: 978-1-78262-967-2
ISSN: 2041-3181

A catalogue record for this book is available from the British Library

Published by The Royal Society of Chemistry,
Thomas Graham House, Science Park, Milton Road,
Cambridge CB4 0WF, UK

Registered Charity Number 207890

For further information see our web site at www.rsc.org

Printed in the United Kingdom by CPI Group (UK) Ltd, Croydon, CR0 4YY, UK

Preface

To act on a bad idea is better than not to act at all. Because the worth of the idea never becomes apparent until you do it. (Nick Cave, 20,000 Days on Earth)

The simulation of enzymatic processes is a well-established field within computational chemistry, as demonstrated by the 2013 Nobel Prize in Chemistry, awarded jointly to Martin Karplus, Michael Levitt and Arieh Warshel for the development of multiscale models for complex chemical systems. It is worth mentioning that this recognition by the Nobel Foundation of the capabilities of computational chemistry was preceded by previous awards to Walter Kohn and John Pople, who won the Nobel Prize in 1998 for their work on density functional theory and computational methods in quantum chemistry. The use of models and simulations to understand and predict chemical processes has been attracting increasing attention, especially in recent years due to the development of high-performance computers. Nowadays, computational chemistry has become an essential tool, complementary to other experimental techniques, to study enzyme reactivity due to their potential applications, in particular in complex systems where molecular details can be elusive to experimental approaches.

This book provides the basic knowledge for postgraduate students and researchers in chemistry, biochemistry and biophysics, who want a deeper understanding of how to get information on complex biological processes at the molecular level through computational simulations. The book is specifically centred on questions concerning the difficulties and recent advances in the computational modelling of enzyme reactivity. Enzymes are biological catalysts that speed up chemical reactions, making them compatible with life. Apart from this catalytic power, often these catalysts show important advantages with respect to non-natural catalysts such as their chemo-, regio- and stereoselectivity, or the ability to work under mild

RSC Theoretical and Computational Chemistry Series No. 9
Simulating Enzyme Reactivity: Computational Methods in Enzyme Catalysis
Edited by Iñaki Tuñón and Vicent Moliner

Published by the Royal Society of Chemistry, www.rsc.org

conditions of temperature and pressure, which makes them the best environmentally-friendly catalysts to speed up the rate of chemical reactions. Nevertheless, although there have been numerous studies that have provided a solid understanding about some of the key factors of these biocatalysts, knowledge about the origin of enzymatic efficiency to catalyse chemical reactions is still not complete.

This book explores the theories, methodologies and applications in simulations of enzyme reactions divided in three sections: Theory, Methods and Applications. This division has been made from a pedagogical point of view although in some cases the three sections are unavoidably connected and some chapters cover aspects related to these three main sections. Our aim has been to gather together some of the most significant researchers in this field to each contribute a closely related topic. The book begins with an overview and perspective of the field written by the Nobel Laureate Arieh Warshel. Arieh Warshel and Ram Prasad Bora provide a perspective on modelling enzymatic reactions and discuss the main findings of simulations about the origin of the enormous catalytic power of enzymes. In doing so, the authors describe the development of the computational strategies for the study of chemical reactions in complex environments, the so-called hybrid quantum mechanics/molecular mechanics (QM/MM) methods where the system is divided into two subsystems described at different levels, as well as the need for careful sampling.

The first block of the book is called *Theory* and tries to provide the basic theoretical tools for the analysis of enzymatic reactions and the interpretation of experimental information for its use in the design of adequate computational simulations. This section devoted to theories begins, as it should, with an experimentalist approach to the problem of the rate of enzymatic processes. Richard L. Schowen briefly introduces or recalls the main ideas of enzyme kinetics and how computational simulations must be designed in connection to experimental approaches. Correct answers need the correct questions too. In the subsequent chapter James T. Hynes and coworkers present the basic principles of transition state theory. The assumptions of this theory are discussed, as are the refinements to account for possible failures, in particular for the study of enzymatic reactions involving the transfer of light particles (proton, hydride and hydrogen atom transfers), where a quantum description of their motion is required. The theory and its deviations are illustrated with applications to a number of enzyme-catalysed reactions. Electron transfer reactions are an important class of enzymatic reactions that are specifically analysed in a separate chapter written by Aurélien de la Lande, Fabien Cailliez and Dennis R. Salahub. Marcus-like approaches to the calculation of the reaction rate for electron transfer reactions in enzymes are presented and some applications are discussed. Apart from the rate constant itself, its variation with the mass of the atoms involved in the reaction, the kinetic isotope effect, provides valuable information where experiments and simulations meet. The next chapter, written by I. W. Williams and P. B. Wilson, analyses methods for calculating kinetic isotope effects and their application to reactions catalysed by enzymes. The chapter emphasises one of the concepts that are omnipresent in the entire book – the importance of averaging in conformationally flexible systems, such as enzymes, in order to obtain magnitudes comparable to experimental determinations. The final chapter in this section, by J. Kästner and coworkers, is devoted to a key concept in the analysis of complex chemical processes: free energy. Free energy not only determines the spontaneity of a process, including obviously enzymatic reactions, but within transition state theory, the activation free energy is also related to the rate of the process. As mentioned above, averaging over important conformations of the system is fundamental to compute properties to be compared to experimental data and, in particular, to obtain free energies. Computational tools applied to reach this goal are reviewed in this chapter, establishing a natural link with the next section of the book.

The second block of the book is devoted to *Methods* or, in other words, to those computational strategies that can be used to calculate the magnitudes introduced in the first section, connecting experimental and theoretical approaches to enzymatic reactivity. Enzymes are proteins that may present different conformations during the catalytic cycle and the selection of the adequate one is a key step in the study of their reactivity. In their chapter, Pedro Sfriso and Modesto Orozco discuss simulations methods designed to study conformational transitions at a reasonable computational cost. The following three chapters of the book are devoted to different families of hybrid QM/MM methods. Keiji Morokuma and coworkers present the ONIOM (Our own N-layer Integrated molecular Orbital and molecular Mechanics) method and recent applications to enzymatic reactions. In the next chapter, Paolo Carloni, Ursula Rothlisberger and coworkers review the principles of QM/MM methods with particular emphasis to hybrid Car–Parrinello/molecular mechanics and continuum-based methodologies. Although all these methods consider a quantum treatment of some of the electrons of the system (while the rest are implicitly described), nuclear motion is generally assumed to be described accurately using classical mechanics. This approach breaks down when dealing with the transfer of light particles. To close this section of the book Dan T. Major and coworkers provide an overview of the different methods available for calculation of nuclear quantum effects in enzyme systems. Applications of these methods are discussed for the widely studied enzyme dihydrofolate reductase (DHFR), an enzymatic system that has attracted the attention of many experimental and theoretical researchers.

The last section of the book, centred on *Applications*, is closely related to those subjects presented in the two previous sections. In their contribution, Kara E. Ranaghan and Adrian J. Mulholland review different applications of QM/MM methods to the study of enzymatic reactivity, highlighting the role played by these methods in understanding fundamental issues related to enzymatic catalysis. The following chapter, by Joan Bertran and Antoni Oliva, is devoted to the study of reaction mechanisms catalysed by ribozymes, biological catalysts that are not formed by a protein structure but by RNA. The large charge associated with these molecules and the presence of counterions introduces additional peculiarities that are analysed in this chapter. Damien Laage and coworkers focus in their contribution on an important aspect of enzymatic catalysis: the role of the solvent and the possibility of exploiting changes in its composition to acquire new catalytic functions in enzymes. The authors show that simulations do not support the simplistic and popular view of the solvent acting as a lubricant of protein motions. The next chapter, by M. Alfonso Prieto and Carme Rovira, is devoted to a special class of enzymes whose study requires specific methodological treatments: metalloenzymes, enzymes that catalyse chemical transformations making use of a metal centre present in the active site. The authors show how density functional theory-based simulations are able to unravel the molecular details in several relevant metalloenzymes. The final chapter in this section tries to complete the circle: if we have been able to understand and rationalise enzymatic reactivity using the tools provided by theoretical and computational chemistry, it might be possible to use all this knowledge and procedures to design new biological catalysts with new functions. In their chapter Agustí Lledós, Jean-Didier Maréchal and coworkers discuss this new and exciting research area at the frontier between organic and inorganic chemistry, protein engineering and structural biology that challenges molecular intuitions and where molecular modelling is a fundamental tool. The chapter analyses pros and cons of current strategies, reviewing the work carried out to date.

Advances in simulating enzyme reactivity are key not only to a better understanding of these complex biological machines but for their application in, for instance, the development of new drugs or new environmentally-friendly catalysts. In this regard, the combination of experimental techniques and computer simulations pave the way to a quicker and deeper understanding of their mode of action. The more we learn about the foundations of processes governing chemical processes in living organisms, including our bodies and cells, the better position we will be in to control them, with corresponding benefits in fields such as biomedicine, biotechnology, pharmacy, *etc*. Thus, one could say that the developments and advances in computer simulations are closely linked to the improvements in quality of life on our planet. We hope that this book will provide a reference for those researchers that want to dive into this amazing field.

Vicent Moliner
Universitat Jaume I, Spain

Iñaki Tuñón
Universitat de València, Spain

Acknowledgements

The editors thank the authors for their contributions, and the editorial staff of the Royal Society of Chemistry for their continuous and excellent cooperation. All the experience gained in the years of research but also in teaching to under- and postgraduate students has been used to edit this book. In this regard, we want to thank our coworkers, our students and our teachers.

RSC Theoretical and Computational Chemistry Series No. 9
Simulating Enzyme Reactivity: Computational Methods in Enzyme Catalysis
Edited by Iñaki Tuñón and Vicent Moliner

Published by the Royal Society of Chemistry, www.rsc.org

Contents

RSC Theoretical and Computational Chemistry Series No. 9
Simulating Enzyme Reactivity: Computational Methods in Enzyme Catalysis
Edited by Iñaki Tuñón and Vicent Moliner

Published by the Royal Society of Chemistry, www.rsc.org

Section I: Theory

Section II: Methods

CHAPTER 1

Perspective on Computer Modelling of Enzymatic Reactions

ARIEH WARSHEL* AND RAM PRASAD BORA

Department of Chemistry, University of Southern California, SGM 420, 3620 McClintock Avenue, Los Angeles, CA 90089, USA
*Email: warshel@usc.edu

1.1 Introduction

Understanding how enzymes work has both fundamental and practical importance, as these remarkable molecules play a key role in controlling and performing most life processes.[1,2] In many respects the most crucial issue is understanding the origin of the enormous catalytic power of enzymes. Although some aspects of this puzzle were elucidated by biochemical and structural studies, the source of the catalytic power of enzymes is still controversial (*e.g.* see references in Warshel,[1] Fersht[2] and Warshel *et al.*[3]). The current discussion is sometimes reduced to statements such as, "the enzyme binds the transition state more strongly than the ground state", without providing any clear idea as to how such extra stabilisation can be provided.

The search for the origin of enzyme catalysis is frequently guided poorly by overlooking the crucial need to select a proper reference state. This issue has been carefully addressed in ref. 3, which defines a reference stat where we have the reaction in a water cage as in the enzyme active site. This unique 'chemistry-filtered' reference selection[3] allows one to focus on the key issues in enzyme catalysis in a well-defined way, asking what is the real

RSC Theoretical and Computational Chemistry Series No. 9
Simulating Enzyme Reactivity: Computational Methods in Enzyme Catalysis
Edited by Iñaki Tuñón and Vicent Moliner

Published by the Royal Society of Chemistry, www.rsc.org

difference in the environment (and its interaction with the reacting substrate) that makes such an enormous difference.

One of the points that we have tried to emphasise is our view that enzyme catalysis is too complex to be resolved experimentally without the use of computer-based models.[4] With this view in mind we can move to the pioneering works in this field and the introduction of the quantum mechanical/molecular mechanical (QM/MM) method[5] in 1976. We will also discuss some of the advances since the inception of the field, emphasising efficient sampling and outlining related problems.

Our perspective will also address different catalytic proposals, focusing on the validation of the electrostatic preorganisation idea and on exploring less consistent ideas. We will also comment on the problems with the relatively slow current advances in enzyme design.

1.2 Defining and Calculating the Catalytic Effect

1.2.1 Using a Logical and Useful Definition

In analysing the origin of the catalytic effect it is crucial to ask catalysis relative to what? Here, as hinted at in the previous section, we must select a reasonable reference state, and this can be done in several ways. For example, we may start with the scale introduced by Wolfenden and coworkers.[6] This scale has established the catalytic power associated with the binding free energy of the transition state (TS) in the enzyme relative to the energy of reaching the TS in the uncatalysed reaction in water, but still left significant misunderstandings about the challenge in rationalising enzyme catalysis. One problem is associated with the fact that the mechanism in the enzyme and in solution can be different (see above) and this difference is a part of the Wolfenden scale. Another serious problem arises from the fact that the real challenge in rationalising enzyme catalysis has not been emphasised by the Wolfenden scale, because it also reflects the binding free energy of the substrate (whose nature is well understood), whereas the real problem is associated with rationalising the large change in free energy upon going from the ES (enzyme–substrate) to the $ES^{\ddagger}$ (TS corresponding to the ES) states (*i.e.* the free energy associated with the k_{cat} and the corresponding $\Delta g^{\ddagger}_{cat}$ of Figure 1.1).

More specifically, it is useful to start with the Figure 1.1(A), where the activation barrier, $\Delta g^{\ddagger}_{enz}$, corresponds to k_{cat}/K_M (more precisely, k_{cat}/K_D), and $\Delta g^{\ddagger}_{cat}$ corresponds to k_{cat} (or more precisely, to the enzyme rate constant for the chemical step). The energetics of the reaction in the enzyme can now be compared to the corresponding energetics of the reaction in solution (Figure 1.1(B)). In this respect, it is useful to consider the free energy profile or the potential of mean force (PMF) for the reaction in water and in the protein, and then to divide the second-order process into the free energy of bringing the fragments to the same cage and the activation barrier of the first-order reactive event. The comparison of the enzyme and solution

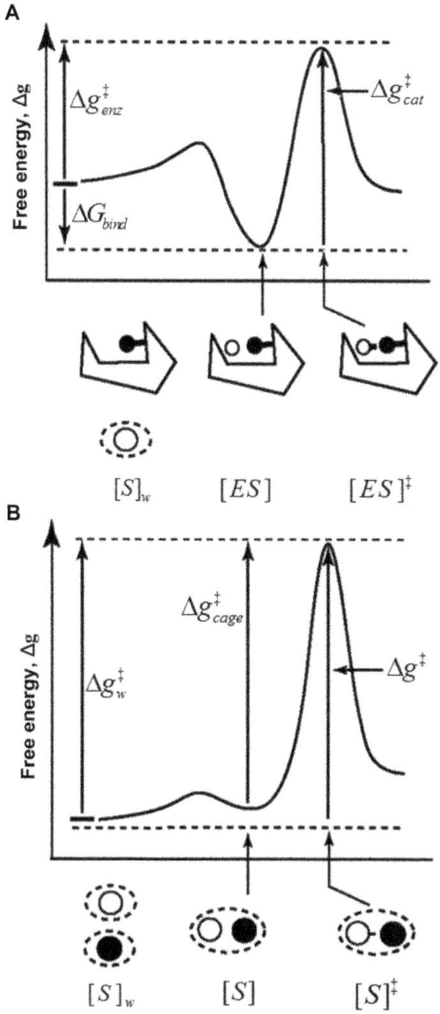

Figure 1.1 A schematic description of the free energy profile for an enzymatic reaction and for the corresponding solution reaction. Figure 1.1(A) describes the free energy $\Delta g^{\ddagger}_{p}$ and $\Delta g^{\ddagger}_{\text{cat}}$ associated respectively with $k_{\text{cat}}/K_{\text{M}}$ and k_{cat}. Figure 1.1(B) describes the energetics of a reference solution reaction.
Reprinted with permission from A. Warshel, P. K. Sharma, M. Kato, Y. Xiang, H. Liu and M. H. Olsson, *Chem. Rev.*, 2006, **106**, 3210–3235.[3] Copyright (2006) American Chemical Society.

reactions can be done by either comparing $\Delta g^{\ddagger}_{\text{cat}}$ to $\Delta g^{\ddagger}_{w}$, or by comparing $\Delta g^{\ddagger}_{\text{cat}}$ to $\Delta g^{\ddagger}_{\text{cage}}$ (see ref. 3).

As pointed out above, the comparison of $\Delta g^{\ddagger}_{\text{cat}}$ and $\Delta g^{\ddagger}_{w}$ should reflect the fact that many enzymatic reactions involve mechanisms different from the corresponding solution reactions (see, for example, ref. 3). Fortunately, this effect is well understood and can be determined by using a proper thermodynamic cycle and in recent years by reliable quantum mechanical calculations.[7,8]

1.2.2 Evaluating Reliable Activation Free Energies by Computational Approaches

Attempts to explore the origin of the catalytic power of enzymes should be based on quantitative methods for calculating the rate constant for reactions starting from the structure of the given enzymes. Obviously the key point in such calculations is the validity of the calculated activation free energies.[9] The general QM/MM strategy provides a generic way of obtaining potential surfaces and, in principle, activation free energies of chemical processes in enzymes. This approach[5] has gained popularity in recent years and has been used in a variety of forms (for reviews see Kamerlin *et al.*[10]). However, implementation of rigorous, *ab initio* QM/MM approaches in quantitative calculations of activation free energies is still extremely challenging. Nevertheless, significant progress has started to be made in recent works.[7,8,11–17] Furthermore, semi-empirical QM/MM studies with reasonable PMF calculations, and in some cases even with least energy paths, can be used to assess the validity of some catalytic proposals.[18–20]

Despite the future promise of well sampled *ab initio* QM/MM evaluations of activation barriers, we prefer to focus here on the empirical valence bond (EVB) method,[1,21,22] because even at this stage it provides what is probably the most effective available way for quantifying the catalytic effect and determining its origin. The EVB method is a QM/MM approach, which describes the system with two or more resonance states (or more precisely, diabatic states) corresponding to classical valence-bond structures (which are basically described as empirical force fields). These diabatic states are allowed to interact with the surroundings through their electrostatic charge distribution and then mixed in an effective Hamiltonian using the same mixing terms (off-diagonal terms) in solution, in the gas phase and in solution (an assumption that has gained major support from our constrained density functional theory calculations (CDFT)).[23]

The free energy surfaces are described as a function of the diabatic energy gap that is taken as the generalised free energy surface where the overall free energy barrier is obtained by a specialised free energy perturbation umbrella sampling approach (FEP/US).[24] This approach allows us to sample the EVB energy surface in an effective way by molecular dynamics (MD) simulations. Now, because trajectories on the reactant surface will reach the TS only rarely, it is usually necessary to run trajectories on a series of potential surfaces ('mapping' potentials) that drive the system adiabatically from the reactant to the product state.[24] In the simple case of two diabatic states the mapping potential (ε_m) can be written as a linear combination of the reactant and product potentials and the FEP/US provides a way to obtain the 'PMF' (or the free energy functional) along the energy gap coordinate.

The FEP/US approach can also be used to obtain the free energy functional of the individual diabatic states for the reactant and product states (the free energy functional) represent microscopic equivalents of the Marcus parabolas in electron-transfer theory.[25] The intersection of this free energy functional provides a quantitative estimate of the reorganisation energy, which will play a key role in our considerations.

The powerful physical picture of intersecting electronic states provided by the EVB treatment is particularly useful for exploring environmental effects on chemical reactions in condensed phases.[26] The ground-state charge distribution of the reacting species (solute) polarises the surroundings (solvent), and the charges of each resonance structure of the solute then interact with the polarised solvent.[1] For example, if ionic and covalent states are used to describe the solute, preferential stabilisation of the ionic state by the solvent will give the adiabatic ground state more ionic character. This allows us, for example, to obtain a very well-defined separation of covalent (charge transfer) and electrostatic effects and thus, to analyse in a clear way some covalent hypotheses. In addition, the EVB method lends itself to proper configurational sampling and converging free energy calculations, which makes it possible to evaluate non-equilibrium solvation effects.[9]

The EVB and other QM/MM methods allow us to simulate chemical reactions in enzyme active sites and solution and to reproduce the corresponding changes in activation barriers. The results of the corresponding calculations of the total EVB catalytic effects in some enzymes are given in Figure 1.2.

The results shown are taken from ref. 3, and recent EVB studies gave similarly good agreement (for example, Kamerlin and Warshel,[22] as well as very challenging cases that involve evaluation of activation entropy (*e.g.* Isaksen *et al.*[27])).

1.2.3 Electrostatic Transition State Stabilisation (TSS)

Our calculations have indicated that in all cases studied the catalytic effect is due to electrostatic effects (see Introduction and ref. 3). With current insight it might be argued that electrostatic effects must have been the most obvious candidates for explaining enzyme catalysis. However, careful studies in the early stages of the field have basically excluded this possibility. That is, early experiments with model compounds in solution (*e.g.* ref. 28 and 29) that explored the role of electrostatic effects (by introducing charged groups to stabilise the TS charge distribution) concluded that such effects must be small (*e.g.* see ref. 28 and 29). Similarly, phenomenological attempts to estimate the magnitude of electrostatic contributions to catalysis[30] also indicated that such effects are small. Thus it was assumed that more or less uniformly (at least in studies that attempted to quantify the catalytic effect) electrostatic effects do not play a very important role. The problem has, however, been that the physical organic chemistry experiments in solution might be rather irrelevant to an enzyme active site. Overall, phenomenological attempts to estimate the electrostatic effects in enzymes have

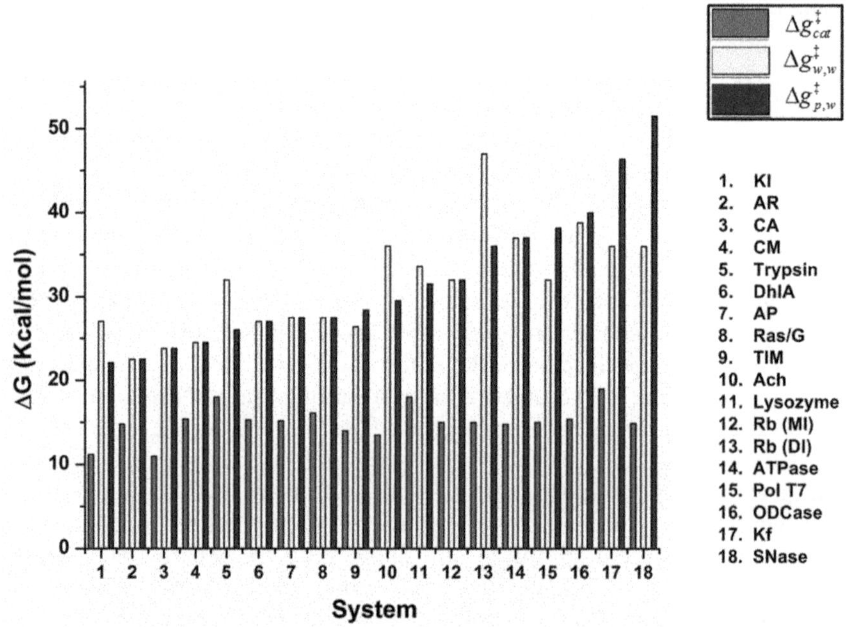

Figure 1.2 The activation free energies of representative enzymatic reactions ($\Delta g^{\ddagger}_{\text{cat}}$) and the corresponding reference solution reaction for the same mechanism as the enzymatic reaction ($\Delta g^{\ddagger}_{p,w}$) and the actual mechanism in water ($\Delta g^{\ddagger}_{w,w}$). The notation of the different enzymatic reactions is defined in ref. 3.
Reprinted with permission from A. Warshel, P. K. Sharma, M. Kato, Y. Xiang, H. Liu and M. H. Olsson, *Chem. Rev.*, 2006, **106**, 3210–3235.[3] Copyright (2006) American Chemical Society.

been very problematic because it is almost impossible to assess the dielectric effects in the protein without a proper computational model. In this respect, it is also important to clarify that the view expressed by the pioneering work of Jencks[31] did not consider electrostatic stabilisation of the TS as a major catalytic effect.

1.3 What was Found by Reliable Computational Studies?

1.3.1 General Findings

Before we consider the growing theoretical support for the role of electrostatic stabilisation in catalysis, it is important to comment about the insight that emerged from mutation experiments starting from around 1984 (*e.g.* ref. 2, 32–37). These mutation experiments have provided major insights, and in many cases pointed towards the importance of electrostatic effects. More recent works have added extensive support to this view (*e.g.* ref. 38, 39). However, since the catalytic effect reflects the overall effect of the enzyme active site, it has been very hard to reach unique experimental conclusions about the overall electrostatic effect. Furthermore, even in the seemingly unique case in which a mutation of an ionised group to a non-polar group leads to a large reduction in k_{cat}, it has been very hard to determine experimentally whether this is an electrostatic effect or some other factor (an excellent example is the D102N mutation of trypsin,[40] discussed in ref. 41).

With the above observation in mind, it seems to us that the use of QM/MM and related approaches provides what is perhaps the best way to convert the structures of enzyme active sites to catalytic contributions. In fact, since 1976, there has been a growing number of molecular orbital–QM/MM (MO–QM/MM) and EVB calculations that identify electrostatic effects as the key factor in enzyme catalysis. This trend has moved the field gradually from a stage of qualitative statements (*e.g.* see ref. 42) to more quantitative conclusions. Here it is useful to consider the studies summarised in ref. 3. This work only considered studies that were done until the works mentioned in ref. 3, where the actual catalytic effect was reproduced, rather than general QM/MM studies. Full quantitative evaluation of the electrostatic effects has been provided at present mainly by EVB studies, because this requires not only calculations of the activation free energy in enzyme and solution but also evaluation of the electrostatic contribution to the binding free energy of the reactant state (RS) and the TS. However, MO–QM/MM approaches have studied the electrostatic interaction energies (*e.g.* ref. 19, 43) and even reported systematic progress in evaluating change in electrostatic free energy along the reaction coordinate (*e.g.* ref. 44). Furthermore, many of the EVB studies evaluated the solvent reorganisation energy and demonstrated that this contribution accounts for a major part of the catalytic effect (see below). There were also some more recent attempts to estimate the reorganisation energies by MO–QM/MM approaches (*e.g.* ref. 45), but these only considered the change in the environment MM energy rather than systematic calculations using eqn (1.1) (see below) or related linear response approximation (LRA) treatments. Unfortunately, the change in the very large total molecular mechanics (MM) energy during the reaction is a rather unstable quantity, which is hard to evaluate in a quantitative way.

In many cases MO–QM/MM calculations can provide clear indications that the electrostatic effects play a major role in catalysis by simply evaluating the contributions of different residues to the activation barrier. Unfortunately, this type of 'mutational' analysis is frequently very qualitative because the simulations do not provide a sufficient dielectric screening. The underestimation of the screening effect is quite problematic, when one deals with ionised protein residues (see discussion in Schutz and Warshel[46]). Moreover, a proper analysis of the catalytic effect should explore the overall electrostatic contribution of the active site rather than just the contribution of some residues. In any case, we also list in ref. 3 studies that explore the electrostatic effects of different residues. In a few cases (*e.g.* ref. 47) we already have QM/MM studies that gradually excluded the electrostatic effect of the enzyme environment and thus established the importance of the overall electrostatic effect.

Some QM/MM and other related studies (see below) have not supported the idea of electrostatic TSS. However, at present (see also Section 1.4), all studies that did not support the electrostatic idea have involved significant inconsistencies. Key examples are: (a) Works that attributed the catalysis to desolvation or the ground-state electrostatic destabilisation. These works did not consider the actual binding of the TS and RS (*e.g.* ref. 48, 49) or could not reproduce the actual catalytic effect by the binding calculations (*e.g.* ref. 20, 49, 50). (b) Works that could not reproduce the catalytic effect without the use of entirely inconsistent entropic cycles and calculations that involved major overestimates based on gas-phase vibrational analysis (*e.g.* ref. 51, also see discussion in ref. 52). (c) Works that promoted the near attack conformations (NAC) proposal of Bruice and coworkers[53] and also were supported by other groups in related forms (*e.g.* ref. 54, 55) might have seemed reasonable. However, a deeper physically-based analysis has proved that the NAC reflects an incorrect analysis of what is actually an electrostatic TSS effect (this point requires one to follow the discussion below and in ref. 56). (d) Finally, it is important to comment here on the idea that enzyme catalysis is due to RS destabilisation and to the decrease of the enzyme self-energy upon moving to the TS. As will be shown below this idea is also based on inconsistent considerations.

To summarise this section, it seems to us that careful considerations of the works mentioned in ref. 3, as well as consistent attempts to identify the origin of large catalytic effects, point towards the conclusion that electrostatic effects are the key factors in enzyme catalysis (this issue will be emphasised and quantified further in the following section).

1.3.2 Quantifying the Source of Electrostatic Contributions to Catalysis

The studies reported above provide a general support to the electrostatic proposal. A more quantitative analysis is provided in Table 3 of ref. 3. As seen from the table we have clear examples of specific cases where most of the catalytic effect is due to electrostatic interactions. What remains to be established is that these effects are associated with TSS and to examine the reasons as to why the protein is able to provide such large effects. These issues can be explored by using the LRA expression (Lee *et al.*[57]):

$$\begin{aligned}\Delta G(Q^{\ddagger}) &= 0.5(\langle U(Q=Q^{\ddagger}) - U(Q=0)\rangle_{Q=Q^{\ddagger}}) + \langle U(Q=Q^{\ddagger}) - U(Q=0)\rangle_{Q=0}) \\ &= 0.5\langle\Delta U\rangle_{Q^{\ddagger}} + \langle\Delta U\rangle_{0}) \end{aligned} \tag{1.1}$$

$$\lambda = 0.5\langle\Delta U\rangle_{Q^{\ddagger}} - \langle\Delta U\rangle_{0}) - \Delta G_{0} \tag{1.2}$$

where U is the solute–solvent interaction potential, Q designates the residual charges of the solute atoms where $Q^{\ddagger}$ indicates the TS charges and $\langle\Delta U\rangle_{Q}$

designates an average over configurations obtained from a MD run with the given solute charge distribution. The quantity λ in eqn (1.2) is the reorganisation energy. The first term in eqn (1.1) is the above-mentioned interaction energy at the TS, where $Q = Q^{\ddagger}$, which is similar in the enzyme and in solution. The second term expresses the effect of the environment preorganisation. If the environment is randomly oriented towards the TS in the absence of charge (as is the case in water), then the second term is zero and we obtain:

$$\Delta G(Q^{\ddagger})_w = \frac{1}{2} \langle \Delta U \rangle_{Q^{\ddagger}} \tag{1.3}$$

where the electrostatic free energy is half of the average electrostatic potential.[58] However, in the preorganised environment of an enzyme, we obtain a significant contribution from the second term and the overall ΔQ $(Q^{\ddagger})$ is more negative than in water. This extra stabilisation leads to the catalytic effect of the enzyme. Another way to see this effect is to realise that in water, where the solvent dipoles are randomly oriented around the uncharged form of the TS, the activation free energy includes the free energy needed to reorganise the solvent dipoles towards the charged TS. On the other hand, the reaction in the protein costs less reorganisation energy because the active site dipoles (associated with polar groups, charged groups and water molecules) are already partially preorganised towards the TS charge.[59] The reorganisation energy is related to the well-known Marcus' reorganisation energy, but it is not equal to it. More specifically, the Marcus' reorganisation energy[60] is related to the transfer from the reactant to the product state, while here we deal with charging the TS. In other words, the suggestion (*e.g.* ref. 61, 62) that the reduction in the protein reorganisation energy will result in catalysis according to the Marcus relationship is problematic, because it implies that the reduction in the reorganisation energy is due to the existence of a non-polar active site. Unfortunately, protein active sites are polar (instead of being non-polar) and having a non-polar active site would drastically destabilise rather than stabilise ionic TSs (see discussion of desolvation models in Warshel *et al.*[63] and references given in this paper). In fact, the source of enzyme catalysis is the preorganisation of a very polar environment (see ref. 9 for a detailed discussion and a demonstration in ref. 3).

Regardless of the above clarification, it is almost always true that the catalytic effect is associated with the reduction of the Marcus' reorganisation energy so that $\lambda_p \leq \lambda_w$. This point and the related role of preorganisation of the electrostatic environment is demonstrated schematically in Figure 1.3 and ref. 3, and quantified for the case of DhlA[64,65] in Figure 10 of ref. 3.

An LRA analysis based on eqn (1.2) is given in Table 3 of ref. 3 for haloalkane dehalogenase (DhlA) and for chorismate mutase (CM) in Table 4 of ref. 3. As seen from these cases and other related studies, the catalytic effect appears to be associated mainly with the electrostatic stabilisation of

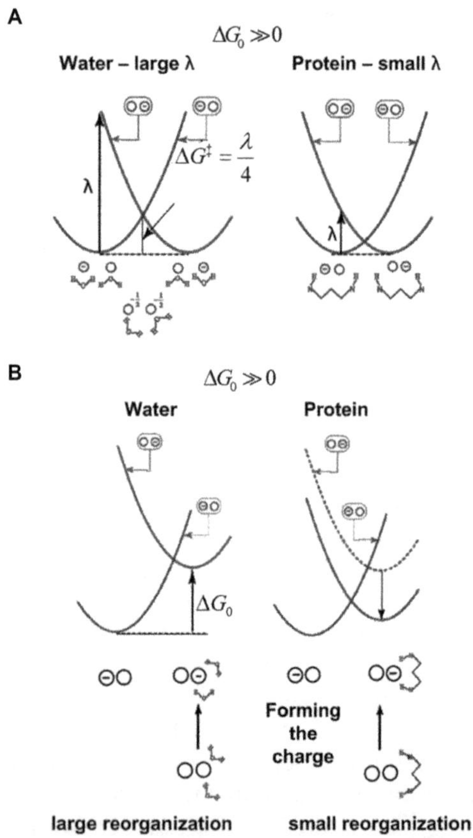

Figure 1.3 Illustrating the nature of the preorganization effect in two limiting cases. (A) In the limit when $\Delta G_0 = 0$, the enzyme has smaller λ because its dipoles are already partially preorganized towards the TS charge distribution. This corresponds to a reduction of the Marcus 'reorganisation energy'. (B) In the case when $\Delta G_0 \gg 0$, the enzyme dipoles are preorganised towards the product charge distribution. In this way, the preorganisation helps to increase the solvation of the product state. Reprinted with permission from A. Warshel, P. K. Sharma, M. Kato, Y. Xiang, H. Liu and M. H. Olsson, *Chem. Rev.*, 2006, **106**, 3210–3235.[3] Copyright (2006) American Chemical Society.

its TS, and a large part of the effect is associated with the preorganisation contribution. Interestingly, even in the case of peptide bond formation by the ribosome (which constitutes a very early stage in the evolution of biocatalysts) it has been found that the preorganisation effect provides the major catalytic effect.[66,67]

The correlation between the reorganisation energy and the catalytic effect has been explored (ref. 68) in a study of the effect of mutations in dihydrofolate reductase (DHFR). This is a very interesting benchmark, because the effects of mutations were used as evidence of the catalytic effect of correlated motions, but our calculations[68] established that the actual

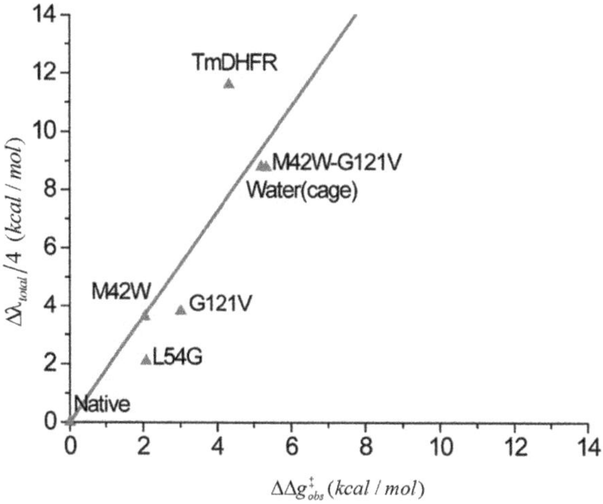

Figure 1.4 Correlation between the calculated reorganisation energy and the observed mutational effects in DHFR.
Reprinted with permission from H. Liu and A. Warshel, *Biochemistry*, 2007, **46**, 6011–6025.[68] Copyright (2007) American Chemical Society.

trend reflects a good correlation between the reorganisation energy and the catalytic effect (see Figure 1.4).

Another important issue that may be mentioned at this point is the Jencks' idea that enzymes use their binding energies to destabilise the substrate and to bring about the positioning of the reacting groups. Actually, a large part of the preorganisation effect is due to the inherent folding energy and not due to the interaction with the substrate. Furthermore, the preorganisation effect results in TSS rather than ground-state destabilisation.

As stated above the catalytic power of enzymes is largely due to the preorganised electrostatic environment of their active sites. Our considerations of the overall energetics of this effect led to the idea that the preorganisation is associated with reduction in the protein folding energy.[1,69] This stability/activity idea was also supported by experimental works[70,71] and electrostatic modelling.[72]

Recent works have tried to argue that enzyme structure can tell us about changes in the protein preorganisation upon mutation[73] and that with such a concept one can show that the changes in the rates in mutants of DHFR are due to reduced dynamics and not to increases in reorganisation energy. Fortunately, due to our ability to actually calculate reorganisation energy and the activation barrier it was possible to show[74] that the rate constant change reflects the change in reorganisation energy. The same conclusions were obtained subsequently by other research groups.[75–77]

In exploring the robustness of our conclusions it is useful to try to evaluate the observed mutational effect. In this respect we consider our evaluation of mutational effects in CM (Figure 6 of ref. 78) or the results of directed evolution in the design of Kemp elimenase,[79] as well as the results of distanced mutations in DHFR,[80] as powerful elements in the validation of our computational methods. All the above analysis requires a very significant benchmark that appears to be provided by the EVB.

1.4 What are the Problems with Other Proposals?

Although we have introduced compelling evidence for the overwhelming importance of electrostatic contributions, it is important to consider other proposals. This issue has been discussed extensively elsewhere (*e.g.* ref. 1, 11, 52, 81, 82), but it seems appropriate to summarise the results of computer modelling of the main alternative proposals.

1.4.1 Ground-state Destabilisation by Steric Strain Does Not Provide a Large Catalytic Effect

The idea that enzyme catalysis is associated with ground-state destabilisation was put forward in the classical studies of lysozyme.[83] Later studies that examined the actual amount of energy associated with steric strain found it to be small, due to the inherent flexibility of proteins.[1,84,85] Nevertheless, the strain proposal has been invoked in several recent studies, which were shown to be inconsistent with energy considerations (see ref. 3); this will be discussed below.[86,87]

One instructive example provided by spectroscopic studies was interpreted as a ground-state destabilisation due to electrostatic effects (electrostatic strain-induced mechanism). In particular, this idea was further elaborated by Anderson.[88] Unfortunately, the logic of ref. 88 was shown to be very problematic, as shown in ref. 3. At any rate, the main effect of the field from the reorganised active site is to stabilise the TS and not to destabilise the RS, and this fact has been established in many detailed computational studies that actually examined this issue (*e.g.* see Table 5 in ref. 3).

One of the systems that challenges the idea that strain is not important to catalysis is the action of B12-containing enzyme catalysis; thus it may be useful to consider the Co–C bond cleavage in coenzyme B12 enzymes. This system involves a radical bond breaking process and yet displays a very large catalytic effect of about 12 orders of magnitude.[89,90] This catalytic effect has been attributed to reactant state destabilisation (RSD) and, in particular, to the distortion of the corrin ring or other strain effects.[49,89–94] In particular, it was suggested that the strain is operated by the so-called mechanochemical trigger mechanism associated with the upward folding of the corrin ring (*e.g.* ref. 95–97). However, recent theoretical studies show that such compression cannot destabilise the Co–C bond (*e.g.* ref. 98, 99). A recent QM/MM study[93] provides an impressive analysis of the system and reproduces the catalytic effects. The decomposition of the catalytic effect resulted in about 8 kcal mol^{-1} electrostatic effect (between the protein and the leaving group) and about 15 kcal mol^{-1} strain in the leaving group. However, decomposition to energy contributions in QM/MM calculations, which do not involve free energy calculations and sufficient sampling and relaxation (*e.g.* see ref. 100), is extremely challenging and can lead to unstable results. A more recent study[101] that used the EVB and a very extensive free energy umbrella sampling calculation found, in agreement with ref. 93, that the catalysis is due to the interaction with the leaving group, but concluded that this effect is almost entirely an electrostatic effect (the catalysis disappears with a hypothetical, fully non-polar leaving group). The study of ref. 101 also used the LRA approach and established that the enzyme does not use RSD and stabilise the substrate more strongly than water does. The enzyme stabilisation of the leaving group increases, however, when the Co–C bond is stretched towards the TS. Interestingly in a more recent work,[102] we succeeded in accounting for a very large electrostatic stabilisation due to entropic effects.

1.4.2 Dynamical Effects Do Not Contribute Significantly to Enzyme Catalysis

The proposal that special 'dynamical' effects play a major role in enzyme catalysis (*e.g.* ref. 103, 104) has become popular in recent years (*e.g.* ref. 105–113). Unfortunately, the only way to explore this proposal is to start with a clear definition of the dynamical effects and then to examine carefully whether the corresponding contributions are different in enzymes and in solution. Although this issue has been analysed in great detail in several recent reviews,[9,52,114] we will consider here some key points as well as some recent works that supported the dynamical proposal.

There are several ways to define dynamical effects, and these will be considered below. However, in order to establish a dynamical contribution to catalysis by a given definition, we must obtain (with the specific definition) different magnitudes of dynamical contributions to the rate constant in the enzyme and in water. Now, in considering different definitions, we may start with the transmission factor, because it is agreed in the chemical physics community (see references in Villa and Warshel[9]) that all the dynamical effects are contained in this factor, which corrects the absolute rate theory for recrossing of the reactive trajectories (see ref. 52 for a clear definition). To the best of our knowledge, all the reported simulation studies going back to the earliest analysis[115] and to subsequent studies (*e.g.* ref. 69) found that the transmission factors are similar in the enzyme and in solution and do not differ much more than unity in the enzyme (*e.g.* ref. 9, 110). Typical values of the transmission factors are 0.8 and 0.6 in enzyme and solution, respectively.[116] These values are too similar to each other to be considered a source for any catalytic effect.

One of the best definitions of dynamical effects is to distinguish between inertial models (where coherent models funnel the binding motion to motion over the TS) and motions where the TS theory (with a minor tunnelling correction) is fully valid. The problems with the inertial model were established by the first realistic simulation[117] of the long-term behaviour of a reacting enzyme. This simulation has shown that once the barrier for the chemical step is sufficiently high, the system loses all its memory about the movement to the bottom of the reaction barrier. Unfortunately, after the clear demonstration that properly defined dynamical effects do not contribute to catalysis, it was argued by the supporters of the dynamical idea that no dynamical proposal was ever put forward by the experimental community and that the theoreticians are those who reject the 'active dynamics' proposal. Thus we wish to clarify that in clear contrast to the implications of ref. 118 and 119 the issues are not semantic at all. That is, a recent review[119] argued that the experimental community talks about stochastic dynamical models (which it called passive dynamics), because this is what Marcus-like models imply. Unfortunately, as we illustrated with explicit examples in our previous review,[120] it is very clear that the majority of these works explicitly refer to a dynamical contribution to catalysis, and stochastic approaches are not dynamical and thus are not part of a correctly formulated dynamical proposal. This problem extends to the extremely puzzling implications that enzymologists knew this and meant that dynamics is related to statistical models. It is obviously hard to prove what someone knew without providing references, but to attribute knowledge of effect, which is not discussed in the literature, is not an accepted way of establishing knowledge and understanding. Additionally, as far as Marcus-like models are concerned (which we introduced consistently to solution and enzymes in the correct adiabatic limit in 1990[121] and 1991[122]), attempts to use such models in analysing tunnelling, in for instance all the arguments and discussions in early and ongoing works,[123–126] seem to clearly be an attempt to promote the problematic real dynamical (gating) proposal, as can be established by the reader, and not an attempt to support the established relationship between the activation free energy and the electrostatic preorganisation.[3,127] Finally, we would also like to clarify that the nature of the catalytic landscape presents a thermodynamic (including entropic) factor rather than a dynamical effect.

Fortunately, the key issues in elucidating the origin of enzyme catalysis and catalytic contributions have already been proposed, and thus what was meant by the proposal can be frequently established, without being sidetracked by the names that either have been or are now being used to describe such proposals. Of course, to eliminate confusion, it is crucial to have clear logical and scientific definitions of what is meant by a dynamical proposal. Here, we will try to examine the validity of the dynamical proposal in cases where it is fully clear that what is meant is a dynamical contribution to the chemical step of catalysis. These proposals will be examined and analysed within the framework of clear, physically-based definitions of dynamical contributions to catalysis, in order to demonstrate that within rigorous frameworks, there is no need to invoke dynamical contributions to rationalise the observed effects, and of course, for example, conformational sampling gives entropic contributions, but these again have nothing to do with dynamics but rather with the available configurational space.

Thus, having presented our arguments, we will leave it to the readers to decide whether, within rigorously defined frameworks, it is still possible to argue that dynamical contributions play a major role in enzyme catalysis. In addition we would like to clarify that in contrast to the implication of ref. 118 and 119 it is not true that the theoretical community objected to the active dynamics proposal. In fact, it was only Warshel who has evaluated the dynamical effects in the early 80s[128] and showed that they do not contribute to catalysis, and only recent changes in the common trend led to new identification of the problems associated with the dynamical proposal (see ref. 129–131).

Another definition can imply that dynamical effects are related to the availability of special coherent motions. In this way, the dynamical proposal implies that enzymes 'activate' a special type of coherent motions, which are not available in the solution reaction. However, if the difference between the reaction in enzyme and in solution can be accounted for by evaluating the corresponding $\Delta g^{\ddagger}$ using non-dynamical Monte Carlo (MC) methods we do not have dynamical contributions. In other words, if the results from MC and MD are identical, then we do not have dynamical contributions to catalysis. Now, careful and systematic studies (*e.g.* ref. 9, 132) have shown that the reactions in both enzymes and solutions involved large electrostatic fluctuations. However, these fluctuations follow the Boltzmann distribution and, thus, do not provide dynamical contributions to catalysis.

It has been suggested (*e.g.* ref. 110) that dynamical effects are associated with the so-called non-equilibrium solvation effects. However, the corresponding analysis has been shown to be very problematic (see ref. 9, 52). Furthermore, it has been clearly demonstrated that the difference between the non-equilibrium solvation effects in the enzyme and that in solution is an integral part of the difference between the corresponding activation free energies.

Apparently, there is no single experimental finding that can be used to consistently support the dynamical hypothesis. Most of the experiments that were used to support this proposal have not compared the catalysed and uncatalysed reaction, and thus have not addressed the issue of catalysis (see discussion in ref. 9). Instructive NMR experiments (*e.g.* ref. 111) demonstrated the involvement of different motions in enzymatic reactions (see also below). The obvious existence of motions that have components along the reaction coordinate does not constitute a dynamical effect unless these motions are shown to be coherent. Probably all the motional effects identified so far are related to entropic factors (*i.e.* to change in the available configurational space) rather than to real dynamical effects.

Despite our previous reviews of the dynamical proposal (*e.g.* ref. 114) we find it useful to consider the most recent work that implied or explicitly supported this idea. We start by recognising that the advance in NMR studies (*e.g.* ref. 111, 133) allows one to probe the interesting nature of the relatively slow protein motions. This, however, does not prove that proteins can "harness thermal motions through specific dynamic networks to enable molecular function" as suggested by ref. 133.

An instructive example of what we see as an over-interpretation of exciting experimental findings is a follow-up[134] to the study of ref. 111. That is, study of the action of cyclophilin[111] found that the protein motions are correlated with the substrate turnover, while the more recent study of ref. 134 found that the same motions still exist in the absence of the substrate. This led to the interesting proposal that both protein structure and dynamics have co-evolved synergistically and that dynamical pre-sampling is "harvested for catalytic turnover". Unfortunately, the authors have not addressed the facts that catalysis must be defined relative to a reference reaction in solution, and that the catalytic effect of virtually every enzyme that has been studied consistently has been found to be associated with electrostatic rather than dynamical effects (this is true also in the present case, *e.g.* ref. 135).

A theoretical work[136] that was considered as a support of the finding of ref. 134, has attempted to evaluate the dynamical contribution from the protein's vibrations to the transmission factor of the erection of cyclophilin, and concluded that the dynamical contribution is significant. This study propagated trajectories from the TS, placing different amount of kinetic energy in the protein normal modes. Unfortunately this work had several problems. First, adding arbitrarily non-Boltzmann energy to specific modes at the TS, or any other state, has no relationship to correct rate theories. One has to prove that these vibrations are populated in a non-Boltzmann way and then to use the correct density matrix or an alternative treatment to examine whether there is any validity to such an assumption. In other words, adding arbitrary kinetic energy in the direction of the product will certainly change the recrossing in any model and, thus, cannot serve as a way of examining the contributions of the protein mode; this challenging problem can perhaps be addressed by starting an assumed coherent mode from the ground state and examining whether it retained coherence in the long time that it takes to reach the TS. Second, the same approach, whether justified or not, should have been performed on the reference solution reaction. Such a study would almost certainly reproduce similar effects in solution and thus correspond to little or no catalytic effect.

Another recent theoretical attempt to support the dynamical proposal[137] used transition path sampling to explore the catalytic reaction of lactate dehydrogenase (LDH). It was concluded that some trajectories in the TS region move in a concerted way and some in a stepwise path, and this was used to imply that the enzyme dynamics help to catalyse the reaction. However, this study also involved several problems. First, no attempt was made to evaluate the activation free energy and no comparison was made to the uncatalysed reaction, in contrast to earlier studies that actually elucidated the role of the reduction in reorganisation energy in the same enzyme.[138] Second, the fact that the reaction path may involve both concerted and stepwise paths has little to do with dynamical effects. It simply reflects the shape of the calculated reaction landscape.

In summary of the discussion, it is useful to recognise that consistent simulation studies found no evidence for dynamical contributions to catalysis. Another related issue is associated with the suggestion that vibrationally enhanced tunnelling (VET) plays a major role in enzyme catalysis (see, for example, ref. 112, 139). Some workers (*e.g.* ref. 113) assumed that there exists here an entirely new phenomenon that makes transition state theory (TST) inapplicable to enzymatic reactions. However, the VET effect is not new and is common to many chemical reactions in solution.[140–142] Moreover, the VET is strongly related to TST. That is, when the solvent fluctuates and changes the energy gap (see ref. 132, 140), the light atom sees a fluctuating barrier that allows in some cases for a greater rate of tunnelling. As shown in ref. 132, these fluctuations are taken into account in the statistical factor of the classical TST and the same is true when quantum effects are taken into account. Thus, the recent finding that the solvent coordinates should be considered in tunnelling studies is not new and does not mean that this effect is important in catalysis.

Hwang *et al.*[143] were the first to calculate the contribution of tunnelling and other nuclear quantum effects to enzyme catalysis. Since then and in particular in the past few years, there has been a significant increase in simulations of quantum mechanical–nuclear effects in enzyme reactions. The approaches used range from the quantised classical path (QCP) (*e.g.* ref. 9, 144, 145), the centroid path integral approach,[146,147] vibrational TST,[148] and the molecular dynamic with quantum transition (MDQT) surface hopping method.[149] Most studies have not yet examined the reference water reaction and thus, could only evaluate the quantum mechanical contribution to the enzyme rate constant, rather than the corresponding catalytic effect. However, studies that explored the actual catalytic contributions (*e.g.* ref. 9, 114, 144, 145) concluded that the quantum mechanical contributions are similar for the reaction in the enzyme and in solution, and thus, do not contribute to catalysis (see also ref. 129, 130).

1.4.3 Correlated Modes Clearly Exist in Proteins, but They Also Exist in Solution

Early studies[108,150,151] have explored the reaction of dihydrofolate reductase by NMR. They found that site-directed mutations of the residues in a loop that undergoes relatively large backbone motions had detrimental effects on catalysis, and they suggested that the dynamics of these residues could be important for catalysis. This suggestion was supported by MD simulations[107,152] that did not examine, however, any of the TSs in the reaction or demonstrate any dynamical effects on the rate constant.

More recent studies (*e.g.* ref. 153–155) have led to growing recognition that the mutational effects in DHFR reflect equilibrium structural effects rather than dynamical effects. However, the focus shifted to discussion of correlated motions (*e.g.* ref. 156, 157) rather than on reorganisation effects identified in ref. 80 and shown in Figure 1.4. This seems to create an impression that here we have a special catalytic effect with new implications beyond the concept of electrostatic TSS. However, the identification of correlated motions does not provide a new view of enzyme catalysis, because the reorganisation of the solvent along the reaction path in solution also involves highly correlated motions.[132,143] Correlated motions of an enzyme do not necessarily contribute to catalysis, and indeed could be detrimental if they increase the reorganisation energy of the reaction. Our EVB and dispersed-polaron approaches described elsewhere (*e.g.* ref. 52) considered the enzyme reorganisation explicitly and automatically assess the complete structural changes along the reaction coordinates. A dispersed-polaron analysis of the type represented in ref. 52, for example, determines the projection of the protein motion on the reaction coordinate and provides a basis for a quantitative comparison with a reference reaction in solution. In other words, our early studies indicated quite clearly that the motions along the reaction coordinate involve many modes in both the enzyme and solution reactions, but could not find any evidence that the existence of coupled modes contribute to catalysis.

One may still wonder about the connection between correlated motions and the effect of mutations on enzyme catalysis. However, the effect of distant mutations in DHFR (see Figure 1.4) is likely to be due to propagation of structural changes to the active site region, as is the case in many allosteric systems (*e.g.* ref. 158, 159). The new active site configuration is then unable to provide the same preorganised environment as the native enzyme. In other words, the mutation can change the curvature of the reaction coordinate and this change can be described as the effect of coupled modes (although such a description is neither predictive nor particularly useful). However, the issue is not the decomposition of the reaction path to the different protein modes but the height of the activation barrier. This barrier is determined by the reorganisation energy, which depends on the sum of the displacement of the different modes upon motion from the reactant to product state. Apparently, the mutations lead to an increase in the distance between the product and reactant states and in fact to larger displacements of the modes that are projected on the reaction coordinate. This means that the coupled modes reduce rather than increase the catalytic effect.

Perhaps the most effective way to classify and quantify the effect of mutation energy is to use allosteric diagrams of the type discussed in ref. 160. In this case, the focus is on the transfer of information due to energy coupling rather than just the correlation between simulated structural changes and the relationship to the active site preorganisation is clearer.

1.4.4 Problems with the Generalised Compression Idea

Bruice and coworkers have advanced the idea that enzymes catalyse reactions by favouring configurations in which the reactants are pushed to a close interaction distance (*e.g.* ref. 53). In most cases that we have studied, the energy associated with moving the reacting fragments from their average configuration in water to the average configuration in the enzymes was small, indicating that the corresponding catalytic effect was relatively minor.[65,161] In one case, where the NAC effect appeared to be large, it was found that the actual catalytic effect was attributable to electrostatic stabilisation of the TS.[56] In other words, the NAC effect evidently has been found to be a consequence rather than the reason for the electrostatic catalytic effect.[56]

The most notable example is CM, whose RS and TS are illustrated in ref. 56, where both the RS and TS of CM have similar charge distributions, and thus the same preorganisation effects that stabilise the RS also stabilise the TS and lead to an apparent NAC effect by making the RS structure closer to that of the TS. However, this is an inherent result of the TSS rather than being the reason for catalysis (see ref. 56).

Interestingly, despite the fact that the NAC effect has been shown to reflect incorrect energy considerations, it is still assumed to be an important effect and a reflection of compression effect. The inconsistency in the compression idea has been established by careful studies (*e.g.* ref. 162).

1.4.5 RSD by Desolvation Effects Does Not Provide Large Catalytic Effects

The idea that enzymes reduce the activation barrier by desolvating and destabilising the ground state of their reacting fragments has been put forward by many workers (*e.g.* ref. 48, 163–165). However, systematic analyses have demonstrated that the TS are solvated much more strongly in many enzymes than in the reference solution system.[1,11,64] It is important to note that the only way to test the desolvation proposal computationally is to calculate the actual binding energies of the reactants in the ground and TSs (see, for example, ref. 64). Most of the computational studies that are claimed to favour the desolvation proposal have not included such calculations.

One of the best illustrations of the problem with the RSD proposal has been given in the case of orotidine 5′-monophosphate decarboxylase (ODCase).[63] Although this case was discussed extensively, it gained additional importance due to an experiment[166] that justifies taking this as a specific general example. Now, the catalytic action of ODCase was first proposed to reflect the desolvation effect.[165] This was shown to involve an incorrect thermodynamic cycle (*e.g.* ref. 63). The elucidation of the structure of this enzyme showed that its active site is extremely polar (highly charged), but this led to a new RSD proposal, where the negatively-charged groups of the protein destabilise the carboxylate of the orotate substrate.[20] This proposal was shown to be inconsistent with the fact that a destabilised orotate will accept protons and become stable.[63] Furthermore, a careful computational study illustrates that the protein works by TSS and not by RSD (see ref. 63 and discussion below). Finally, studies by Wolfenden and coworkers[167,168] have provided strong evidence against the RSD proposal. These studies demonstrated that mutations of Asp96 and other residues that were supposed to destabilise the orotate led to weaker rather than stronger binding. As predicted in ref. 63, this result is inconsistent with the RSD, because destabilisation of the RS should result in a reduction of the binding energy. The strongest support has been provided in a very unique way by the experiments of Amyes *et al.*,[166] once these experiments have been analysed in a physically consistent way as has been done in ref. 3. In fact, the analysis can be used to disprove Jencks' proposal that enzymes work by using binding energies to destabilise the ground state of the reactive part of the substrate.

More major and fundamental problems with the analysis of ODC and the analysis of Wu *et al.*[20] that serve as a basis for the RSD proposal are provided in ref. 3.

1.4.6 Entropy Contributions of Bringing the Reactants Together are Unlikely to Account for Large Catalytic Effects

The idea that enzyme catalysis is associated with the entropy loss upon substrate binding was advanced in the early work of Jencks and coworkers[163,169] and has gained some support in recent computational studies.[170,171] However, Villa *et al.*[172] have argued that this proposal is based on an incomplete thermodynamic cycle. The entropic contribution probably cannot be large because the activation entropy in solution is usually much smaller than one might assume. This reflects the fact that the formation of the TS does not lead to loss of many degrees of freedom.[172] Problems with the entropic proposal have also emerged from experimental studies of cystidine deaminase by Wolfenden and coworkers,[173] as established by Åqvist and coworkers.[174]

1.4.7 Allosteric Control of Catalytic Activity is Also Associated with Electrostatic Effects

Allosteric effects control many enzymatic processes where interaction with another protein or with effectors drastically changes the catalytic activity of the given enzymes. So far, all the systems that have been explored by consistent simulations are found to be controlled by electrostatic effects.

The activation of Ras by GTPase activating protein provides a general example of a molecular switch that controls cell differentiation (*e.g.* see discussion in ref. 175). Through our simulation studies of this system, we have shown that the binding of GTPase activating protein leads to a major electrostatic stabilisation of the TS for the guanosine triphosphate (GTP) hydrolysis by both the so-called arginine finger[176] and by the transfer to a catalytic configuration, where the p-loop and other dipolar motifs stabilise the product of the hydrolysis reaction (see ref. 159, 175).

Another example is provided by the transition from the inactive chymotrypsinogen to the active chymotrypsin. This transition involves the cutting of the single bond between residues 15 and 16.[177] The new amino terminus at Ile-16 then forms a salt bridge with Asp 194, and this leads to a large shift of the main chain dipoles and the formation of the preorganised oxyanion hole.[178] The energetics of a related structural change due to the Gly-216/ Gly-226 mutation to alanine's was explored by EVB calculations[178] and shown to reflect changes in electrostatic TSS.

Even in the case of haemoglobin we were able to show[158] that a significant fraction of the allosteric effect is associated with the change in interaction between the charge shift upon oxygen binding and the change in protein tertiary structure (see discussion in ref. 158).

The fidelity of DNA replication by DNA polymerases is controlled by the active site (where the incorporation reaction is catalysed) and by the binding site of the incoming nucleotide that already includes the template base (*e.g.* see discussion in ref. 179). The high fidelity is guaranteed by the fact that the rate of incorporation of an incoming wrong nucleotide, W, is drastically slower than the corresponding rate of the right nucleotide, R (see ref. 180). Now, the origin of this control can be quantified by considering the interplay between the binding site of the incoming base and the stabilisation of the TS in the chemical site. Our previous studies[181,182] already indicated that the binding of the incoming base is determined by the preorganisation energy provided by the base binding site (which includes the template base); now the remaining challenge is to show that the TS stabilisation by the preorganised active site is anti-correlated with the preorganisation in the base binding site. This point has been explored in our preliminary studies, when we generated an 'interaction matrix' to describe the interaction between the TS and the protein groups as well as the interaction of the base of the incoming nucleotide with its surroundings.[160] Using such diagrams for the R and W systems (at the corresponding relaxed TS structures) provides an instructive decomposition of the allosteric effect that controls replication fidelity. In particular, taking the difference between the R and W matrices helps to identify the residues that are involved in the transfer of information from the base site to the TS site. Without going into the details (which will be addressed elsewhere) we note that the transfer of information between the base site and the chemical active site is controlled by electrostatic energies.

1.5 Conclusions and Perspectives

The exploration of the catalytic power of enzymes requires one to use well-defined references and the most unique reference is our 'chemistry-filtered' reference state that involves the same mechanism in the corresponding enzymatic reaction. With this reference we can address the challenge of evaluating the enormous catalytic power of some enzymes and relating it to the nature of the active site environmental effect.

The evaluation of the catalytic effect requires models that convert the active site structure to activation free energy and such models started with our 1976 QM/MM model[5] and evolved over the last 40 years. This review describes some of the directions taken in using and developing QM/MM methods. It appears that many studies are still overlooking the need for significant sampling and the need for calibration and validation on reference reactions in solutions. One way to see this point is to examine whether the given model can evaluate p*K*a in a protein active site, as large errors in such a calculation would also be reflected in calculations of activation barriers.

Using QM/MM methods that give converging and calibrated results, and in particular the EVB method, has indicated that enzyme catalysis is due mainly to electrostatic effects. These effects are due to polar preorganisation of the active site environment that stabilises the TS much more that the corresponding environment in water. We also clarified and demonstrated that the preorganisation effect involves a reduction in the folding energy and results in an inverse activity/stability correlation.

In order to verify the proposal that the catalytic effect is primarily due to electrostatic effects, it is important to demonstrate that the contributions from other factors and proposals are relatively small. This was done in the present review by considering various proposals and summarising studies that established the problems with those proposals. Thus, although it is reasonable to assume that evolution has exploited many possible catalytic effects, it appears that, with the exception of the electrostatic preorganisation effects, most of the mechanisms that have been proposed cannot lead to significant catalytic effects. Of course, our findings cannot be extrapolated to enzymes that have not yet been studied. But the only way to examine the feasibility of a proposed effect is to assess its magnitude in a variety of known enzymes, and the finding that a particular effect is relatively unimportant in all of these test cases indicates that this effect cannot contribute significantly to catalysis.

In summary, our studies and those of others have provided clear support for the view that electrostatic effects are the most important factor in enzyme catalysis.[59,183] It also appears that the issue in studies of enzyme catalysis is not the reformulation of TST, but the ability to evaluate the activation free energy in a reliable way including, if needed, quantum corrections. We believe that the accelerated increase in theoretical studies will provide increasing support to the electrostatic proposal and that the ability of such theoretical studies to reproduce experimental observations will lend credibility to their ability to dissect the overall catalytic effects to their key components and thus to establish the origin of enzyme catalysis.

The EVB can be parameterised conveniently in many cases by studying the solution reaction, but in some cases the surface of the solution reaction is very complex and requires careful *ab initio* calculations of the free energy surface in solution. A good example is GTP hydrolysis in solution, which requires major effort[184] in an attempt to distinguish between a mechanism with one nucleophilic water (1W) and with an additional water (2W) where the proton of the nucleophilic water is transferred to a second water and then to the phosphate oxygen. Calibrating the EVB surface on the *ab initio* surface provides a powerful way to study the reaction in G proteins.[8]

The advances in QM/MM studies are likely to continue in the near future and are likely to help in rational enzymes, and some progress in this direction has been reported. However, most attempts to use computer-aided enzyme design involved approaches that cannot reproduce the barriers in known cases (see ref. 185) and thus are not expected to provide realistic predictions for effective catalysis. Furthermore, even approaches that reproduce observed activation barriers are still facing a major challenge in taking non-active enzymes and reaching the same enormous catalytic effect presented in enzymes that were developed by evolution. Accomplishing this task would clearly be a way to demonstrate real understanding of enzyme catalysis.

Acknowledgements

This work was supported by the National Institutes of Health (NIH) grants GM 24492 and U19CA105010, and NSF grant MCB-1243719. We thank the High Performance Computing Center (HPCC) at the University of Southern California (USC) for computer time. We also thank Extreme Science and Engineering Discovery Environment (XSEDE) (which is supported by National Science Foundation grant number ACI-1053575) for computer time.

References

1. A. Warshel, *Computer Modeling of Chemical Reactions in Enzymes and Solutions*, Wiley-Interscience, 1991.
2. A. Fersht, *Structure and Mechanism in Protein Science: A Guide to Enzyme Catalysis and Protein Folding*, W. H. Freeman and Co., New York, 1999.
3. A. Warshel, P. K. Sharma, M. Kato, Y. Xiang, H. Liu and M. H. Olsson, *Chem. Rev.*, 2006, **106**, 3210–3235.
4. A. Warshel, *Angew Chem., Int. Ed.*, 2014, **53**, 10020–10031.
5. A. Warshel and M. Levitt, *J. Mol. Biol.*, 1976, **103**, 227.
6. A. Radzicka and R. Wolfenden, *Science*, 1995, **267**, 90–93.
7. N. V. Plotnikov and A. Warshel, *J. Phys. Chem. B*, 2012, **116**, 10342–10356.
8. B. R. Prasad, N. V. Plotnikov, J. Lameira and A. Warshel, *Proc. Natl. Acad. Sci. U. S. A.*, 2013, **110**, 20509–20514.
9. J. Villa and A. Warshel, *J. Phys. Chem. B*, 2001, **105**, 7887–7907.
10. S. C. L. Kamerlin, M. Haranczyk and A. Warshel, *J. Phys. Chem. B*, 2009, **113**, 1253–1272.
11. A. Shurki and A. Warshel, *Adv. Protein Chem.*, 2003, **66**, 249–313.
12. M. H. M. Olsson, G. Hong and A. Warshel, *J. Am. Chem. Soc.*, 2003, **125**, 5025–5039.
13. M. Strajbl, G. Hong and A. Warshel, *J. Phys. Chem. B*, 2002, **106**, 13333–13343.
14. Y. Zhang, H. Liu and W. Yang, *J. Chem. Phys.*, 2000, **112**, 3483–3492.
15. M. L. Wang, Z. Y. Lu and W. T. Yang, *J. Chem. Phys.*, 2004, **121**, 101–107.

16. Z. Y. Lu and W. T. Yang, *J. Chem. Phys.*, 2004, **121**, 89–100.
17. E. Rosta, M. Klahn and A. Warshel, *J. Phys. Chem. B*, 2006, **110**, 2934–2941.
18. J. J. Pavelites, J. L. Gao, P. A. Bash and A. D. Mackerell, *J. Comput. Chem.*, 1997, **18**, 221–239.
19. B. Szefczyk, A. J. Mulholland, K. E. Ranaghan and W. A. Sokalski, *J. Am. Chem. Soc.*, 2004, **126**, 16148–16159.
20. N. Wu, Y. R. Mo, J. L. Gao and E. F. Pai, *Proc. Natl. Acad. Sci. U. S. A.*, 2000, **97**, 2017–2022.
21. A. Warshel and R. M. Weiss, *J. Am. Chem. Soc.*, 1980, **102**, 6218–6226.
22. S. C. L. Kamerlin and A. Warshel, *Faraday Discuss.*, 2010, **145**, 71–106.
23. E. Rosta and A. Warshel, *J. Chem. Theory Comput.*, 2012, **8**, 3574–3585.
24. J. K. Hwang, G. King, S. Creighton and A. Warshel, *J. Am. Chem. Soc.*, 1988, **110**, 5297–5311.
25. R. A. Marcus, *Angew. Chem., Int. Ed. Engl.*, 1993, **32**, 1111–1121.
26. J. K. Hwang, S. Creighton, G. King, D. Whitney and A. Warshel, *J. Chem. Phys.*, 1988, **89**, 859–865.
27. G. V. Isaksen, J. Aqvist and B. O. Brandsdal, *Plos Comput. Biol.*, 2014, 10.
28. B. M. Dunn and T. C. Bruice, *Adv. Enzymol. Relat. Areas Mol. Biol.*, 1973, **37**, 1–60.
29. T. H. Fife, S. H. Jaffe and R. Natarajan, *J. Am. Chem. Soc.*, 1991, **113**, 7646–7653.
30. J. A. Thoma, *Theor. Biol.*, 1974, **44**, 305–317.
31. W. P. Jencks, *Adv. Enzymol. Relat. Areas Mol. Biol.*, 1975, **43**, 219–410.
32. S. J. Gardell, C. S. Craik, D. Hilvert, M. S. Urdea and W. J. Rutter, *Nature*, 1985, **317**, 551–555.
33. G. Winter, A. R. Fersht, A. J. Wilkinson, M. Zoller and M. Smith, *Nature*, 1982, **299**, 756–758.
34. A. Kuliopulos, P. Talalay and A. S. Mildvan, *Biochemistry*, 1990, **29**, 10271–10280.
35. P. Carter, B. Nilsson, J. P. Burnier, D. Burdick and J. A. Wells, *Proteins: Struct., Funct., Genet.*, 1989, **6**, 240–248.
36. S. J. Benkovic, C. A. Fierke and A. M. Naylor, *Science*, 1988, **239**, 1105–1110.
37. H. M. Wilks, K. W. Hart, R. Feeney, C. R. Dunn, H. Muirhead, W. N. Chia, D. A. Barstow, T. Atkinson, A. R. Clarke and J. J. Holbrook, *Science*, 1988, **242**, 1541–1544.
38. T. K. Harris, G. Wu, M. A. Massiah and A. S. Mildvan, *Biochemistry*, 2000, **39**, 1655–1674.
39. N. C. Horton, C. Otey, S. Lusetti, M. D. Sam, J. Kohn, A. M. Martin, V. Ananthnarayan and J. J. Perona, *Biochemistry*, 2002, **41**, 10754–10763.
40. S. Sprang, T. Standing, R. J. Fletterick, R. M. Stroud, J. Finermoore, N. H. Xuong, R. Hamlin, W. J. Rutter and C. S. Craik, *Science*, 1987, **237**, 905–909.
41. A. Warshel, G. Naray-Szabo, F. Sussman and J. K. Hwang, *Biochemistry*, 1989, **28**, 3629–3637.

42. Computational Approaches to Biochemical Reactivity, in *Understanding Chemical Reaction*, ed. G. Naray-Szabo and A. Warshel, 1997, vol. 19.
43. M. Roca, S. Marti, J. Andres, V. Moliner, M. Tunon, J. Bertran and A. H. Williams, *J. Am. Chem. Soc.*, 2003, **125**, 7726–7737.
44. T. Ishida and S. Kato, *J. Am. Chem. Soc.*, 2003, **125**, 12035–12048.
45. A. Soriano, E. Silla, I. Tunon, S. Marti, V. Moliner and J. Bertran, *Theor. Chem. Acc.*, 2004, **112**, 327–334.
46. C. N. Schutz and A. Warshel, *Protein: Struct., Funct., Genet.*, 2001, **44**, 400–417.
47. K. E. Ranaghan, L. Ridder, B. Szefczyk, W. A. Sokalski, J. C. Hermann and A. J. Mulholland, *Org. Biomol. Chem.*, 2004, **2**, 968–980.
48. L. S. Devi-Kesavan and J. L. Gao, *J. Am. Chem. Soc.*, 2003, **125**, 1532–1540.
49. M. Garcia-Viloca, J. Gao, M. Karplus and D. G. Truhlar, *Science*, 2004, **303**, 186.
50. S. Fischer, S. Michnick and M. Karplus, *Biochemistry*, 1993, **32**, 13830–13837.
51. B. Kuhn and P. A. Kollman, *J. Am. Chem. Soc.*, 2000, **122**, 2586–2596.
52. A. Warshel and W. W. Parson, *Q. Rev. Biophys.*, 2001, **34**, 563–679.
53. S. Hur and T. C. Bruice, *J. Am. Chem. Soc.*, 2003, **125**, 1472–1473.
54. H. Guo, Q. Cui, W. N. Lipscomb and M. Karplus, *Proc. Natl. Acad. Sci. U. S. A.*, 2001, **98**, 9032–9037.
55. S. Hay, L. O. Johannissen, M. J. Sutcliffe and N. S. Scrutton, *Biophys. J.*, 2010, **98**, 121–128.
56. M. Strajbl, A. Shurki, M. Kato and A. Warshel, *J. Am. Chem. Soc.*, 2003, **125**, 10228–10237.
57. F. S. Lee, Z. T. Chu, M. B. Bolger and A. Warshel, *Protein Eng.*, 1992, **5**, 215–228.
58. A. Warshel and S. T. Russell, *Q. Rev. Biophys.*, 1984, **17**, 283–422.
59. A. Warshel, *Proc. Natl. Acad. Sci. U. S. A.*, 1978, **75**, 5250–5254.
60. R. A. Marcus, *J. Chem. Phys.*, 1956, **24**, 966–978.
61. W. J. Albery, *Annu. Rev. Phys. Chem.*, 1980, **31**, 227–263.
62. L. I. Krishtalik, *J. Theor. Biol.*, 1980, **86**, 757–771.
63. A. Warshel, M. Strajbl, J. Villa and J. Florian, *Biochemistry*, 2000, **39**, 14728–14738.
64. M. H. M. Olsson and A. Warshel, *J. Am. Chem. Soc.*, 2004, **126**, 15167–15179.
65. A. Shurki, M. Strajbl, J. Villa and A. Warshel, *J. Am. Chem. Soc.*, 2002, **124**, 4097–4107.
66. P. K. Sharma, Y. Xiang, M. Kato and A. Warshel, *Biochemistry*, 2005, **44**, 11307–11314.
67. S. Trobro and J. Åqvist, *Proc. Natl. Acad. Sci. U. S. A.*, 2005, **102**, 12395–12400.
68. H. Liu and A. Warshel, *Biochemistry*, 2007, **46**, 6011–6025.
69. A. Warshel, F. Sussman and J. K. Hwang, *J. Mol. Biol.*, 1988, **201**, 139–159.

70. B. K. Shoichet, W. A. Baase, R. Kuroki and B. W. Matthews, *Proc. Natl. Acad. Sci. U. S. A.*, 1995, **92**, 452–456.
71. B. M. Beadle and B. K. Shoichet, *J. Mol. Biol.*, 2002, **321**, 285–296.
72. A. H. Elcock, *J. Mol. Biol.*, 2001, **312**, 885–896.
73. G. Bhabha, J. Lee, D. C. Ekiert, J. Gam, I. A. Wilson, H. J. Dyson, S. J. Benkovic and P. E. Wright, *Science*, 2011, **332**, 234–238.
74. A. J. Adamczyk, J. Cao, S. C. Kamerlin and A. Warshel, *Proc. Natl. Acad. Sci. U. S. A.*, 2011, **108**, 14115–14120.
75. E. J. Loveridge, E. M. Behiry, J. Guo and R. K. Allemann, *Nat. Chem.*, 2012, 292–297.
76. L. Y. P. Luk, J. J. Ruiz-Pernia, W. M. Dawson, M. Roca, E. J. Loveridge, D. R. Glowacki, J. N. Harvey, A. J. Mulholland, I. Tunon, V. Moliner and R. K. Allemann, *Proc. Natl. Acad. Sci. U. S. A.*, 2013, **110**, 16344–16349.
77. Y. Fan, A. Cembran, S. Ma and J. Gao, *Biochemistry*, 2013, **52**, 2036–2049.
78. M. Roca, A. Vardi-Kilshtain and A. Warshel, *Biochemistry*, 2009, **48**, 3046–3056.
79. M. P. Frushicheva, J. Cao, Z. T. Chu and A. Warshel, *Proc. Natl. Acad. Sci. U. S. A.*, 2010, **107**, 16869–16874.
80. H. B. Liu and A. Warshel, *Biochemistry*, 2007, **46**, 6011–6025.
81. M. J. Field, *J. Comput. Chem.*, 2002, **23**, 48–58.
82. S. Marti, M. Roca, J. Andres, V. Moliner, E. Silla, I. Tunon and J. Bertran, *Chem. Soc. Rev.*, 2004, **33**, 98–107.
83. C. C. F. Blake, L. N. Johnson, G. A. Mair, A. C. T. North, D. C. Philips and V. R. Sarma, *Proc. R. Soc. London, Ser. B*, 1967, **167**, 378–388.
84. A. Warshel and M. Levitt, *J. Mol. Biol.*, 1976, **103**, 227–249.
85. M. Levitt, in *Peptides, Polypeptides and Proteins*, ed. E. R. Blout, F. A. Bovey, M. Goddman and N. Lotan, Wiley, New York, 1974, pp. 99–113.
86. N. A. Khanjin, J. P. Snyder and F. M. Menger, *J. Am. Chem. Soc.*, 1999, **121**, 11831–11846.
87. O. Tapia, J. Andres and V. S. Safont, *J. Chem. Soc., Faraday Trans.*, 1994, **90**, 2365–2374.
88. V. E. Anderson, *Arch. Biochem. Biophys.*, 2005, **433**, 27–33.
89. J. Halpern, S. H. Kim and T. W. Leung, *J. Am. Chem. Soc.*, 1984, **106**, 8317–8319.
90. C. D. Garr, J. M. Sirovatka and R. G. Finke, *J. Am. Chem. Soc.*, 1996, **118**, 11142–11154.
91. M. K. Geno and J. Halpern, *J. Am. Chem. Soc.*, 1987, **109**, 1238–1240.
92. R. Padmakumar, R. Padmakumar and R. Banerjee, *Biochemistry*, 1997, **36**, 3713–3718.
93. K. P. Jensen and U. Ryde, *J. Am. Chem. Soc.*, 2005, **127**, 9117–9128.
94. J. L. Gao, *Curr. Opin. Struct. Biol.*, 2003, **13**, 184–192.
95. H. A. O. Hill, J. M. Pratt and R. P. Williams, *J. Chem. Br.*, 1969, **5**, 169–172.
96. J. H. Grate and G. N. Schrauzer, *J. Am. Chem. Soc.*, 1979, **101**, 4601–4611.

97. J. Halpern, *Science*, 1985, **227**, 869–875.
98. K. P. Jensen and U. Ryde, *J. Mol. Struct.: THEOCHEM*, 2002, **585**, 239–255.
99. J. M. Sirovatka, A. K. Rappe and R. G. Finke, *Inorg. Chim. Acta*, 2000, **300**, 545–555.
100. M. Klahn, S. Braun-Sand, E. Rosta and A. Warshel, *J. Phys. Chem. B*, 2005, **109**, 15645–15650.
101. P. K. Sharma, Z. T. Chu, M. H. M. Olsson and A. Warshel, *Proc. Natl. Acad. Sci. U. S. A.*, 2007, **104**, 9661–9666.
102. P. Schopf, M. J. L. Mills and A. Warshel, *Proc. Natl. Acad. Sci. U. S. A.*, 2015, **112**, 4328–4333.
103. G. Careri, P. Fasella and E. Gratton, *Annu. Rev. Biophys. Bioeng.*, 1979, **8**, 69–97.
104. M. Karplus and J. A. McCammon, *Annu. Rev. Biochem*, 1983, **52**, 263–300.
105. A. Kohen, R. Cannio, S. Bartolucci and J. P. Klinman, *Nature*, 1999, **399**, 496–499.
106. J. Basran, M. J. Sutcliffe and N. S. Scrutton, *Biochemistry*, 1999, **38**, 3218–3222.
107. J. L. Radkiewicz and C. L. Brooks, *J. Am. Chem. Soc.*, 2000, **122**, 225–231.
108. C. E. Cameron and S. J. Benkovic, *Biochemistry*, 1997, **36**, 15792–15800.
109. H. J. C. Berendsen and S. Hayward, *Curr. Opin. Struct. Biol.*, 2000, **10**, 165–169.
110. E. Neria and M. Karplus, *Chem. Phys. Lett.*, 1997, **267**, 23–30.
111. E. Z. Eisenmesser, D. A. Bosco, M. Akke and D. Kern, *Science*, 2002, **295**, 1520–1523.
112. A. Kohen and J. P. Klinman, *Chem. Biol.*, 1999, **6**, R191–R198.
113. M. J. Sutcliffe and N. S. Scrutton, *Trends Biochem. Sci.*, 2000, **25**, 405–408.
114. M. H. M. Olsson, W. W. Parson and A. Warshel, *Chem. Rev.*, 2006, **106**, 1737–1756.
115. J. K. Hwang and A. Warshel, *Biochemistry*, 1987, **26**, 2669–2673.
116. M. Roca, J. Andres, V. Moliner, I. Tunon and J. Bertran, *J. Am. Chem. Soc.*, 2005, **127**, 10648–10655.
117. A. V. Pisliakov, J. Cao, S. C. L. Kamerlin and A. Warshel, *Proc. Natl. Acad. Sci. U. S. A.*, 2009, **106**, 17359–17364.
118. A. Kohen, *Acc. Chem. Res.*, 2015, **48**, 466–473.
119. J. P. Klinman and A. Kohen, *Annu. Rev. Biochem.*, 2013, **82**, 471–496.
120. S. C. L. Kamerlin and A. Warshel, *Proteins: Struct., Funct., Bioinf.*, 2010, **78**, 1339–1375.
121. A. Warshel and Z. T. Chu, *J. Chem. Phys.*, 1990, **93**, 4003–4015.
122. J.-K. Hwang, Z. T. Chu, A. Yadav and A. Warshel, *J. Phys. Chem.*, 1991, **95**, 8445–8448.
123. A. Kohen, R. Cannio, S. Bartolucci and J. P. Klinman, *Nature*, 1999, **399**, 496–499.
124. A. Kohen and J. P. Klinman, *Chem. Biol.*, 1999, **6**, R191–R198.

125. D. Antoniou and S. D. Schwartz, *Proc. Natl. Acad. Sci. U. S. A.*, 1997, **94**, 12360–12365.
126. S. C. Sharma and J. P. Klinman, *J. Am Chem. Soc.*, 2008, **130**, 17632–17633.
127. A. Warshel, *Proc. Natl. Acad. Sci. U. S. A.*, 1978, **75**, 5250–5254.
128. A. Warshel, *Proc. Natl. Acad. Sci. U. S. A.*, 1984, **81**, 444–448.
129. D. R. Glowacki, J. N. Harvey and A. J. Mulholland, *Nat. Chem.*, 2012, **4**, 169–176.
130. I. Tuñón, D. Laage and J. T. Hynes, *Arch. Biochem. Biophys.*, 2015, **582**, 42–55.
131. N. Boekelheide, R. Salomon-Ferrer and T. F. Miller, *Proc. Natl. Acad. Sci. U. S. A.*, 2011, **108**, 16159–16163.
132. A. Warshel, *Proc. Natl. Acad. Sci. U. S. A.*, 1984, **81**, 444–448.
133. G. Bouvignies, P. Bernado, S. Meier, K. Cho, S. Grzesiek, R. Bruschweiler and M. Blackledge, *Proc. Natl. Acad. Sci. U. S. A.*, 2005, **102**, 13885–13890.
134. E. Z. Eisenmesser, O. Millet, W. Labeikovsky, D. M. Korzhnev, M. Wolf-Watz, D. A. Bosco, J. J. Skalicky, L. E. Kay and D. Kern, *Nature*, 2005, **438**, 117–121.
135. G. H. Li and Q. Cui, *J. Am. Chem. Soc.*, 2003, **125**, 15028–15038.
136. P. K. Agarwal, *J. Am. Chem. Soc.*, 2005, **127**, 15248–15256.
137. J. E. Basner and S. D. Schwartz, *J. Am. Chem. Soc.*, 2005, **127**, 13822–13831.
138. A. Yadav, R. M. Jackson, J. J. Holbrook and A. Warshel, *J. Am. Chem. Soc.*, 1991, **113**, 4800–4805.
139. W. W. Cleland and M. M. Kreevoy, *Science*, 1994, **264**, 1887–1890.
140. A. Warshel, *J. Phys. Chem.*, 1982, **86**, 2218–2224.
141. D. Borgis and J. T. Hynes, *J. Chem. Phys.*, 1991, **94**, 3619–3628.
142. E. D. German and A. M. Kuznetsov, *J. Chem. Soc. Faraday Trans.*, 1981, **77**, 397–412.
143. J. K. Hwang, Z. T. Chu, A. Yadav and A. Warshel, *J. Phys. Chem.*, 1991, **95**, 8445–8448.
144. J.-K. Hwang and A. Warshel, *J. Am. Chem. Soc.*, 1996, **118**, 11745–11751.
145. I. Feierberg, V. Luzhkov and J. Aqvist, *J. Biol. Chem.*, 2000, **275**, 22657–22662.
146. M. J. Gillan, *J. Phys. C Solid State Phys.*, 1987, **20**, 3621–3641.
147. G. A. Voth, *Adv. Chem. Phys.*, 1996, **93**, 135–218.
148. C. Alhambra, J. C. Corchado, M. L. Sanchez, J. L. Gao and D. G. Truhlar, *J. Am. Chem. Soc.*, 2000, **122**, 8197–8203.
149. S. R. Billeter, S. P. Webb, P. K. Agarwal, T. Iordanov and S. Hammes-Schiffer, *J. Am. Chem. Soc.*, 2001, **123**, 11262–11272.
150. G. P. Miller and S. J. Benkovic, *Biochemistry*, 1998, **37**, 6327–6335.
151. J. R. Schnell, H. J. Dyson and P. E. Wright, *Annu. Rev. Biophys. Biomol. Struct.*, 2004, **33**, 119–140.
152. T. H. Rod, J. L. Radkiewicz and C. L. Brooks, *Proc. Natl. Acad. Sci. U. S. A.*, 2003, **100**, 6980–6985.

153. J. B. Watney, P. K. Agarwal and S. Hammes-Schiffer, *J. Am. Chem. Soc.*, 2003, **125**, 3745–3750.
154. S. Hammes-Schiffer, *Curr. Opin. Struct. Biol.*, 2004, **14**, 192–201.
155. I. F. Thorpe and C. L. Brooks, *J. Am. Chem. Soc.*, 2005, **127**, 12997–13006.
156. P. K. Agarwal, S. R. Billeter, P. T. R. Rajagopalan, S. J. Benkovic and S. Hammes-Schiffer, *Proc. Natl. Acad. Sci. U. S. A.*, 2002, **99**, 2794–2799.
157. K. F. Wong, T. Selzer, S. J. Benkovic and S. Hammes-Schiffer, *Proc. Natl. Acad. Sci. U. S. A.*, 2005, **102**, 6807–6812.
158. A. Warshel and R. M. Weiss, *J. Am. Chem. Soc.*, 1981, **103**, 446–451.
159. T. M. Glennon, J. Villa and A. Warshel, *Biochemistry*, 2000, **39**, 9641–9651.
160. Y. Xiang, P. Oelschlaeger, J. Florián, M. F. Goodman and A. Warshel, *Biochemistry*, 2006, **45**, 7036–7048.
161. K. E. Ranaghan and A. J. Mulholland, *Chem. Commun.*, 2004, 1238–1239.
162. S. C. L. Kamerlin, P. K. Sharma, Z. T. Chu and A. Warshel, *FEBS Lett.*, 2010, **584**, 2759–2766.
163. W. P. Jencks, *Catalysis in Chemistry and Enzymology*, Dover Publication, New York, 1986.
164. J. Crosby, R. Stone and G. E. Lienhard, *J. Am. Chem. Soc.*, 1970, **92**, 2891–2900.
165. J. K. Lee and K. N. Houk, *Science*, 1997, **276**, 942–945.
166. T. L. Amyes, J. P. Richard and J. J. Tait, *J. Am. Chem. Soc.*, 2005, **127**, 15708–15709.
167. B. G. Miller, G. L. Butterfoss, S. A. Short and R. Wolfenden, *Biochemistry*, 2001, **40**, 6227–6232.
168. B. G. Miller and R. Wolfenden, *Annu. Rev. Biochem.*, 2002, **71**, 847–885.
169. M. I. Page and W. P. Jencks, *Proc. Natl. Acad. Sci. U. S. A.*, 1971, **68**, 1678–1683.
170. R. V. Stanton, M. Perakyla, D. Bakowies and P. A. Kollman, *J. Am. Chem. Soc.*, 1998, **120**, 3448–3457.
171. P. A. Kollman, B. Kuhn, O. Donini, M. Perakyla, R. Stanton and D. Bakowies, *Acc. Chem. Res.*, 2001, **34**, 72–79.
172. J. Villa, M. Strajbl, T. M. Glennon, Y. Y. Sham, Z. T. Chu and A. Warshel, *Proc. Natl. Acad. Sci. U. S. A.*, 2000, **97**, 11899–11904.
173. M. J. Snider, S. Gaunitz, C. Ridgway, S. A. Short and R. Wolfenden, *Biochemistry*, 2000, **39**, 9746–9753.
174. M. Kazemi, F. Himo and J. Åqvist, *Proc. Natl. Acad. Sci. U. S. A.*, 2016, **113**, 2406–2411.
175. A. Shurki and A. Warshel, *Proteins: Struct., Funct., Bioinf.*, 2004, **55**, 1–10.
176. M. R. Ahmadian, P. Stege, K. Scheffzek and A. Wittinghofer, *Nat. Struct. Biol.*, 1997, **4**, 686–689.
177. J. Kraut, *Annu. Rev. Biochem.*, 1977, **46**, 331–358.
178. A. Warshel and F. Sussman, *Proc. Natl. Acad. Sci. U. S. A.*, 1986, **83**, 3806–3810.

179. J. Florian, M. F. Goodman and A. Warshel, *J. Am. Chem. Soc.*, 2003, **125**, 8163–8177.
180. J. Florian, M. F. Goodman and A. Warshel, *Proc. Natl. Acad. Sci. U. S. A.*, 2005, **102**, 6819–6824.
181. J. Florian, M. F. Goodman and A. Warshel, *J. Phys. Chem. B*, 2002, **106**, 5739–5753.
182. J. Florian, A. Warshel and M. F. Goodman, *J. Phys. Chem. B*, 2002, **106**, 5754–5760.
183. A. Warshel, *Acc. Chem. Res.*, 1981, **14**, 284–290.
184. N. V. Plotnikoy, B. R. Prasad, S. Chakrabarty, Z. T. Chu and A. Warshel, *J. Phys. Chem. B*, 2013, **117**, 12807–12819.
185. M. P. Frushicheva, M. J. L. Mills, P. Schopf, M. K. Singh, R. B. Prasad and A. Warshel, *Curr. Opin. Chem. Biol.*, 2014, **21**, 56–62.

Section I: Theory

CHAPTER 2

Fundamentals of Enzyme Catalysis: Determination of Rate Constants

RICHARD L. SCHOWEN

University of Kansas, 962 East 1338 Road, Lawrence, KS 66046, USA
Email: rschowen@ku.edu

2.1 Introduction

This chapter addresses methods for the extraction of rate constant values from simulation studies. Such an extraction is useful because rate constants are one of the chief products of experimental studies of enzymes and enzyme catalysis. Thus their determination also in the course of theoretical simulations is desirable, not merely to establish coherence between theoretical and experimental work, but more generally to aid in establishing the mechanisms that underlie the experimental kinetics.

Much biochemical history remains embedded in the notation and language of enzyme kinetics. Two recent books, that of Cook and Cleland[1] and that of Stein[2] can be recommended for those needing or desiring refreshment of memory or more detail than is presented here. The older volume of Segel[3] is ideal for its depth and scope ("where you look up whether you have it right"). The excellent didactic volume of Warshel,[4] on the subject of computer modelling, is a treasure.

RSC Theoretical and Computational Chemistry Series No. 9
Simulating Enzyme Reactivity: Computational Methods in Enzyme Catalysis
Edited by Iñaki Tuñón and Vicent Moliner

Published by the Royal Society of Chemistry, www.rsc.org

2.2 The Elements of Enzyme Kinetics, in Particular Rate Constants

2.2.1 Rate Constants Experimentally Determined

Box 2.1 introduces the main ideas by means of a simple example. The chemical equation shows an elementary mechanism that illustrates the

Box 2.1 Elements of enzyme kinetics.

Equation 2.1:

$$\mathrm{E+S} \underset{k_2}{\overset{k_1}{\rightleftharpoons}} \mathrm{ES} \underset{k_4}{\overset{k_3}{\rightleftharpoons}} \mathrm{EP} \xrightarrow[(k_6,\text{ omitted at low [P]})]{k_5} \mathrm{E+P}$$

v = velocity = d[P]/dt = −d[S]/dt; e_o = [E] + [ES] + [EP] = constant; e_o/v = reciprocal specific velocity. T_{IJ} is the transition state for step I (forward) and step J (reverse).

Equation 2.2

Term in e_o/v	Term corresponds to this conversion
$e_o/v = 1/k_1[\mathrm{S}]$ +	E + S to T_{12}
$k_2/k_1k_3[\mathrm{S}]$ +	E + S to T_{34}
$k_2k_4/k_1k_3k_5[\mathrm{S}]$ +	E + S to $T_{5(6)}$
$1/k_3$ +	ES to T_{34}
k_4/k_3k_5 +	ES to $T_{5(6)}$
$1/k_5$.	EP to $T_{5(6)}$

$$e_o/v = \{1/k_1 + k_2/k_1k_3 + k_2k_4/k_1k_3k_5\}\{1/[\mathrm{S}]\} + \{1/k_3 + k_4/k_3k_5 + 1/k_5\} = \{K_M/k_{cat}\}\{1/[\mathrm{S}]\} + \{1/[k_{cat}]\}$$

Equation 2.3, the Michaelis–Menten equation

$$e_o/v = \{K_M/k_{cat}\}\{1/[\mathrm{S}]\} + \{1/[k_{cat}]\} \text{ or } v/e_o = \{k_{cat}[\mathrm{S}]\}/\{K_M + [\mathrm{S}]\}$$
$$\text{where } k_{cat} = k_3k_5/(k_3 + k_4 + k_5)$$

$$K_M/k_{cat} = 1/k_1 + k_2/k_1k_3 + k_2k_4/k_1k_3k_5 \quad k_{cat}/K_M = k_1k_3k_5/(k_3k_5 + k_2k_5 + k_2k_4)$$

$$(K_M/k_{cat})(k_{cat}) = \{k_3k_5 + k_2\,(k_4 + k_5)\}/\{k_1\,(k_3 + k_4 + k_5)\}$$
$$= K_M = \{[\mathrm{E}][\mathrm{S}]\}/\{[\mathrm{ES}] + [\mathrm{EP}]\}$$

Thus K_M = [free enzyme][S]/[bound enzyme]

Note the steady-state requirements for the species [E], [ES] and [EP] are: $k_1[\mathrm{E}][\mathrm{S}] = k_2[\mathrm{ES}] + k_5[\mathrm{EP}]$; $k_1[\mathrm{E}][\mathrm{S}] = k_1[\mathrm{E}][\mathrm{S}] + k_4[\mathrm{EP}] = \{(k_2 + k_3)[\mathrm{ES}]\}$; slightly tedious algebra yields the definitions of K_M above.

Thus in this general sense, a small value for K_M indicates a high affinity of the enzyme for ligand(s).

best-known phenomena: complex formation between enzyme and substrate, transformation of the substrate complex to product complex, perhaps through intermediates, culminating in product release; the applicability of steady-state kinetics to the reaction under relevant conditions.

The Cleland School of enzyme kinetics[1] designates as the kinetic mechanism the order in which reactants combine with the enzyme and the products depart from the enzyme, and has provided a colourful language for the various possibilities (sequential, ping-pong, *etc.*). These scholars distinguish the kinetic mechanism from the chemical mechanism, which describes the nature and sequence of the chemical events, such as the making and breaking of chemical bonds that occur in the conversion of reactants to products, and the identification and characterisation of transition states and intermediate states.

It is often the case that important information on the kinetic mechanism can be deduced from the kinetic law that describes the relationship of the reaction rate to the rate constants for individual steps (microscopic rate constants), the rate constants as they aggregate in a kinetic law (phenomenological, or sometimes observed, rate constants), and to the concentrations of substrates, products, inhibitors, *etc.*

The specific velocity of the enzyme-catalytic reaction, v/e_o where v is the reaction velocity and e_o is the total concentration of enzyme in all forms, *i.e.* free or complexed, can be expressed in reciprocal form as in Box 2.1. Recall that because the enzyme is a catalyst and undergoes no permanent change in the course of reaction, the value of e_o is constant. Then the expression readily falls into a sum of terms, six in this example. As Box 2.1 shows, three of these terms contain the substrate concentration and three do not. The older among us recognise the linear equation of the Lineweaver–Burk double-reciprocal plot, once a staple of enzymological data reduction. Each term is a reciprocal rate constant that links a particular reactant state to a particular transition state, as shown in the box. Thus the reactant state E + S is linked to the transition state T_{12} for the first step (the transition state is necessarily the same forward and reverse reactions), and to the transition state T_{34} for the second step, and to the transition state $T_{5(6)}$ for the third step. The parentheses in the subscript remind us that if P remains at low concentrations it will not rebind to the enzyme, so k_6 need not be brought in. The reactant state ES is then linked to each of the two transition states that follow it, T_{34} and $T_{5(6)}$, and finally the reactant state EP is linked to $T_{5(6)}$.

How many of these individual or microscopic rate constants can be determined experimentally is a matter of both human ingenuity and the nature of each particular experimental system, but it is prudent and desirable for a theoretical study to produce values of each rate constant whenever it is possible and practicable. Box 2.1 shows how the constants sort (*via* the Michaelis–Menten equation[1,2]) into one composite first-order rate constant k_{cat} (units s^{-1}) and one composite second-order rate constant k_{cat}/K_M (units $M^{-1}\,s^{-1}$). These composite rate constants, or phenomenological rate constants, are those that emerge directly from kinetic studies.

The ratio of k_{cat} to k_{cat}/K_M gives a quantity K_M with units of M. Quite a few buckets of ink, not to say blood, have been spilt over the interpretation of K_M values, but little of it is important here. For our purposes k_{cat}/K_M can be considered merely a phenomenological second-order rate constant, defined by the microscopic rate constants as in Box 2.1. It is, however, true, as shown at the bottom of Box 2.1, that K_M is the ratio of the free-enzyme concentration (concentration of enzyme bearing no ligand[s]) to the sum of concentrations of all bound forms of the enzyme (forms bearing ligand[s]), multiplied by the concentration of substrate [S], so that in this sense it measures the tendency of the substrate and the enzyme to remain free of each other (in the same sense that an actual dissociation constant reflects the tendency of the two bonding partners to dissociate and remain free of each other).

2.2.2 Comparison of Experimental Rate Constants with Theoretically Computed Values

Box 2.1 serves to emphasise that in a steady-state rate expression for a one-substrate enzyme reaction, the reciprocal velocity resolves into a sum of terms, each of which links a single reactant state to a single transition state. In the simple case shown, the reactant state E + S is linked by one of the three additive terms to (i) the first transition state T_{12}, (ii) the second transition state T_{34}, and (iii) the third transition state $T_{5(6)}$. The sum of the three terms (eqn (2.3)) is the reciprocal of an experimentally accessible quantity, k_{cat}/K_M.

If it is desired to calculate values from a free-energy surface or potentials of mean force for the free energies of activation of the three contributing aggregate rate constants then one can in principle obtain the value of (i) k_1 for the binding step of E to S to form ES, (ii) k_1k_3/k_2, for overall equilibrium formation of ES from E + S (k_1/k_2) followed by the chemical reaction step (k_3), and finally (iii) $k_1k_3k_5/k_2k_3$, for the overall equilibrium conversion from E + S to the product-release step (k_5). These might be obtained by calculating each of the rate constants k_1, k_2, k_3, k_4 and k_5, it is a rare case where all of these constants will have been measured experimentally – although that could change in the future.

Let us pause to take note that other experimental data than individual rate-constant measurements might be at hand to help. For example, it might be so that one of the three components of this observed rate constant k_{cat}/K_M would dominate over the others. The second of the terms, for the overall chemical conversion from E + S to EP, might have the largest value of the three terms, as might have been demonstrated by kinetic isotope effects.

Note that if the chemical conversion has a much larger reciprocal rate constant than either the bonding step or the product-release step, the aggregate rate constant will become equal to that for the overall chemical conversion: the largest reciprocal rate constant will belong to the smallest rate constant. This rate constant is then said to determine the rate and is therefore equal to the observed rate constant.

A further sometimes confusing point is that kineticists are accustomed to say "the chemical conversion is the rate-determining step". A glance at Box 2.1 might then lead one to conclude that "the rate constant for the rate-determining step is that for the chemical conversion, namely k_3". At second glance k_3 cannot become equal to k_{cat}/K_M: the former is a first-order rate constant, the latter is a second-order rate constant. The resolution of the conundrum is that a very slow chemical conversion will lead to a small value of the second-order rate constant k_{cat}/K_M that is equal to k_1k_3/k_2, which is indeed a second-order rate constant.

Such confusion as there is probably arises from the conversational use of expressions like "the proton-transfer step is the rate-determining step", which generates a mental image of a reactant state Donor:H + :Acceptor that is transformed in a simple unitary reaction to a product state Donor: + H:Acceptor, such that if k_3 is described as "the rate constant of the proton-transfer step", then the erroneous conclusion seems inevitable.

Some clarification may arise from examination of the situation with respect to the first-order rate constant k_{cat}. Eqn (2.2) in Box 2.1 shows that there are three contributing rate constants isolated in the reciprocal specific velocity. The first point to notice is that only microscopic rate constants that follow after the full formation of the complex ES are present. Second, of the three terms that contribute, two have ES as the reactant state with one leading to the adjacent transition state T_{34} and the other by the equilibrium conversion of ES to EP (equilibrium constant k_3/k_4) followed by conversion of EP to $T_{5(6)}$. The third contributing rate constant is simply k_5 linking EP to the final transition state in the reaction sequence. The observed rate constant k_{cat} is therefore composed of rate constants for the 'chemical step', conversion of ES to EP, for the equilibrium conversion of ES to EP with following product release, and for the product-release step itself.

In the case of both k_{cat}/K_M and k_{cat}, there is no way to know in general which of the component processes of substrate binding, chemical conversion and product release will be the rate-determining process; only experimental and/or theoretical studies can illuminate this point for a given system. Indeed, if two or more of the component aggregate rate constants are of similar magnitude, then both will determine the rate together. Such a situation might be thought unlikely on the basis of experience with the reactions of organic and inorganic chemistry, but the opposite is the case. Albery and Knowles[22] showed long ago that the molecular evolution of enzymes by random mutations followed by selection for high catalytic activity will tend to increase the free energy of reactant states and decrease the free energy of transition states along the reaction pathway, resulting in lowered barriers and faster reaction, thus higher catalytic power.

If one or a few of the component reaction barriers are substantially higher in free energy than the others during the early stages in the evolution of an enzyme, then mutations that reduce these barriers will be preferred strongly (because the resulting increases in the rates of passage over these high barriers will be carried undiminished into the rate of product formation, while mutations in lower barriers, because the rate is much less sensitive to the height of lower barriers, will have negligible influence on the rate). This situation will continue, with reactant states being raised and transition states being lowered in free energy, until all barriers become of equal height, all reactant states of equal energy and thus necessarily, all transition states are of equal energy. Thereafter mutations that lower all or many barriers will be selected in order to bring all states down together. The limit is presented by the combination of E and S to form ES, with rate constant k_1 in Box 2.1. When it has been reduced to the height of the barrier to diffusion of E and S together then enzyme mutation can (probably) go no further. An enzyme like triosephosphate isomerase that has reached this limit is known as a perfect enzyme.

The importance of these matters for the simulating theorist is that many enzymes are likely to have a number of important contributions to the phenomenological constants in the kinetic law. The theorist who compares his calculations to such experimental data will need to address such features. Furthermore, it is probably apparent that the mechanism shown in Box 2.1 is as close to an utterly minimal example as one could imagine. As the complexity of a mechanism increases, so does the kinetic law, so that K_M and K_M-like constants and other novelties multiply floridly.

2.2.3 A Note on Other Approaches

As explained by Cook and Cleland[1] and by Stein,[2] steady-state kinetic studies are not the only possibility. Transient kinetics methods are constantly being used for determination of microscopic rate constants and elegant isotopic methods, also well described by the same authors, can reveal much.

2.3 Typical Components of a Simulation Study of Enzyme Catalysis

2.3.1 Structural and Other Background

As with all scientific studies, enzymic simulation approaches begin with a suitable problem and when the determination of rate constants is to be involved, then a well characterised enzyme (structurally, kinetically, mechanistically, . . .) is an ideal starting point, and the situation is commonly 'the more the better'. Minimally, a high-resolution X-ray diffraction or neutron-diffraction structure of the enzyme is needed, along with good structural data for substrates, products, as well as any intermediates and cofactors thought to participate; structural data for complexes of any of the latter species with the enzyme again would be ideal.

2.3.2 Selection of QM and MM Regions and Methods

QM calculations are relatively expensive in time and money. MM calculations, in contrast, make use of carefully established potential-energy functions for small distortions of bond distances and angles, non-bonded interactions, electrostatic interactions and other features (like the widely used CHARMM force field, 'Chemistry at Harvard Molecular Mechanics') and are fast and cheap. Thus, while an enzyme-catalysed chemical reaction might ideally be subjected to a QM treatment of the entire system, practicality dictates that the luxury of quantum mechanics be indulged only when truly needed, *i.e.* in the part of the system where bonds are forming and breaking. Obviously, the QM region encompasses the active site structures that participate in the catalytic chemistry, whether enzyme, substrate, cofactor or solvent fragments. However, quite generally the QM region is much smaller than the MM region.

2.3.3 The Border of the QM Region and its Embedding in the MM Region

Essentially the entire power of modern quantum chemistry is at the disposal of the theorists. A great many theorists have in fact simplified the lives of others by making excellent programs for the requisite computations available on websites. Of course, the higher the level of the QM method, the better, most of us would have thought. The expense involved in treating even part of a system with very high-level approaches is high, however. The bright side is that QM approaches at even very low levels, if cleverly deployed, appear in practice to yield very satisfactory results.

As the Truhlar group has taught us the QM region may be thought of as embedded in the surrounding MM region and the group has distinguished both electrostatic embedding and mechanical embedding. Figure 2.1 shows, as an example for reference, the QM region of a simulation study of the action of a bacterial zinc-metallo-β-lactamase in catalysing the hydrolysis of two β-lactam antibiotics.[5] As the caption describes, the link-atom approach was used at the indicated points to allow the crossing of a bond between the QM and MM regions. Lin and Truhlar[6] have reviewed the various ways that have been used to achieve this and arrived at the verdict that the mechanical approach is simple and fast, while the electrostatic approach is more complicated but that both link-atom approaches and others are likely to be pursued in the future.

The entire MM region is treated according to an analytical force-field procedure such as CHARMM or one of its competitors. One of the valuable properties of these procedures is that the potential energy functions of the system, which are sums of the contributing functions for the structural variables such as internal coordinates or Cartesian coordinates, can be analytically differentiated to the first degree, which then yields the forces acting on each atom (and thereby guides the system in finding reaction pathways), and analytically differentiated to the second degree to produce

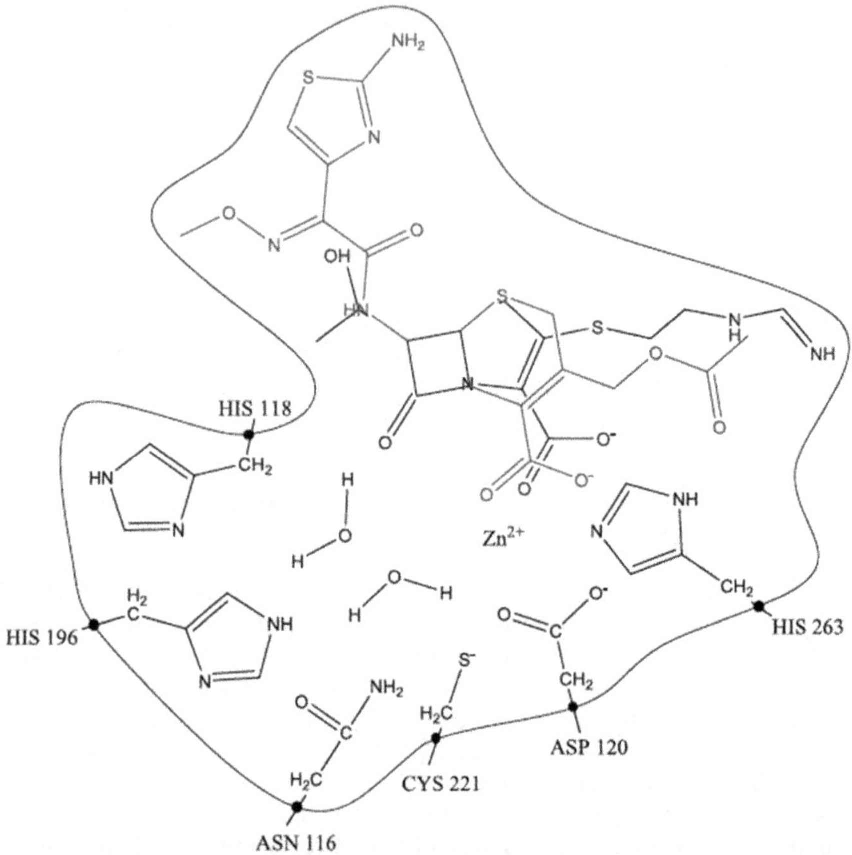

Figure 2.1 The QM region for a QM/MM study of the hydrolysis of two antibiotics, shown in black (imipenem) and grey (cefotaxime), by the zinc metallo-β-lactamase, CphA from *Aeromonas hydrophila.* The QM region thus consists of 94 (imipenem) or 104 (cefotaxime) atoms. The MM region contained all remaining atoms of the enzyme, counterions and waters of crystallisation, as well as the water molecules occupying a box of $80 \times 80 \times 100$ Å after the deletion of a central cavity to accommodate the enzyme–substrate system. The block spots on the boundary between QM and MM regions are 'link atoms' that represent a scheme to address the fact that a bond of each of the six enzyme residues that lie partly in the QM and partly in the MM region must cross the boundary line. The simplest approach may be to place a hydrogen nucleus on the boundary and bond it to the electron pair of the QM atom. This leaves a charge on the MM atom nearby but this can be artificially set to zero or otherwise altered if desired.
Reprinted from C. Meliá *et al.*, Theoretical studies of the hydrolysis of antibiotics catalyzed by a metallo-β-lactamase, *Arch. Biochem. Biophys.*, **582**, 116–126.[5] Copyright 2015 with permission from Elsevier.

the Hessian matrix of the second derivatives of the potential energy with respect to the spatial coordinates, which give the curvature of the potential-energy surface.

2.3.4 Establishing the Potential-energy Surface

Thus in the MM region, relatively straightforward computations yield at any point on the surface the value of the potential energy, its first derivative with respect to structural coordinates (the tangents to the surface), and its second derivatives with respect to the structural coordinates (the curvatures of the surface along each coordinate). These results produce what amounts to an avalanche of information on the system and its physical behaviour.

The values of the potential energy itself allow the construction of the familiar maps of potential-energy surfaces (it is probably good to remind oneself that the 3D maps one commonly encounters are typically potential energy *versus* two coordinates such as a solvent coordinate and a ligand coordinate, while the highly multidimensional maps at issue in the actual computations have the energy plotted *versus* $3N$ coordinates where N is the number of atoms in the system (enzyme plus substrate plus relevant solvent molecules), and to trace out on it pathways such as reactive trajectories or routes from reactant state through transition state to product state.

When a point is located at a stable state on the surface, representing a structure such as that of the enzyme bound to a reactant molecule, a product molecule, or an intermediate-state molecule, then the point must be at a local minimum along all internal coordinates, *i.e.* the structure must exist at the bottom of an energy well (*i.e.* the structure is at a local minimum of energy, with the tangents at the minimum all zero (from the first derivatives) and the curvatures from the second derivatives all positive, showing that the energy rises as the system departs in any direction from the energy minimum.

If a point instead exhibits all gradients equal to zero and all curvatures positive except for one and only one curvature which is negative, then that point is a saddle point on the PES: the coordinate with negative curvature will experience a decrease in energy when the system departs from the saddle point in either direction. Along all other coordinates, departure from the saddle point in either direction will lead to an increase in energy, so that there is an energy minimum at the saddle point for these $3N-7$ coordinates; along the reaction coordinate and it alone, there is an energy maximum at the saddle point. The saddle point structure of the system is a potential transition structure and an ensemble of such structures may be said to populate a transition state.

In principle, there may be points that lie at maxima of energy along two or more of the internal coordinates; this kind of point is analogous to the highest point of a peak in potential energy in a three-dimensional space, as opposed to the saddle point generated by one and only one coordinate of negative curvature. Murrell and Laidler[9] expressed the important fact that no minimum-energy pathway (MEP) can pass through a stationary point that is a peak in the potential energy because there must always be pathways passing parallel to a line through the maximum of the peak but which are formed of points lower down the slopes that go up to the maximum and which thus constitute competitive but lower-energy pathways.

2.3.5 Establishing the Reaction Path or Swath

As noted above, an ambition for the final PES is that along the MEP for the reaction (in terms of internal coordinates q for the reactive assembly), the PES will provide at any point q accurate values of three entities: (i) the potential energy $V(q)$, (ii) the gradients $g(q)$ (first derivatives of the energy), which identify stationary points on the surface and guide the simulation forward along the MEP, and (iii) the curvatures along each internal coordinate perpendicular to the MEP $f(q)$ (second derivatives of the energy), which provide harmonic force constants for the degrees of freedom normal to the MEP that allow the construction of partition functions and calculation of free energies. The points $V(q)$, taken together with the first and second derivatives of V, therefore define not a pathway in the narrow sense but a swath (Oxford English Dictionary, 1971: "the space covered by a sweep of the mower's scythe; the width of grass or corn so cut").

With not only the energy but also the gradients which permit the identification of stable points at energy minima AND also the Hessian matrix of second derivatives, one is in a position to construct from the properties available including the vibrational energy levels (easily including zero-point energy) and frequencies, thus the partition functions at each point.

2.3.6 Development of a Free-energy Surface

The partition functions are the route to thermodynamic functions and a free-energy surface. Sections through it such as potentials of mean force and free-energy profiles follow readily. Again as an example[10] from the groups of Soliman and Williams for reference, Figure 2.2 shows a free-energy surface for a step in the xylanase-catalysed hydrolysis of a phenyl ester at the anomeric carbon. One reaction coordinate, ξ_1, is that for nucleophilic displacement by a water molecule at the anomeric centre of the *p*-nitrophenyl leaving group, while the second reaction coordinate, ξ_2, is that for transfer to the leaving-group oxygen of a proton from an enzymic residue. The free-energy surface is beautifully informative and suggests, as the caption emphasises, that the displacement is prior to the proton transfer and complete before the proton transfer begins.

2.3.7 Calculation of Rate Constants

Once the free-energy surface is in hand, the calculation of rate constants may be as simple as reading from the surface the free energy of activation and using a simple relationship to obtain the rate constant, or use an expression like those in the next section to perform the calculation.

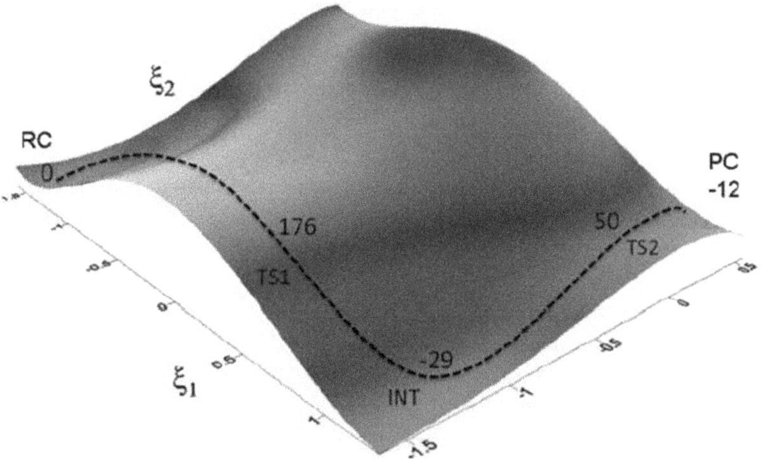

Figure 2.2 A free-energy landscape for the action of the *endo*-1,4-β-xylanase of *Bacillus circulans* in the glycosylation step for the nucleophilic attack by Glu78 on the anomeric carbon of the substrate, assisted by protonation of the oxygen of the *p*-nitrophenyl leaving group by Glu172. The variable on the left axis is the reaction coordinate for nucleophilic displacement at the anomeric carbon and the variable on the right axis is the reaction coordinate for proton transfer to the leaving-group oxygen. The two reactions emerge from the QM/MM simulation as occurring in two largely uncoupled reactions, the nucleophilic displacement first and the proton transfer faster but endergonic. Note that the protonation of the *p*-nitrophenoxide leaving group is endergonic under the circumstances. Reproduced from ref. 10 with permission from the Royal Society of Chemistry.

In many cases, the simulation procedures described elsewhere in this volume are used. To illustrate typical general approaches, a study of the thymidylate synthase mechanism by Kanaan *et al.*[7,8] will be alluded to all too briefly. This enzyme possesses the capacity to accelerate a reaction involving the cofactors methylene-tetrahydrofolate and dihydrofolate and as substrate dUMP (2′-deoxyuridine-5′-monophosphate) and as product dTMP (2′-deoxythymidine-5′-monophosphate). An enzymic cysteine residue assists in the process. In an early step, the cysteine S forms a covalent bond to the deoxyuridine ring at C6 and displaces electron density to C5, where it serves to accept the methylene group from methylene-tetrahydrofolate. The next step is the focus of this study: a hydride ion moves from C8 of the liberated tetrahydrofolate to the methelenyl side chain to generate the methyl group of the product dTMP, and the displaced electron density occasions the fission of the C6–S bond, thus generating the free product dTMP. One question is the degree to which the hydride transfer reaction and C–S bond fission reaction are coupled into a single process (are they concerted?).

In short, the QM region consisted of most atoms of the folate moiety, most atoms of dUMP, and most atoms of the cysteine residue and a nearby arginine residue thought to stabilise electron density of the sulfur atom in transition states for the formation and fission of the C–S bond. Four carbon atoms lay on the QM/MM boundary and were adapted for their roles in transferring information between the two regions by building onto the carbon atoms hybridised s + p orbitals such that some of the hybrid orbitals existed in the QM and some in the MM region (a procedure known as the generalised hybrid orbital or GHO method). The QM calculations were performed with the Austin Model 1 (AM1) approach, while CHARMMc22 was used for MM calculations within a 20 Å radius of the QM centre; outside this sphere, the structures were held rigid. Solvating waters were treated by the TIP3P force field.

A potential-energy surface was calculated and two-dimensional potentials of mean force (free-energy maps) were generated. The two distinguished reaction coordinates corresponded to the H-transfer contribution and the C–S fission. The results corresponded to a common transition state for the two processes (thus concerted) but with very different degrees of progress between reactant-like structures and product-like structures (thus asynchronous). The H-transfer process appeared to be about half-complete but the bond fission reaction was in its earliest stages.

Kinetic isotope effects are treated in general and in detail elsewhere in this volume. In the present example, they were calculated by first obtaining classical-mechanical free energies of activation. Then from the simulation results, instantaneous normal-mode analysis was carried out on 1200 [configurations] in the reactant zone, 800 in the [transition-state] zone, and 800 in the product zone. These led to a quasiclassical potential of mean force (PMF) and quasiclassical free energy of activation, and the estimation of tunnelling corrections in a heroic set of calculations. These results did not exactly reproduce the experimental isotope effects but the most difficult of the results to reproduce in past work, namely their independence of temperature, has been in fact reproduced.

2.4 Analytical Expressions for Rate Constants

Once the free-energy swath is established connecting the reactant region by way of the neighbourhood of the saddle point to the product region, there may be many ways to obtain a rate constant under conditions of interest. Table 2.1 contains four examples of such expressions.

2.4.1 The Stable States Picture[11,12]

The Grote–Hynes model is somewhat distinct from the others. Like all the examples, the stable states picture develops a MEP on a PES and then a PMF or free-energy swath along the MEP connecting the reactant-state energy minimum with the product-state energy minimum by way of the neighbourhood of the saddle point. In contrast to the others, however, Grote and Hynes did not erect a single, infinitesimally thick surface at the saddle point. Instead, they envisioned the conversion of a stable reactant state (stable in the sense that the state resides at a multidimensional minimum along all

Table 2.1 A few examples of analytical expressions for rate constants.[a]

Hynes *et al.*[b] Stable states picture	$k = \frac{\beta}{2\hbar}\langle C^2\rangle\left(\frac{\pi}{\beta\Delta G_{tot}}\right)^{1/2} x_{sc}\exp(-\beta\Delta G_{tot})$	Two dividing surfaces, one between reactant state and saddle point, one between the latter and product state. Fluxes from time correlation functions.
Truhlar and Garrett[c] Variational transition-state theory	$k^{CVT} = k(T)\frac{k_B T}{hC^o}\exp\{-[G_T^0(\text{CVT}) - G_T^0(R)]/RT\}$ $\Delta G_T^0 = \max_s \Delta G_T^0(GT, s)$	Separability of reaction coordinate assured by eliminating recrossing (maximum barrier height); quantisation along the reaction coordinate then allows tunnelling corrections for that coordinate.
Hammes-Schiffer *et al.*[d] Proton-coupled electron transfer	$k^{EPT} = \sum_\mu P_\mu \sum_\nu \frac{\lvert V^{e1}S_{\mu\nu}\rvert^2}{h}\sqrt{\frac{\pi}{\lambda k_B T}}\exp\left[-\frac{\left(\Delta G_{\mu\nu}^0 + \lambda\right)^2}{4\lambda k_B T}\right]$	Variant for electron/proton transfer shown, the electron and proton are both fully quantised and reach different termini in the product state.
Klinman *et al.*[e] Quantum statistical mechanical rate theory	$k_{tun} = \sum_v P_v \sum_w \frac{1}{2\pi}\lvert V_{e1}\rvert^2\sqrt{4\pi^3/\lambda RT\hbar^2}\exp^{-(\Delta G^0 + E_{vib} + \lambda)^2(4\lambda RT)} \times (\text{F.C. term})_{v,w}$	Klinman and Kohen[23] and others who employ this expression usually cite Kuznetsov and Ulstrup.[16]

[a] k = rate constant, k_B = Boltzmann constant, T = temperature in Kelvin, $\beta = k_B T/h$, h = Planck's constant, $h/2\pi = \hbar$ with diagonal bar, G = Gibbs free energy.

[b] Hynes *et al.*[11,12] $\langle C^2\rangle$ = proton coupling, χ_{SC} is a usually small correction to reactant–product coupling and energetic splitting deriving, in an H-transfer reaction, from solvent-driven fluctuations in the donor–acceptor distance.

[c] Truhlar and Garrett[13] CVT = canonical variational transition-state theory or transition state, κ = transmission coefficient, G_T^0 = Standard Gibbs free energy at temperature T, R = reactant state or gas constant, GT = generalised transition state, s = distance along a minimum-energy pathway from the saddle point.

[d] Hammes-Schiffer *et al.*[14,15] EPT = electron, proton transfer; P_μ, Boltzmann probability of occupancy of μth reactant vibronic energy level, $V^{el}S_{\mu\nu}$, product of electronic coupling and overlap integral between μth (reactant) and νth (product) vibronic energy levels, λ the reorganization energy.

[e] Klinman *et al.*[23] Most symbols the same as in Hammes-Schiffer *et al.* (F.C. term)$_{\mu\nu}$ = Franck–Condon terms between μth (reactant) and νth (product) vibronic levels.

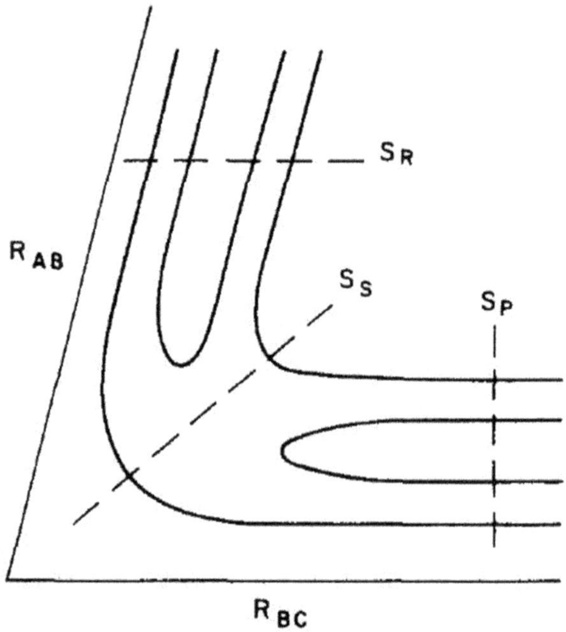

Figure 2.3 The contours represent potential energy plotted against mass-weighted (isoinertial) internal coordinates for the reaction of A+B-C to generate A-B+C. The reactant-state valley is that which 'points north'; the valley that 'points east' is the product-state valley. The three dividing surfaces of the Grote–Hynes theory are that labelled S_R which divides the reactant state on the 'north side' of the surface from the reactant-side intermediate region on the 'south side' of S_R; S_S, which divides the reactant side of the intermediate region on the 'north-west side' of S_S from the product-side of the intermediate region on the 'south-east side' of S_S and S_P, which divides the adjacent part of the intermediate region to the 'west' of S_S from the product-state valley to the 'east' of S_S. Reproduced with permission from R. F. Grote and J. T. Hynes, *J. Chem. Phys.*, 1980, **73**, 2715.[11] AIP Publishing LLC.

internal coordinates) overall into a stable product state, by transmission through, first, a dividing surface that separates the reactant state from an intermediate region and, second, through another dividing surface that separates the intermediate region from the stable product state (see Figure 2.3).

The expression shows the expected resemblance to transition-state theoretical approaches but contains the square of the proton coupling between reactant and product states, ⟨C2⟩ which controls the state-to-state tunnelling. For a fuller and more interesting treatment, the reader cannot do better than to consult the next chapter in this volume.

2.4.2 Variational Transition-state Theory

The expression of Truhlar and Garrett[13] for the canonical variational transition-state theory (CVT) shows a familiar transition-state theoretical form as well but in the version shown, the transition state can be moved away from the saddle point at $s = 0$ and variationally adjusted along the reaction path s until the free energy of activation is maximised. This is done in order to select that dividing surface between reactants and products that minimises the flux into the product state through making the free-energy barrier a maximum (see Section 2.5.1).

2.4.3 Hammes-Schiffer *et al.* and Klinman *et al.*

These two entries have some resemblances but also some important differences. Both are based on the concept that the reaction (for Hammes-Schiffer *et al.*, proton-coupled electron transfer, PCET, in the most general sense but here the expression is for the case of electron-proton transfer; for Klinman *et al.*, the reaction types can be quite general but commonly involve proton transfer and sometimes electron transfer as well) can be described in a framework of state-to-state particle transfers between reactant-state vibronic or vibrational states. The detailed model for Klinman *et al.* is a version of that described by Kuznetsov and Ulstrup[16] who attributed the basic idea, that chemical proton-transfer reactions can be regarded as quantum-mechanical electronic-vibrational transitions, to Dogonadzhe, Kuznetsov and Levich. The model used by Hammes-Schiffer *et al.* makes use of her NEO (nuclear electronic orbital) formalism, in which the Born–Oppenheimer approximation is not applied.[14,15]

2.5 An Instructive Example: Rate Constants from the Multiconfigurational Molecular Mechanics Approach QM/MM–MCMM

D. G. Truhlar and his coworkers have described in several publications[17–19] the approach of multiconfigurational molecular mechanics, abbreviated MCMM. This approach is characterised by a powerful focus on developing QM/MM simulation approaches to calculate accurate values of rate constants while keeping the costs of computation, in both time and money, within practical limits.

The full method is known as QM/MM–MCMM, which is usable for "studies of very large systems" such as enzymes and enzyme-catalysed reactions. (If the readers of this volume are not aware already, they will soon discover that the field is replete with lengthy acronyms not as intuitively decipherable as was probably intended.) Although enzyme-catalysed reactions are of course the focus of the present volume, we will describe this application of QM/MM–MCMM to smaller systems to make the rather complex procedure as transparent as we are capable of making it. The method is targeted on eventually producing rate constants that can fruitfully be compared with experimental rate constants in order to illuminate their physical underpinnings. At present, however, the rate constants computed in QM/MM–MCMM studies are to be compared to 'direct dynamics' calculations, in which high-level calculations of the dynamics along the MEP are used to establish a standard set of accurate rate constants as targets toward which to work.

2.5.1 Elements of the QM/MM–MCMM Approach

The main elements of this approach to using simulation methods to obtain values of rate constants are several:

1. Obtaining important points on the PES makes use of an empirical valence-bond method with two resonance contributing structures, one reactant-like and one product-like, for QM parts of the calculations.
2. In addition, MM functions are liberally employed wherever appropriate, even in the QM region.
3. Third, mass-weighted internal coordinates are used to describe the atomic positions and will facilitate low-order expansions of the energy about calculated points.
4. Fourth, the technique of Shepard interpolation, in which each new point is made a weighted average of all or some previous points, with the weights employed in averaging strongly emphasising the nearest of the previously calculated points.
5. The construction of the MEP from transition state forward to the product state and backward to the reactant state according to the Page–McIver procedure,[20] the conversion from a PES along the reaction swath to a free-energy surface by constructing the partition functions and from them the thermodynamic functions, and thence the rate constants, *etc.*

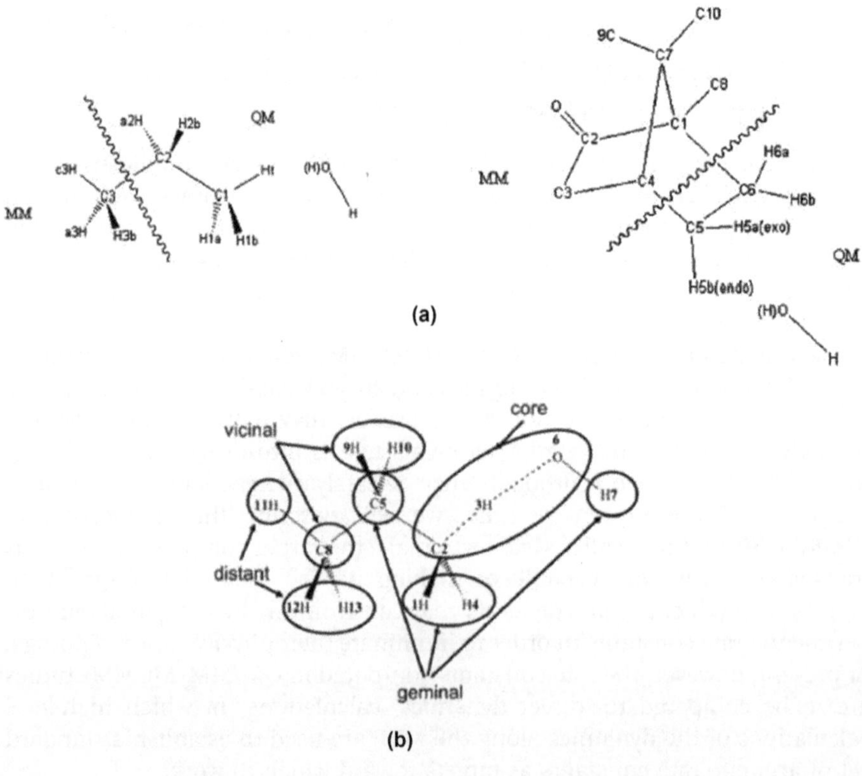

6. Application of the variational transition-state theory to obtain accurate rate constants. A more extensive discussion of some of the features of QM/MM–MCMM follows.

2.5.2 The Empirical Valence-bond Technique for the QM Region

In the bimolecular reactions for which the reactive assemblies are shown in Figure 2.4, the reactant-like structure will be a van der Waals complex at a shallow energy minimum in the reactant valley of the PES, containing the hydroxyl-radical H-acceptor and the organic H-donor. The product-like

Figure 2.4 (a) The QM–MM regional segregation of the system of H-atom transfer to hydroxyl radical from propane (top left) and camphor (top right). In each structure, the wavy surface divides the MM region where the energy will be described by the mechanical functions of the MM formalism and where no bonds will be made or broken. The QM region contains the hydroxyl radical that will accept a hydrogen atom, forming a bond between the oxygen and the parenthetical hydrogen atom to generate a water molecule in the product state; the same hydrogen is seen still attached to the C atom of the co-reactant, as will be the case for the reactant-like structure. During the simulation the H will in each case move from the organic structure to the hydroxyl oxygen, leaving a carbon radical behind. Reproduced with permission from H. Lin, J. Pu, T. V. Albu and D. G. Truhlar, *J. Phys. Chem. A*, 2004, **108**, 4112.[21] Copyright (2004) American Chemical Society. (b) The higher-efficiency scheme of Lin *et al.*[19] The *core* region is so labelled; it contains only the three atoms C_2–H–O undergoing bond formation and bond breaking. This region must always be treated as a QM region. The three *geminal* atom groups contain atoms bonded to core atoms; they may be treated as QM or MM depending on the circumstances, although QM should give higher quality results (just as MM gives faster computation) particularly near the saddle point. Near the reactant or product state, MM should suffice. The two sets of *vicinal* atoms are bonded to germinal atoms and should inflict still less damage from losing QM status but could be treated as QM if desired. The two groups of *distant* atoms would be elevated to QM status only if all atoms were treated as QM. The strategic computational application of Shepard interpolation with variable assignments of atoms to QM and MM can be illustrated by the use of high-level electronic structure data to determine the Hessian matrix elements for the saddle point of structure (b). At the saddle point itself (Shepard point 0) the 10 atoms labeled core, germinal and vicinal are all reassigned as core atoms. QM data across all 10 atoms (excluding only the distant H11, H112 and H13) are used to interpolate the Hessian elements for the saddle point. Then Shepard points 1–10, located nearby, are subjected to stepwise transfer out of the core of first three vicinal atoms at supplementary Shepard points 1, 2 and 3, then the four geminal atoms at Shepard points 4, 5, 6 and 7, with Shepard points 8, 9 and 10 having only the three atoms shown as core atoms treated to interpolation of QM data and all others treated by MM methods. Reproduced with permission from H. Lin, Y. Zhao, O. Tishchenko and D. G. Truhlar, *J. Chem. Theory Comput.*, 2006, **2**, 1237.[19] Copyright (2006) American Chemical Society.

structure will be a similar van der Waals complex, now in the product valley of the PES, containing the atoms of the product water molecule and the organic radical product. The Hamiltonian matrix for the QM calculation will have two diagonal elements, V_{11} for the reactant-like structure and V_{22} for the product-like structure. Because each of the structures contains only stable, fully bonded constituents, the energies of both can be accurately estimated from the appropriate MM functions.

The two off-diagonal elements $V_{12} = V_{21}$, the resonance integral, give the resonance energy or the increment by which the reactant and product and states between them are altered in their energies as a result of configuration interaction between the contributing structures. There are thus two PESs relevant to the chemical reaction (H-atom transfer in the systems of Figure 2.4); the lower-energy surface (the ground state) and the higher-energy surface, an electronically excited state. In the work under discussion, the reaction was taken to occur on the ground-state PES.

2.5.3 The Case of the Resonance Integral

The identical off-diagonal resonance integrals $V_{12} = V_{21}$ describing the configurational interaction between the two contributing states in the QM calculation just described emerges as difficult and techniques for modelling it have been explored. For example, in a paper on increasing computational efficiency in rate-constant calculations, Lin *et al.*[21] made use of a protocol in which $V_{12}(q)$ is interpolated for a new point on the MEP by the Shepard procedure.

The Shepherd interpolation procedure consists of constructing a weighted sum of all previously calculated values of the desired quantity – say, V_{12} – where the weights are normalised values of the reciprocal distance along the MEP from the new point to each other existing point on the surface, the reciprocal distances being raised to the fourth power. In an abundance of caution, because this approach tends to 'blow up' at greater distances, a further modifying function to exclude unsatisfactory values of the resonance integral accompanies each 'Shepard point' contribution into the interpolation. This weighting powerfully favours the nearest points and diminishes the role of the more distant and seems to have proven effective in arriving at a PES that leads to good rate constants.

2.5.4 Identification and Characterisation of Stationary Points

At any point q in the multidimensional space of internal coordinates, one can obtain the energy V, the first derivative of V, or gradient, and the second derivative of V, or curvature, from which vibration frequencies can be calculated (the matrix of second derivatives is called the Hessian matrix). For stable states like reactants, products and intermediates, the gradients will be zero at the point when all $3N-6$ internal coordinates describe the minimum-energy structure. In addition, the second derivatives of V should all be positive at the minimum of energy and give the vibrational force constants for the stable structures. Transition states are characterised by having all gradients equal to zero at the saddle point structure, as with stable structures; the force constants or curvatures will be positive for all the vibrational coordinates except for that which corresponds to motion along the MEP that connects reactants, products and saddle point. The saddle point lies at a maximum of energy and the MEP curves downward toward both reactant and product states: thus the curvature or force constant at the saddle point is negative for the motion along this normal coordinate. The vibrational 'frequency' for this motion is proportional to the square root of the force constant and is an imaginary number. The saddle point coordinates thus define the structure of the transition state.

2.5.5 Minimum-energy Pathways

One might be tempted to strike out from the reactant state in a simulation by watching for high-energy members of the Boltzmann distribution and following their progress toward the saddle point of the transition state, at a sufficient number of points pausing to calculate the free energy in the swath about the point. It soon emerges that such events are far too rare to be useful. Instead one must find the saddle point and progress away from it toward either reactants or products, choosing the pathway of steepest descent; such a procedure was followed by Truhlar *et al.*[17–19] At any point, the potential energy, its gradient and the curvatures are available, and thus all that is needed for the conversion of a potential-energy surface to a free-energy surface. The free-energy difference between saddle point and the reactant point is the free energy of activation which after consideration of tunnelling is all that is necessary to calculate the rate constant.

2.5.6 Toward Good, Cheap Hessians

Expense in conducting simulations of large-scale systems such as enzymic reactions, insofar as the main data available from PESs are the value of the potential energy $V(q)$, the first derivative or gradient $g(q)$, and the Hessian matrix of second derivatives, the vibrational force constants $f(q)$. The last are the highest in cost and also the most necessary in that accuracy is sensitive to quality in cases like vibration frequency and tunnelling, both quantum phenomena. In their quest for efficiency in computing, Truhlar and his coworkers have sought ways of obtaining Hessian matrices short of full-scale high-level QM calculations in all cases. One approach was described by Lin *et al.*, who combined the use of successive Shepard interpolations of Hessian matrix elements with high-level QM calculations for the beginning 'Shepard points' incorporating the sets of (i) all atoms involved in bonds that make or break at the saddle point (core atoms); (ii) all atoms directly bonded to core atoms (geminal atoms) and all atoms directly bonded to geminal atoms (vicinal atoms). Then a series of supplementary Shepard points in the neighbourhood are selected and interpolations are carried at each of these. For those closest, the vicinal atoms are removed one by one from QM treatment and demoted to MM treatment. Following this, at more distant supplementary Shepard points, the geminal atoms are similarly, one by one, demoted to MM treatment. An example is explained in Figure 2.4 above.

2.6 Good Hessians Give Good Rate Constants

When the operators are satisfied, they may bring themselves to compare the rate constants calculated from the VTST approach in this 'on the fly' method with those from direct dynamics (DD) as the theoretical gold standard. In the reaction of HO with CH_4 in the gas phase at 300, 600 and 600 K, the DD results were 1013 k (cm^3 $molecule^{-1}$ s^{-1}) = 1.80, 988 and 7650 at the three respective temperatures. The highest level of calculation (MCMM-10) of the technique yield 1.59, 900 and 7280. This must have been very satisfying.

Acknowledgements

I am far too grateful to be able to express it adequately to Iñaki Tuñón and Vicent Moliner, two excellent scientists and gentlemen of high honour for a level of kind and indeed saintly patience and collegiality with me as I struggled like a mewling infant to prepare this article. I fear now it recalls Aesop's story of the "mountain who labored and brought forth a mouse".

References

1. P. F. Cook and W. W. Cleland, *Enzyme Kinetics and Mechanism*, Taylor & Francis Group, LLC, New York and Oxford, 2007.
2. R. L. Stein, *Kinetics of Enzyme Action – Essential Principles for Drug Hunters*, John Wiley & Sons, Hoboken, 2011.
3. I. H. Segel, *Enzyme Kinetics – Behavior and Analysis of Rapid-Equilibrium and Steady-State Systems*, Wiley-Interscience, New York, London, Sydney, & Toronto, 1975.
4. A. Warshel, *Computer Modeling of Chemical Reactions in Enzymes and Solutions*, John Wiley & Sons, 1991.
5. C. Meliá, S. Ferrer, V. Moliner and J. Bertran, *Arch. Biochem. Biophys.*, 2015, **582**, 116.
6. H. Lin and D. G. Truhlar, *Theor. Chem. Acc.*, 2007, **117**, 185.
7. N. Kanaan, M. Roca, I. Tuñón, S. Martí and V. Moliner, *J. Phys. Chem. B*, 2010, **114**, 13393.
8. N. Kanaan, S. Ferrer, S. Martí, M. Garcia-Viloca, A. Kohen and V. Moliner, *J. Am. Chem. Soc.*, 2011, **133**, 6692.
9. J. H. Murrell and K. J. Laidler, *Trans. Faraday Soc.*, 1968, **64**, 371.
10. M. E. S. Soliman, J. J. R. Pernía, I. R. Greig and I. H. Williams, *Org. Biomol. Chem.*, 2009, **7**, 5236.
11. R. F. Grote and J. T. Hynes, *J. Chem. Phys.*, 1980, **73**, 2715.

12. D. Borgis and J. T. Hynes, *J. Chem. Phys.*, 1991, **94**, 3619.
13. D. G. Truhlar and B. C. Garrett, *Annu. Rev. Phys. Chem.*, 1984, **35**, 159.
14. S. Hammes-Schiffer, *J. Am. Chem. Soc.*, 2015, **137**, 8860.
15. A. Sirjoosingh, M. V. Pak, C. Swalina and S. Hammes-Schiffer, *J. Chem. Phys.*, 2013, **139**, 034103.
16. A. M. Kuznetsov and J. Ulstrup, *Can. J. Chem.*, 1999, 77, 1085.
17. Y. Kim, J. C. Corchado, J. Villa, J. Xing and D. G. Truhlar, *J. Chem. Phys.*, 2000, **112**, 2718.
18. T. V. Abu, J. C. Corchado and D. G. Truhlar, *J. Phys. Chem. A*, 2001, **105**, 8465.
19. H. Lin, Y. Zhao, O. Tishchenko and D. G. Truhlar, *J. Chem. Theory Comput.*, 2006, **2**, 1237.
20. M. Page and J. W. McIver, Jr., *J. Chem. Phys.*, 1988, **88**, 922.
21. H. Lin, J. Pu, T. V. Albu and D. G. Truhlar, *J. Phys. Chem. A*, 2004, **108**, 4112.
22. W. J. Albery and J. R. Knowles, *Biochemistry*, 1976, **15**, 5631.
23. J. P. Klinman and A. Kohen, *Annu. Rev. Biochem.*, 2011, **82**, 471.

CHAPTER 3

A Transition State Theory Perspective for Enzymatic Reactions: Fundamentals and Applications

JAMES T. HYNES,*[a,b] DAMIEN LAAGE,*[b] IÑAKI TUÑÓN[c] AND VICENT MOLINER[d]

[a] Department of Chemistry and Biochemistry, University of Colorado, Boulder, CO 80309-0215, USA; [b] Ecole Normale Supérieure-PSL Research University, Chemistry Department, Sorbonne Universités-UPMC Univ. Paris 06, CNRS UMR 8640 PASTEUR, 24 rue Lhomond, 75005 Paris, France; [c] Departament de Química Física, Universitat de València, Spain; [d] Departament de Química Física i Analítica, Universitat Jaume I, Spain
*Email: chynes43@gmail.com; damien.laage@ens.fr

3.1 Introduction

The character of this chapter has been dictated by the fact that there are a significant number of reviews of the theoretical aspects of enzyme reactions already in existence,[1–6] including a very recent one by several of us last year on the much discussed issue of dynamical effects in enzymatic catalysis.[7] In view of this, we have decided to focus on issues that have received much less exposure in the review literature. Namely, we address in some detail the more molecular aspects of the transition state theory (TST) approach.

RSC Theoretical and Computational Chemistry Series No. 9
Simulating Enzyme Reactivity: Computational Methods in Enzyme Catalysis
Edited by Iñaki Tuñón and Vicent Moliner

Published by the Royal Society of Chemistry, www.rsc.org

This includes attention to: what TST actually says, why it works when it does, and what deviations from it – even if not enormous numerically – have to tell us about the molecular events occurring during the passage through the transition state (TS) region for the reaction. The latter events are of course intimately related to the reaction mechanism in the critical TS region; it is they that decide whether the reaction is successful or not. We also focus on theoretical ideas and models rather than computational methods, which are discussed elsewhere in this book. However, computational results are discussed in connection with those theoretical aspects, as revealing their validity and value or the lack thereof.

Even though our chosen structure represents in some ways a limited goal, we will still need to restrict ourselves to a few representative enzymatic reactions in order to expose the ideas in an intelligible way. This has resulted in a definite (and we hope understandable) slant towards the work and perspective of the present authors, but we do attempt to mention other related work in the literature as we proceed. But we confess at the outset that we have probably missed some relevant contributions, for which we ask our colleagues' pardon.

Our focus is of course on enzymatic reactions, but as in our earlier effort,[7] we consider, along with ref. 8 for example, that the analogue solution reactions provide useful independent sources of significant insight, in addition to playing their more usual 'bystander' role as providing a numerical comparison reference in order to quantitatively assess the enzyme's catalytic impact. Most of this is expounded upon in the context of particular enzyme reactions.

The outline of the rest of this chapter is as follows. Section 3.2 exposes first the essential features of the TST approach and theoretical descriptions of deviations from the TST predictions. The discussion is not extensive but we hope sufficient for the purpose; the interested reader is referred to our more recent detailed exposition.[7] Section 3.3 is primarily concerned with enzymatic reactions of SN2 nucleophilic substitution character, an important reaction class. We discuss three enzyme examples in detail, with what we hope is illuminating connection with an aqueous solution example. The two SN2 example systems do not span the character of different types of departure from TST, and we remedy this lacuna by discussing an enzymatic addition reaction. Section 3.4 discusses the additional issues which arise when a quantum particle such as a proton or hydride ion is transferred in the reaction, together with several enzymatic reaction examples. We provide some concluding remarks in Section 3.5.

3.2 TST and Allied Theories for Enzyme Reactions

3.2.1 Assumptions and Structure of TST

The theoretical basis most important for our (and any) enzyme rate constant discussion is the TST. This theory often provides the sole basis for the rate assessment of an enzyme or other reaction, and even in those cases where the description goes beyond TST, the theory still provides the reference value for the rate. The clearest understanding of TST is provided by the 'no-recrossing rule' version, due to E. Wigner.[9] In this version, the TST rate constant (for the forward reaction) is given by the equilibrium average of the one-way flux across the transition state surface:

$$k_{\mathrm{TST}} = \langle J_{+}^{\ddagger} \rangle_{\mathrm{R}} \tag{3.1}$$

where the average is normalised with respect to the reactants R, which reflects its focus on the forward reaction. If the TS – traditionally labelled by the symbol ‡ – is defined by the surface $x = x^{\ddagger}$, then $J_{+}^{\ddagger}$ is the one-way flux:

$$J_{+}^{\ddagger} = \frac{p}{m}\theta_{+}(p)\delta(x - x^{\ddagger}) \tag{3.2}$$

Here p and m are the x coordinate's momentum and mass at the TS, $\theta_{+}(p)$ is the step function restricting p to have only positive values so that trajectories cross the TS surface $x = x^{\ddagger}$ in the direction from reactants R to products P, and finally the delta function $\delta(x - x^{\ddagger})$ restricts x to the TS surface. Because the specification of the reaction coordinate at the TS still leaves many other 'orthogonal' degrees of freedom unspecified, one can speak of an 'ensemble' of trajectories crossing the TS surface. A simple example may be of use here. In the famous collinear $H + H_2$ reaction[10,11] treated classically, the reaction coordinate at the TS is the unstable antisymmetric stretch motion involving old bond rupture and new bond formation. The orthogonal degree of freedom – which provides the surface aspect of the TS in this simple example – is the symmetric stretch motion where the central H remains fixed and the two other Hs expand and contract. The ensemble of trajectories at the TS surface would present the antisymmetric stretch motion at various conditions of this orthogonal coordinate.

The fundamental TST assumptions – within the context of the idea that the rate constant for a system in chemical equilibrium is the same as in a non-equilibrium kinetics experiment – are (i) that there is classical mechanical motion for the nuclei, and (ii) that the entire distribution of trajectories at the TS, *i.e.* crossing the TS surface from the side of reactants to the side of the products, successfully continue to become (stable) products, *i.e.* there is no recrossing of the TS surface $x = x^{\ddagger}$ (Figure 3.1). Note that this is quite a different statement than saying that all trajectories starting from the reactant region satisfy the above no-crossing condition; indeed, most of those trajectories fail even to reach the TS (see, for example, ref. 7). Eqn (3.1) is already restricted to those trajectories having attained the improbable situation of being at the TS, *i.e.* which are activated at a cost of an activation free energy, as now discussed.

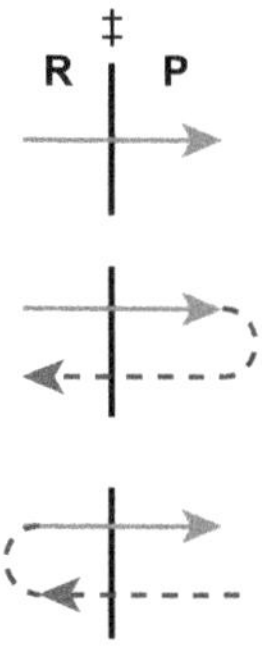

Figure 3.1 Crossing and recrossing trajectories at the TS surface. The TST assumption is that all trajectories are like the topmost trajectory, leading successfully from the side of reactants (R) to the side of products (P), *i.e.* successful crossings of the TS surface, which we label RP. The remaining two trajectory types – which we can label as RR and PR, respectively – are in violation of the TST no-recrossing assumption.

Two alternate forms of eqn (3.1) are frequently encountered. The first (*e.g.* ref. 12):

$$k_{\mathrm{TST}} = \frac{k_{\mathrm{B}}T}{h}\frac{Q^{\ddagger}}{Q_{\mathrm{R}}}\exp\left[-\Delta V^{\ddagger}/k_{\mathrm{B}}T\right] = \frac{k_{\mathrm{B}}T}{h}\left(C^{0}\right)^{1-n}\exp\left[-\Delta G^{\ddagger}/k_{\mathrm{B}}T\right] \tag{3.3}$$

involves TS and reactant partition functions $Q^{\ddagger}$ and Q_{R}, the TS activation potential energy $\Delta V^{\ddagger}$ and activation free energy $\Delta G^{\ddagger}$ – the feature just mentioned in the preceding paragraph. Other factors here are a reference concentration factor C^{0} to guarantee correct dimensions, and the famous Eyring prefactor involving the ratio of the thermal energy and Planck's constant h (see ref. 7 and 12 concerning the correct and incorrect interpretations of this prefactor). Another form of eqn (3.1) and one that is more convenient for our purposes is:

$$k_{\mathrm{TST}} = \frac{\omega_{\mathrm{R}}}{2\pi}\exp\left[-\Delta G^{\ddagger}/k_{\mathrm{B}}T\right] \tag{3.4}$$

Here ω_{R} is a collision frequency of the reactants for a bimolecular reaction (with appropriate units), or is a vibrational frequency in a reactant free energy well for a unimolecular reaction. This form ensures the important feature that the activation free energy $\Delta G^{\ddagger}$ refers to the same number of degrees of freedom in the TS and in the reactants, a feature not shared by eqn (3.3).[13]

The explicit appearance of the exponential activation free energy factor in eqn (3.3) and (3.4) emphasises the point made above that even the attempt at crossing of the TS surface is an improbable event for any barrier height much greater than the thermal energy $k_{\mathrm{B}}T$. It also emphasises the now common feature that calculations of free energy are made along some presumed reaction coordinate, especially for enzymatic reactions and condensed-phase reactions. Such a coordinate will typically change its character on the way from reactants to the TS and on to products (the collinear $H + H_2$ reaction mentioned above is a clear example); it may or may not include aspects of the environment of the basic reacting molecular system, *e.g.* the enzyme active site. Such free energy curves generally involve the assumption that at each point along that coordinate, all other degrees of freedom are equilibrated to it; this condition is rarely met, but usually it is only in the TS region that this assumption is relevant for a rate constant discussion (see ref. 7 for a detailed discussion).

3.2.2 TS Surface Recrossing Corrections to TST

Of course, the TST no-recrossing assumption can be violated in enzyme reactions, *i.e.* trajectories such as those illustrated in Figure 3.1 can occur. The resulting departure of the reaction rate constant from its TST value is conveniently measured by the transmission coefficient κ, defined by the ratio k/k_{TST}, so that:

$$k = k_{\mathrm{TST}}\kappa \tag{3.5}$$

Because those recrossing classical trajectories do not represent successful R to P transitions (in contrast to what TST assumes), in fact they do not contribute to the rate constant, and so κ is reduced to a value below unity. Here we need to stress an important point: the κ value – and thus the departure from TST – depends *via* eqn (3.5) on the definition that is adopted for the TST rate constant, which in turn depends upon the choice that is made for reaction coordinate x in the TS neighbourhood. Such a choice must always be made, because we never know in advance the true, exact, reaction coordinate (and probably never will know it in any real reaction case). Thus, κ could be unity (or very close to it) for one clever or fortunate choice of x and be very small for another less inspired or fortuitous choice, with a corresponding difference in the assessment of the validity of TST and the importance of recrossing. This issue is coupled to what is often called 'variational transition state theory', as we will discuss later. In a simulation context, the transmission coefficient κ for an enzyme reaction can be determined *via* trajectories started at the selected reaction TS; one efficient formula useful for this purpose is the asymptotic limit (usually rapidly established) of:[14]

$$\kappa(t) = \frac{\langle J_+^{\ddagger} \cdot \theta(x(+t))\rangle - \langle J_+^{\ddagger}\theta(x(-t))\rangle}{\langle J_+^{\ddagger}\rangle} \tag{3.6}$$

Here $J_+^{\ddagger}$ is the initially positive flux at $t = 0$ at the TS, and $\theta(x)$ is a step function equal to one on the product side of the reaction coordinate x ($x > x^{\ddagger}$) and zero on the reactant side ($x < x^{\ddagger}$). Several applications of this formula will be detailed in this chapter.

All the above discussion has assumed classical nuclear motion, an assumption clearly inappropriate for proton, hydride and hydrogen atom transfer reactions, and special discussions are required (Section 3.4).

Theoretical descriptions of κ for condensed-phase reactions began with Kramers,[13,15,16] and these ideas have been carried over to enzyme reactions. Kramers described the reaction system's motion in the environment by a Langevin equation involving a frictional force $-\zeta v$, in which $v = \dot{x}$ is the reaction coordinate velocity and ζ is the friction constant. This frictional force can alter the reaction coordinate momentum, *e.g.* reversing a trajectory crossing the TS from R to P, resulting in TS recrossing. Kramers' result for κ is:

$$\kappa_{KR} = [(\zeta/2m\omega_b)^2 + 1]^{1/2} - (\zeta/2m\omega_b) \tag{3.7}$$

where the barrier frequency ω_b is the square root of the magnitude of the TS free energy barrier's negative curvature, divided by the reaction coordinate's effective mass m. κ_{KR} steadily reduces the rate constant from its TST value as the friction increases. The friction constant ζ is proportional to the time integral of the time correlation function of the environment's "random", non-systematic force F on the reaction coordinate at the TS:

$$\zeta \propto \int_0^\infty \mathrm{d}t \langle FF(t) \rangle \tag{3.8}$$

which is larger the larger and longer-lived is the molecular level environment's force on the reaction coordinate. Grote and Hynes (GH) generalised Kramers' result[17] to take into account the very important feature that the environment's impact on the TS barrier recrossing should depend not on the long-term scale friction constant ζ, but instead on the time-dependent friction $\zeta(t)$ (whose time-integrated area is ζ). This reflects the feature that the relevant environmental influence exerted on the reaction system is that exerted during the (often very) short timescale of the passage in the barrier neighbourhood. This timescale is approximately $\omega_b{}^{-1}$, the barrier frequency's inverse; longer timescale opposition by the environment is irrelevant for the barrier recrossing. For many modest to high barrier reactions, this timescale is very short, just a few fs or few tens of fs. The GH transmission coefficient:

$$\kappa_{GH} = \left\{ \kappa_{GH} + \left[\hat{\zeta}(\kappa_{GH}\omega_b) \middle/ m\omega_b \right] \right\}^{-1} \tag{3.9}$$

where $\hat{\zeta}(\omega_b \kappa_{GH})$ is the Laplace transform of $\zeta(t)$, is generally much closer to the TST prediction $\kappa_{TST} = 1$ than is κ_{KR}. As already implied in the discussion above, κ_{GH} predicts smaller environmental dynamical effects on the rate because it is usually difficult for the environment to exert a significant influence on the relevant very short space and timescales of the reaction system's passage through the TS region.

Because GH theory accurately accounts for simulation results for enzyme reactions[18,19] (and more generally for assorted solution reaction classes[20]), we will employ this perspective in some relevant sections of what follows. We can already stress that one of the theory's lessons – noted above and to be confirmed within – is that deviations from TST are not often very large; with a reasonable choice of reaction coordinate, TST is typically an excellent description of the enzymatic reaction rate constant.

Here we devote a limited discussion to one of the limits or regimes of GH theory which will be seen to be useful in characterising enzymatic reaction dynamics. Although this was originally described for a solvent environment,[14,21,22] we couch the discussion in enzymatic terms.

In the 'frozen environment' limit,[14,18,21,22] the enzyme free energy barrier frequency ω_b is sufficiently high and the coupling between the enzyme reaction system and its environment – the active site and even beyond – is not too strong. In this case, there is no time for the friction $\zeta(t)$ to depart from its initial value $\zeta(t=0)$ during any recrossing events. Then the reaction proceeds without any dynamics in the environment during the short barrier passage duration of the trajectories. Of course, the environment is not 'frozen' during the much longer time duration passage from the reactant region to the TS;[23] otherwise no reaction would ever occur! This limit can also be characterised in several useful ways, including a variational transition state perspective,[22] as described below. In the opposite limiting regime – the 'polarisation cage' limit[21] – where the reaction system–environment coupling is instead strong, no successful reaction can occur until environmental dynamics occurs. Here the environmental dynamics is accordingly critical. While there are examples in solution where such dynamics have an influence,[21] enzymatic examples of this limit remain to be discovered.

3.3 Classical Enzyme Reactions

In this section, we will elaborate the application of the theories of Section 3.2 to a series of enzyme reactions. These examples stress both the capability of TST to explain the observed rate constants of chemical steps in enzymatic processes and how, using the theoretical tools explained above, it is possible to obtain a deeper insight into the detailed reaction mechanisms. In this section, we restrict attention where the reaction does not involve the transfer of a quantum moiety, *i.e.* a proton or hydride ion. Such reactions are addressed in Section 3.4.

3.3.1 TST Analysis of an Enzymatic Inverse Menshutkin Reaction: Catechol *O*-methyltransferase

We begin with the application of the TST approach to the methyl transfer reaction from the cationic cofactor S-adenosyl-L-methionine (SAM) to the catecholate anion, catalysed by catechol *O*-methyl transferase (COMT).[24] This reaction (see Figure 3.2) is a direct bimolecular SN2 process that can be

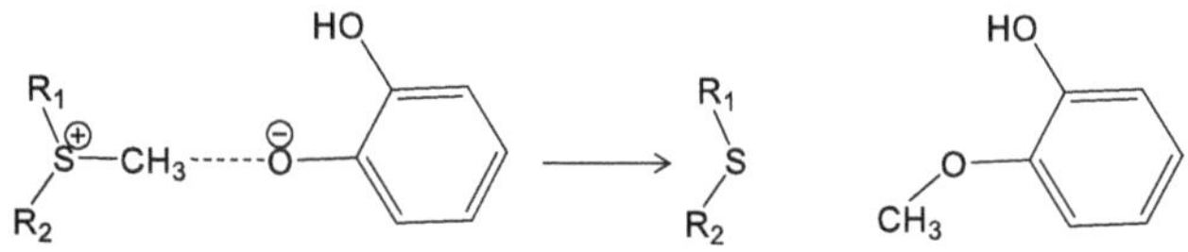

Figure 3.2 The SN2 methyl transfer reaction catalysed by the enzyme catechol *O*-methyl transferase.

considered as an inverse Menshutkin reaction, in which ionic reactants are transformed to neutral products. Our goal here will be to establish some general major features for enzyme reactions with this example. A full description can be found elsewhere.[18,25]

3.3.1.1 *Reaction Coordinate and Free Energy Profile*

A priori, a reasonable choice for the reaction coordinate x is the antisymmetric combination of the distances defining the bonds to be broken (S–C) and formed (C–O). With this coordinate, a TS 'ensemble' (corresponding to a TS surface perpendicular to x) can be defined as those structures found at the maximum of the corresponding free energy profile. According to eqn (3.3) or (3.4), the rate is largely determined by the free energy difference between this ensemble and that of the reactants.

The free energy profile can be obtained from the probability distribution of the selected reaction coordinate (x) as:[26]

$$G(x) = C' - kT\ln\rho(x) \tag{3.10}$$

where C' is an integration constant. The probability distribution can be obtained from adequate simulations of the system for which the reaction coordinate is monitored. Because the reaction coordinate involves bond breaking and forming, these simulations must consider at least two key aspects:

1. The potential energy function must be able to describe correctly the energy changes associated with changes in the bonding pattern. This can be accomplished by means of hybrid potentials, where one part of the system (the reacting fragment) is described using quantum mechanics (QM) and the associated electrons are explicitly considered. The rest of the system (the environment) is described using simpler classical force fields or molecular mechanics (MM), where the explicit electrons are ignored. Hybrid QM/MM methods are reviewed in Chapters 8 and 9. Another possibility is to employ empirical valence bond (EVB) approaches, where the (usually approximately) correct quantum description of the bond breaking and forming is obtained by combining two or more classical force fields that correspond to the different electron distributions considered to occur during the reaction.

2. Because of the large energy changes associated with the bond breaking and forming, straightforward simulations of a reacting system such as that presented in Figure 3.2 would fail to explore, *i.e.* sample, the entire range of values of the reaction coordinate x: simulations started around x values corresponding to the reactants' or products' valleys would practically remain in the starting valley without spontaneously crossing to the other one. In particular, the TS region would certainly not be sampled. This problem can be alleviated by adding a biasing potential that forces the system to visit the desired values of the coordinate such as:

$$V_{\text{bias}}(x) = \frac{1}{2}K(x - x_{\text{ref}})^2 \tag{3.11}$$

where K is a selected force constant and the location coordinate x_{ref} is changed to explore different regions of the whole x coordinate distribution in different independent simulations or windows. This is the umbrella sampling technique.[27] The full distribution corresponding to the original unbiased system can be finally recovered from the biased windows by using statistical techniques such as the weighted histogram analysis method.[28] This and other statistical procedures for obtaining free energy profiles are explained in Chapter 6 of this book.

The free energy profile obtained as a function of the selected reaction coordinate x for the SN2 methyl transfer enzymatic reaction is shown in Figure 3.3, with the computational details available in ref. 18 and 25. The TS ensemble, or dividing surface, is defined by the value of x at the maximum of the profile, $x^{\ddagger}$. As we described in detail in Section 3.2, the rate constant according to TST is governed by the free energy difference between the TS and the reactants. Further, in TST every trajectory crossing from the

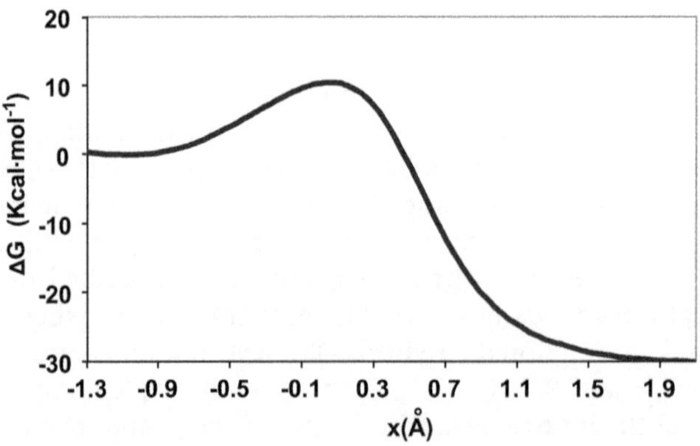

Figure 3.3 Free energy profile for the SN2 methyl transfer reaction catalysed by the enzyme catechol *O*-methyl transferase. The reaction coordinate x corresponds to the anti-symmetric combination of the distances defining the two bonds to be broken (S–C) and formed (C–O).

reactants (R) side and crossing the dividing surface should commit to products (P) – an RP trajectory in Figure 3.1 – without ever recrossing. As also discussed in that section, TST can overestimate the rate constant because some trajectories (see Figure 3.1) may recross the dividing surface leading from reactants to reactants – an RR trajectory – or recross to the products side having ultimately originated from the products side (rather than the reactants side) – a PP trajectory.

3.3.1.2 Deviations from TST

The validity of the no-recrossing TST hypothesis can be tested, for example by analysing the fate of the trajectories initiated at configurations selected from the TS ensemble. For each selected configuration, random Maxwell–Boltzmann velocities are assigned and trajectories are then followed forward and backward in time by just reversing the sign of these velocities. Then each configuration can be assigned a probability to commit to the products (p), or the reactants (1 − p). An ideal TS surface should correspond to an isocommitor surface of value 0.5 without any trajectory recrossing. In practice, even when a fairly or a good reaction coordinate has been selected, such as x in the present case, one obtains a probability distribution of commitor values such as that presented in Figure 3.4. While the distribution is peaked at 0.5 – which demonstrates that the reaction coordinate choice is certainly a reasonable one – some configurations show a clear preference to produce PP trajectories ($p > 0.5$) while others preferentially result in RR trajectories ($p < 0.5$).

The distribution of commitor values obtained for configurations belonging to the TS dividing surface defined by $x^{\ddagger}$ demonstrates that the fate of a particular reaction is not completely determined by the value of this coordinate. Other degrees of freedom play a role in determining if a particular initially crossing trajectory is directly reactive or not. For example, for the present reaction case, environmental configurations stabilising the

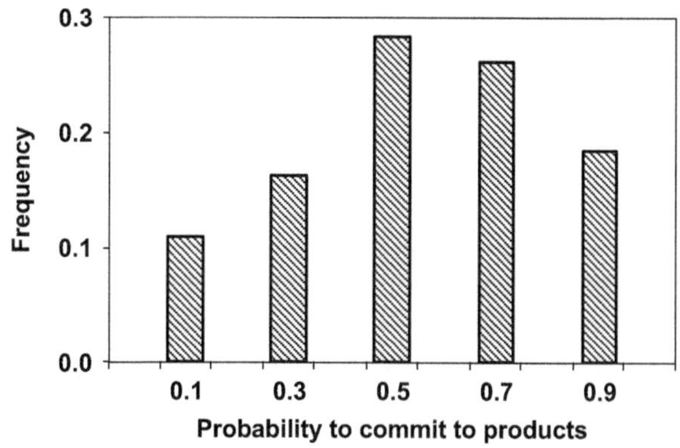

Figure 3.4 Commitor distribution for configurations selected at $x^{\ddagger}$ for the SN2 methyl transfer reaction catalysed by the enzyme catechol *O*-methyl transferase.

charge separation characterising the reactants state favour the evolution of the system from $x^{\ddagger}$ to the R basin, resulting in a larger probability to observe an RR trajectory. Otherwise if the environment favours the charge annihilation observed in the products state, the system will avoid the reactants basin, and then the probability to observe a PP trajectory increases. Although we will return later to this point, this observation explicitly indicates that a transmission coefficient must be taken into account to correct the TST rate constant value. Another implication is that the selection of the reaction coordinate may be improved by incorporating other degrees of freedom in its definition. In the latter connection, the variational version of TST seeks to find TS dividing surfaces with minimal recrossing to make the rate constant estimations as accurate as possible.[29] However, in complex systems such as enzyme reactions, this may prove difficult. The situation is further complicated by the feature that non-linear dynamics aspects can make the fate of a trajectory a non-local property that cannot be completely determined from the properties of the dividing surface.[30]

As we have already emphasised, the transmission coefficient κ in eqn (3.5) in Section 3.2.2 accounts for the recrossing reduction of the TST rate constant to give the correct reaction rate constant. As noted in that section, it can be determined *via* trajectories initiated at the TS which in the present SN2 reaction (Figure 3.2) is defined by the antisymmetric reaction coordinate choice; this evaluation was accomplished *via* the equation:[14]

$$\kappa = \frac{\sum_{i=1}^{N} v_i \cdot Q_i}{\sum_{i=1}^{N} |v_i|} \tag{3.12}$$

where all the trajectories were appropriately obtained from an equilibrium distribution. Here v_i is the initial velocity associated with the reaction coordinate x in trajectory i and Q_i is equal to 1 for RP trajectories, 0 for RR or PP and −1 for PR trajectories. Eqn (3.12) is equivalent to eqn (3.6) for the longer times during which the transmission coefficient value is established. The resulting estimation was $\kappa_{MD} = 0.83 \pm 0.03$.[31] The close proximity of this result to unity indicates that TST, with the given choice of reaction coordinate, is an excellent approximation to the actual rate constant (for the modelled reaction).

As discussed in detail in Section 3.2, the GH theory and its limiting cases can be applied to predict reaction transmission coefficients κ. With the same choice of the antisymmetric S–C–O stretch as the reaction coordinate, the departure from TST determined by the GH theory transmission coefficient was evaluated as well, and found to be in excellent agreement with the MD result: $\kappa_{GH} = 0.89 \pm 0.03$. In contrast, the Kramers theory result $\kappa_{KR} = 0.10$ clearly strongly underestimated the transmission coefficient, because it includes the full-time integrated frictional effect, *i.e.* the full impact of environmental dynamics. The latter is not relevant on the TS region short timescale, with an inverse barrier frequency of ~30 fs.[31] Finally, the frozen environment (FE) limit of GH theory gave the transmission coefficient result $\kappa_{FE} = 0.86 \pm 0.02$, in good agreement with both the full GH and MD results. This is an example then of a reduction to the enzymatic rate constant not involving dynamics. We will discuss another enzymatic example of this FE limit later and connect it to configurational effects on the reaction rate.

3.3.1.3 Contributions to the Catalytic Effect

The simulation study of this SN2 reaction establishes several important points regarding catalysis.[18,25] First, the catalytic effect is traditionally assessed by comparison with the aqueous phase reaction, and we consider that here. Using the antisymmetric S–C–O stretch as the natural choice of reaction coordinate, it was found that, within the framework of TST, the generated free energy curve provided an activation free energy $\Delta G^{\ddagger}$ for the enzymatic reaction which was lower than the aqueous solution value by ~6 kcal mol^{-1}, a significant reaction acceleration by the enzyme.[25]

Second, one could ask about the role of κ in the catalytic effect. In this connection, the transmission coefficient results could be related to the concept of active site preorganisation previously applied to the COMT enzymatic reaction:[32] in contrast to the solution case, the protein structure provides a reaction site that is already organised to favour the reaction. This provides an important reduction of the free energy barrier included in a TST perspective, as noted above. But in addition, this preorganisation reduces the coupling of environmental motions with the reaction coordinate at the TS, because the environment suffers smaller changes during the TS, barrier top passage, as now explained. In solution, water molecule dipoles must be reoriented in order to follow the charge flow taking place in the reaction (see Figure 3.2); this involves a free energy cost not required in the enzyme. But in contrast to the finding for the enzymatic reaction above, this also means that environmental dynamics are important in solution, an aspect reflected in reduction of the rate by almost a factor of 2: $\kappa_{MD} = 0.62 \pm 0.04$. This is a dynamical effect, because the frozen environment transmission coefficient – which gave an excellent account for the enzyme reaction – underestimates the aqueous solution transmission coefficient: $\kappa_{FE} = 0.44 \pm 0.09$. The difference in these transmission coefficients and the associated lack or presence of environmental dynamics for the two environments is important for a detailed comprehension of the reaction microscopics, although it is of minor numerical importance compared to the exponential impact of the difference in the enzyme and solution phase activation barrier $\Delta G^{\ddagger}$ values, a feature captured by TST.

3.3.2 Analysis of Haloalkane Dehalogenase. A Conventional SN2 Reaction

We now turn to a second prototypical reaction to be analysed in the TST context: the SN2 nucleophilic reaction between dichloroethane and Asp124

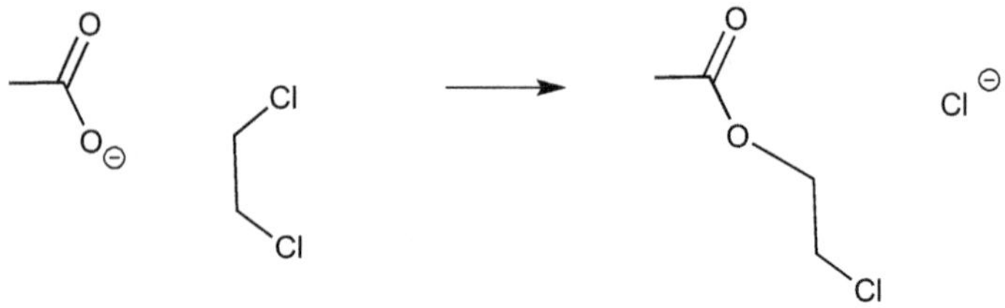

Figure 3.5 The SN2 reaction catalysed by the enzyme haloalkane dehalogenase (in the enzyme the nucleophile is the side chain of residue Asp124).

catalysed by DhlA, a haloalkane dehalogenase from *Xanthobacter autotrophicus* (see Figure 3.5).[33] The counterpart SN2 reaction in aqueous solution, where a molecule of acetate is employed as the nucleophile, is also analysed. A very extensive comparison of various molecular aspects between the reaction in these two media offers an opportunity to comprehend the role of the enzymatic *versus* aqueous solution environments in this important chemical reaction class.

3.3.2.1 *Free Energy Profiles and the Catalytic Effect*

TST can be applied to the DhlA-catalysed SN2 reaction displayed in Figure 3.5, using as a reasonable choice for the reaction coordinate x the antisymmetric combination of the bonds to be broken (C–Cl) and formed (C–O). With this coordinate, a TS 'ensemble' (corresponding to a TS surface perpendicular to x) can be defined as those structures found at the maximum of the corresponding free energy profile. According to eqn (3.3) or (3.4), the rate is largely determined by the free energy difference between this ensemble and that of the reactants, and we first examine that feature. The free energy profiles for the SN2 nucleophilic reaction between dichloroethane and Asp124/acetate in DhlA/aqueous solution are presented in Figure 3.6, with the computational details available in ref. 34.

The maxima of the free energy $G(x)$ profiles in the two media, which define the TS ensembles along the selected antisymmetric reaction coordinate described above, are located at $x^{\ddagger} \approx 0.2–0.3$ Å for the uncatalysed and enzymatic reactions. Thus, at the TS the new C–O bond distance is shorter than that of the initial C–Cl bond. With the assumption of accurate potentials, the difference in the activation free energies for the reaction in the two environments, as we will now see, can account, in the TST perspective, for the rate acceleration observed experimentally in the enzyme reaction relative to the uncatalysed reaction:

$$\frac{k_{\mathrm{cat}}}{k_{\mathrm{uncat}}} \sim e^{\frac{\Delta G^{!}_{\mathrm{uncat}} - \Delta G^{!}_{\mathrm{cat}}}{RT}} \tag{3.13}$$

More precisely, the computed difference in the activation free energies from the QM/MM free energy profiles is 10.9 kcal mol^{-1}; according to eqn (3.13),

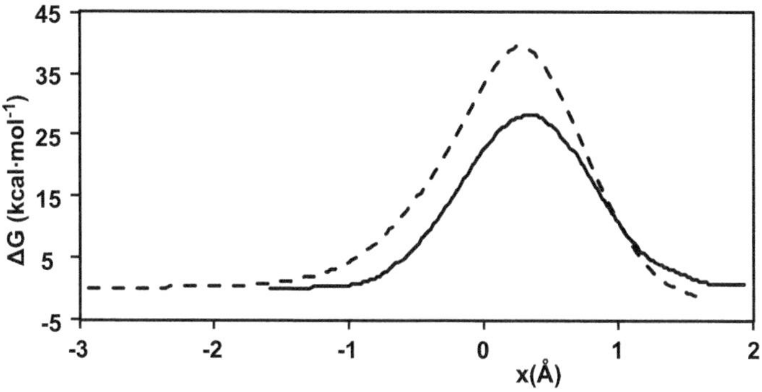

Figure 3.6 Free energy profiles for the SN2 reaction catalysed by DhlA (normal line) and the counterpart process (acetate and dichlorotehane) in aqueous solution (dashed line).

this results in a rate increase at 300 K of $\sim 10^8$, close to the ratio between experimental rate constants (see ref. 35 and references therein). Evidently, TST in its simplest version does a good job of reproducing experimental rate constant magnitude trends. Thus, the origin of the catalysis can be explained if the reason for the free energy barrier diminution in the enzyme compared to aqueous solution can be unveiled; see below.

We pause to emphasise here that the free energy profile is an equilibrium concept, obtained by assuming that all the remaining degrees of freedom of the system can be equilibrated at any value of x. The actual path may deviate from this (but we emphasise that TST does not assume that this path is followed). For example, the actual path for an aqueous SN2 reaction is much more complex (see ref. 23). Instead TST assumes, as described in Section 3.2, that there is an equilibrium distribution for successful trajectories at the TS, which introduces the activation free energy into the TST rate constant expression, for example eqn (3.3) and (3.4).

Returning to the main thread, the success of TST here indicates that the most important contribution to catalysis comes from variations in the equilibrium interactions established between the reacting system and the environment at the reactants state and at the TS. The rate constants of SN2 reactions in solution that are similar to that analysed here – in which a charge initially located in the reactants is ultimately transferred to the products – are known to be diminished when the polarity of the solvent increases (this is a specific illustration of the well-known Hughes–Ingold rules[36]). Water molecules, for example, preferentially stabilise concentrated charges, as is the case here in the reactants. For ease of discussion, we will focus on the nucleophilic oxygen atom (where the negative charge resides in the reactants) and the chlorine atom on the reactant dichloroethane, which carries that negative charge in the products (see Figure 3.5). At the TS, the charge is distributed (primarily) between these two atoms, and this delocalisation results in less stabilisation by interactions with water molecules. (This situation is similar to that for, for example, a simpler $Cl^- + MeCl \rightarrow ClMe + Cl^-$ reaction.[38]) This differential equilibrium solvation effect results in a decrease of the free energy difference between the TS and the reactants as the effective polarity of the environment decreases (because the interactions biasing the reaction in favour of the reactants compared to the TS are diminished; in the gas phase the barrier would be minimum), and an increase of the rate constant results. This is exactly the case for the transition from the aqueous solvent to the enzyme environment here. The main difference between the water solvent and the protein is that – despite the fact that some water molecule(s) are located in the active site – the latter presents a largely covalent structure. In fact, one of the reacting fragments (Asp124) is covalently bonded to the rest of the enzyme.

But beyond this general effect, there are some addition-specific enzymatic effects, not present in the solution, which preferentially stabilise the TS and as a result lower the barrier further. This aforementioned covalent structure results in a less flexible and more organised environment, and this environment preferentially stabilises the TS charge distribution $[O^{\delta-} - C^{\delta+} - Cl^{\delta-}]$, in the following fashion. The hydrogen bond established between the carboxylate group of Asp124 and the amide hydrogen atom of Trp125 plays a key role in this stabilisation.[37] While the carboxylate group is desolvated at the TS in solution (at a free energy cost), this hydrogen bond is in fact reinforced at the enzymatic TS. In general, the enzyme preorganisation phenomenon – an active site structure already prepared to accommodate the TS – is a central concept in the rationalisation of enzymatic catalytic efficiency origin.[3]

3.3.2.2 *TS Recrossing and Deviations from TST*

Even though the calculation of the free energy profile undoubtedly presents a challenge, discussed elsewhere in this book (see Chapter 6), it should be stressed how simple the TST rate constant is, keeping in mind the complexity of the trajectories attempting the reaction. This is not at all to say that the molecular details of the fate of the trajectories reaching the TS region from the reactant and attempting to reach the product from the TS region are not interesting. As stressed before in Section 3.2, TST assumes that all the trajectories initiated at the reactants side at the TS surface and crossing this surface successfully continue to the products side without recrossing (RP trajectories, see Figure 3.1). However this assumption does not always hold and some trajectories recross the TS surface to come back to the reactants side (RR trajectory). In addition, trajectories which TST would consider to have arisen from the reactants side, instead arise from the products side and recross towards the products side (PP trajectory). This would be incorrectly counted as contributing to the forward rate constant in TST. Figure 3.7 shows a schematic representation of all these three scenarios depending on the stabilising interactions established by

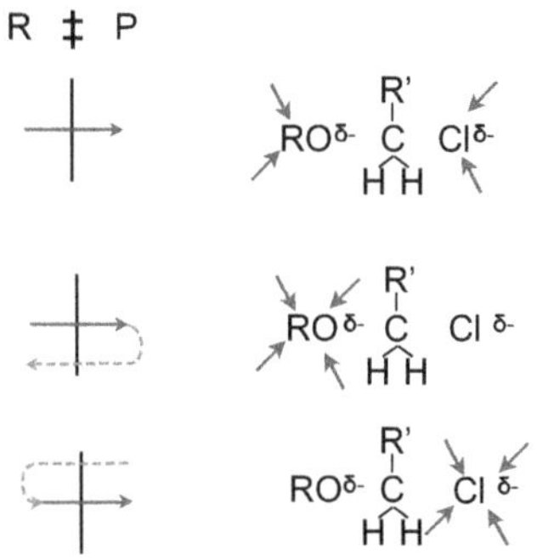

Figure 3.7 Trajectory types for the SN2 reaction between Asp/acetate and dichloroethane, together with their associated schematic enzyme/water hydrogen-bonding patterns of the transition state SN2 substrate/solute.

the environment (represented by arrows) with the nucleophile and the leaving group.

In the remainder of our discussion, the DhlA analysis can be most readily understood by first summarising the results for a related but formally simpler reaction: the symmetric $Cl^- + MeCl'$ SN2 process in water.[14] This reaction, for which one of the very first free energy reaction profiles was calculated,[38] has been employed as a model reaction to thoroughly analyse the validity of TST and any necessary corrections to this theory. Just as for the COMT SN2 reaction studied in Section 3.3.1, a reasonable reaction coordinate for this system is the antisymmetric combination of the distances associated with the Cl–C bond to be formed and the C–Cl′ bond to be broken. This coordinate, denoted as usual by x, is negative on the reactants side and positive on the products side. Because of the reaction's symmetry, the TS dividing surface is defined by $x = 0$, which corresponds to the local maximum of the free energy profile $G(x)$ traced as a function of this coordinate.

A detailed analysis of the $Cl^- + MeCl'$ SN2 reaction in water showed that trajectories started at the dividing surface $x = 0$ display recrossings which diminish the rate constant – in the patterns shown in Figure 3.1 – that violate the TST assumption depending on the solvation pattern found for the TS configuration.[14] The fate of a particular trajectory initiated at the TS dividing surface is largely determined by which of the two chlorine atoms (Cl or Cl′) is better solvated by the solvent water molecules at that particular configuration. The transmission coefficient κ value, which reduces the rate constant from its TST prediction, was found to be ~ 0.5, so that TST, while not perfect, is not very seriously numerically inaccurate. These features are also found for the DhlA-catalysed SN2 reaction; for example, the $Cl^- + MeCl'$ SN2 recrossing patterns are also found for DhlA-catalysed SN2 reaction as illustrated in Figure 3.7. We now turn to a detailed discussion of recrossing-related issues for the DhlA-catalysed SN2 reaction and its aqueous solution analogue.

The transmission coefficient for the SN2 reaction was determined along the lines outlined in Section 3.2. Trajectories were initiated at the TS location $x^{\dagger}$, in the unstable antisymmetric O–C–Cl stretch reaction coordinate (the carbon-chlorine distance minus the carbon-oxygen distance), for both the reaction between acetate/Asp124, respectively and dichloroethane in aqueous solution and for the reaction in the DhlA active site. It was determined that a significant fraction recrossed the TS surface. In solution, only 42% of the trajectories were of directly successful RP type, while in the enzyme this percentage was increased to 59%. The time-dependent κ was obtained from the analysis of the time evolution of the trajectories *via* eqn (3.6) of Section 3.2.2.

The results obtained for the time-dependent transmission coefficient in the enzyme and in aqueous solution are presented in Figure 3.8. The shape of $\kappa(t)$ shows a fast decay in both media: the fate of the reaction is completely defined after the first 20 and 30 fs in solution and in the enzyme, respectively. After this short time period, the transmission coefficient reaches a plateau; the computed values of κ are 0.56 ± 0.03 and 0.77 ± 0.03 in solution and in the enzyme, respectively. This shows that, while deviations from TST exist, they are not too large, indicating that a reasonable choice of the reaction coordinate has been made. Here TST gives a quite good rate constant result in solution and an excellent description for the enzymatic reaction rate.

Even though TST performs well and even quite well for the SN2 reaction, it is important for the molecular level understanding of catalysis to comprehend the recrossing events that make the actual rates differ (somewhat) from the TST predictions. As discussed in Section 3.2, a limit of GH theory – the FE limit – is obtained when the timescale of the barrier passage is short and the coupling between the reacting system and the environment is not too strong. Then the reaction barrier crossing attempts of the TS surface occur in an essentially FE in the TS region; on this short timescale, the time-dependent friction does not deviate from its initial value $\zeta(t=0)$. These short timescale

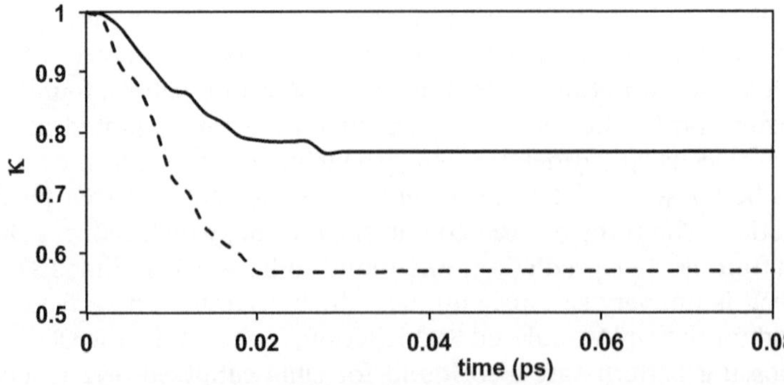

Figure 3.8 Time-dependent transmission coefficient $\kappa(t)$ for the SN2 reaction in aqueous solution (dashed line) and in the DhlA enzyme (solid line). The asymptotic lines give the transmission coefficient κ.

and modest coupling conditions are found to hold in the SN2 reaction between Asp/acetate and dichloroethane. The initial friction is substantially larger in solution than in the enzyme: the values obtained for the SN2 reaction in solution and in the active site of DhlA are 950 and 605 cm^{-1}, respectively. The larger coupling in solution indicates that the effective short-time friction acting on the reaction coordinate motion is larger, so that a larger fraction of trajectories recross the TS surface. The FE limit equation of GH theory gives transmission coefficient values of 0.57 ± 0.02 and 0.82 ± 0.02 in solution and in the enzyme, respectively.[37] These results agree, within the error bars, with those derived from the analysis of the MD trajectories initiated at $x^{\ddagger}$ and presented above, supporting the validity of the FE picture.

3.3.2.3 *Environmental Conformations Responsible for TS Surface Recrossings*

With the FE limit established for our DhlA-catalyzed reaction, we can confidently discuss the reaction recrossings' origin. This origin is to be sought in the role that coordinates, other than our selected reaction coordinate x, play in the chemical reaction's TS surface passage step. Because the FE situation applies, we can discuss this origin in terms of static configurations as opposed to the dynamics of those configurations. We are guided here by the discussion for the $Cl^- + MeCl'$ SN2 reaction[14] and comments on that reaction in ref. 7. For the present SN2 reaction, static environmental coordinates on the TS surface play the key role in determining the fate of the trajectory started at a particular configuration of the system. In the equilibrium distribution at $x^{\ddagger}$, some of the environmental configurations characterising the interactions with the reacting system – which has the approximate charge distribution $O^{\delta-} - C^{\delta+} - Cl^{\delta-}$ – will favour the charge being on the nucleophilic O atom or being on the leaving Cl moiety. Figure 3.7 indicates the recrossing patterns. RR recrossings occur when the environment–reaction system interactions pattern favours the R side of the TS, *i.e.* the charge localised on the nucleophilic O atom. PP recrossings occur when this interactions pattern favours the P side of the TS, *i.e.* the charge localised on the leaving Cl atom. The final situation is that in which there is an essentially symmetric interaction pattern such that there is no bias either way, and successful $R \rightarrow P$ trajectories occur, just as the TST assumption would have it. This implies that in the enzyme, the preorganisation of the TS has increased the sufficiently symmetric interaction pattern compared to the solution case, so that there is less environmentally-induced recrossing, and the transmission coefficient is thus larger.

3.3.2.4 *Improving the Reaction Coordinate: Variational TST*

Another insight available from this enzymatic SN2 study concerns the issue of altering the initial choice of reaction coordinate to include the environmental motions to generate a better reaction coordinate, *i.e.* such that recrossing is reduced and the transmission coefficient is increased.[35] The discussion throughout Section 3.3.2.2 has involved the choice of the unstable antisymmetric O–C–Cl combination of the distances associated with the bond to be broken (C–Cl) and the bond to be formed (C–O), as usual labelled here as x. However, as discussed in Section 3.3.2.3, the pattern of reaction system–surroundings interactions can also determine the fate of a particular trajectory crossing the TS. These interactions can be associated with a 'solvent coordinate', hereafter called s. This coordinate s can been related to a collective property of the environment,[39] in this case the electrostatic potential created on the nucleophilic and leaving groups.[35] (Other, related, choices of this coordinate could be made.[22])

In order to motivate the specific choice of a new reaction coordinate, two-dimensional free energy surfaces $G(x,s)$ obtained as a function of x and s for the SN2 reaction between Asp/acetate and dichloroethane in aqueous solution and in the enzyme are presented in Figure 3.9. The analysis of the recrossing discussed in Section 3.3.2.3 revealed that these events could be quite closely associated with extra, albeit comparatively small, barriers in x, which are not apparent in the equilibrium free energy curves shown in Figure 3.6. The presence of such a barrier is indicated by curved arrows in the left panel of Figure 3.9, *i.e.* a trajectory is reflected by the barrier. To pursue the point, in the lower half of the figure, RR recrossing trajectories arise from s-dependent barriers in x located slightly past the nominal TS surface as determined by x (the vertical dotted lines in Figure 3.9) so that they are encountered by the trajectory coming from R and are reflected back to R. (PP trajectories arise from barriers located to the left of the TST surface in the upper half of the picture; see ref. 22 for this phenomenon for the $Cl^- + MeCl'$ SN2 reaction.) In an orthogonal x–s coordinate system, the original $x^{\ddagger}$ surface can be rotated to a new TS surface lying along the peaks of these barriers so as to avoid these barrier-induced recrossings and to minimise the flux across the new surface.[22] Figure 3.9 shows a qualitative description of this effect, where the solid lines represent the improved TS surfaces in the x–s coordinate system. In this variational transition state theory (VTST) perspective,[40] recrossing by the trajectories of the optimised TS surface is reduced and the description of the rate *via* TST in the new coordinate system is improved.

The two-dimensional free energy surfaces in Figure 3.9 also emphasize some key enzymatic catalysis aspects. In order to appreciate this, we first note that there is a smaller rotation needed to reach the optimised x–s TS from $x^{\ddagger}$ in the enzyme than in solution (which is also related to the larger value of the transmission coefficient). Since it was established in Section 3.3.2.2 that the FE limit applies for the reaction in both media, this means that in the enzyme, the distribution of environmental configurations existing in the TS surface are more conducive, or less interfering, for the successful passage through the surface. An alternative was to say this is that the amount of environmental rearrangement needed to reach that TS surface

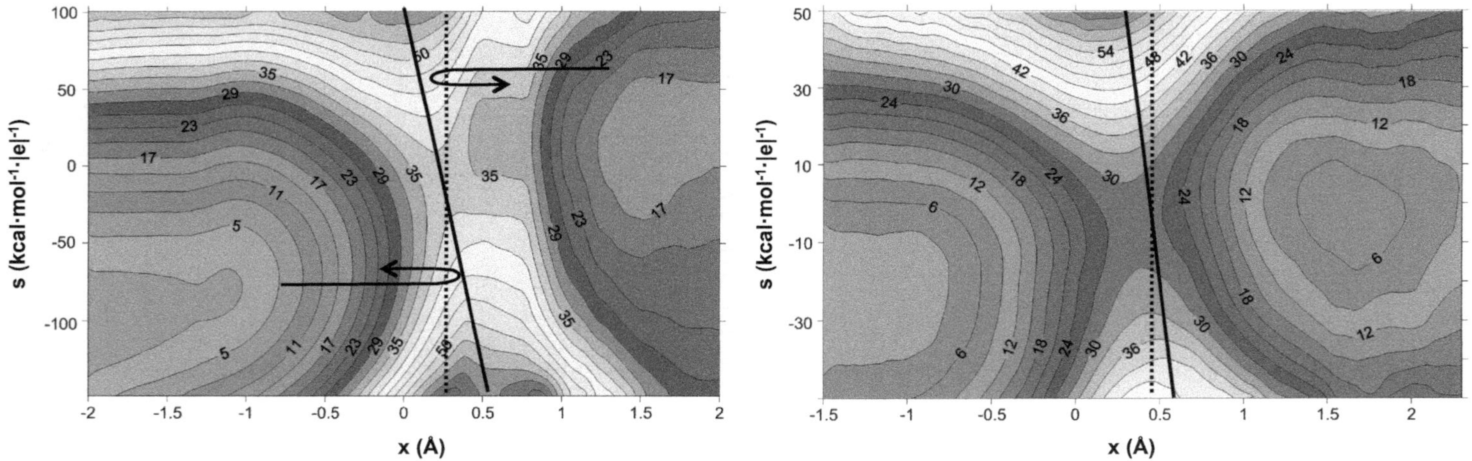

Figure 3.9 Free energy surfaces for the SN2 reaction between Asp/acetate and dichlorotehane in aqueous solution (left) and in DhlA (right). The x-coordinate is the O–C–Cl stretch while the solvent coordinate s is obtained from the anti-symmetric combination of the electrostatic potential on the nucleophile and leaving atoms. Arrows indicate posible trajectories recrossing the TS surface characterized as $x = x^{\ddagger}$ (dotted line). An optimised dividing surface as a funcition of x and s is indicated by a continuous line. Iso(free)energetic lines are given in kcal mol^{-1}.

is smaller in the enzyme than in aqueous solution: the differences in the s scale clearly show that reaching the TS from the reactants valley involves a smaller reorganisation of the environment in the enzyme than in solution. The electrostatic environment provided by the enzyme is already significantly prepared to accommodate the TS charge distribution and thus needs to be reorganised substantially less than the solvent does in order to reach the TS surface configurations.[3]

3.3.3 Beyond the FE Limit: The Michael Addition Catalysed by Chalcone Isomerase

As our final classical enzymatic reaction example, we consider the addition reaction catalysed by chalcone isomerase (CHI). The CHI enzyme catalyses the transformation of chalcones *via* an intramolecular Michael addition of a deprotonated hydroxyl group to the α,β-double bond, as shown in Figure 3.10.[41]

This reaction provides an example of strong coupling between the reaction coordinate and the remaining degrees of freedom, where as a result the FE limit fails. Here the environment definition – which clearly depends on the definition of the reaction coordinate – will also involve other degrees of freedom of the reaction system.

3.3.3.1 Free Energy Profiles and Transmission Coefficients

The CHI reaction was studied computationally using the O–Cβ stretch as the reaction coordinate (see Figure 3.10).[19] The corresponding free energy profiles for the enzymatic reaction and the counterpart process in aqueous solution are shown in Figure 3.11, with the computational details available in ref. 19. The activation free energy is noticeably smaller in CHI than in aqueous solution and this diminution explains essentially all of the rate enhancement observed for the catalysed process. The position of the TS dividing surface along the selected coordinate ($x^{\ddagger}$) is found to be slightly earlier in the enzyme than in solution (1.84 and 1.81 Å, respectively) corresponding, in both cases, to a quite advanced formation of the new C–O bond.

The involvement of the dynamics of the environment in the enzymatic reaction – *i.e.* all degrees of freedom other than the O–Cβ stretch – was

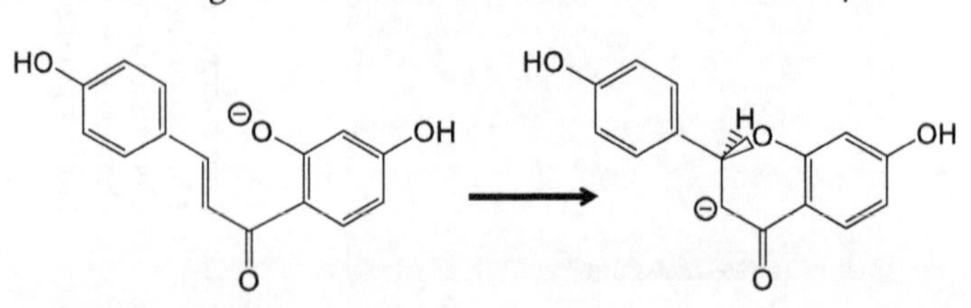

Figure 3.10 The intramolecular Michael addition reaction catalysed by the enzyme chalcone isomerase (CHI).

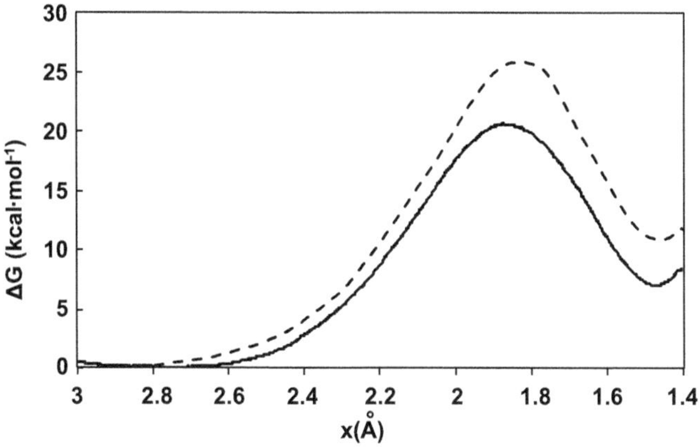

Figure 3.11 Free energy profiles for the intramolecular Michael addition reaction catalysed by the enzyme chalcone isomerase (continuous line) and in aqueous solution (dashed line).

signalled by the finding that the MD transmission coefficient $\kappa_{MD} = 0.87 \pm 0.03$, determined from trajectories initiated at the TS – was noticeably greater than predicted by the frozen environment limit of GH theory $\kappa_{FE} = 0.56$. The fact that the full GH theory result $\kappa_{GH} = 0.82 \pm 0.02$ is in good agreement with the MD transmission coefficient indicates that the omission of environmental dynamics is responsible for the failure of the frozen environment description κ_{FE}. (As a side remark, Kramers theory severely underestimates the transmission coefficient: $\kappa_{KR} = 0.03$, incorrectly predicting – in contrast to GH theory – a large deviation from TST.) Again, GH theory predicts that TST is an excellent approximation for the enzymatic rate constant. In the next section, we use the theory to gain further insight on the reaction dynamics.

3.3.3.2 *Improving the Reaction Coordinate*

Even though we have just seen that there is not extensive recrossing in the CHI reaction, this reaction has been used to illustrate an interesting application of GH theory:[19] the identification of the degrees of freedom coupled to the selected reaction coordinate,[23] which in the CHI case is the O–Cβ stretching. The magnitude of the barrier frequency associated with the motion along the O–Cβ unstable vibration was found to be 715 cm^{-1}. The Fourier transform of the friction kernel displayed in Figure 3.12 shows that different motions contribute to the friction, some of which are effective in diminishing the transmission coefficient below unity. Recall from eqn (3.9) that it is essentially the frequency component of the time-dependent friction that is effective in this sense.

Some motions, such as C–H stretches, mainly involving the Cβ and Cα atoms, appear at frequencies higher than 3000 cm^{-1} and so can be considered to be fast enough to be equilibrated with the O–Cβ stretching.

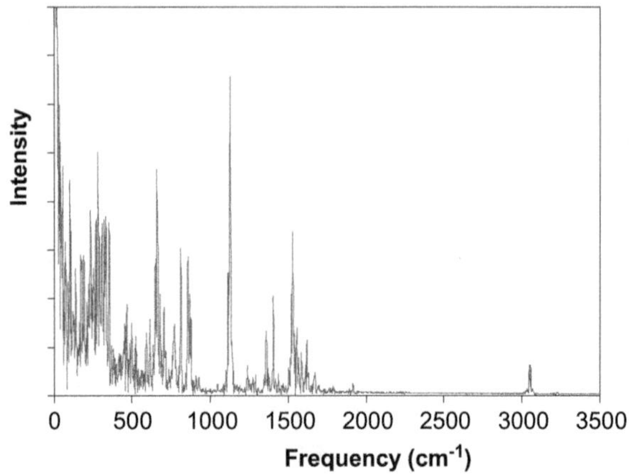

Figure 3.12 The (relaxed) frequency spectrum of the friction, in arbitrary units, for the enzymatic CHI reaction (see the text).

Other motions are clearly slower than the chosen reaction coordinate and thus can be considered to be frozen during the barrier crossing event. Collective protein motions and hydrogen bonds reorientation appear in this region of the friction spectrum; but these only have an effect on the transmission coefficient if they strongly coupled to it, and most of these motions are not so coupled. Finally, some environmental motions that are strongly coupled to the O–Cβ reaction coordinate were found to occur on the barrier crossing timescale, with frequencies similar to that of the selected reaction coordinate so that their dynamics is relevant for the reaction event. Most of these coupled motions are in fact found to be located in the molecular reacting system and not in the protein environment. In particular, the hybridisation change at the Cβ atom due to the nucleophilic addition leads to an out-of-plane motion of the bonded hydrogen atom and the lengthening both of the Cβ–Cα bond and of the carbonyl double bond (part of the charge is delocalised on the carbonyl oxygen atom). An important reason that this dynamical coupling involves to some extent the motion of some active site residues is that the changes suffered by the substrate must be accommodated in the protein active site. One could envisage including these aspects in a more complicated reaction coordinate, if desired. In a more general perspective, one can expect more applications of this particular methodology in the future to readily identify and assess the impact of key motions dynamically coupled to a selected enzymatic reaction coordinate.

3.4 Enzyme Reactions Involving Quantum Nuclear Motion

As we remarked in Section 3.2, standard TST is a classical theory: the reaction coordinate's passage through the TS surface is treated as classical nuclear motion. Transfer reactions for a proton H^+, hydride H^- or hydrogen atom H (and their isotopes) require a quantum treatment for the proper nuclear motion description.[42–57] The first two of these are charge transfers and usually have strong coupling to a polar/charged environment such as is common in an enzyme active site. We focus on these two reaction types here, but use mainly proton transfer (PT) language. Many aspects of our remarks also apply to the PT portion of proton-coupled electron transfers.[55–57] Our goal here is to provide an introductory perspective for the enzymatic quantum particle transfer reactions described in detail subsequently in this section.

3.4.1 A Two-dimensional Perspective

There are assorted available approaches for quantum transfer reactions.[42–51] Here we employ the theoretical perspective described in the group of one of us:[45–49] The rapid quantum nuclear motions can be quantised at fixed values of the remaining slower and largely environmental coordinates. It is the latter, and not the former (as is usually imagined), that provide the reaction coordinate, which accordingly is classical rather than quantum. For solution reactions, this reaction coordinate is typically a solvent coordinate reflecting the solvent's nuclear electrical polarisation in interaction with the reacting solute system.[45–49] This coordinate can also be related to an energy gap coordinate associated with the different interaction of the solvent's configuration with either of reactant and product configurations, especially with their different charge distributions (see eqn (3.14) below). Related considerations can be used to define an environmental reaction coordinate for enzymatic reactions.[50] In the following, we will use the symbol s for the environmental coordinate, and for the purposes of this introduction, we assume that it is an energy gap coordinate.

Our principal goal in this introduction will be to discuss the two PT regimes, the adiabatic and non-adiabatic (or tunnelling) regimes. To this end, we focus our discussion on a PT acid–base reaction:

$$AH\cdots B \rightarrow A^{-}\cdots HB^{+} \tag{3.14}$$

occurring in an H-bonded complex. The fundamental picture in either regime, characterised in Figure 3.13, is that the quantum PT event itself – when it occurs – is rapid compared to the much slower environment's reorganisation which is necessary (but perhaps not sufficient) for a successful transfer attempt. It follows that this reorganisation must occur prior to the attempt, which in turn means that the environment (and other 'slow' coordinates not including the proton) constitutes the reaction coordinate and determines (the major portion of) the activation free energy barrier.

3.4.2 Adiabatic PT

In the adiabatic regime,[45,48] the proton adiabatically follows (*i.e.* rapidly adjusts to) the slower environmental coordinate s (Figure 3.13a) and at the

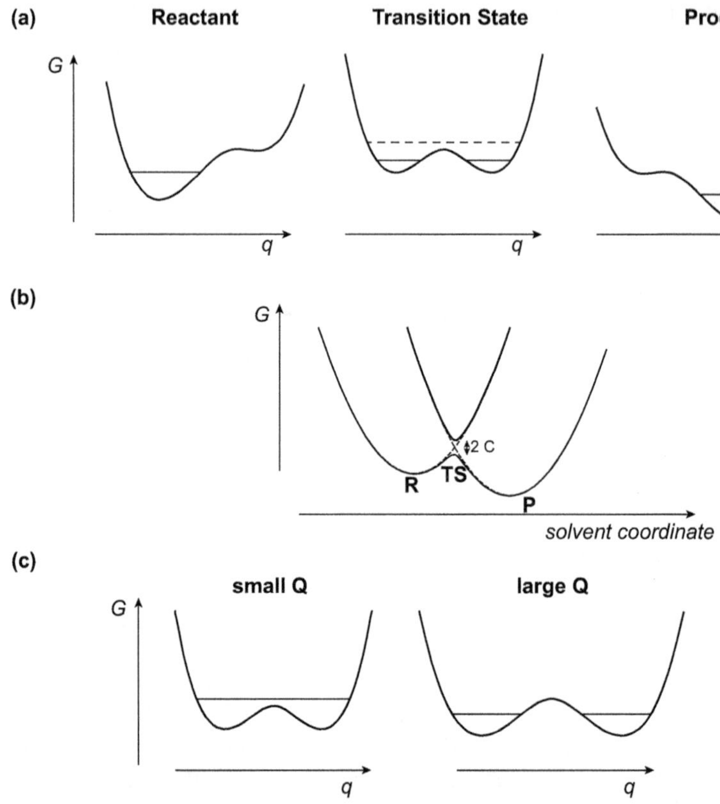

Figure 3.13 Schematic free energy (G) curves for the PT reactions (eqn (3.14)) in a polar environment, involving the proton coordinate q and the hydrogen-bond coordinate Q. (a) Proton potentials at different values of the environmental solvent coordinate s, with the proton vibrational levels shown. At the TS value of s, there are two different situations. For the non-adiabatic tunnelling limit there are two different levels (solid lines) shown below the proton barrier: these are the localised, non-adiabatic, levels. For the different, adiabatic level, there is one level (dashed line), which is above the proton barrier. (b) Free energy curves in s; the reactant and product curves cross in the tunnelling regime, but are split by twice the proton coupling C to produce upper and lower curves in the adiabatic regimes. (c) Illustration that the adiabatic regime is favoured by small Q (larger C), while the tunnelling regime is favoured by larger Q (smaller C). The exponential behaviour of C is especially important in the latter regime. See also ref. 45, 47, for example. The non-adiabaticity here is of nuclear origin, *i.e.* the proton motion, due to the weak coupling ⟨C2⟩. The electronic motion governing the proton potentials in panel (a) is instead adiabatic, because electronic coupling in proton (and hydride) transfer systems is very strong (~1 eV). Some tunnelling rate formulas in the literature have an additional weak electronic coupling transmission coefficient factor, which is inappropriate.[52]

TS for s – reached from the reactants at a free energy cost $\Delta G^{\ddagger}$ (Figure 3.13b) – the zero point energy (ZPE) level for the proton lies above the barrier in the proton coordinate q. It is crucial to note that this is a bound vibrational motion. It starkly contrasts with the common view of unstable, above the barrier classical, supposed reaction coordinate, motion of the proton. The effective free energy barrier involves the difference $\Delta ZPE^{\ddagger}$ of the ZPEs at the environmental reaction coordinate's TS $s^{\ddagger}$ and reactant values; see eqn (3.15) below. (A telling contrast with the standard picture can be seen for the symmetric reaction. In the standard view, there would be no $ZPE^{\ddagger}$ contribution because there is no proton motion in the orthogonal coordinate; but in the adiabatic PT perspective there is such a contribution since the proton motion is a non-reactive, stable quantum vibration). Quantum kinetic isotope effects, KIEs, also enter *via* this same difference $\Delta ZPE^{\ddagger}$ just discussed. Beyond this, all of our previous remarks of Section 3.2 concerning TST apply to this regime, with the classical environmental coordinate being the reaction coordinate. The rate constant is (compare eqn (3.4) and (3.5)):

$$k_{AD} = k_{AD,TST}\kappa = \frac{\omega_R}{2\pi}\exp\left[-(\Delta ZPE^{\ddagger} + \Delta G^{\ddagger})/k_B T\right]\kappa \tag{3.15}$$

Here ω_R is the frequency in the R well of the free energy curve in Figure 3.13b and κ is the transmission coefficient determined by any TS surface recrossing in the environmental coordinate s. Evaluations of κ for adiabatic PT indicate that GH theory provides an accurate description, and that – as for classical reactions – deviations for TST so defined are small.[49]

3.4.3 Non-adiabatic PT

We next consider the non-adiabatic, tunnelling regime.[45–47,52] In contrast to the adiabatic regime, the proton ZPE level lies below the barrier in the proton coordinate, at the TS $s^{\ddagger}$ of the environmental coordinate (Figure 3.13a). We assume for simplicity here that both the first two (diabatic, uncoupled) levels lie below this barrier (see the lower two levels in the middle panel of Figure 3.13a), and that when they are coupled, they are separated by the tunnelling splitting of twice the coupling $C(Q)$; this shows up in Figure 3.13b. At $s^{\ddagger}$, the overlap of the (non-adiabatic) proton wave functions localised in the R and P wells at $s^{\ddagger}$ is approximately an exponential function of the H-bond vibrational coordinate Q, *i.e.* the donor–acceptor A–B separation, and this is reflected in a corresponding exponential dependence for the coupling $C(Q)$. (Figure 3.13c gives a schematic illustration of the important impact of Q on the PT potential and levels.) In the tunnelling regime, the coupling is small, so that the activation free energy $\Delta G^{\ddagger}$ is, to a very good approximation, given by the intersection of the non-adiabatic (proton-localised) R and P curves (Figure 3.13b). The picture is that, when $s^{\ddagger}$ is reached, at the cost of $\Delta G^{\ddagger}$, a successful (incoherent) tunnelling through the proton potential barrier (Figure 3.13a) occurs with a probability proportional to $\langle C^2 \rangle$, where the brackets indicate an appropriate average over the Q motion, discussed further in a moment. This defines a quantum tunnelling transmission coefficient κ_T proportional to $\langle C^2 \rangle$. This of course differs from the transmission coefficient κ, which is – as per usual – associated with classical recrossing, here in the environmental coordinate, of the TS surface at $s^\ddagger$. The rate constant is thus of the form:

$$k_{NA} = k_{NA,TST}\kappa_T\kappa = \frac{\omega_R}{2\pi}\exp\left[-\Delta G^\ddagger / k_B T\right]\kappa_T\kappa \quad (3.16)$$

Here no ZPE difference enters, in contrast to the adiabatic case in eqn (3.15). The exponential sensitivity of the coupling to the Q motion is important here: as Q decreases or the A–B 'bond' is compressed, the PT transfer barrier is lowered, and its width is diminished, so that the tunnelling rate constant can increase dramatically. Further, due to the well-known strong sensitivity of tunnelling to the mass, it is also the principal source of a KIE, the ratio between the rate constants for the transfers of a proton and of a heavier isotope, *e.g.* deuteron.

3.4.4 Examples of Enzyme Reactions Involving Quantum Nuclear Motion

In the following, we discuss some studies of paradigmatic enzymatic reactions involving PT, hydride transfer or both. Several of these studies were originally carried out from a classical perspective together with *a posteriori* corrections to account for the quantum nature of the transferred particle. In other cases, the studies were carried out with a perspective closer to that we have just presented here. But in all cases, these efforts are useful to stress the distinctive and characteristic features of reactions involving quantum transfer discussed in Sections 3.4.1 to 3.4.3 with respect to those that can be satisfactorily described within a classical framework in Section 3.3.

3.4.4.1 PT in Triosephosphate Isomerase

We first consider the paradigm PT reaction catalysed by the triosephosphate isomerase (TIM) enzyme. TIM catalyses the isomerisation of dihydroxyacetone phosphate (DHAP) into glyceraldehyde-3-phosphate (GAP) (see Figure 3.14), which is an important step in the glycolytic pathway in muscles. TIM has sometimes been designated as a catalytically 'perfect' enzyme because of its extremely high efficiency:[58] it accelerates the reaction by a factor of more than 10^9 with respect to the aqueous solution conditions. While different catalytic mechanisms have been suggested, the rate-limiting step is now established to be a PT reaction to form the enediolate and involving a glutamate residue (see Figure 3.14). (A remarkable feature is that the PT occurs from a C–H bond.)

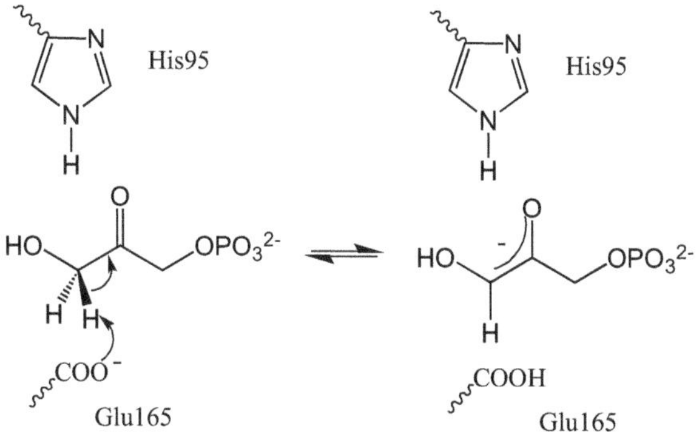

Figure 3.14 Proton transfer reaction catalysed by triosephosphate isomerase.

The first computational studies[59] of this catalytic mechanism considered the antisymmetric stretch coordinate, *i.e.* the difference in the classical distances between the transferred proton and the donor and acceptor, respectively. As the antisymmetric coordinate changes, the energy along the reaction path is determined by relaxing the nearby active site atoms and by adding corrections to include nuclear quantum effects (see Chapter 10). Tunnelling probabilities are calculated by a semiclassical treatment for the transmission probability of incoming wavepackets through a barrier. A subsequent study[60] calculated a value of 0.47 for the classical transmission factor at the TS along the classical antisymmetric coordinate. This value suggests that TST should provide an adequate description of the rate constant and that the antisymmetric stretch coordinate is an important part of the reaction coordinate at the TS.

However, this type of approach and this choice of reaction coordinate have two main limitations: (i) the assumption that active site atoms can instantly adapt to the rapidly evolving charge distribution in the substrate and (ii) the classical description of the motion along the reaction coordinate even when it involves the displacement of the proton. It should be emphasised that even in the absence of tunnelling – when the proton ZPE level is above the barrier (see Figure 3.13), the motion of the proton is not classical at room temperature, and quantisation would offer an improved description.

3.4.4.2 *Hydride Transfer in Formate Dehydrogenase*

To illustrate the importance of both an explicit description of the quantum nature of the hydride (or proton) and of the succession of rearrangements occurring in the active site and preceding the actual transfer of the quantum particle, we now turn to an enzyme-catalysed hydride-transfer reaction.

For this purpose, we focus on the formate dehydrogenase (FDH) enzyme, but several other enzymes catalysing hydride-transfer reactions have been studied, including for example the paradigm dihydrofolate reductase,[61–65] and the alcohol,[66–68] and lactate dehydrogenases.[67–71] FDH catalyses the

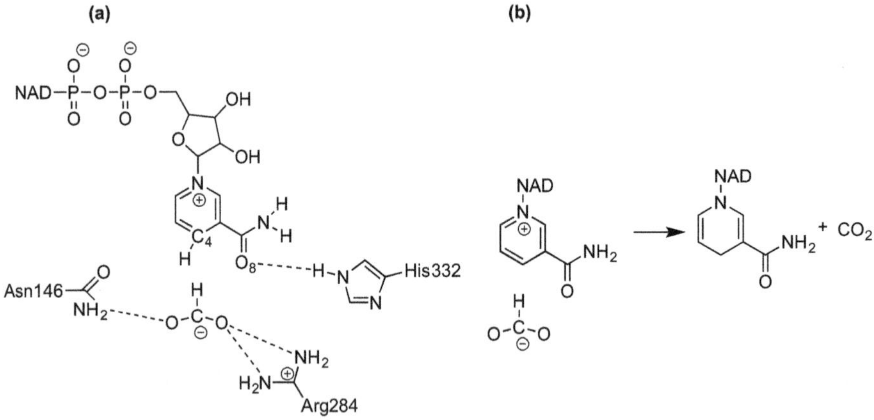

Figure 3.15 Active site of formate dehydrogenase (a) and its catalysed hydride transfer reaction (b).

hydride transfer from a formate $HCOO^-$ anion to a nicotinamide adenine dinucleotide (NAD^+) cofactor, yielding carbon dioxide and NADH (Figure 3.15). This hydride-transfer reaction, which is used for example for cofactor regeneration, is the rate-limiting step in the FDH catalytic mechanism.

Theoretical studies of the mechanism have shown that several geometric rearrangements take place in the active site before the hydride transfer.[72,73] While using the antisymmetric stretch coordinate – defined as the difference between the distances from the hydride to the donor and acceptor carbon atoms – is an intuitive choice for the reaction coordinate, it was shown to yield a transmission factor of 0.46 in the enzyme when this coordinate is treated classically.[72] The large recrossing probability shown by this value suggests that additional coordinates and/or quantisation should be considered.

As already largely recognised for proton and hydride-transfer reactions in solution,[47] a critical coordinate is the distance between the donor and acceptor sites. This distance (here denoted by Q) affects the height of the potential energy barrier along the hydride coordinate. At short Q, the barrier is lower than the hydride ZPE and the transfer is adiabatic, in the language of Section 3.4.2. In contrast, at larger Q separations, the barrier is above the hydride ZPE and the transfer requires tunnelling through the barrier, *i.e.* a non-adiabatic transfer in the language of Section 3.4.3. In the reaction catalysed by FDH, a contraction of the C4–C distance (see Figure 3.15) was observed at the TS.[72,73] If a broad distribution of Q distances are found at the TS, the reactive pathways may thus include both adiabatic and tunnelling transfers and the rate constant would have to be averaged over these conformations (see Figure 3.13c).

An additional rearrangement was found to occur in the hybridisation of the donor and acceptor carbons. In the reactants, both the donor carbon in $HCOO^-$ and the acceptor carbon in the NAD^+ cofactor have an sp2 hybridisation, while in the products the donor carbon has an sp hybridisation leading to the linear configuration of CO_2, and the acceptor carbon becomes sp3-hybridised. These changes imply local geometric rearrangements that occur before the actual hydride transfer.[72]

It is interesting to note that for the acceptor cofactor carbon atom, C4, the change in hybridisation implies the displacement of the hydrogen atom already bonded to this carbon. Although the frequency of this angular motion is slightly lower than that of the hydride stretch motion, the two coordinates would thus need to be simultaneously treated quantum mechanically. The importance of this hydrogen atom is confirmed by experimental secondary KIE measurements which showed that deuterating the acceptor carbon site has a significant effect on the reaction rate constant ($k_H/k_D = 1.2$ as reported in ref. 73).

In addition to these local geometrical rearrangements, and similarly to the SN2 reactions discussed above, an essential part of the reaction coordinate arises from changes in the polarisation of the environment, which stabilise differently the reactant and product charge distributions. In the hydride transfer catalysed by FDH, when the reaction proceeds, the negative charge on the $HCOO^-$ reactant is destabilised by the elongation of the hydrogen bond donated by Arg284, while the growing negative charge on the O8 oxygen atom in the cofactor is increasingly stabilised by His332.[72]

A comprehensive description of the enzyme-catalysed hydride-transfer reaction should thus include not only the antisymmetric hydride stretch coordinate to be treated at the quantum level, but also the donor–acceptor distance which governs the height of the hydride barrier and the (non-)adiabatic character of the transfer, the donor/acceptor hybridisation changes (which also may require some quantum treatment), and finally the critical reorganisation impact of the local electrostatic potential. The latter can be described, for example, by the electrostatic potential difference used for the SN2 reaction in Section 3.3.2.4, or by the instantaneous energy gap between resonant valence bond states used for the hydride transfer catalysed by dihydrofolate reductase (see, for example, ref. 61, 63, 64). Regarding the quantum treatment of the proton or hydride coordinate, several approaches have been used for enzymatic reactions, ranging from semiclassical corrections[59,74] to mixed quantum/classical simulations where only the proton/hydride motions are quantum (see, for example, ref. 61), and to path integral and ring-polymer molecular dynamics where all nuclear quantum effects are included.[63] These methods are presented in detail in Chapter 10.

3.4.4.3 Hydride Transfer in Liver Alcohol Dehydrogenase

As a final quantum enzymatic reaction example, we make brief mention of a theoretical analysis[67] of a model of catalysis by the enzyme liver alcohol dehydrogenase (LADH). In particular, in this study, which involves a number of the concepts we have discussed in Sections 3.4.1 to 3.4.3, a detailed investigation was made of the hydride transfer from a zinc-bound benzyl

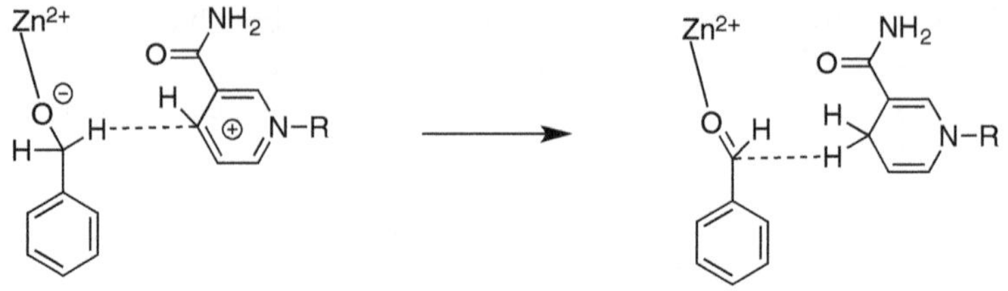

Figure 3.16 Schematic representation of hydride transfer in LADH.

alkoxide to the coenzyme NAD^+ to produce benzaldehyde and the reduced form NADH (see Figure 3.16). The adiabatic and non-adiabatic nuclear quantum effects of the transferring hydrogen were included with a mixed quantum/classical approach in which a three-dimensional wave function represents the transferring hydrogen moiety. A collective energy gap coordinate, rather than a coordinate involving the hydride motion, was selected as the reaction coordinate.

A TST perspective for the reaction was presented for a free energy curve for what we would term an 'adiabatic' hydride transfer in the language of Section 3.4.3 (free energy, including the ZPE of the transferring hydride, *versus* the energy gap reaction coordinate). The recrossing of the TS surface in this representation was found to be almost unity: $\kappa = 0.95$. Indications of some tunnelling contributions to the rate were also discussed, with an important role for these played by the hydride donor–acceptor distance (a distance that also influences the free energy barrier).

3.5 Concluding Remarks

Here we have presented a discussion of the basic principles and ideas of TST for kinetic rate constants, together with six different examples of its applications to the study of enzymatic reactions: three for classical particle transfer and three for quantum particle transfer. We have addressed the basic fundamental assumptions of TST; we have also discussed the transmission coefficient that incorporates the correction to the TST prediction for the reaction rate, due in particular to the failure to some degree of TST's assumption that there is no recrossing of the TS surface.

We have emphasised the importance for TST of the selection of a reaction coordinate and the associated generation of free energy surfaces. Different choices of the reaction coordinate will give different realisations of TST and different activation free energies. In addition, we stress that TST is designed to describe the reaction rate constant, and not address detailed reaction path all the way from reactants to products.

Several examples of enzyme reactions are given in which there is deviation from TST, and that deviation can be accurately described in terms of a time-dependent friction exerted by the environment on the reaction coordinate. We have also discussed how such deviations can be reduced by a new choice of reaction coordinate, a special illustration of variational TST. Further, we have pointed out that such deviations from TST – while often quite revealing for the molecular level aspects of the enzymatic reactions in the TS surface neighbourhood – are typically numerically quite small. In addition, the origin of the enzyme's catalytic acceleration of the reaction compared to the solution rate is generally accurately given by TST, because the effect is heavily dominated by the activation free energy component of TST.

TST, in its usual form, assumes classical nuclear motion of the reaction coordinate, and we give three reaction examples where this assumption is a reasonable one, *i.e.* transfer of a heavy particle in an SN2 reaction. However this assumption does not hold for enzymatic reactions involving the transfer of light particles such as a proton, a hydride or a hydrogen atom. In such cases a quantum description of the particle motion is required. Two basic approaches are discussed here. The first is that classical TST is applied and then quantum corrections are added. In the second approach, the particle motion is quantised and a different, classical coordinate (related to the reaction environment) is selected as the reaction coordinate. In the latter approach, to which we give special attention, TST can be directly employed. Our three chosen enzymatic examples of quantum transfer illustrate both approaches.

The basic conclusions of the admittedly limited – but we believe nonetheless very relevant – discussion of this chapter are that, in general, TST and its variants can be successfully used to obtain rate constants for enzymatic reactions, to understand important aspects of their mechanisms, and to identify the sources of the catalytic effect itself. Alternately stated, in our view TST has to be regarded as the first theory of choice for discussions of enzymatic rate constants – whether classical or quantum – and the origin of the enzyme's catalytic effect. Even when there are deviations from TST, the theory still provides the basic perspective in terms of which of those deviations can be usefully interpreted.

Acknowledgements

This research was supported in part by the USA NSF Grant CHE-1112564 (JTH) and by the Spanish Project from Ministerio de Economía y Competitividad CTQ2015-66223-C2 (IT and VM) and by Generalitat Valenciana project PROMETEOII/2014/022 (VM). The research leading to these results has received funding from the European Research Council under the European Union's Seventh Framework Program (FP7/2007-2013)/ERC Grant Agreement No. 279977 (DL).

References

1. M. Garcia-Viloca, J. Gao, M. Karplus and D. G. Truhlar, *Science*, 2004, **303**, 186.
2. S. Martí, M. Roca, J. Andrés, V. Moliner, I. Tuñón and J. Bertran, *Chem. Soc. Rev.*, 2004, **33**, 98.

3. A. Warshel, P. K. Sharma, M. Kato, Y. Xiang, H. Liu and M. H. M. Olsson, *Chem. Rev.*, 2006, **106**, 3210.
4. S. Hammes-Schiffer, *Acc. Chem. Res.*, 2006, **39**, 93.
5. D. R. Glowacki, J. N. Harvey and A. J. Mulholland, *Nat. Chem.*, 2012, **4**, 169.
6. S. D. Schwartz, *Top. Curr. Chem.*, 2013, **337**, 189.
7. I. Tuñón, D. Laage and J. Hynes, *Arch. Biochem. Biophys.*, 2015, **582**, 42.
8. C. Walsh, *Enzymatic reaction mechanisms*, W. H. Freeman, San Francisco, 1979.
9. E. Wigner, *Trans. Faraday Soc.*, 1938, **34**, 29.
10. M. Karplus, R. N. Porter and R. D. Sharma, *J. Chem. Phys.*, 1965, **43**, 3259.
11. E. A. McCullough and R. E. Wyatt, *J. Chem. Phys.*, 1971, **54**, 3578.
12. H. S. Johnston, *Gas Phase Reaction Rate Theory*, Ronald, New York, 1966.
13. J. T. Hynes, The Theory of Reactions in Solution, *Theory of Chemical Reaction Dynamics*, CRC Press, Boca Raton, FL, 1985, vol. 4, pp. 171–234.
14. J. P. Bergsma, B. J. Gertner, K. R. Wilson and J. T. Hynes, *J. Chem. Phys.*, 1987, **86**, 1356.
15. H. A. Kramers, *Physica*, 1940, **7**, 284.
16. P. Hänggi, P. Talkner and M. Borkovec, *Rev. Mod. Phys.*, 1990, **62**, 251.
17. R. F. Grote and J. T. Hynes, *J. Chem. Phys.*, 1980, **73**, 2715.
18. M. Roca, V. Moliner, I. Tuñón and J. T. Hynes, *J. Am. Chem. Soc.*, 2006, **128**, 6186.
19. J. J. Ruiz-Pernía, I. Tuñón, V. Moliner, J. T. Hynes and M. Roca, *J. Am. Chem. Soc.*, 2008, **130**, 7477.
20. J. T. Hynes, Crossing the transition state in solution, in *Solvent Effects and Chemical Reactivity*, Kluwer, Amsterdam, 1996; pp. 231–258.
21. G. Van der Zwan and J. T. Hynes, *J. Chem. Phys.*, 1982, **76**, 2993.
22. B. J. Gertner, J. P. Bergsma, K. R. Wilson, S. Lee and J. T. Hynes, *J. Chem. Phys.*, 1987, **86**, 1377.
23. B. J. Gertner, R. M. Whitnell, K. R. Wilson and J. T. Hynes, *J. Am. Chem. Soc.*, 1991, **113**, 74–87.
24. F. Takusagawa, M. Fujioka, A. Spies and R. L. Schowen, in *Comprehensive Biological Catalyzis*, ed. M. Sinnot, Academic Press, San Diego, CA, 1998, vol. 1, pp. 1–30.
25. M. Roca, S. Martí, J. Andrés, V. Moliner, I. Tuñón, J. Bertran and I. H. Williams, *J. Am. Chem. Soc.*, 2003, **125**, 7726.
26. G. M. Torrie and J. P. Valleau, *Chem. Phys. Lett.*, 1974, **28**, 578.
27. G. M. Torrie and J. P. Valleau, *J. Comput. Phys.*, 1977, **23**, 187.
28. S. Kumar, D. Bouzida, R. H. Swendsen, P. A. Kollman and J. M. Rosenberg, *J. Comput. Chem.*, 1992, **12**, 1011.
29. D. G. Truhlar, W. L. Hase and J. T. Hynes, *J. Phys. Chem.*, 1983, **87**, 2664–2682.
30. R. G. Mullen, J. E. Shea and B. Peters, *J. Chem. Phys.*, 2014, **140**, 041104.
31. R. Castillo, M. Roca, A. Soriano, V. Moliner and I. Tuñón, *J. Phys. Chem. B*, 2008, **112**, 529.
32. A. Warshel, *J. Biol. Chem.*, 1998, **273**, 27035.

33. J. P. Schanstra, J. Kingma and D. B. Janssen, *J. Biol. Chem.*, 1996, **271**, 14747.
34. A. Soriano, E. Silla, I. Tuñón, S. Martí, V. Moliner and J. Bertran, *Theor. Chem. Acc.*, 2004, **112**, 327.
35. R. García-Meseguer, S. Martí, J. J. Ruiz-Pernía, V. Moliner and I. Tuñón, *Nat. Chem.*, 2013, **5**, 566.
36. E. D. Hughes and C. K. Ingold, *J. Chem. Soc.*, 1935, 244.
37. A. Soriano, E. Silla, I. Tuñón and M. F. Ruiz-López, *J. Am. Chem. Soc.*, 2005, **127**, 1946.
38. J. Chandrasekhar and W. L. Jorgensen, *J. Am. Chem. Soc.*, 1985, **107**, 2974.
39. M. H. M. Olsson and A. Warshel, *J. Am. Chem. Soc.*, 2004, **126**, 15167.
40. D. G. Truhlar and B. C. Garrett, *Annu. Rev. Phys. Chem.*, 1984, **35**, 159.
41. J. M. Jez and J. P. Noel, *J. Biol. Chem.*, 2002, **277**, 1361.
42. J. Ulstrup, *Charge Transfer Processes in Condensed Media*, Springer-Verlag, Berlin, New York, 1979.
43. R. R. Dogonadze, A. M. Kuznetsov, M. G. Zaqaraya and J. Ulstrup, A Quantum Theory of Low-Temperature Chemical and Biological Rate Processes, *Tunneling in Biological Systems*, Academic Press, New York 1979, pp. 145–171.
44. A. Kuznetsov, *Charge transfer in chemical reactions kinetics*, Press polytechniques et universitaires romandes, Lausanne, 1997.
45. P. M. Kiefer and J. T. Hynes, Theoretical Aspects of Proton Transfer Reactions in a Polar Environment, in *Hydrogen Transfer Reactions*, Wiley-VCH, Weinheim, 2007, vol. 1, pp. 307–348.
46. D. Borgis and J. T. Hynes, *Chem. Phys.*, 1993, **170**, 315.
47. P. M. Kiefer and J. T. Hynes, *J. Phys. Org. Chem.*, 2010, **23**, 632.
48. K. Ando and J. T. Hynes, *J. Phys. Chem. B*, 1997, **101**, 10464.
49. A. Staib, D. Borgis and J. T. Hynes, *J. Chem. Phys.*, 1995, **102**, 2487.
50. S. Hammes-Schiffer and S. R. Billeter, *Int. Rev. Phys. Chem.*, 2001, **20**, 591.
51. D. G. Truhlar, *J. Phys. Org. Chem.*, 2010, **23**, 660.
52. P. M. Kiefer and J. T. Hynes, *J. Phys. Chem. A*, 2004, **108**, 11809.
53. S. Hammes-Schiffer, *ChemPhysChem*, 2002, **3**, 33.
54. D. G. Truhlar, J. Gao, C. Alhambra, M. Garcia-Viloca, J. Corchado, M. L. Sánchez and J. Villà, *Acc. Chem. Res.*, 2002, **35**, 341.
55. R. I. Cukier and D. G. Nocera, *Annu. Rev. Phys. Chem.*, 1998, **49**, 337.
56. S. Hammes-Schiffer, *Acc. Chem. Res.*, 2001, **34**, 273.
57. S. Hammes-Schiffer and A. V. Soudackov, *J. Phys. Chem. B*, 2008, **112**, 14108.
58. J. R. Knowles and W. J. Albery, *Acc. Chem. Res.*, 1977, **10**, 105.
59. Q. Cui and M. Karplus, *J. Am. Chem. Soc.*, 2002, **124**, 3093.
60. M. Wang, Z. Lu and W. Yang, *J. Chem. Phys.*, 2004, **121**, 101.
61. P. K. Agarwal, S. R. Billeter and S. Hammes-Schiffer, *J. Phys. Chem. B*, 2002, **106**, 3283.
62. M. Garcia-Viloca, D. G. Truhlar and J. Gao, *Biochemistry*, 2003, **42**, 13558.

63. N. Boekelheide, R. Salomón-Ferrer and T. F. Miller III, *Proc. Natl. Acad. Sci. U. S. A.*, 2011, **108**, 16159.
64. A. J. Adamczyka, J. Caoa, S. C. L. Kamerlin and A. Warshel, *Proc. Natl. Acad. Sci. U. S. A.*, 2011, **108**, 14115.
65. D. Doron, A. Kohen, K. Nam and D. T. Major, *J. Chem. Theory Comput.*, 2014, **10**, 1863.
66. C. Alhambra, J. Corchado, M. L. Sánchez, M. Garcia-Viloca, J. Gao and D. G. Truhlar, *J. Phys. Chem. B.*, 2001, **105**, 11326.
67. S. R. Billeter, S. P. Webb, P. K. Agarwal, T. Iordanov and S. Hammes-Schiffer, *J. Am. Chem. Soc.*, 2001, **123**, 11262.
68. Q. Cui, M. Elstner and M. Karplus, *J. Phys. Chem. B*, 2002, **106**, 2721.
69. A. J. Turner, V. Moliner and I. H. Williams, *Phys. Chem. Chem. Phys.*, 1999, **1**, 1323.
70. J. E. Basner and S. D. Schwartz, *J. Phys. Chem. B*, 2004, **108**, 444.
71. S. Ferrer, I. Tuñón, S. Martí, V. Moliner, M. Garcia-Viloca, A. González-Lafont and J. M. Lluch, *J. Am. Chem. Soc.*, 2006, **128**, 16851.
72. M. Roca, M. Oliva, R. Castillo, V. Moliner and I. Tuñón, *Chem. - Eur. J.*, 2010, **16**, 11399.
73. A. Vardi-Kilshtain, D. T. Major, A. Kohen, H. Engel and D. Doron, *J. Chem. Theory Comput.*, 2012, **8**, 4786.
74. D. G. Truhlar, J. Gao, M. Garcia-Viloca, C. Alhambra, J. Corchado, M. L. Sanchez and T. D. Poulsen, *Int. J. Quantum Chem.*, 2004, **100**, 1136.

CHAPTER 4

Electron Transfer Reactions in Enzymes: Seven Things that Might Break Down in Vanilla Marcus Theory and How to Fix Them if They Do

AURÉLIEN DE LA LANDE,*[a] FABIEN CAILLIEZ*[a] AND DENNIS R. SALAHUB*[b,c]

[a] Laboratoire de Chimie Physique, UMR8000, CNRS/Univ. Paris-Sud, Univ. Paris Saclay, 91405 Orsay, France; [b] Department of Chemistry, Centre for Molecular Simulation, Institute for Quantum Science and Technology, University of Calgary, 2500 University Drive, N.W., Calgary, Alberta, Canada, T2N 1N4; [c] College of Chemistry and Chemical Engineering, Henan University of Technology, No. 100, Lian Hua Street, High-Tech Development Zone, Zhengzhou 450001, P.R. China
*Email: aurelien.de-la-lande@u-psud.fr; fabien.cailliez@u-psud.fr; dennis.salahub@ucalgary.ca

4.1 Introduction

Scheme 4.1, two intersecting parabolas, represents the Marcus theory of electron transfer. It is, arguably, as recognisable to chemists, physicists and biologists as the Volkswagen logo is to the general public. This powerful two-state theory has spawned literally thousands of detailed computations and

RSC Theoretical and Computational Chemistry Series No. 9
Simulating Enzyme Reactivity: Computational Methods in Enzyme Catalysis
Edited by Iñaki Tuñón and Vicent Moliner

Published by the Royal Society of Chemistry, www.rsc.org

Scheme 4.1 The Marcus diagram.

simulations, including some from our own groups. More importantly, it has provided the conceptual framework for the interpretation of tens, if not hundreds, of thousands of electron transfer experiments. Rudy Marcus is a true giant of physical chemistry, on whose shoulders stand a legion of adopters and adapters; a scientific hero of the first magnitude. His Nobel Prize was more than 20 years ago and he is still going strong at the age of 92, publishing at a rate and an impact that would shame researchers 40 years his junior, or more! In terms of bibliometrics, he has published over 450 papers that have been cited nearly 50 000 times. His most cited paper has more than 6000 citations and he has six papers with more than 1000. His h-index is a staggering 94 (94 papers with more than 94 citations). We respectfully dedicate this review chapter to Professor Marcus.

Marcus theory (MT) was designed in the 1950s to describe redox reactions of ions in solution. While one might wonder what such a theory is doing in a book on enzymes, we hope to show in this review chapter how more complex theories have evolved over the past 60 years from the 'paper and pencil' chemical physics of Marcus' first papers to the current supercomputer simulations of electron transfer reactions in complex environments. While this review is not meant to be a work on science history, we think that an overview of the roots of MT will help put the modern methods into proper perspective.

'Marcus theory' may mean different things to different people, depending on the particular variant of MT they have in mind. In this chapter, we take Marcus' work as it stood in 1965 to represent the standard or 'vanilla' MT. After describing the salient features of vanilla MT, the concepts and approximations involved, we then attempt to provide a pedagogical review from the perspective of features that can break down in real applications, focusing on enzymes, and we discuss available fixes. The chapter is meant to provide a rigorous graduate-level review of both theoretical and computational issues that will allow graduate students to perform and understand real simulations in a biological context, avoiding pitfalls as much as possible.

While our bibliography is extensive it is inevitably far from exhaustive. We have tried to find good pedagogical examples from the literature and we apologise in advance to the authors of the countless great papers that we have not included.

In Section 4.2 we describe vanilla MT in terms of its underlying hypotheses and its defining characteristics. In bullet form these are:

- Theory intended for ions in solution
- Born–Oppenheimer (BO) approximation
- Condon approximation (horizontal Frank–Condon factors)
- Two-state model
- Diabatic or adiabatic cases
- Non-adiabatic, small gap, electron transfer (ET) described with Landau–Zener theory
- Electrostatic free energy of states having non-equilibrium polarisation
- Polarisable continuum model for the solvent
- Electrostatically unsaturated (linear) response of solvent to the ET
- Global reaction coordinate
- Free energy functions are parabolas
- Rates depend on the solvent reorganisation energy and on the electronic coupling between the two states
- An inverted region is predicted (rate is a non-monotonic function of the driving force)

All of these fundamental assumptions are subject to breaking down in real systems, proteins in particular. In what follows we will examine seven aspects of vanilla MT that can potentially break down, see if they do or do not, and see how they may be ameliorated in various 'extended Marcus theories'. The seven potential trouble spots we have identified, again in bullet form (see Appendix A for a chronology of some important contributions to ET theory), are:

- The linear response approximation may break down due to electrostatic issues (Section 4.4)
- The ergodic hypothesis may break down (Section 4.4)
- Quantised inner-sphere vibrational modes (Section 4.5)
- Nuclear quantum effects arising from the outer sphere (protein, solvent, Section 4.5)
- The fluctuating chemical structure of the intervening medium determines the electronic coupling (Section 4.6)
- Non-Condon effects give rise to inelastic tunnelling (Section 4.6)
- The two-level approximation is questionable for chains of ET (Section 4.7)

In what follows we have attempted to group these pitfalls and extensions of MT within a logical framework that will hopefully provide a coherent view of the field. In Section 4.3 we take a modern view and discuss the relation between microscopic and macroscopic concepts and simulations, introduce the energy gap as the reaction coordinate and explain how to use it in simulations. In Section 4.4 we discuss non-ergodic effects, ET involving strong modification of the environment or of the donor (D)/acceptor (A) complex (trouble spots 1 and 2). In Section 4.5 we discuss the formulation of quantum ET theory, either to account for the quantisation of the internal modes of the redox cofactors or to describe nuclear quantum effects in a general, tractable formalism (trouble spots 3 and 4). Section 4.6 deals with the electronic coupling strength and its fluctuations within pathway and tunnelling current approaches, non-Condon effects, the effects of inelastic tunnelling, electron-nuclear entanglement and decoherence (trouble spots 5 and 6). In Section 4.7 we examine methods that go beyond the two-state model, incoherent hopping, flickering resonance (trouble spot 7). Finally in Section 4.8 we sum up and offer some perspectives for future work.

4.2 Vanilla MT

We start the lead-up to MT, somewhat arbitrarily, with a contribution from Libby[1] on the theory of electron exchange in aqueous solution. Libby started by considering H_2^+ in solution and examined the frequency of exchange of the electron between the two protons. He realised that two timescales were involved, fast for the actual quantum mechanical electron exchange and slower for the rearrangement of the heavier water molecules in the solvent ("... the hydration energy is not transmissible as rapidly as the electrons"). He therefore constructed a two-part theory in which the quantum mechanics of the ET itself was treated in vacuum and the Franck–Condon principle was applied to the aqueous solutions, modulating the exchange frequency. For example, in the case of Fe^{2+} to Fe^{3+} exchange, the ferrous ion becomes a ferric ion, but still in a ferrous environment and *vice versa.* Energy of hydration is stored in the process and it eventually has to leak across by slower relaxation of the solvent. Hence a barrier to the ET is created. Libby then estimated the barrier from electrostatics:

$$\Delta H^{\ddagger} = \frac{e^2}{2rD}(2Z + 1) \tag{4.1}$$

where e is the charge of the electron, Z is the charge of the ion with the smaller charge, r is its ionic radius and D is the dielectric constant of water. Libby considered both simple atomic ions and also complex ions such as $Fe(CN)_6^{3-}$ and $Fe(CN)_6^{4-}$ presaging the notion of outer-sphere and inner-sphere ET.

We mention next the 1953 paper of Noel Hush,[2] another giant of ET theory. Although this paper deals with a chemical reaction rather than ET, as shown in Figure 4.1, it is within the framework of two-state models and shows that the idea of intersecting energy curves along a reaction coordinate (a particular cut on an energy surface) was very much a part of the zeitgeist of the 1950s.

In one of two landmark papers in 1956, Marcus[3] laid the groundwork for his theory by considering that a fast ET reaction leaves the surrounding solvent in an electrostatic configuration that was not at equilibrium with the new electronic state. He realised that, in order to conserve energy, the

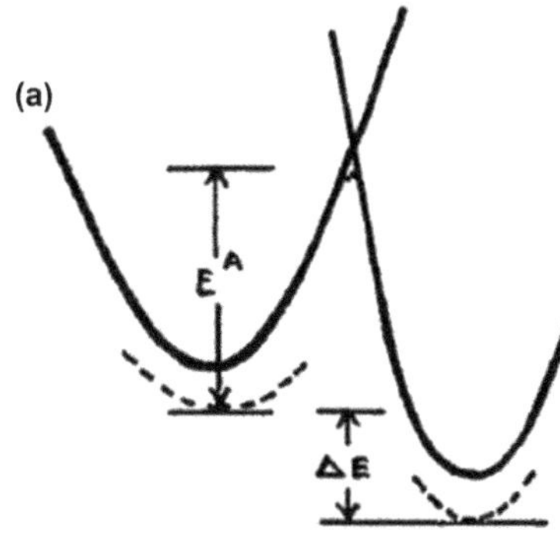

Figure 4.1 Reaction coordinate (abscissa) *vs.* potential energy (ordinate). (a) and (b) are the diabatic reactant and product states.
Reproduced from ref. 2 with permission from John Wiley and Sons. Copyright © 1953 Interscience Publishers, Inc., New York.

change in redox state proposed by Libby could only take place at the crossing point of the two diabatic curves. He then went on to derive the proper equations for various properties of systems with non-equilibrium electrostatic configurations, including the free energy, energy and entropy. Here, we outline some of the key concepts and equations. The thing to keep in mind is that the ET is fast and the solvent reorganisation is slower.

The electric field, **E**, is given by the gradient of the electrostatic potential and this holds also for non-equilibrium polarisation. For equilibrium situations, two vectors are required, **E** and the dielectric displacement, which is defined by the charge distribution and involves the dielectric constant of the medium. For non-equilibrium situations three vectors are required, **E**, the polarisation strength, **P**, and $\mathbf{E}_c$, the electric field strength which the charge distribution would exert if it were in a vacuum rather than a polarised medium. Marcus then defines the state of polarisation of a non-equilibrium state in terms of two types of polarisation, **U**-type (atomic and orientation polarisation) and **E**-type (electronic polarisation). The total polarisation $\mathbf{P}(\mathbf{r})$ is the vector sum of these two:

$$\mathbf{P}(\mathbf{r}) = \mathbf{P}_u(\mathbf{r}) + \mathbf{P}_e(\mathbf{r}) \tag{4.2}$$

He then calculates the free energy of the non-equilibrium state by finding a reversible path for reaching that state involving a two-stage charging process. The free energy calculation requires the electrostatic potential, $\psi(\boldsymbol{r})$, which Marcus gives as:

$$\psi(\boldsymbol{r}') = \int \frac{\rho(\boldsymbol{r})\mathrm{d}V}{|\boldsymbol{r}-\boldsymbol{r}'|} + \int \frac{\sigma(\boldsymbol{r})\mathrm{d}S}{|\boldsymbol{r}-\boldsymbol{r}'|} + \int \boldsymbol{P}(\boldsymbol{r}) \cdot \nabla_r \frac{1}{|\boldsymbol{r}-\boldsymbol{r}'|}\mathrm{d}V \tag{4.3}$$

where $\rho(\boldsymbol{r})$ is the charge density, $\sigma(\boldsymbol{r})$ is the charge density per unit area at a surface element dS of an interface of the system and the polarisation is the sum of **E**- and **U**-types.

Eqn (4.3) is then used to calculate the free energy G of the two-step charging process, the final result being:

$$G=\frac{1}{2}\int\left\{\frac{\mathbf{E}_c\cdot\mathbf{E}_c}{4\pi}-\mathbf{P}\cdot\mathbf{E}_c+\mathbf{P}_u\cdot\left(\frac{\mathbf{P}_u}{\alpha_u}-\mathbf{E}\right)\right\}\mathrm{d}V \tag{4.4}$$

where α_u is the polarisability. Energy and entropy components were also derived.

In his second landmark paper of 1956, Marcus[4] uses the above results to derive equations for the rates of redox reactions for the slightly overlapping (weak electronic coupling) case. An important notion at the outset is that two electronic wave functions are involved, one for reactants and one for products. In an intermediate activated complex the two wave functions mix and, importantly, the total energy of a hypothetical system having the electronic configuration of the reactants has to have the same total energy as the hypothetical system having the electronic configuration of the products (hence the importance of the crossing point (or crossing seam) in the Marcus diagram – it represents the activated complex). Marcus then brings in the argument about non-equilibrium intermediate states, rapid ET and slow solvent reorganisation, and discusses the role of solvent fluctuations which can lead to formation of the activated complex. It is assumed that the reactants can be "treated as a sphere which in turn may be surrounded by a concentric spherical region of saturated dielectric" (an inner sphere), "outside of which the medium is dielectrically unsaturated" (which will allow a linear response treatment) (the outer sphere). He then goes on to minimise the free energy (using eqn (4.4)) subject to the constraint that the reactant and product configurations have the same energy. When all of the dust had settled, for this model of spherical ions in a polarisable continuum, the free energy of activation, $\Delta G^{\ddagger}$, is given in terms of ionic charges, ionic radii, static and optical dielectric constants, and the change in the electronic entropy on ET (see eqn (38)–(40) of ref. 4). $\Delta G^{\ddagger}$ is then used within a kinetics scheme to derive the final expression for the overall bimolecular rate constant:

$$k_{\mathrm{b}}\cong Z\exp\left(-\frac{\Delta G^{\ddagger}}{k_{\mathrm{B}}T}\right) \tag{4.5}$$

where Z is the collision number.

Marcus then went on to apply his theory to the rates of isotopic exchange reactions[5] and organic redox reactions.[6] In these applications the following three equations were used, in addition to eqn (4.5):

$$\Delta G^{\ddagger}=m^2\lambda+e_1^*e_2^*/D_{\mathrm{s}}r \tag{4.6}$$

$$2m+1=-\left[\Delta G^0+\frac{(e_1e_2-e_1^*e_2^*)}{D_{\mathrm{s}}r}\right]\Big/\lambda \tag{4.7}$$

$$\lambda = \left(\frac{1}{2a_1} + \frac{1}{2a_2} - \frac{1}{r}\right)\left(\frac{1}{D_{\text{op}}} - \frac{1}{D_{\text{s}}}\right) \tag{4.8}$$

which contain the same information as the aforementioned eqn (38)–(40) of ref. 4, although in a rearranged form and with minor changes in nomenclature.

In eqn (4.6) to (4.8) the *e*s are charges on reactants and products in the two states, *a*s are the effective radii of the reactants, D_{op} is the square of the refractive index and D_{s} is the static dielectric constant. *r* is the distance between the centres of the reactants in the activated complex. We now know λ as the reorganisation energy (see below). Eqn (4.8) gives the reorganisation energy for Marcus' spherical electrostatic continuum model.

Hush returns to the scene in 1958[7] with a paper on adiabatic rate processes at electrodes, focusing on the energy–charge relationships. He argues that only adiabatic or near-adiabatic paths are important for thermally-activated reactions and criticises the conventional approach in which an approximate potential energy path for the reaction is calculated from intersecting curves for the initial and final states, the very Franck–Condon restrictions that are at the heart of Marcus' 1956 papers. He does, however, paint a similar picture to Marcus as concerns the formation of a transition state in which the reaction field of the dielectric is explicitly considered. He proposes to replace the conventional representation of the energy of the system as a function of displacement along a reaction coordinate in space by calculation of the energy as a function of the electron density on the solution side of the interface (*cf.* Section 4.3, in which the contemporary choice of reaction coordinate, the diabatic energy gap, which depends on the electron density, is introduced). He then goes on to calculate the free energy as a function of charge, as it varies adiabatically from reactants, through the transition state and on to the products, showing that the activation energy is greater than in a constant-charge curve-intersection treatment (the dielectric is doing its work). We refer the interested reader to ref. 7 for details. (Interestingly, Hush makes no reference to Marcus' 1956 work, perhaps because of the long delay between submission of his manuscript and publication. His manuscript was received on 29 November 1955, just shortly after the reception dates of Marcus' papers, 11 July and 28 July 1955.)

In Marcus' 1960 paper[8] we see most of the ingredients that define vanilla MT. We see a two-state diagram of intersecting diabatic curves, potentially split by the electronic coupling to form two adiabatic curves, plotted against a reaction coordinate representing atomic coordinates (Figure 4.2).[9] Electronic coupling between the two diabatic states opens a gap and, depending on the magnitude of the coupling, the ET reaction may proceed either adiabatically, for strong coupling, or non-adiabatically, for weak coupling. Fluctuations of the solvent bring the system to the intersection allowing the reaction to occur. The possibility is noted of an inverted behaviour, a decrease of the rate constant with increasing driving force, the negative of the free energy change of the reaction. He also introduced the

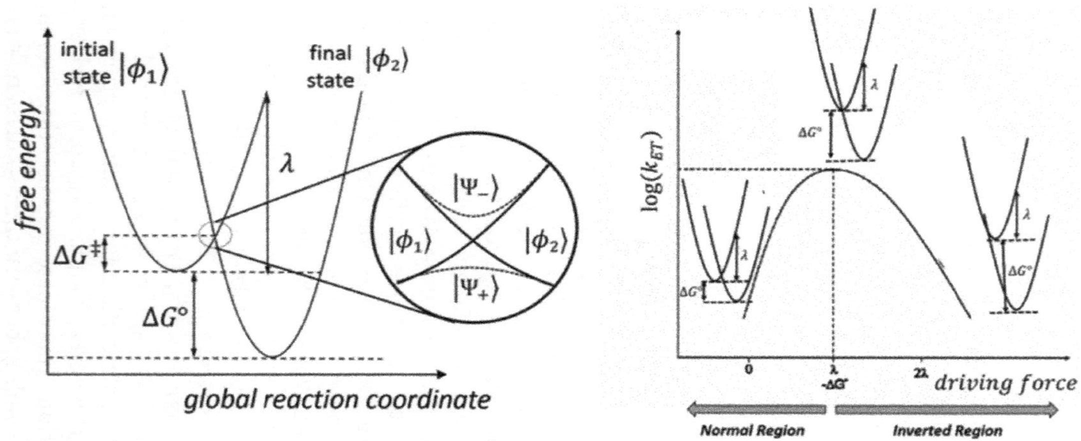

Figure 4.2 Left, canonical two-level Marcus theory free energy profile. The diabatic states $|\phi_1\rangle$ and $|\phi_2\rangle$ are shown as full lines. Reproduced with permission from ref. 9 with permission from the PCCP Owner Societies. Right, non-monotonic shape of ET rate with driving force, defining the normal and inverted regions. The asymmetry of the curve will later be explained by nuclear tunnelling (see Section 4.5).

notion of contributions to the reorganisation energy of bond-length and bond-angle changes in the inner sphere, on which he expanded in his 1963 paper.[10]

In 1961, Hush continued his work on adiabatic theory with a paper[11] on outer-sphere ET in solution using his probability density parameter as the reaction coordinate which he calls λ (it corresponds to the m in eqn (4.6) and (4.7)...). Once the appropriate correspondences are made he finds that his work and Marcus' lead to general conclusions that are very similar.

We will take Marcus' 1965 paper[12] to define vanilla MT. He presents there a unified treatment for homogeneous and electrode reactions, treats the inner-sphere motions as classical vibrations and generalises the treatment of the outer sphere to non-spherical cavities.

While the above overview of vanilla MT is sketchy, we hope it will give a graduate student a notion of the main concepts behind MT, a sense of science history and an inkling of some of the complex equations that are involved in classical electrostatics and statistical mechanics. Not all students will have the desire or, for many, the ability, to go through such classic chemical physics reasoning and mathematics but we wish those who do have the taste for it a hearty 'bonne lecture'! We now wish to fast forward about 20 years to the time when the contemporary version of the Marcus diagram and the main concepts it embodies became pervasive. Many reviews of MT have converged on the diagram shown in Figure 4.2.[13,14]

Figure 4.2 shows two intersecting parabolas, plotted against a global reaction coordinate. The parabolic shape is a result of the linear response approximation used in a polarisable continuum treatment with a dielectrically unsaturated medium. The two parabolas have the same curvature and they are simply displaced with respect to each other both horizontally and vertically. Because of these approximations, high-school algebra shows that there is a special relationship between the free energy change of the reaction, ΔG^0, the free energy of activation, $\Delta G^\ddagger$, and the reorganisation energy, λ (the free energy required to move the system from the coordinates of the reactants to the coordinates of the products but without effectuating the ET):

$$\Delta G^\ddagger = \frac{(\Delta G^0 + \lambda)^2}{4\lambda} \tag{4.9}$$

Figure 4.2 also shows that coupling between the two diabatic curves leads to the opening of a gap, ΔE, and the mixing of the diabatic wave functions $|\phi_1\rangle$ and $|\phi_2\rangle$ to form adiabatic wave functions $|\Psi_+\rangle$ and $|\Psi_-\rangle$. If the gap is large enough then the usual BO adiabatic transition state theory may be applied, leading to the adiabatic rate equation:

$$k = \kappa \cdot A\sigma^2 \cdot \exp\left(-\frac{\Delta G^\ddagger}{k_\mathrm{B}T}\right) \tag{4.10}$$

where the transmission coefficient κ is taken as unity (as soon as the crossing point is reached products are produced) and $A\sigma^2$ has dimensions of the collision frequency. In the contrary case of small coupling, which is more interesting for eventual use for long-distance ET in proteins, κ is much lower than unity. The Landau–Zener theory provides an estimate of population exchange probability (P_{LZ}) between states $|\phi_1\rangle$ and $|\phi_2\rangle$ after one passage through the crossing seam:

$$P_{LZ} = 1 - \exp\left(\frac{-2\pi H_{DA}^2}{\hbar \dot{\Delta E}}\right) \tag{4.11}$$

where H_{DA} is the electronic coupling matrix element $\langle\phi_1|H_{el}|\phi_2\rangle$, $\dot{\Delta E}$ is the time derivative of the energy gap and $\hbar$ is the reduced Planck constant. Both are assumed to be constant during the passage through the crossing seam. To obtain the overall transmission coefficient, it is necessary to consider multiple passages of the system through the intersection seam.[14] Under the steady-state assumption one then finds $\kappa = 2P_{LZ}/(1 + P_{LZ})$ for ET in the normal region and $\kappa = 2P_{LZ}/(1 - P_{LZ})$ for the inverted region.[14]

According to eqn (4.9) the maximum rate occurs when the driving force of the reaction, $-\Delta G^0$, is equal to the reorganisation energy, λ. Beyond this point, further increase of the driving force actually leads to a decrease in the rate, in the so-called abnormal or inverted region. The eventual experimental discovery of the inverted region[15] represents a great triumph for the MT.

4.3 Relation Between Microscopic and Macroscopic Concepts and Molecular Simulation

4.3.1 Microscopic Derivation of the Marcus Activation Free Energy

As described in Section 4.2, the original derivation of the Marcus activation energy (eqn (3.9)), connecting $\Delta G^{\ddagger}$ and ΔG^0, relies on a macroscopic description of the solvent and on the linear response approximation (LRA). This approximation leads to parabolic shapes of the free energies of the diabatic states as a function of the solvent polarisation, which plays the role of a global reaction coordinate for the ET reaction. In this section, we will derive the Marcus activation energy using a more detailed, and modern, microscopic description of matter that emerged in the early 1980s, especially with the work of Warshel and collaborators.[16–18]

In his seminal 1982 paper,[16] Warshel proposed that the energy gap $\Delta E = E_2 - E_1$ is a more appropriate reaction coordinate than the polarisation at the microscopic level. In particular, this definition implies a degeneracy of the two electronic states at the activated complex. We note in passing that this definition of the coordinate already emerged from works of Hopfield[19] and Zusman[20] prior to its use by Warshel. The activation free energy is defined as the locus where the free energy curves $G_1(\varepsilon)$ and $G_2(\varepsilon)$ of states 1 and 2 intersect. The free energy in state x is connected to the probability $p_x(\varepsilon)$ of observing a given value ε of ΔE when the system is in the electronic state x:

$$G_x(\varepsilon) = -k_B T \ln p_x(\varepsilon) + G_x^0 \tag{4.12}$$

where G_x^0 is the 'full' free energy of state x, which is given in the canonical ensemble by:

$$G_x^0 = -k_B T \ln Z_x = -k_B T \ln\left(\int \exp(-\beta E_x) \mathrm{d}\Gamma\right) \tag{4.13}$$

where $\beta = 1/k_B T$ and the integration is over the whole of the phase space Γ accessible to the system. The probability of observing a given value ε of the energy gap in the electronic state x is given by:

$$p_x(\varepsilon) = \frac{\int \exp(-\beta E_x)\delta(\Delta E - \varepsilon)\mathrm{d}\Gamma}{\int \exp(-\beta E_x)\mathrm{d}\Gamma} = \frac{\int \exp(-\beta E_x)\delta(\Delta E - \varepsilon)\mathrm{d}\Gamma}{Z_x} \tag{4.14}$$

where δ is the Dirac delta function.

We note here a very important relation connecting the free energy of the two electronic states for the case that the system is ergodic (we will come back to this ergodicity hypothesis in Section 3.4):

$$G_2(\varepsilon) - G_1(\varepsilon) = \varepsilon \tag{4.15}$$

This relation is obtained directly by combining eqn (4.12) and (4.14). We emphasise that eqn (4.15) does not rely on the form of $p_x(\varepsilon)$ but only on the ergodicity hypothesis.

The next step in deriving the Marcus relation is to use the LRA which, at the microscopic level, leads to a Gaussian distribution of the energy gap:

$$p_x(\varepsilon) = \frac{1}{\sigma_x\sqrt{2\pi}} \exp\left(-\frac{\left(\varepsilon - \langle \Delta E \rangle_x\right)^2}{2\sigma_x^2}\right) \tag{4.16}$$

$\langle \Delta E \rangle_x$ is the average value of the energy gap in electronic state x and σ_x^2 its variance. This Gaussian statistics hypothesis for ΔE can be understood by the fact that the energy gap is a function of a large number of variables (the position of every charged or polarizable particle of the system). Even if the contribution of one particular part of the system to ΔE may not be Gaussian, the sum of all the contributions is expected to follow a normal distribution, as a consequence of the central limit theorem.

Combining eqn (4.12) and (4.16), one obtains the free energy of diabatic state x as a parabolic function of the energy gap:

$$G_x(\varepsilon) = G_x^0 + \frac{\left(\varepsilon - \langle \Delta E \rangle_x\right)^2}{4\lambda_x} + \frac{k_B T}{2} \ln(4\pi k_B T \lambda_x) \tag{4.17}$$

In this equation, λ_x is the reorganisation energy defined here from the fluctuations of the energy gap:

$$\lambda_x = \frac{\sigma_x^2}{2k_B T} \tag{4.18}$$

In order to fulfil eqn (4.15) for every value of the energy gap, the curvatures of the parabolic functions $G_1(\varepsilon)$ and $G_2(\varepsilon)$ have to be equal, leading to $\lambda_1 = \lambda_2 = \lambda$. Applying eqn (4.15) for $\varepsilon = \langle \Delta E \rangle_1$ and $\varepsilon = \langle \Delta E \rangle_2$ further leads to the two following relations:

$$\Delta G^0 = G_2^0 - G_1^0 = \frac{\langle \Delta E \rangle_1 + \langle \Delta E \rangle_2}{2} \tag{4.19}$$

$$\lambda = \frac{\langle \Delta E \rangle_1 - \langle \Delta E \rangle_2}{2} = \lambda^{St} \tag{4.20}$$

which gives an alternative definition of the reorganisation energy often referred to as the Stokes reorganisation energy λ^{St}, because it is linked to the Stokes shift between emission and absorption spectra. The 'Marcus' reorganisation free energies λ_x^M are also defined (see Section 4.2) by the reversible work necessary to bring the system from its optimum configuration in state x (x being 1 or 2) to the optimum configuration of the other state involved in the ET (resp. 2 or 1). Using this definition, one gets:

$$\lambda_1^M = G_1(\langle \Delta E \rangle_2) - G_1(\langle \Delta E \rangle_1) = \frac{(\langle \Delta E \rangle_2 - \langle \Delta E \rangle_1)^2}{4\lambda} = \lambda \tag{4.21}$$

and a similar result for λ_2^M. As a consequence, the LRA leads to the equality between all the possible definitions of the reorganisation energy: $\lambda = \lambda_1 = \lambda_2 = \lambda_1^M = \lambda_2^M = \lambda^{St}$.

The two parabolic free energy curves cross when $\varepsilon = 0$, and thus the activation free energy is given by: $\Delta G^{\ddagger} = G_1(0) - G_1(\langle \Delta E \rangle_1)$. Noting that $\langle \Delta E \rangle_1 = \lambda + \Delta G^0$ from eqn (4.19) and (4.20), one obtains the Marcus expression of the activation free energy as given in eqn (4.9).

4.3.2 ET Theories and Molecular Simulations

For a long time, experiment has been the only way to (in)validate MT. Unfortunately, and despite the great ingenuity of experimentalists, it is not easy to design experiments in which all the parameters entering the MT can be fine-tuned and controlled. The advent of powerful computer and simulation algorithms have allowed one (i) to explore various situations using model systems, (ii) to compute parameters entering the MT, (iii) to go beyond the harmonic approximation of quantum mechanical calculations, (iv) to examine the validity of the approximations of the MT (and of its derivatives), and (v) to help to build new ET theories.

In the 1980s, Warshel made a great contribution by developing semi-classical molecular dynamics (MD) simulations dedicated to the theoretical study of ET, in which the system is described at the microscopic level.[16] The first ET simulations[18,21–25] aimed to test the LRA, the validity of which is fundamental to the derivation of the MT activation free energy. The principle is to build the free energy profiles from the probability distributions of the energy gap $p_1(\varepsilon)$ and $p_2(\varepsilon)$ using eqn (4.12) and (4.14). These distributions can in principle be computed from molecular simulations (MD or Monte Carlo) performed on each diabatic state by 'simply' counting the occurrence of a given value of ε during the simulation. However, in cases where the activation free energy is high, more involved simulation protocols such as free energy perturbation (FEP) and umbrella sampling are needed to provide reliable results. As an illustration of these difficulties, we discuss here the example of a simple model system first studied by Carter and Hynes.[24] In 1989, they studied model ion-pair (IP) and neutral-pair (NP) species in a model polar solvent, composed of rigid dipolar molecules with a dipole moment of 2.4 D, a rough caricature of methyl chloride. They performed two equilibrium MD simulations (one for each diabatic state) and thus derived free energy profiles for the two states as a function of the energy gap. The two curves could be fitted by parabolic expressions but with different curvatures, thus lending support to suggestions of Kakitani and Mataga,[26] but contradictory to the LRA. However, a few months later, Tachiya[27] showed that the interpretation proposed by Kakitani and Mataga violated the fundamental ergodicity relation given in eqn (4.15). In 1990, King and Warshel[25] re-examined the same system as Carter and Hynes using FEP simulations, in addition to long (for that time) equilibrium MD simulations on the two diabatic states. Their main results are shown on Figure 4.3. Using the same methodology as Carter and Hynes, they arrived at similar conclusions, that is: one can fit the free energy curves for the two states by two parabolas with different force constants. These are shown by dashed curves on the left graph of Figure 4.3. However, the use of these parabolic fits to extrapolate the free energy curves far from their minima leads to strong deviations with respect to the 'true' results extracted from the FEP simulations (symbols and full line in Figure 4.3). The FEP results on the other hand are in close agreement with the MT: the free energy curves for the two diabatic states have similar curvatures (left graph) and $\Delta G^{\ddagger}/\lambda$ is a quadratic function of $\Delta G^0/\lambda$ (right graph), as predicted by eqn (4.9), thus validating the LRA. We think this example sheds light on the danger of insufficient sampling, which is a central concern in molecular simulation.

Validation of the LRA was obtained for ET between monoatomic donors and acceptors[21,22,25] or benzene-type molecules[18] immersed in non-polarisable water molecules, ET between monatomic solutes taking into account solvent polarisability.[25] Concerning ETs in proteins the LRA was verified, for example, in the bacterial photosynthetic reaction centre,[23,28] in copper proteins[29,30] and in haem proteins.[31]

Since the first studies in the 1980s, molecular simulation has become an indispensable tool to test and accompany the development of ET theories.

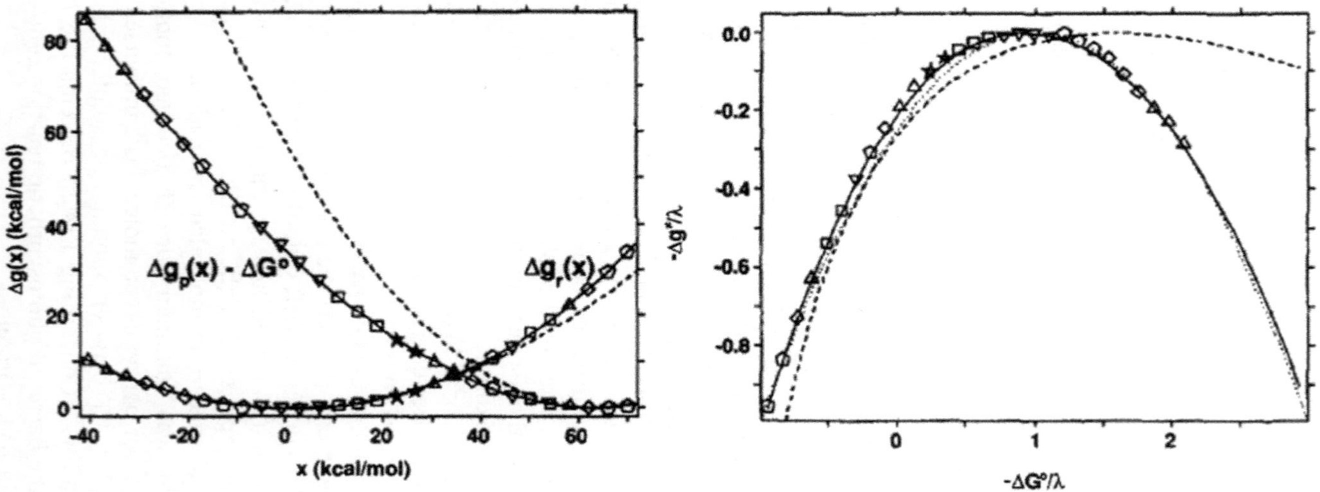

Figure 4.3 Left, free energy curves as a function of the energy gap for a two-atom solute model ET-system immersed in a simple dipolar solvent obtained from FEP simulations. Symbols represent the simulation results, the solid line is a fourth-order polynomial fit of the two curves. The dashed lines are parabolic fits of MD simulation data performed on each diabatic state. Right, relation between the activation free energy and the reaction free energy for the same system. Symbols and solid line are the results of FEP simulations, the dotted line (almost coincident with the solid line) is the prediction of the MT, and the dashed line is the prediction from the results on MD simulations on the diabatic states only. Both figures are Fig. 4 and Fig. 6 of ref. 25. Reprinted from G. King and A. Warshel, *J. Chem. Phys.*, 1990, **93**, 8682–8692, with the permission of AIP Publishing.

The increase in computational power and the improvement of algorithms has allowed more complex systems to be addressed, situations where the validity of the MT can be tested, especially ET reactions occurring in proteins. Various research groups have specialised in the study of such issues and develop or use dedicated simulation tools.[24,32–37] The interested reader is referred to various recent reviews on simulations of ET.[38–40] In the course of this chapter, many examples will be taken from the molecular simulation literature as illustrations.

4.4 Beyond the LRA

4.4.1 What May Cause the LRA to Break Down?

The LRA is a fundamental assumption of vanilla MT. It is characterised by parabolic expressions for the free energy of the two electronic states exhibiting the same curvature or, equivalently, a Gaussian distribution of the energy gaps (see Section 4.3). Until now, three main reasons have been put forward to explain the breakdown of the LRA in ET systems. The first is that there may be a large change in the polarisability of the donor/acceptor moieties upon electron transfer.[36,41] The second arises from a large modification of the charge transfer (CT) complex during the process of ET. This happens particularly when ET is correlated with a change in the coordination of a transition metal[42,43] or when covalent bonds are broken/formed, like in PCET reactions. Finally, deviation from the LRA may be due to the breaking of the ergodicity hypothesis, for example in the case of fast ETs (in the ns–ps range).[36]

4.4.2 Change of the Polarisability of the Acceptor/Donor Moieties

In cases where the donor and the acceptor in the CT complex are neutral in one of the electronic states and charged in the other, it is expected that the polarisability of the CT will be strongly affected upon ET. The primary charge separation in the photosynthetic reaction centre (PRC) is a typical example of such a situation. In this system, after photoexcitation of the so-called special pair (denoted P), an electron hops to a close bacteriochlorophyll (denoted B_x): $[P^*,B_x] \rightarrow [P^+,B_x^-]$, prior to further CTs. Two possible, approximately symmetric, ET paths could in principle be anticipated due to the presence of two bacteriochlorophylls B_M and B_L close to P. Actually, ET is much faster along the 'L-branch' than along the 'M-branch'. Le Bard and Matyushov investigated in depth the mechanism of the different potential ETs in the bacterial PRC, with the use of MD simulations, taking into account the polarisability of the special pair P. They especially checked the statistics of the energy gaps.[36,44] Figure 4.4 shows the free energies of the $[P^*,B_M]$ (F_1^M) and $[P^+,B_M^-]$ (F_2^M) states, extracted from MD simulation data, as a function

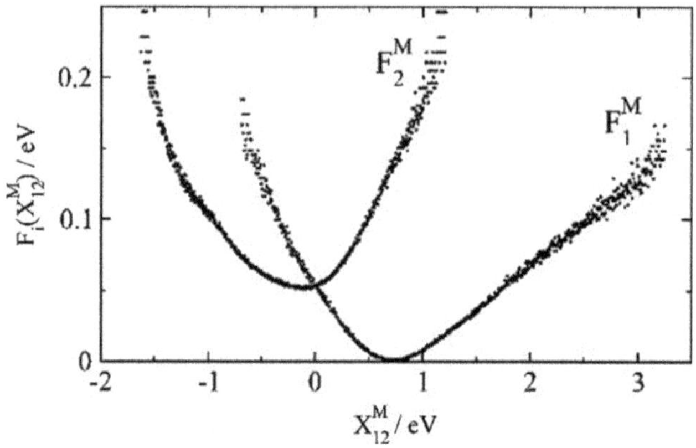

Figure 4.4 Free energies of the [P*,B_M] ($F_1{}^M$) and [P^+,B_{M^-}] ($F_2{}^M$) diabatic states in the bacterial PRC, as a function of the energy gap between the two states ($X_{12}{}^M$) obtained from MD simulations.
Reprinted with permission from D. N. Le Bard and D. V. Matyushov, *J. Phys. Chem. B*, 2009, **113**, 12424–12437.[36] Copyright (2009) American Chemical Society.

of the reaction coordinate (that is the diabatic energy gap, denoted $X_{12}{}^M$ on Figure 4.4). $F_1{}^M$ is clearly non-parabolic, exhibiting high asymmetry and a quasi-linear behaviour when the energy gap is greater than ~1.5 eV. This behaviour can be nicely recovered in the framework of the so-called Q-model, described below.

The Q-model for electron transfer was developed by Matyushov and Voth.[41,45] This model is aimed at describing the fact that the force constant of the collective mode that drives the ET has a different value in the two electronic states. Considering that the solute/solvent coupling is linear, the energy of the system in the electronic state x (1 or 2) is given by:

$$E_x(q) = E_x^0 - C_x q + \frac{1}{2}\zeta_x q^2 \tag{4.22}$$

where E_x^0 is the vacuum energy of the state x, C_x is the strength of the solute/solvent coupling and ζ_x is the force constant of the collective mode along the coordinate q. The standard Marcus model corresponds to the case where $\zeta_1 = \zeta_2$. In that case, the energy gap is a linear function of q, whereas in the more general case, the energy gap assumes a quadratic form with respect to q. Based on this model, Matyushov and Voth[45] derived an analytical expression for the free energy as a function of the diabatic energy gap ε:

$$e^{-\beta G_x(\varepsilon)+\beta G_x^0} = \left(\frac{1}{1-e^{-\beta\lambda_x\alpha_x^2}}\right)\sqrt{\frac{\lambda_x|\alpha_x|^3}{|\varepsilon-\varepsilon_0|}}\,e^{-\beta\left(|\alpha_x||\varepsilon-\varepsilon_0|+\lambda_x\alpha_x^2\right)}\,I_1\left(2\beta\sqrt{|\alpha_x|^3\lambda_x|\varepsilon-\varepsilon_0|}\right) \tag{4.23}$$

In eqn (4.23), I_1 is the first-order modified Bessel function. The definition of each free energy curve requires three parameters (ε_0, α_x and λ_x). α_x measures the variation of the force constant of the collective mode upon ET:

$$\alpha_x = \frac{\zeta_x}{\zeta_2 - \zeta_1} \tag{4.24}$$

λ_x is the reorganisation energy in state x defined from the variance of the energy gap (eqn (4.18)), given by:

$$\lambda_x = \frac{1}{2\zeta_x}\left(\frac{C_x}{\alpha_x} - (C_2 - C_1)\right) \tag{4.25}$$

Finally, the parameter ε_0 is obtained from:

$$\varepsilon_0 = \Delta G^0 - \lambda_1 \frac{\alpha_1^2}{\alpha_2^2} = \left(E_2^0 - E_1^0\right) - \frac{(C_2 - C_1)^2}{2(\zeta_2 - \zeta_1)} \tag{4.26}$$

As a whole, the Q-model needs three parameters, because of the two relations connecting α_x and λ_x:

$$\alpha_2 = 1 + \alpha_1 \tag{4.27}$$

$$\alpha_1^3 \lambda_1 = \alpha_2^3 \lambda_2 \tag{4.28}$$

For most of the values of the energy gap ε, one has $2\beta\sqrt{|\alpha_x|^3 \lambda_x |\varepsilon - \varepsilon_0|} \gg 1$, and eqn (4.23) reduces to:

$$G_x(\varepsilon) = G_x^0 + \left(\sqrt{|\alpha_x||\varepsilon - \varepsilon_0|} - |\alpha_x|\sqrt{\lambda_x}\right)^2 \tag{4.29}$$

However, when $|\varepsilon - \varepsilon_0| \gg \lambda_x |\alpha_x|$, eqn (4.23) can be approximated by:

$$G_x(\varepsilon) = G_x^0 + |\alpha_x| \left|\varepsilon - \Delta G^0 + \lambda_x \frac{\alpha_1^2}{\alpha_2}\right| \tag{4.30}$$

and the free energy thus exhibits a linear dependence with respect to the energy gap. Figure 4.5 shows free energy curves obtained with the Q-model for various values of the parameters. One clearly sees that highly asymmetric curves can be obtained, which can be applied to the observation of Le Bard and Matyushov for ET $[P^*, B_M] \rightarrow [P^+, B_M^-]$ in the bacterial PRC.[36,44] We finish this description of the Q-model by noting that if $|\alpha_x| \gg 1$ (which corresponds to $\zeta_1 \approx \zeta_2$), one gets $\lambda_1 \approx \lambda_2 = \lambda$ from eqn (4.25) and the classical Marcus parabolic free energy expressions are recovered. This can be observed on the third graph of Figure 4.5 (the one with the higher value of α_1), where the free energy curves are nearly parabolic.

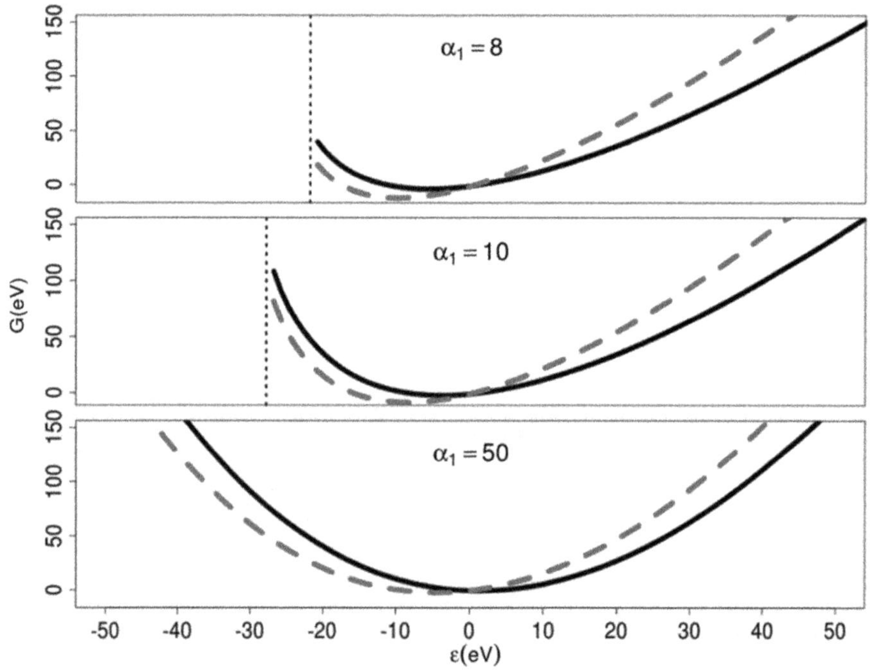

Figure 4.5 Some realisations of the Q-model for various values of the α_1 parameter. The other parameter values are the same in the three graphs: $\Delta G^0 = -0.5$ eV and $\lambda_1 = 3$ eV. Black solid line: $G_1(\varepsilon)$; red dashed line: $G_2(\varepsilon)$. The dotted line indicates $(-\varepsilon_0)$. It is not seen on the third graph because $(-\varepsilon_0) \approx 150$ eV in that case.

4.4.3 Modification of the 'Solvation State' upon ET

A second source of deviation from the LRA is a large reorganisation of the CT complex associated with the ET. This may happen when covalent bonds are broken/formed during the ET process. Such situations are of paramount importance in biological systems. We will restrict our discussion to cases where the 'solvation state' of the acceptor/donor moieties is very different before and after the ET. We use here the generic term 'solvation state' to describe an ensemble of configurations of the surroundings of the A/D moieties (solvent and/or amino acids) that are strongly favoured in one electronic state (and strongly disfavoured in the other). We will use as examples the oxido-reduction of Cu^{2+}/Cu^{+} and Ag^{+}/Ag^{0} systems, that have recently been studied using first-principles molecular dynamics (FPMD) simulations.[42,43] In both systems, the free energy curves for the two electronic states, depicted on Figure 4.6, are clearly non-parabolic.

Vuilleumier and coworkers showed that the Q-model described in the preceding section is not able to rationalise the results for these systems. They attributed this to the very different solvation state of the oxidised and reduced forms (see Figure 2 of ref. 43 for the Ag^{+}/Ag^{0} system), and derived a

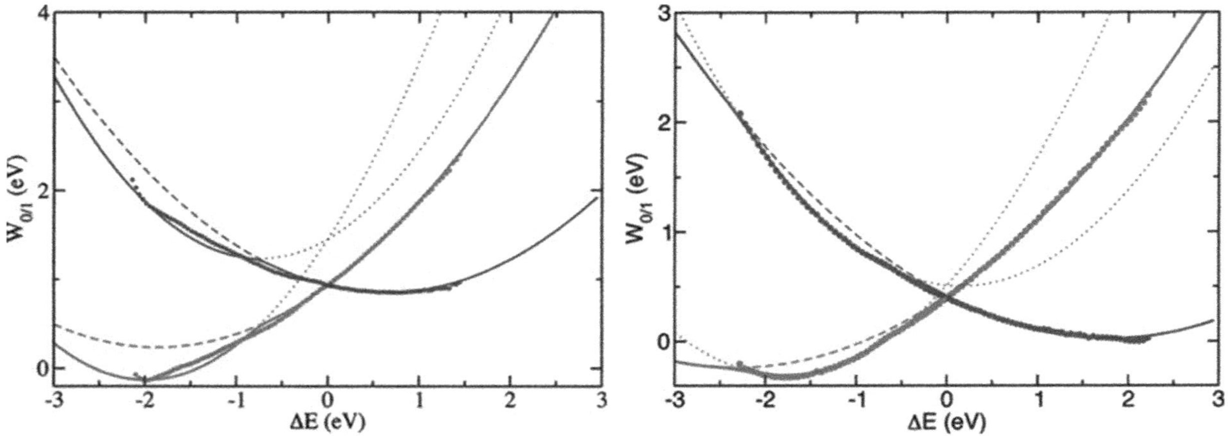

Figure 4.6 Free energy curves for oxidised (red, with minima for negative ΔE) and reduced (blue, with minima for positive ΔE) states of Ag (left graph) and Cu (right graph). Dots are outputs of the FPMD simulations. The continuous lines correspond to the TGS model. The dotted and dashed lines correspond to the two solvation states S_1 and S_2. Both figures are reproduced from ref. 43. Reprinted with permission from R. Vuilleumier, K. A. Tay, G. Jeanmairet, D. Borgis and A. Boutin, *J. Am. Chem. Soc.*, 2012, **134**, 2067–2074.[43] Copyright (2012) American Chemical Society.

two-solvation-state model (hereafter referred to as the two Gaussian solvation (TGS) model) to account for it.

In the TGS model, the configuration space of the solvent can be decomposed into two distinct configuration spaces S_1 and S_2 that correspond to the accessible configurations in the electronic states 1 and 2, respectively. The union of S_1 and S_2 encompasses the full configuration space. In each of the solvation states, the LRA applies, *i.e.* the fluctuations of the energy gap are supposed to be Gaussian, so that we can write:

$$G_{S,1}(\varepsilon) = G^0_{S,1} + \frac{(\varepsilon - \lambda_S - \Delta G^0_S)^2}{4\lambda_S} + \frac{k_B T}{2}\ln(4k_B T\lambda_S) \tag{4.31a}$$

$$G_{S,2}(\varepsilon) = G^0_{S,2} + \frac{(\varepsilon + \lambda_S - \Delta G^0_S)^2}{4\lambda_S} + \frac{k_B T}{2}\ln(4k_B T\lambda_S) \tag{4.31b}$$

These expressions correspond to the results one would obtain if the full configuration space were limited to the solvation state S. The parameters $G^0_{S,1}$, $G^0_{S,2}$, λ_S and $\Delta G^0_S = G^0_{S,2} - G^0_{S,1}$ have the same definitions as in the standard MT (except that the integrations are not performed over the full configuration space but only on the configuration space corresponding to the solvation state S). The probability that the energy gap is ε when the system is in the electronic state x is given by the sum of its probabilities to be in the solvation state S_1 or in the solvation state S_2:

$$p_x(\varepsilon) = p_{1,x}(\varepsilon) + p_{2,x}(\varepsilon) = \frac{\exp(-\beta G_{1,x}(\varepsilon))}{Z_x} + \frac{\exp(-\beta G_{2,x}(\varepsilon))}{Z_x} \tag{4.32}$$

As a consequence, the free energy for electronic state x is:

$$G_x(\varepsilon) = -k_B T \ln p_x(\varepsilon) + G^0_x = -k_B T \ln\left(\exp(-\beta G_{1,x}(\varepsilon)) + \exp(-\beta G_{2,x}(\varepsilon))\right) \tag{4.33}$$

with $G^0_x = -k_B T \ln Z_x$ the full free energy of the electronic state x. As a whole, the definition of the two diabatic free energy curves $G_1(\varepsilon)$ and $G_2(\varepsilon)$ finally requires five parameters: λ_1, ΔG^0_1, λ_2, ΔG^0_2, and $\Delta_S G^0_1 = G^0_{1,1} - G^0_{0,1}$. $\Delta_S G^0_1$ measures the free energy difference between the two solvation states, the electronic state being fixed to 1. This last parameter allows one to constrain the total free energy difference to $\Delta G^0 = G^0_2 - G^0_1$.

Figure 4.6 shows that the TGS model gives free energy curves that are consistent with the results extracted directly from the FPMD simulations for the Ag^+/Ag^0 and Cu^{2+}/Cu^+ systems. This agreement is due to the fact that the microscopic environment of the copper and silver particles in aqueous solution varies greatly with their oxidation state. This may not be true when transition metals are located within biological matrices that are clearly less flexible than water. Yet numerous enzymes catalyse oxido-reduction reactions thanks to active sites containing transition metal ions. Their coordination spheres must offer a certain level of lability to bind external ligands to allow cycling between various metal ion redox states. So far there have been relatively few studies of ET from or towards such flexible redox cofactors. The two-solvation-state model could be appropriate to study ET processes in such systems. This formalism may also prove to be useful when ET is coupled to proton transfer or to bond breaking. Note, however, that extensions of the MT have been proposed to deal with such problems (see, for example, ref. 46–48).

4.4.4 Non-ergodic Effects

Failure of the LRA can also be observed in cases where the ET happens faster than some motions of the system. Such situations, not uncommon in biological systems, arise either when the ET is ultra-fast, leading to the dynamical arrest of low-frequency vibrational modes, or when the ET is coupled with a slow relaxation of the system, for example large conformational changes. The consequence is that the phase space accessible to the system in the course of the ET reaction is reduced, leading to a breakdown of ergodicity, and notably to the breakdown of eqn (4.15). Matyushov and coworkers extensively studied non-ergodic effects in ET reactions arising especially in biological systems, and developed corrections for the MT to take these into account. In the rest of this section, we will review their contributions to the issue of broken ergodicity in ET reactions.

We first describe discontinuous ergodicity breaking. In this case, introduced for the study of the glass transition, the entire configuration space is split into two complementary regions, called components.[49] Within the timescale of ET, the system is constrained to stay in only one of the components because of a high activation energy barrier separating the two regions of the configuration space. Ergodicity breaking arises when the global equilibrium states of the initial and final electronic states are located in different components so that at the end of the ET process, the system lies in a metastable state, that will further evolve prior to ending up in the equilibrium state of the final electronic state. A schematic picture of this is given in Figure 4.7.

The reaction is described with the use of two reaction coordinates P and q, to which the energy gap is linearly coupled: $\Delta E = C_P P + C_q q$. P is a fast coordinate that drives the ET, whereas q is a slow reaction coordinate, *e.g.* a conformational change of the protein. The initial equilibrium state Σ_1 is centred around (P_1, q_1) and the final equilibrium state Σ_2 is centred around (P_2, q_2). However, there exists a large activation energy barrier along the q coordinate, separating the configuration space into two components. During the ET, the system is constrained to stay in the left component and the actual final state is Σ_2', centred around (P_2, q_1). It is important to note that if the time τ_c required for the transition along the q coordinate is greater than the

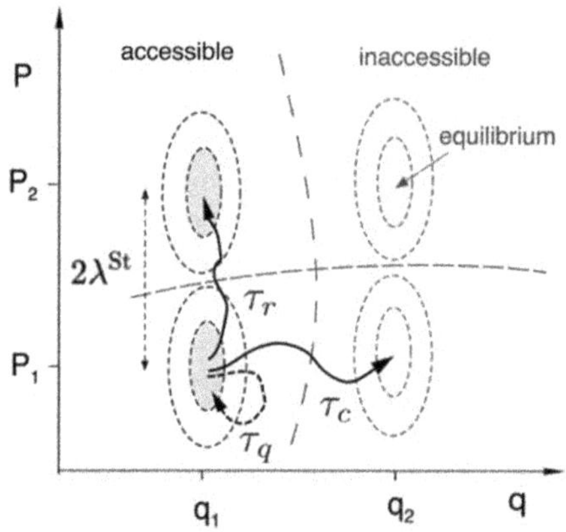

Figure 4.7 Schematic representation of discontinuous ergodicity breaking. The phase space is represented as a function of a fast (P) and a slow (q) reaction coordinate. The two equilibrium states are around (P_1, q_1) and (P_2, q_2). The barrier separating the minima along the q variable is much higher than the one along the P coordinate, so that transition along the q coordinate is not achieved during the ET reaction time ($\tau_r \ll \tau_c$). As a consequence, the equilibrium state around (P_2, q_2) is never reached, and the system ends up in the metastable state (P_2, q_1). However, relaxation along the q coordinate is still faster than the ET rate: $\tau_q \ll \tau_r$. Reproduced from ref. 52.

ET reaction time τ_r, 'local' relaxation of the slow coordinate (characteristic time τ_q) is still faster than τ_r, so that we have $\tau_q \ll \tau_r \ll \tau_c$.

If one supposes full relaxation of the vibrational modes into each component, the theoretical treatment of the ET kinetics is then similar to the MT developed in Section 4.3, except that the integrations are not performed over the full configuration space but only over the allowed component. As a consequence, the energy gap distribution in the initial state Σ_1 is a Gaussian distribution centred around (P_1, q_1) with a width given by: $\sigma_1^2 = C_P\sigma_P^2 + C_q\sigma_q^2$, giving a reorganisation energy computed from the fluctuations of the energy gap (eqn (4.18)):

$$\lambda_1 = \frac{\sigma_1^2}{2k_BT} = \frac{C_P\sigma_P^2}{2k_BT} + \frac{C_q\sigma_q^2}{2k_BT} = \lambda_P + \lambda_q \tag{4.34}$$

λ_P and λ_q are reorganisation energy contributions due to the coupling of the energy gap respectively with the P and q coordinates. Because the coordinate q is constrained to stay close to q_1, the Stokes reorganisation energy (eqn (4.20)) is only:

$$\lambda^{St} = \frac{\langle \Delta E \rangle_{\Sigma_1} - \langle \Delta E \rangle_{\Sigma_2'}}{2} = \frac{C_P}{2}(P_1 - P_2) = \lambda_P \tag{4.35}$$

We thus have $\lambda^{St} < \lambda_1$, which contradicts MT. Under the approximation of Gaussian fluctuations, the free energy of diabatic state 1 is given by:

$$G_1(\varepsilon) = G_1(q_1) + \frac{(\varepsilon - \Delta E_1)^2}{4\lambda_1} + \frac{k_B T}{2}\ln(4\pi k_B T \lambda_1) \tag{4.36}$$

where $\Delta E_1 = \langle \Delta E \rangle_{\Sigma_1} = C_P P_1 + C_q q_1$ is the average value of the diabatic energy gap when the system is constrained to stay in the Σ_1 region and $G_1(q_1)$ is the free energy of electronic state 1 computed over the accessible component where q stays close to q_1. Under the hypothesis of similar fluctuations of the energy gap in the final state Σ'_2 ($\lambda_1 = \lambda_2 = \lambda$), $G_2(\varepsilon)$ is given by:

$$G_2(\varepsilon) = G_2(q_1) + \frac{(\varepsilon - \Delta E_2)^2}{4\lambda} + \frac{k_B T}{2}\ln(4\pi k_B T \lambda) \tag{4.37}$$

with $\Delta E_2 = \langle \Delta E \rangle_{\Sigma'_2} = C_P P_2 + C_q q_1$.

The system being non-ergodic, eqn (4.15) does not hold anymore, and the only relation that connects the two free energy surfaces is their crossing for $\varepsilon = 0$. Using this and the definition of λ^{St} (eqn (4.35)), one arrives at:

$$\Delta E_1 = \frac{\lambda}{\lambda^{St}}\Delta G' + \lambda^{St} = \chi_G \Delta G' + \lambda^{St} \tag{4.38}$$

$\Delta G' = G_2(q_1) - G_1(q_1)$ is the free energy difference between states Σ_1 and Σ'_2. The activation free energy is obtained by setting $\varepsilon = 0$ in eqn (4.36):

$$\Delta G^{\ddagger} = G_1(0) - G_1(q_1) = \frac{\Delta E_1^2}{4\lambda} = \frac{(\chi_G \Delta G' + \lambda^{St})^2}{4\lambda} \tag{4.39}$$

In the case $\chi_G = 1$, one recovers the standard Marcus relationship.

Two main aspects of discontinuous ergodicity breaking in the context of spontaneous ($\Delta G^0 < 0$) biological ET merit discussion: thermodynamics and kinetics. Concerning thermodynamics, ergodicity breaking leads to a decrease of the heat released to the medium. Indeed, this is measured by the variation in free energy during the reaction, which is lower in magnitude in non-ergodic ET ($\Delta G^0 < \Delta G' < 0$), because the system ends up only in a metastable state. Concerning kinetics, the effect is less evident. The ratio between the activation free energy in the non-ergodic and ergodic case (assuming fluctuations are similar in both cases) is:

$$R = \frac{\Delta G^{\ddagger}_{ne}}{\Delta G^{\ddagger}} = \frac{(\chi_G \Delta G' + \lambda^{St})^2}{(\Delta G^0 + \lambda)^2} \tag{4.40}$$

For a given value of ΔG^0 and λ, which defines the ergodic case, the value of this ratio depends on the particular values of χ_G and $\Delta G'$ (a value $R < 1$ corresponds to an acceleration of the reaction). Large values of χ_G

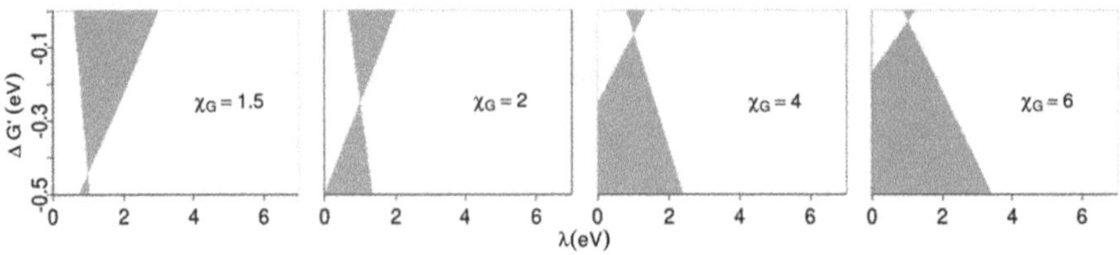

Figure 4.8 Ratio between non-ergodic and ergodic activation free energies for various combinations of λ, $\Delta G'$ and χ_G values. ΔG^0 has been set to −0.5 eV. Grey (resp. white) regions correspond to ratio values greater (resp. lower) than 1.

correspond to $\lambda^{St} \ll \lambda$, which tends to lower the activation free energy, but may also lead to $\chi_G \Delta G' \ll \Delta G^0 < 0$, which acts in the opposite direction. Figure 4.8 shows numerical calculations of R for various values of χ_G observed in some ET systems and realistic ΔG^0 and λ values. One can see that ergodicity breaking leads to acceleration of the ET kinetics in a large number of cases, especially for values of $\chi_G \approx 2$ to 4. MD studies have been used to determine χ_G values in different proteins, showing that it can largely exceed unity.[36,50,51] For example, χ_G values around 2–3 are found for all ET steps in the bacterial PRC, with λ values around 2 eV, while λ^{St} is in the range 0.7–1.1 eV.[52] Experimental evidence of discontinuous ergodicity breaking requires access to both λ and λ^{St}, which can be obtained from spectroscopic analysis. The former can be extracted from the bandwidth of optical transitions and the latter from the Stokes displacement between the absorption and emission spectra. There are only a few available examples to date. We mention here the example of the mStrawberry fluorescent protein for which a χ_G value of around 1.8 has been determined.[53,54]

The framework of discontinuous ergodicity breaking requires that the relaxation of the vibrational modes is still effective within the timescale of the ET reaction. This condition may not be fulfilled for ultra-fast ET reactions that occur for example in the PRC or in the DNA-photolyase/cryptochrome family, where the ET occurs on the sub-nanosecond timescale. In such situations, the phase space accessible to the system is greatly reduced and the slow frequency vibrational modes are frozen during the reaction.

An adequate formalism to study ultra-fast ETs is that of the dynamically restrained ensemble, introduced by Matyushov.[55] In a classical equilibrium ensemble, the phase space is composed of the collection of dynamic variables $Q(t)$ that the system is allowed to visit during a (supposedly) infinite time. It is mathematically equivalent to making a time Fourier transform of the variables $Q(t) \rightarrow Q(\omega) = Q_\omega$ and performing the integration over the ensemble of the variables $Q(\omega)$. In a dynamically restrained ensemble, the coordinates Q_ω for which the frequency is lower than the ET rate k are frozen during the reaction to a value Q_ω^f, and thus integration is only performed over the variables corresponding to $\omega > k$. The effective free energy of the initial diabatic state is thus given, up to an additive constant, by:

$$G_1(\varepsilon) = -k_B T \ln\left(\int \exp(-\beta E_1(Q_\omega))\delta(\Delta E - \varepsilon) \prod_{\omega < k} \delta(Q_\omega - Q_\omega^f) \prod_\omega dQ_\omega\right) \tag{4.41}$$

Note that the values Q_ω^f depend on the initial state of the system. For example, if the initial electronic state is produced by photoactivation, the Q_ω^f values follow their Boltzmann distribution in the initially equilibrated ground state. For each set of Q_ω^f values, one gets a single effective free energy profile (and associated properties) and a subsequent average over the Q_ω^f values should be made.

Assuming a Gaussian distribution for the energy gap during the ET (MD simulation results on ultra-fast ETs in the PRC and in (6,4)-photolyase support this assumption),[36,56,57] the non-ergodic free energy surface for diabatic state 1 is given by:

$$G_1(\varepsilon) = \frac{(\varepsilon - \Delta E_1(k))^2}{4\lambda(k)} \tag{4.42}$$

Values of the reorganisation energy $\lambda(k)$ and of the average energy gap $\Delta E_1(k)$ are now a function of the rate constant k of the ET reaction. The reorganisation energy can be obtained from the spectral density (defined in Section 4.5) of the energy gap but restraining the integration to frequencies that are higher than k:

$$\lambda(k) = \frac{2}{\pi}\int_k^\infty \frac{J(\omega)}{\omega} d\omega \tag{4.43}$$

This results in an effective reduction of the reorganisation energy with respect to the ergodic reorganisation energy $\lambda(0)$. Matyushov and coworkers defined a non-ergodicity factor $f_{ne}(k) = \lambda(k)/\lambda(0)$, which measures the fraction of the fluctuation spectrum accessible on the ET reaction timescale.[58]

Among the components of the energy gap, only the Coulombic part ΔE_1^C of the solute/solvent interaction is affected by the freezing of low-frequency modes. Le Bard and Matyushov proposed a simple expression that connects the non-ergodic energy gap to the ergodic average energy gap ΔE_1:[56,58]

$$\Delta E_1(k) = \Delta E_1 - (1 - f_{ne}(k))\Delta E_1^C \tag{4.44}$$

The two quantities $\lambda(k)$ and $\Delta E_1(k)$ are then used to compute the ET rate constant, with whatever 'flavour' of the MT, depending on the particular problem under study. In the simplest case, one can use for example:

$$k_{ET} = \frac{2\pi}{\hbar}\frac{1}{\sqrt{4\pi\lambda(k)k_B T}}\exp\left(-\frac{(\Delta E_1(k))^2}{4\lambda(k)}\right) \tag{4.45}$$

where in the exponential, the effective activation free energy is obtained using the resonance condition between the two diabatic states ($\varepsilon = 0$) in eqn (4.42). Eqn (4.45) can be solved self-consistently (starting for example with an initial value of $\lambda(0)$) in order to compute the actual ET rate.

The concept of continuous ergodicity allows the reconciliation of experimental and theoretical data (obtained from MD simulations), as for example in the case of the ps initial charge separation reaction in bacterial photosynthesis. Le Bard and Matyushov obtained reorganisation energies of 2.36 eV when computed from long (on the ns timescale) equilibrium MD simulations,[44] whereas the experimental value is 0.24 eV.[59] On the other hand, when the reorganisation energy is computed from eqn (4.43) using the right ps-reaction rate one gets a value $\lambda(k) = 0.36$ eV.

4.5 Quantum Theories of Electron Transfer

4.5.1 The Fermi Golden Rule

The theory of ET developed by Marcus in the 1950s was essentially based on a classical treatment of the nuclear degrees of freedom. The theory allows the description of adiabatic reactions as well as non-adiabatic reactions thanks to the transmission coefficient κ. Although the theory works well for reactions taking place at room temperature, many interesting experimental trends cannot be explained by the classical theory. Among them are (i) the temperature dependence of certain ET rates that increase with decreasing temperature,[60,61] (ii) the observation of vibrational coherences by femtosecond pump-probe, for example in the primary charge separation step of the PRC[62] or (iii) the non-symmetric shape of the $k_{\mathrm{ET}} = f(\Delta G^0)$ curves.[15] A quantum mechanical approach to ET is needed to account for these effects. Quantum theories of non-adiabatic ET were first established by Levich and coworkers in the 1960s and further developed by various groups. However, earlier crucial theoretical developments appeared in the field of radiationless transitions in solids.[63,64]

We consider a two-level system with electronic states $|\phi_1\rangle$ and $|\phi_2\rangle$ and assume weak coupling between them. They may be seen as first-order solutions (diabatic states) of the time-independent Schrödinger equation with an electronic Hamiltonian deprived of non-adiabatic coupling operators. The energy gap between the energy minima of the two states is $\Delta E = E_2 - E_1$. The initial and final electronic states are associated with a manifold of vibrational states denoted $|v\rangle$ and $|w\rangle$, respectively that are similarly obtained by solving the nuclear Schrödinger equation. The set of quantum numbers $\{n_1^1,\ldots n_k^1,\ldots,n_N^1\}$ of the N vibrators leads to the vibrational energy ε_v (and similarly for ε_w). In case the modes are harmonic $\varepsilon_{v/w} = \sum_k (n_k^{1/2} + 1/2)\hbar\omega_k^{1/2}$. The decay probability of the vibronic state $|1v\rangle$ is denoted by W_{1v} and is given by Fermi's Golden Rule (FGR). Summing the contributions from all accepting modes, $|2w\rangle$, W_{1v} reads:

$$W_{1v} = \frac{2\pi}{\hbar}\sum_{w} \left|H_{1v,2w}\right|^2 \delta(E_1 + \varepsilon_v - E_2 - \varepsilon_w) \tag{4.46}$$

$$H_{1v,2w} = \langle v|\langle 1|H|2\rangle|w\rangle \tag{4.47}$$

The delta function ensures energy conservation. $|H_{1v,2w}|$ is the coupling between states $|1v\rangle$ and $|2w\rangle$. If one makes the hypothesis that it does not depend on the nuclear configuration (the Condon approximation) one may write:

$$|H_{1v,2w}| \approx |\langle 1|H|2\rangle||\langle v|w\rangle| \equiv |H_{DA}||\langle v|w\rangle| \tag{4.48}$$

and the electronic part (H_{DA}) can be factored out of the sum of eqn (4.46). The overall rate is obtained by summing over all the initial states $|1v\rangle$ weighted by their Boltzmann probabilities (P):

$$k_{ET} = \frac{2\pi}{\hbar}|H_{DA}|^2 \sum_{v}\sum_{w} P(\varepsilon_v)|\langle v\,|\,w\rangle|^2 \delta(E_1 + \varepsilon_v - E_2 - \varepsilon_w) \tag{4.49}$$

The terms $|\langle v|w\rangle|^2$ are known as Franck–Condon factors. Besides the two-level approximation we made the hypothesis that no interferences exist among reactive channels, that is, we assume the energy spacing between the initial vibronic states is much larger than the width of the accepting modes. To ensure an irreversible decay of the initial state there should be a dense manifold of accepting states, which is a reasonable assumption for ET in the condensed phase. Other hypotheses underlie eqn (4.49) and we refer the reader to ref. 65 for a comprehensive description. Contrary to the classical picture, according to which electron tunnelling only occurs at the intersection of potential energy surfaces, the appearance of nuclear overlap terms opens the possibility for nuclear tunnelling. Let's examine the $|\langle v|w\rangle|$ terms more closely. If no mode is displaced or mixed between the two electronic states, *i.e.* there is no electron-nuclear coupling, $|\langle v|w\rangle|$ will equal 1 or 0 if the sets of quantum numbers $\{n_k^1\}$ and $\{n_k^2\}$ are identical or differ by at least one vibrational quanta, respectively. The nuclear modes would only be spectators and, unless the electronic energies are strictly equal ($\Delta E = 0$), no ET will take place because of energy conservation. This highlights the central role of the electron–nuclear coupling as previously identified by Marcus at the macroscopic level. Now, if some modes are displaced or mixed, $|\langle v|w\rangle|$ will be lower than unity whatever the vibrational levels in the two states. For the sake of illustration, let's consider a unidimensional system characterised by a unique mode of mass m and frequency ω with a displacement of q. Introducing the dimensionless displacement, $\Delta_c = (m\omega/\hbar)^{1/2}q$, the FC between the ground vibrational state and the nth final excited state reads:[66]

$$|\langle 0|n\rangle| = (S^n/n!)\exp(-S) \tag{4.50}$$

$$S = \Delta_c^2/2 \tag{4.51}$$

Typically S is comprised between 0 and 1 so that the larger the displacements or the mixing the larger the FC terms, hence the contribution to the ET rate. Furthermore, because of the delta function in eqn (4.49), one anticipates that a large number of modes will favour ET by creating a continuum of relaxation paths.

Although the FGR formulation is valuable to understand quantum effects in ET reactions, this formalism is not particularly convenient for actual rate calculations as soon as the system involves thousands of active modes, a situation commonly encountered for biological ET. Indeed explicit calculation of all FC terms entering eqn (4.49) becomes intractable with current computational methodologies (even though noticeable progress has been made in recent years[67,68]). A powerful alternative strategy starts by replacing the delta of eqn (4.46) by its Fourier transform, which leads, after a few algebraic manipulations, to the following closed expression:

$$k_{\mathrm{ET}} = \frac{|H_{\mathrm{DA}}|^2}{\hbar}\int_{-\infty}^{+\infty} \mathrm{e}^{-i\Delta E\tau} f(\tau)\mathrm{d}\tau \tag{4.52}$$

The function f is the Fourier transform of the non-radiative decay probability.[63] It takes manageable analytic forms if one assumes harmonic oscillators with similar frequencies (spin-boson model). In this case Levich and his school[69–71] derived quantum rate expressions for ET in polar solvents, assuming internally rigid redox cofactors. Kestner *et al.*[65] further developed the theory to incorporate the effect of distortions of the first coordination sphere of the cofactors. For space reasons we will not describe these models here but refer the reader to original publications or to excellent review papers.[72,73] We instead focus on two formulations that turn out to be extremely useful from conceptual and practical points of view. One is the mixed quantum classical formulation of ET where the quantum treatment is preserved only for a subset of vibrational modes; the other is the reformulation of ET theory based on the dispersed polaron or spin-boson model.

4.5.2 Mixed Quantum Classical Formulations

Marcus had introduced the effect of inner-sphere deformation using a classical harmonic model. However bond elongation/compression are likely to vibrate at frequencies well above the thermal energy (around 200 cm^{-1} at 300 K), and thus require a quantum treatment. Attempts to incorporate inner-sphere contributions at the quantum level date back to the 1960s with the work of Sutin.[74] Let us take the specific example of the $Fe(H_2O)_6{}^{2+} \rightarrow Fe(H_2O)_6{}^{3+}$ exchange reaction. Sutin considers the reorganisation of the inner coordination shell of the reactants using a harmonic model. Under the hypothesis that inner-sphere motions are rate limiting he finds that the reaction barrier is roughly triangular, an approximation that allows the nuclear tunnelling factor of about 40 to be calculated. Clearly, in these inner-sphere reactions, tunnelling cannot be ignored. Additional support for the importance of tunnelling came from a Franck–Condon type of calculation. He considered a pair of vibrational states, assuming the electron exchange rate is proportional to an electronic factor multiplied by a vibrational factor. The vibrational factor is the square of the overlap integral between the initial and final states ($|\langle v|w\rangle|$). Combining this with appropriate Boltzmann factors, Sutin estimates a nuclear tunnelling factor of about 300 when the reorganisation energy is around 6 kcal mol^{-1}.

In the 1970s, Efrima and Bixon[66,75] and Ulstrup and Jortner[76] proposed a more general formalism to include inner-sphere contributions at the quantum mechanical level, while retaining a classical description for the outer-sphere contribution. Outer-sphere modes have very low frequencies (a few tens of reciprocal centimetres) for which quantum effects are negligible even at low temperature. They proposed the following expression for the ET rate:

$$k_{\mathrm{ET}} = \frac{2\pi}{\hbar}\frac{1}{\sqrt{4\pi\lambda_{\mathrm{o}}k_{\mathrm{B}}T}}H_{\mathrm{DA}}^2\sum_{v,w}|\langle v\,|\,w\rangle|^2\exp\left(-\frac{(\Delta E+\Delta E_{\mathrm{vib}}+\lambda_{\mathrm{o}})^2}{4\lambda_{\mathrm{o}}k_{\mathrm{B}}T}\right) \tag{4.53}$$

v, w thus only refer to the inner-sphere nuclear modes. The overall ET process can be seen as a sum of ET channels, each being associated with a Marcus-like rate with a dressed coupling $(H_{\mathrm{DA}}|\langle v|w\rangle|)^2$ and a reaction energy shifted by the vibrational energy $\Delta E_{\mathrm{vib}} = \varepsilon_w - \varepsilon_v$. Note that the reorganisation energy λ_{o} only refers to the classical outer-sphere contribution. A great merit of this formalism is to account for the non-symmetric shape of the ET rate dependence on the reaction energy. To better see this let us assume that all the N inner-sphere modes have the same frequency ω and reduced displacements Δ_{c}. It can be shown that under the supplementary assumption that only the ground vibrational state is populated, eqn (4.53) reduces to:

$$k_{\mathrm{ET}} = \frac{2\pi}{\hbar}\frac{1}{\sqrt{4\pi\lambda_{\mathrm{o}}k_{\mathrm{B}}T}}H_{\mathrm{DA}}^2\exp(-2NS)\sum_{k=0}^{\infty}\frac{N^k}{k!}S^k\exp\left(-\frac{(\Delta E+\hbar\omega k+\lambda_{\mathrm{o}})^2}{4\lambda_{\mathrm{o}}k_{\mathrm{B}}T}\right) \tag{4.54}$$

which illustrates the impact of quantised inner-sphere modes especially for very exothermic reactions. According to vanilla MT, ET rates are predicted to reach a maximum when $\Delta G^0 = \lambda$ and then to drop symmetrically around this point (Figure 4.9). However nuclear quantum effects reduce the effective barrier by allowing nuclear tunnelling (transfer allowed below the classical

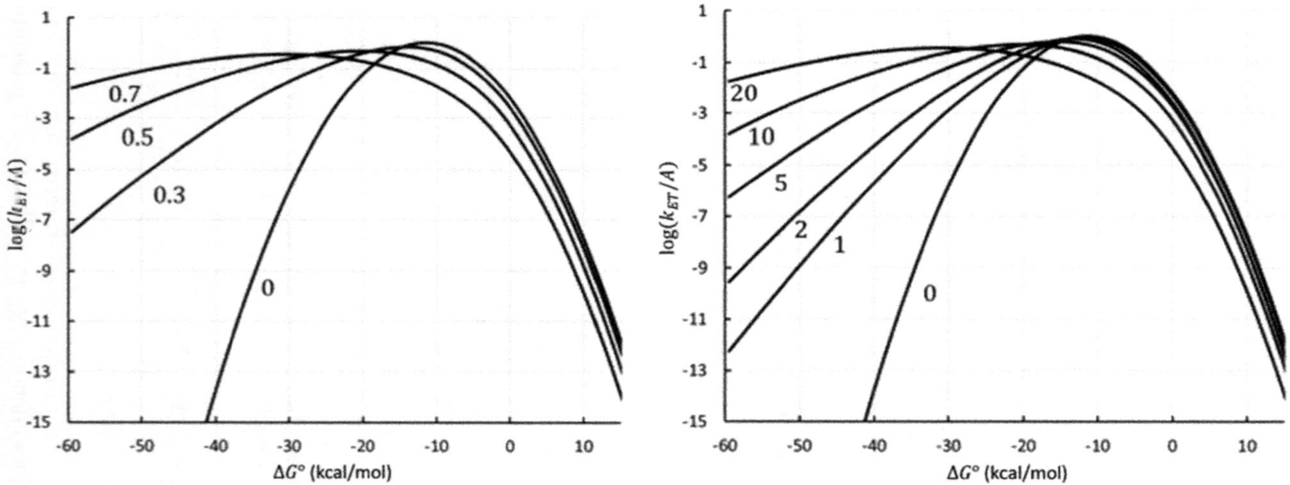

Figure 4.9 Mixed quantum-classical nuclear factor computed by eqn (4.54) as a function of the reaction energy with $A = 2\pi H_{DA}^2 / (\hbar\sqrt{4\pi\lambda_o k_B T})$. Left: effect of the amplitude of mode displacement Δ (0, 0.3, 0.5, 0.7) with $N = 10$. Right: influence of the number of displaced modes N with $\Delta = 0.5$. For both graphs $\lambda = 4000$ cm^{-1}, $\omega = 1500$ cm^{-1}, $T = 284$ K.

turning point) and by reducing the effective barrier by enabling transfer to vibrationally excited states. These effects are more pronounced in the inverted region where the parabolas are interlocked than in the normal region and may produce strong asymmetry in the plots of Figure 4.9. Nuclear quantum effects are all the more pronounced when the active modes are strongly displaced and when they are numerous. Note also some noticeable effects in the normal region. The abnormal inverted effect has been spectacularly confirmed experimentally in the pulse-radiolysis measurements conducted by Miller and coworkers for both inter- and intramolecular ET.[15]

4.5.3 Spectral Density as a Key Ingredient of ET Rates

Kubo and Toyozawa[63] showed that the quantum rate equation for a multidimensional harmonic bath reads:

$$k_{\mathrm{ET}} = \frac{H_{\mathrm{DA}}^2}{\hbar} \int_{-\infty}^{+\infty} \exp[i\omega_{12}t + \gamma(t)]\mathrm{d}t \tag{4.55}$$

$$\gamma(t) = \sum_j \Delta_j^2[(\bar{n}_j + 1/2)(\cos\omega_j t - 1) + (i/2)\sin\omega_j t] \tag{4.56}$$

$$\bar{n}_j = \frac{1}{\exp(\hbar\omega_j \,/\, k_{\mathrm{B}}T) - 1} \tag{4.57}$$

where $\bar{n}_j$ is the mean number of vibrational quanta for the harmonic oscillator j of frequency ω_j at thermal equilibrium. ω_{12} is the difference between the minimum electronic energies of the two diabatic states. At first glance this elegant formalism cannot be applied to real systems such as proteins, because the vibrations may be largely anharmonic. However, Warshel made a crucial step by showing that if the LRA applies and if the reorganisation is distributed over a large number of nuclear modes, the response of the system to interactions that fluctuate around each frequency becomes identical to that given by a multidimensional harmonic system.[18] One can then map the fluctuations of the classical, anharmonic system onto the equivalent harmonic one. The key ingredients entering these equations (Δ_j and ω_j) can be obtained from classical MD simulations. Indeed the power spectrum of the autocorrelation function (ACF) of ΔE exhibits frequencies ω_j corresponding to modes that are displaced or mixed upon ET, the amplitudes of which are proportional to $\omega_j\Delta_j^2$.

$$J(\omega) = \left| \int_{-\infty}^{+\infty} \langle \delta\Delta E(0)\delta\Delta E(t)\rangle \exp i\omega t \,\mathrm{d}t \right| \tag{4.58}$$

J reduces to a cosine transform because the energy gap ACF obtained from classical MD simulation is an odd function of time. This so-called dispersed polaron model is remarkable in the sense that quantum corrections to the rate expression can be extracted from classical MD simulations. It is also remarkable that temperature dependence can be caught through eqn (4.57) using ω_j values obtained from MD simulations carried out at room temperature for which the classical approximation is acceptable. Δ_j and ω_j can be extracted from J and injected in eqn (4.55) to (4.57). One may also use J directly in rate expressions:

$$k_{\mathrm{ET}} = \frac{H_{\mathrm{DA}}^2}{\hbar}\int_{-\infty}^{+\infty} \mathrm{d}R \exp\left\{-[1/2k_{\mathrm{B}}T + iR/k_{\mathrm{B}}T]\Delta G^0 - \frac{2}{\pi\hbar}\int_0^{+\infty}\mathrm{d}\omega \frac{J(\omega)}{\omega^2}\frac{\cosh(\hbar\omega/2k_{\mathrm{B}}T) - \cosh(iR\hbar\omega/k_{\mathrm{B}}T)}{\sinh(\hbar\omega/2k_{\mathrm{B}}T)}\right\} \tag{4.59}$$

The rate can be evaluated by numerical quadrature. However the double integral makes this a difficult task and one often prefers to use approximate analytical expressions based for example on the steepest descent method.[77–79] The dispersed polaron model is analogous to the spin-boson model which was developed by Chandler and coworkers[37] or Schulten.[79] Both models were applied to ET in the PRC[23,79] or to CT in DNA,[80] accounting, for example, for the temperature dependence of ET rates.

Alternatively J may serve as input to quantum dynamics propagations. These methods have the potential to account for interference between nuclear trajectories. Active developments are ongoing on this front. The path integral,[81] the multi-layer MCTDH (multiconfigurational time-dependent Hartree)[82,83] or the hierarchical equations of motion (HEOM) methodologies[84] are all appealing methods for propagating quantum dynamics. Many of these formalisms have also been developed for modelling excitation energy transfers in proteins.[85–88] For example, Sim and Makri conducted path integral calculations for ET within the PRC.[89] Our groups recently combined *ab initio* MD simulations to HEOM approaches to model CT in cryptochromes.[90]

4.5.4 Quantum Entanglement Between Electronic and Vibrational Degrees of Freedom

In their 2011 paper McKemmish *et al.*[91] shed light on the quintessential quantum property, entanglement. They studied a simple model involving two coupled intersecting potential energy surfaces and calculated the quantum entanglement of the electronic and vibrational degrees of freedom. From quantum information theory they adopted the simplest kind of system, a bipartite pure state:

$$|\psi\rangle = \sum_{i,j} c_{ij}|a_i\rangle|b_i\rangle \tag{4.60}$$

where $|a_i\rangle$ and $|b_i\rangle$ form an orthonormal basis for subsystem A and B, here the electronic and vibrational degrees of freedom. The von Neumann

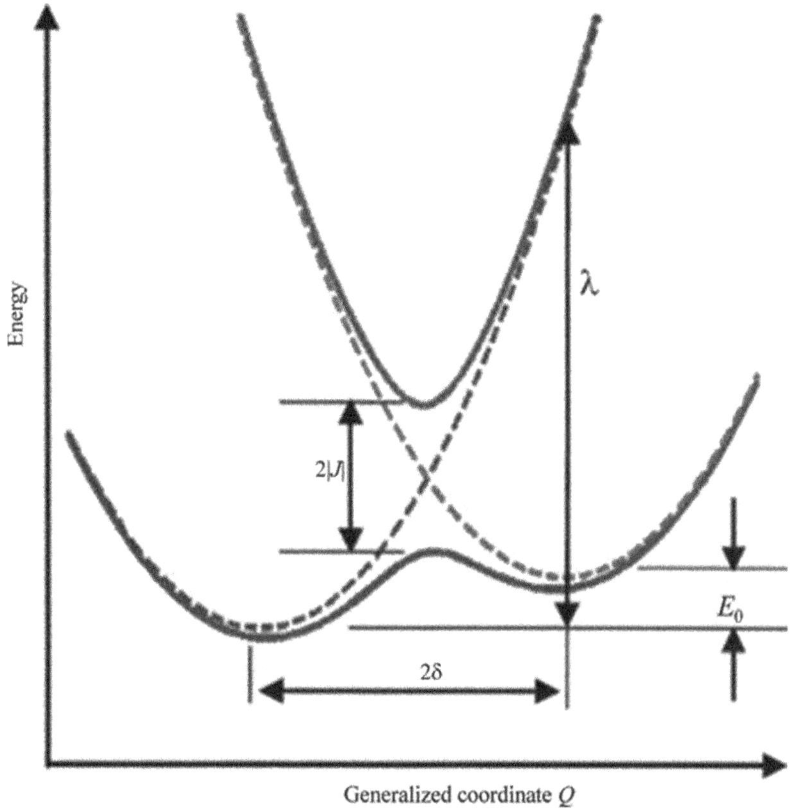

Figure 4.10 The chemical model used to describe electron-vibration entanglement. Reprinted from L. K. McKemmish, R. H. McKenzie, N. S. Hush and J. R. Reimers, *J. Chem. Phys.*, 2011, **135**, 244110–244121,[91] with the permission of AIP Publishing.

entropy of the reduced density matrix is taken as the measure of entanglement (see below). Figure 4.10 shows the parameters for their model.

They considered six molecular examples, ammonia, diboronitrogen (DBN), benzene, the Creutz–Taube (CT) charge transfer ion, the bacterial photosynthetic reaction centre radical cation (PRC) and pyridine excited triplet states (3PYR). Taken together, the six examples span a range of chemistry that can point out the need to go beyond vanilla MT, depending on the Hamiltonian parameters (see Table 1 of ref. 91 for the parameter values of all six examples).

Returning to the question of entanglement, the exact, entangled wavefunctions of the system are expanded in terms of unentangled basis states expressed diabatically as a product of an electronic wavefunction, $\phi(r)$, and a nuclear wavefunction, $\chi(R)$:

$$|\psi_{CA}(r, R)\rangle = |\phi(r)\rangle|\chi(R)\rangle \tag{4.61}$$

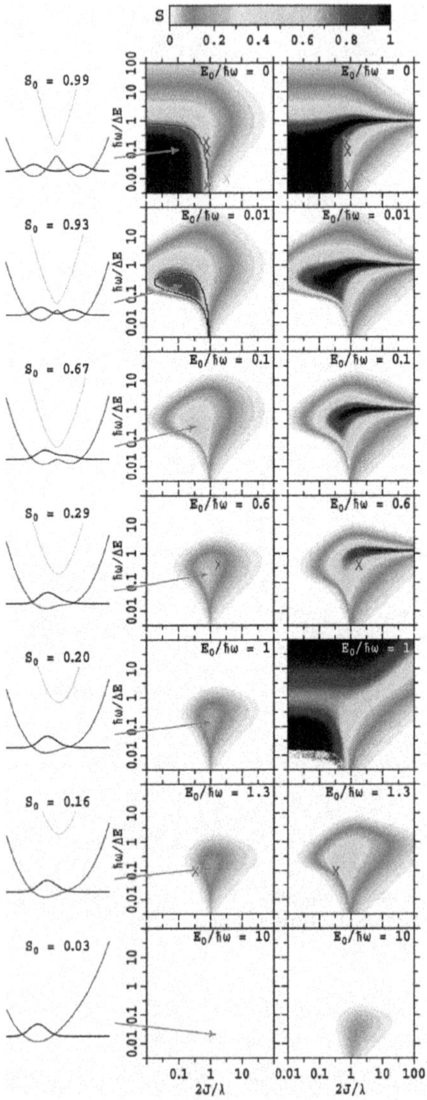

Figure 4.11 Persistent *versus* degeneracy-induced entanglement. Center and right: electron nuclear entanglement (von Neumann entropy) is shown as a contour plot for the vibronic ground state (S_0, centre) and the first vibronic excited state (S_1, right) wavefunctions *vs* $2|J|/\lambda$ and $\hbar\omega/\Delta E$ at various values of $E_0/\hbar\omega$. The black lines denote regions in which the ground-state vibrational probability density is bimodal or unimodal, while the crosses indicate parameter values relevant to red – ammonia; green – benzene, blue – CT, purple – DBN, 3PYR, PRC. Left: the Born-Oppenheimer potential-energy surfaces (purple and green) and the ground-state vibrational probability density (black) at the indicated parameters. Reprinted from L. K. McKemmish, R. H. McKenzie, N. S. Hush and J. R. Reimers, *J. Chem. Phys.*, 2011, **135**, 244110–244121,[91] with the permission of AIP Publishing.

McKemmish *et al.* call this representation the crudely adiabatic approximation, which differs from the BO approximation in that no nuclear dependence of the electronic wavefunction is allowed. The exact wavefunctions are then obtained as:

$$|\psi_{\text{exact}}(r,R)\rangle = \sum_{ij} c_{ij}|\phi_i(r)\rangle|\chi_j(r)\rangle. \tag{4.62}$$

With this expression, the density matrix can be calculated and hence the von Neumann entropy (see eqn (11)–(13) in ref. 91). Figure 4.11 provides an overview of the results for a broad range of parameters.

The figure shows various potential surface types, ranging from those with a single minimum to those having two separate minima. If we look, for example at the central column, for the vibronic ground state, significant entanglement is found only for the cases in the upper part of the figure in which the ground vibronic state has a bimodal density profile, corresponding to two separate local minima. The authors also point out the existence of two different types of entanglement, fragile degeneracy-induced entanglement and persistent entanglement. For even slight asymmetry ($E_0/\hbar\omega = 0.01$) the entanglement is greatly reduced, being restricted mostly to a region within $0.02 < 2|J|/\lambda < 0.5$ and $0.1 < \hbar\omega/\Delta E < 1$. The Creutz–Taube ion, for example, falls in a region of persistent entanglement.

Very recently[92,93] this work has been extended to include a treatment in terms of an adiabatic basis, which elucidates the roles of the various non-adiabatic corrections to the BO approximation. Among other findings, it is found that the entanglement developed by BO breakdown correlates simply with the diameter of the cusp introduced by the BO approximation, and a hierarchy appears between the various BO breakdown correction terms, with the first-derivative correction being more important than the second-derivative correction, which is more important than the diagonal correction. We refer the reader to ref. 92 and 93 for details.

4.6 Dynamical Effects on ET Kinetics

The question whether the protein medium simply acts as an average barrier for electron tunnelling or if the detailed chemical composition of protein and its dynamics matter in determining the propensity for tunnelling has long been a subject of controversy in the field of biological ET (see, for example, ref. 94 and 95). In this section we will examine the factors determining the electronic coupling which enters the Marcus rate expressions. As we shall see below, H_{DA} depends subtly on the chemical structure of the medium separating D from A. Because protein matrices are animated by thermal motions spanning timescales of several orders of magnitude (femto- to milliseconds or more), many of these motions may induce electronic coupling fluctuations, and in turn may affect the mechanism of CT.

This feature may cause some underlying hypotheses of the MT to break down, in particular, the BO and Condon approximations. In MT, H_{DA} is assumed to be independent of the nuclear coordinates; this hypothesis is, in fact, strongly questionable for CTs taking place in proteins.

The eventuality of fluctuating electronic coupling can be traced back to the work of Dogonadze and coworkers.[71] On the other hand Ratner and Madhukar[96] were some of the first authors to devise practical mathematical frameworks to assess the importance of fluctuations of the transfer integral ET kinetics. These early works opened the floodgates to a continuous flow of theoretical, computational and experimental developments aimed at quantifying the impact on ET kinetics of electronic coupling fluctuations, of non-Condon effects, and of inelastic tunnelling.[19,97–100] We now review some of these key models and concepts. First we will present theoretical models that connect the strength of H_{DA} to the chemical structure of the intervening medium. We then describe various limit regimes, the appearance of which depends on the amplitude of electronic coupling fluctuations but also on the speed of these variations.

4.6.1 The Chemical Structure of the Bridge Determines H_{DA}

4.6.1.1 *Tunnelling Through a Square Barrier*

ETs between cofactors separated by more than a few Ångstroms are fundamentally governed by quantum tunnelling. Let's start by looking at a model potential consisting of two Dirac delta functions in one dimension (x).[101] The potential is set to zero for all x except for $\pm a$ where it takes values of $-g$. We define by convention the donor (acceptor) potential to be centred at $x = a(-a)$. The unperturbed (diabatic) donor and acceptor states are obtained by solving the Schrödinger equation (SE) for each simple Dirac delta potential. They are given by $\phi_{D/A}(x) = \sqrt{\kappa} \cdot \exp(-\kappa|x \mp a|)$ with $\kappa = m_e g/\hbar^2$, m_e being the mass of the electron. The electron that is bound to the donor or acceptor site is described by a wavefunction that penetrates into the forbidden region but decays exponentially with x. The transfer integral H_{DA} corresponds to half the energy gap between the ground and first excited adiabatic states obtained by solving the SE taking the full Hamiltonian. This gives:

$$H_{DA} = \frac{-g^2}{2\hbar^2} \exp\left[-2a\sqrt{-2m_e E_0 / \hbar^2}\right] = H_{DA}^0 \exp[-\beta R_{DA}] \tag{4.63}$$

One sees that the electronic coupling decreases exponentially with the distance separating the two potential wells ($2a$ here). The rate of decay depends on the barrier height E_0. This simple model can be generalised to proteins if one assumes that the intervening medium can be seen as a homogenous square potential. An upper limit to the barrier height may be set as half the $\sigma \rightarrow \sigma^*$ band gap of the surrounding material (≈ 3 eV).[19] Higher values would mean that the electron is transported inside the conduction band, which seems unlikely for ET through proteins. Values of 2 eV are commonly used, which gives a decay factor β of 0.72 Å^{-1}. The prefactor can be estimated by taking the resonance integral at short distance (H^0_{DA}). Standard values borrowed from the extended-Hückel theory suggest prefactors around 2.7 eV.

4.6.1.2 Electron Tunnelling Pathways in Proteins

One may anticipate that protein-mediated tunnelling is far too complex to be modelled by a simple square potential. Real biomolecules are three-dimensional entities composed of many nuclei and electrons. The barriers through which the electron tunnels are likely to exhibit complex profiles depending on the presence of covalent bonds of various nature along the bridge. Brute force approaches to the problem of electron tunnelling in realistic models involve calculating the diabatic states by quantum chemistry methods and extracting the coupling between them. We refer the reader to appropriate references for details on these methodologies.[102,103] In this chapter we prefer to emphasise the physical models that have been developed to relate the amplitude of the electronic coupling to the chemical structure of the intervening medium. Despite the complexity of the problem, is it possible to extract the essential physics that governs tunnelling in proteins? We will focus on the tunnelling pathway model developed in the mid-1980s by Hopfield, Beratan and Onuchic.[104–106]

In a first step it is necessary to simplify the problem. This can be done with renormalisation techniques through the examination of the typical energy scale involved in electron tunnelling compared to that of other physico-chemical events.[107] In most biological ETs of interest electron tunnelling takes place between valence molecular orbitals (*e.g.* π/π^* MOs of organic cofactors or between bonding/antibonding metal–ligand MOs). It is thus reasonable to isolate the motion of valence electrons from core electrons, and isolate tunnelling from core/valence excitations. The former can be considered as moving in the pseudopotential field created by the latter. Similarly one can make the assumption that hopping interactions between bonding orbitals are much smaller than typical excitation energies. These approximations justify a one-electron tunnelling model. Furthermore if we restrict our attention to systems for which donor and acceptor MOs are well separated in energy from bridge MOs, the tunnelling process can be recast into a simple two-state tunnelling process; no intermediate electronic states with the electron or hole localised on the bridge is involved. Regarding couplings between tunnelling and vibrational modes we shall make the BO approximation for slow (low-frequency) modes: the electronic coupling depends parametrically on slowly varying nuclear positions. For high-frequency modes one could renormalise the electronic coupling and tunnelling energy, as was done in a previous section to include quantised nuclear modes in the rate expression. Thanks to the above hypotheses, a tight-binding Hamiltonian was established by Onuchic and Beratan,[105,106] giving an expression for h_{DA} through a single pathway connecting D to A:

$$h_{DA} = \frac{\beta_A \beta_D \beta_1}{(E - \alpha_L^1)(E - \alpha_R^1) - \beta_1^2} \prod_i^{N_B} \varepsilon_B^i \prod_j^{N_S} \varepsilon_S^j \prod_k^{N_H} \varepsilon_H^k \tag{4.64}$$

β_D (β_A) is the hopping interaction between the donor (acceptor) state and the first (last) bond of the bridge (Figure 4.12). β_1 is the interaction between the orbitals constituting the first bond. α_L^1 (α_R^1) is the energy of the left (right) hybrid orbital on the first bond. N_B, N_S and N_H are the number of through-covalent bond, through-space and through-hydrogen bond jumps in the pathway, respectively. Each type is associated with distinct decay factors ε. In principle, they should depend on the chemical natures of the bonds (C–C, C–H, C=O, N–H... O, O–H...O, ...) or on the angular polarisations of hybrid orbitals involved in tunnelling. Actually the analysis of ET in synthetic D-bridge-A molecules or in proteins for which experimental ET rates were available led the authors of the model to prefer a simpler formulation, which does not distinguish the nature of the covalent bonds and neglects angular dependences. The following sets of formulae of renormalised decay factors on experimental rates are now used:[106]

$$\varepsilon_B^i = \varepsilon_c = 0.6 \quad \text{for covalent bonds} \tag{4.65}$$

$$\varepsilon_H^k = \varepsilon_c^2 \exp[-1.7(R^H - 2.8)] \quad \text{for hydrogen bonds} \tag{4.66}$$

$$\varepsilon_S^j = \varepsilon_c \exp[-1.7(R^S - 1.4)] \quad \text{for through-space jumps} \tag{4.67}$$

In fact eqn (4.64) gives the h_{DA} for a single pathway connecting D to A. There is not of course a unique pathway connecting D to A. In practice, graph theory search algorithms are needed to find pathways that maximise h_{DA}.[108]

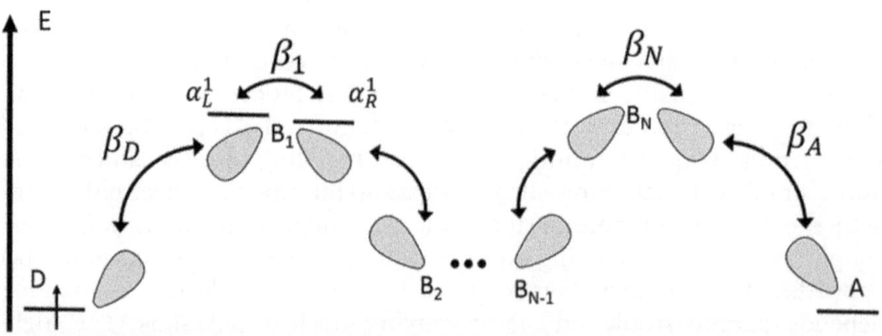

Figure 4.12 Molecular orbital representation illustrating the key energy parameters entering the pathway model (eqn (4.64)) for the electron transfer from D to A through BN bridge atoms *via* a superexchange mechanism.

As a consequence of the exponential forms of the individual decays with the segment lengths there is generally a particular pathway that dominates and $H_{DA} \approx \max(h_{DA})$. On the contrary, if multiple pathways of similar strength coexist, the situation is more delicate. Indeed, real pathways are associated with phases leading to constructive or destructive interferences. However because no phase information is explicitly included in the PM, information on interference effects are lost. If constructive interferences are assumed H_{DA} should be the sum of h_{DA}s, but this assumption is not necessarily correct and the model produces incorrect answers. Actually, because the PM parameters have been calibrated against experimental data that *de facto* account for interference effects, one may assume that the PM incorporates some interference effects in an effective way. Implementations of the pathway model are available in programs like HARLEM,[109] VMD[110] or CUBY.[111]

For illustrative purposes, we consider the copper enzyme PHM (peptidylglycine α-hydroxylating monooxygenase) the catalytic cycle of which involves a long-range ET.[112] PHM contains two copper active sites between which the long-range ET occurs. A pathway model search on wild-type PHM reveals that the most efficient pathway leaves from His108, then passes through a copper ligand of CuH, a water molecule, the carboxylate termination of the enzymatic substrate, and finally arrives at the CuM site.[113] The best pathway involves several covalent bonds but also two hydrogen bonds due to a water molecule that bridges the CuH complex to the substrate. A glutamine residue Gln170 further stabilises the water molecule position. In the Gln170Glu mutant this structural organisation is lost because of the introduction of a negatively charged residue and the medium rearranges, leading to different pathways involving multiple water molecules (Figure 4.13).

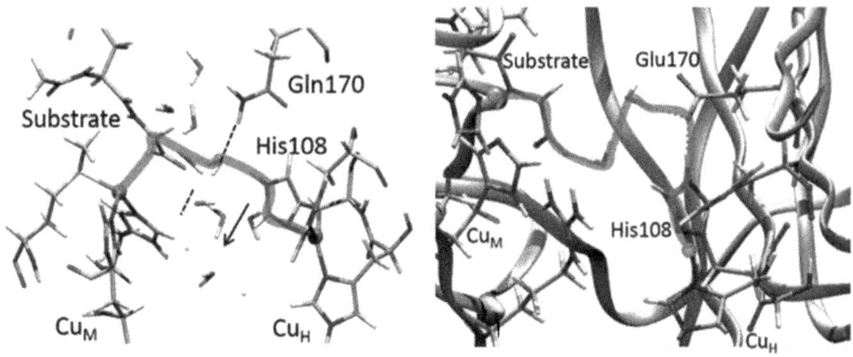

Figure 4.13 Representative tunnelling pathways accounting for electron transfer between the CuM and CuH sites of the enzyme PHM. Left: wild type enzyme. Right: Gln170Glu mutant for which the most efficient pathways involve two water molecules.

4.6.1.3 Interatomic Tunnelling Density Currents

The PM model provides a description of tunnelling that is based on energy considerations. A complementary understanding can be reached in terms of electron density currents (EDC). Early EDC analyses were reported by Bardeen[114] for ETs in metal oxides and by Redi and Hopfield[115] for biological ET with an application to the bacterial photoreactive centre. In the 1990s, Stuchebrukhov[102,116] (and references therein) significantly developed the theory of EDC to make it compatible with modern electronic structure theory. He worked out equations to calculate H_{DA} and to analyse the tunnelling currents from MOs that were expressed using the linear combination of atomic orbitals theory. The essence of the model will now be given. We start by considering a D–B–A system at fixed nuclear geometry. Let's further assume that the nuclear positions correspond to a situation of resonance of the electronic diabatic states ϕ_1 and ϕ_2 characterising the CT process. The time-dependent electronic wave function (ψ_{el}) may be expressed as a linear combination of states ϕ_1 and ϕ_2:

$$\psi_{el}(t) = C_1(t)\phi_1\{R\} + C_2(t)\phi_2\{R\} \tag{4.68}$$

The dependence on time is held by the coefficients C_1 and C_2. They are obtained by resolution of the time-dependent Schrödinger equation. This leads to:

$$C_1(t) = \cos\left(\frac{H_{DA}}{\hbar}t\right) \tag{4.69}$$

$$C_2(t) = -\sin\left(\frac{H_{DA}}{\hbar}t\right) \tag{4.70}$$

The coefficients undergo coherent Rabi oscillations at a pulsation $H_{DA}/\hbar$. The stronger the coupling, the faster the oscillations. As usual, the probability amplitude to find the electron on the A site is obtained by projecting ψ_{el} on the ϕ_2 diabatic state. Starting with the initial conditions $C_1(0) = 1$ and $C_2(0) = 0$, that is, assuming the electron is initially localised on the donor site the Rabi probability equals $\sin^2(H_{DA}t/\hbar)$ (recall $\Delta E = 0$). It becomes tangible that population exchanges take place faster with higher H_{DA}. Knowing ψ_{el}, it is possible to calculate the expectation value of the current density operator which is defined as:

$$\boldsymbol{j}(\boldsymbol{r},t) = \frac{\hbar}{2m_e i}\left(\psi_{el}^{*}\nabla\psi_{el} - \psi_{el}\nabla\psi_{el}^{*}\right) \tag{4.71}$$

Inserting eqn (4.68) into eqn (4.71) one can write the current density as a product of a term depending on the structure of the intervening medium ($\boldsymbol{J}$) and a time-dependent term:

$$\boldsymbol{j}(\boldsymbol{r},t) = \boldsymbol{J}(\boldsymbol{r})\sin\left(\frac{2H_{DA}}{\hbar}t\right) \tag{4.72}$$

$$J(r) = \langle \phi_1 | j | \phi_2 \rangle = \frac{\hbar}{2m_e} (\phi_1^* \nabla \phi_2 - \phi_2 \nabla \phi_1^*) \tag{4.73}$$

Application of the Gauss theorem shows that the integration of the electron density flux through any surface (Ω) separating the donor from the acceptor equals the electronic coupling H_{DA}.

$$H_{DA} = -\hbar \int_{\Omega} J(r) d\Omega \tag{4.74}$$

Equation (4.74) thus represents a means to evaluate the electronic coupling from electronic structure calculations. The calculation of the matrix elements of the density current operator is rather involved, although various reasonable hypotheses can be made to simplify the equations.[117] In particular one often finds that only a single pair of MOs really change shape upon CT. These describe the tunnelling electron while all other MOs describe spectator electrons and remain similar before and after ET. This one-electron approximation and the use of semi-empirical quantum chemistry techniques have frequently been used to compute H_{DA} in various proteins. A further great merit of the current density approach to tunnelling is to provide valuable insights about the structure of the tunnelling flux. Equation (4.74) can be recast as a sum of interatomic currents J_{ab}:

$$H_{DA} = \sum_{a \in \Omega} \sum_{b \notin \Omega} J_{ab} \tag{4.75}$$

The notation $a \in \Omega$ means all the atoms belonging to the region delimited by the Ω surface and including the electron donor. Connections between the tunnelling current and the pathway models are straightforwardly established: the predominant tunnelling pathway corresponds to that of strongest electron density flux. If the matrix of interatomic currents J_{ab} is known, the strongest pathway can be sought with graph theory algorithms. An advantage of the interatomic current description is that interferences among pathways appear explicitly. They manifest themselves through the signs of the interatomic currents. Two pathways interfering constructively will correspond to current densities flowing in the same direction whereas destructive interferences will result in annihilating fluxes of opposite directions. Researchers have developed several mathematical indicators to decipher interference effects in biological ET.[118–120]

For illustration purposes, we present here an example taken from the analysis of electron wiring in the Respiratory Complex 1 of the *Thermus thermophiles* bacterium reported by Hayashi and Stuchebrukhov.[121] The protein features a flavin mononucleotide as initial electron donor and a chain of seven iron–sulfur clusters that form a molecular wire. We focus here on the transfer between the first two clusters. Figure 4.14 depicts the main atoms involved in tunnelling from N3 to N1B. The darker the atom colour the more intense the electronic flux through it. The alanine residue 63 was

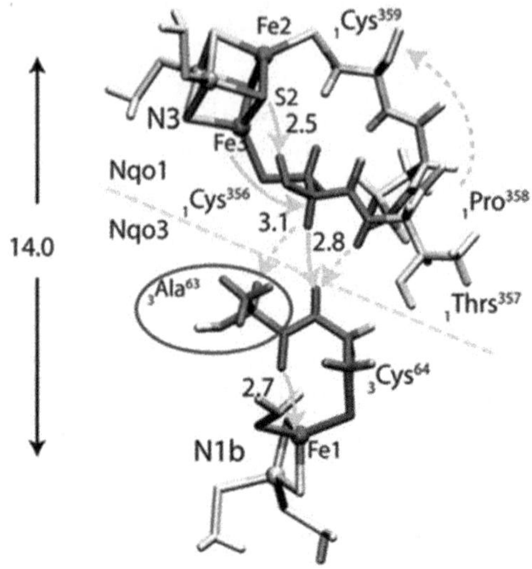

Figure 4.14 Tunnelling currents involved in one of the inter iron–sulfur clusters electron transfer in the Respiratory Complex 1. Interatomic distances are given in Å. The full and dashed arrows represent primary and secondary interatomic fluxes. The stronger the density flux through an atom, the darker its colour.
Reproduced with permission from T. Hayashi and A. A. Stuchebrukhov, *Proc. Natl. Acad. Sci.*, 2010, **107**, 19157–19162.[121]

found to be the main mediator of tunnelling. It is obvious that multiple parallel tunnelling routes contribute to tunnelling, which contrasts somewhat with the PM assumption of a singly emerging pathway.

4.6.2 ET Mechanism and Electronic Coupling Fluctuations

4.6.2.1 *Two Important Characteristic Times*

The value of the electronic coupling parameter is extremely sensitive to the structure of the intervening medium. Even in rather rigid proteins nuclear motions are sufficient to cause significant H_{DA} fluctuations and to introduce a new dependence on ET kinetics, which may be temperature dependent. Now, one may wonder whether a non-constant H_{DA} definitely jeopardises the application of the FGR or if simple corrections can be added to fix the theory? The answer to this question depends on the speed at which the electronic coupling fluctuates compared to the energy gap. Recall that population exchanges between the electronic states occur solely when the diabatic energy gap is close to zero. To make the explanation more concrete let us introduce two characteristic times: τ_{coh} and τ_{FC}. They reflect how fast H_{DA} and ΔE change in time.[122] The subscript 'coh' is used to emphasise that fluctuations in the electronic coupling due to thermal fluctuations cause a loss of coherence of tunnelling routes. The 'FC' subscript stands for

Franck–Condon, that is, τ_{FC} is the time during which the system remains in a region of allowed electronic transition.[107] The definition of τ_{FC} and τ_{coh} allows us to define two limit regimes. In the fast coupling regime, that is, when $\tau_{coh} \ll \tau_{FC}$, H_{DA} changes while the system remains in the FC region, making the Condon approximation inapplicable.[122] The separation of the electronic and nuclear motions that enabled us to write the rate constant as a product of an electronic (H_{DA}^2) and a nuclear (ρ_{FC}) factor is not valid and new formalisms are required. This will be the topic of Section 4.6.3.

On the contrary, in the slow coupling regime ($\tau_{FC} \ll \tau_{coh}$), H_{DA} can be considered to be constant once the system enters a region of allowed electronic transition. The Condon approximation can be preserved. However it is necessary to account for the fact that H_{DA} will be different each time the system reaches resonance. This can be done by replacing H_{DA}^2 in eqn (4.49) by its average value:

$$k_{ET} = \frac{2\pi}{\hbar} \langle H_{DA}^2 \rangle \sum_{v} \sum_{w} P(\varepsilon_v) |\langle v | w \rangle|^2 \delta(E_1 + \varepsilon_v - E_2 - \varepsilon_w) \tag{4.76}$$

What are the typical values for τ_{FC} and τ_{coh}? Numerical simulations are again of great help. τ_{coh} can be extracted from the analysis of the (normalised) autocorrelation functions $C_{H_{DA}}(t) = \langle H_{DA}(t) H_{DA}(0) \rangle / \langle H_{DA}(0)^2 \rangle$. τ_{coh} is approximately the time for $C_{H_{DA}}$ to drop to $1/e$. MD simulations and quantum chemistry calculations provide powerful tools to evaluate τ_{coh}. Note that the sign of H_{DA} depends on the relative phases of the electronic wave functions ϕ_1 and ϕ_2. It is therefore necessary to use the same phase convention for both states. τ_{coh} has been evaluated at a few tens of femtoseconds in the photoreactive centre,[118] in ruthenium-modified azurins[122,123] and more recently in cryptochromes.[90]

Neria and Nitzan,[124] Prezhdo[125] and Lockwood *et al.*[126] devoted various important papers to unravel the molecular mechanisms determining τ_{FC}. In the high-temperature limit τ_{FC} characterises the loss of phase and overlap of the nuclear wavefunctions evolving on the diabatic potential energy surfaces. Borrowing the vocabulary dedicated to the study of open quantum systems, one would qualify this process as decoherence of the electronic wavefunction caused by the interaction with its environment (the so-called bath composed in the present context of the nuclei[127]). Classical and *ab initio* MD simulations suggest timescales of a few femtoseconds for this electronic decoherence process in biomolecules.[57,128,129] Onuchic[107] also proposed a simple relationship relating τ_{FC} to the reorganisation energy in the high-temperature limit: $\tau_{FC} = \hbar / \sqrt{2\lambda k_B T}$. This expression leads to values consistent with results from MD simulations. In fact τ_{FC} always falls within a few femtoseconds for common λ values.

To summarise, H_{DA} decays in general on a longer timescale than ΔE and most protein ETs are expected to operate under the Condon approximation. We now consider various dynamical effects on ET kinetics in the slow coupling regime.

4.6.2.2 Dynamical Effects

When addressing the impact of electronic coupling fluctuations several questions come to mind. For instance is tunnelling controlled by the average structure of the intervening medium or by the structural fluctuations? At which distance do the structural details of the medium cease to be important? What is the role played by interference among pathways, be they destructive or constructive? Much effort has been devoted by researchers to bring partial answers to these questions. This is an area where the confrontation of experimental and computational data proved to be particularly fecund in validating, or not, theoretical models.

To help the discussion, it is useful to write $\langle H_{DA}^2\rangle = \langle H_{DA}\rangle^2 + \sigma^2$ with $\sigma^2 = \langle (H_{DA} - \langle H_{DA}\rangle)^2\rangle$. If $\langle H_{DA}\rangle^2 \gg \sigma^2$, electronic coupling fluctuations around the average value are small and tunnelling is dominated by the average structure, *i.e.* $\langle H_{DA}^2\rangle \approx \langle H_{DA}\rangle^2$. In the opposite limit, when protein fluctuations lead to a large variety of alternative pathways, the variance of H_{DA} dominates over the average value. The so-called coupling parameter $C = \langle H_{DA}\rangle^2/\langle H_{DA}^2\rangle$ was introduced by Balabin and Onuchic[130] to quantify the prevalence of one limit regime. C approaches unity for single-pathway mediated tunnelling and is usually much below 0.1 for multiple pathway regimes.

For illustrative purposes let us consider some results obtained by the groups of Gray and Beratan on ruthenium-modified cytochrome b562. Since the 1970s Gray and coworkers[131] have developed an extensive research programme to measure protein-mediated tunnelling rates. One of the most powerful techniques developed in that group is based on the modification of natural metalloproteins (cytochromes, blue copper proteins, *etc.*) to make them suitable for intra-protein ET. The idea is to introduce histidine residues at a specific location on the protein surface to enable binding of a ruthenium complex. Upon electronic excitation of the Ru complex, ET may take place with the other metal centre naturally present in the protein. UV-visible time-resolved measurements are used to follow the kinetics of ET. By varying the position of the histidine residue, one can collect several rates, each differing from the other due to the different locations of the Ru complex on the protein. This strategy proved extremely valuable to understand how protein structures mediate electron tunnelling.

In the case of cytochrome b562 the measured ET rates reveal two groups of proteins. The first one, encompassing a majority of mutants (His 15, 19, 63, 70, 86, 89, 92), conform with an average exponential law relating the rate to the donor–acceptor distance (Figure 4.15).[132] A decay of 1.3 $Å^{-1}$ characterises this relationship. The second group is comprised of two mutants (His 12 and 73) that dramatically fall below the average curve. For Ru-modified cytochrome b562 ET is tunnelling limited (*i.e.* $(\Delta G^0 + \lambda)^2/4\lambda \approx 0$). So, differences between the two groups of mutants are expected to stem from the $\langle H_{DA}^2\rangle$ term. Beratan and coworkers conducted MD simulations combined with extended-Hückel theory calculations of the electronic coupling on

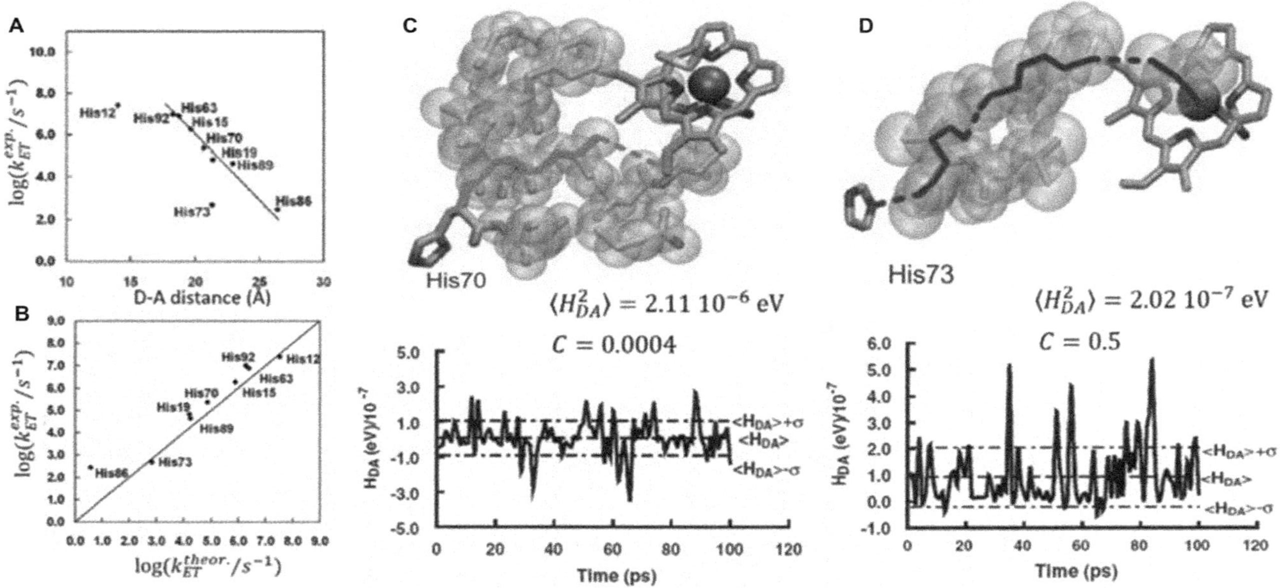

Figure 4.15 Competition between single-pathway and multiple interfering pathway regimes in Ru-modified cytochrome b562. A: experimental tunnelling rates. B: theoretical *vs.* experimental rates. C: H_{DA} fluctuations and tunnelling pathways for the His70 variant (multiple pathway regime). D: same as C for His73 variant (single pathway regime).
Adapted from T. R. Prytkova, I. V. Kurnikov and D. N. Beratan, *Science*, 2007, **315**, 622–625.[133] Reprinted with permission from AAAS.

these mutants.[133] They were able to rationalise the experimental trends for both groups of proteins (Figure 4.15). It appears that the two groups correspond to the mean coupling limit and to the coupling fluctuations limit regime. As seen on Figure 4.15 for the His70 mutant which is representative of the first group, H_{DA} fluctuates around zero and can take either positive or negative values. The coherence parameter is well below 0.1. This is characteristic of tunnelling mediated by multiple interfering pathways. Structurally, this is a consequence of the relative orientation of the Ru complex with respect to the haem edge. Several pathways with different entry points on the haem moiety can be active depending on the (fluctuating) geometry of the bridge atoms. In the second group, the coupling parameter lies above 0.5. Most pathways run through the haem axial ligand which constitutes a bottleneck for the tunnelling electron. The electronic coupling is essentially governed by an average structure, which results, in this case, in slower ET. The take-home message is that the distinction between the groups of proteins is genuinely a dynamical effect. Take the His70 and His73 variants, for example. The D–A distance is almost the same in both cases (≈ 21 Å) but the tunnelling rates are highly different because of electronic coupling fluctuations (the prefactor of eqn (4.76)) that strongly favour the tunnelling in the His70 variant.

Balabin *et al.*[134] further explored the interplay between average structure- and fluctuation-governed ET as a function of donor–acceptor distance. These authors considered an ensemble composed of 24 Ru-modified proteins (cytochrome c, cytochrome b562, myoglobin, azurin) representative of intra-protein ET and a collection of water-mediated inter-cytochrome b5 ET. Numerical simulations indicate that there is a critical distance (R_c) corresponding to $C = 0.5$ beyond which tunnelling is essentially mediated by coupling fluctuations (horizontal dotted line on Figure 4.16). R_c was estimated to be 1.9 ± 1.4 Å and to be 6.8 ± 1.2 Å for water- and protein-mediated ETs, respectively. In general, predominance of coupling fluctuations tends to increase with increasing R_{DA} because for larger distances the variability of tunnelling pathways is higher (except of course for some particular systems as exemplified in the previous example or on Figure 4.16 for $\log(C) > -0.7$). The authors further addressed the question whether at long range, when out-of-equilibrium structures determine $\langle H_{DA}^2 \rangle$ (*i.e.* small C values), the detailed composition of ET pathways continues to be important; or if the intervening medium can be seen as a homogeneous medium on average. One may eventually guess that two proteins exhibiting similar (but large) R_{DA} would produce similar $\langle H_{DA}^2 \rangle$. Rigorous statistical analyses indicate this is not the case: even at long distance H_{DA} fluctuations retain the marks of the underlying structure of the protein. At least for the set of proteins engineered by the Gray group, electron tunnelling cannot be accounted for by a simple homogeneous medium and the identification of tunnelling pathways is needed.

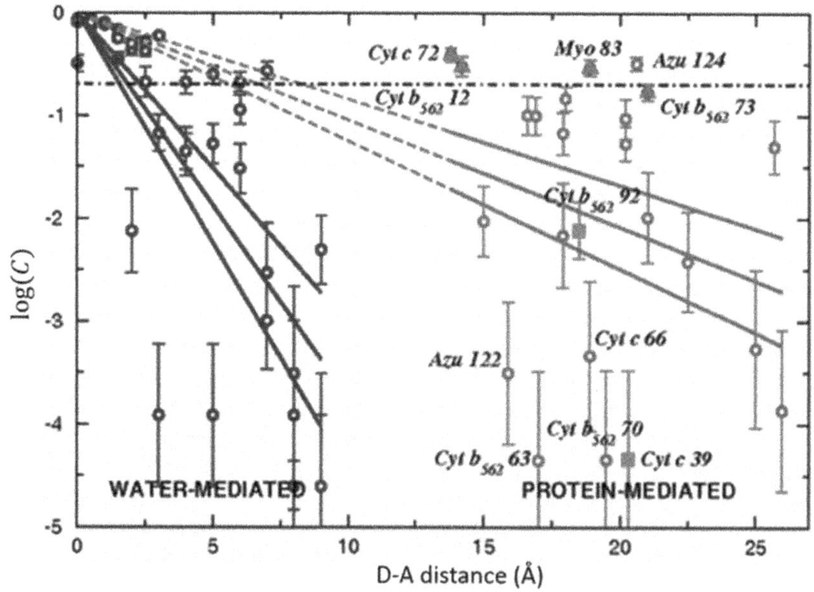

Figure 4.16 Distance dependence of the coherence parameter for water-mediated and protein-mediated ET.
Reproduced from ref. 134. Reprinted figure with permission from I. A. Balabin, D. N. Beratan and S. S. Skourtis, *Phys. Rev. Lett.*, 2008, **101**, 158102.[134] Copyright 2008 by the American Physical Society.

4.6.2.3 Gated Electron Transfer

In the previous sub-section we considered situations where relatively small coupling affected ET kinetics. This was a consequence of the fact that the D and A moieties were kept at almost constant distance so that coupling fluctuations mainly arose from around average fluctuations of the intervening medium. The term gated ET refers to situations where conformations that are efficient for ET are not the most favoured on thermodynamic grounds. This may happen, for example, for inter-protein ET when the most stable complex is not well suited for ET and needs to be activated. To account for such situations the following kinetic scheme may be of help:

$$[A_{ox}-B_{red}]\underset{k_{-x}}{\overset{k_x}{\rightleftharpoons}}[A_{ox}-B_{red}]^*\underset{k_{-ET}}{\overset{k_{ET}}{\rightleftharpoons}}[A_{red}-B_{ox}]$$

The pre-step (x) may be a conformational change or a chemical reaction, for instance. When $K_x \equiv k_x/k_{-x} \gg 1$ and $k_x \gg k_{ET}$ the system is referred to as a 'true' ET, that is, the observed overall rate is that of the ET step.[135] When $k_x \ll k_{ET}$ the system is gated by the realisation of the pre-step. Finally when $k_x \gg k_{ET}$ but $K_x \ll 1$ the system is under a coupled regime and the observed rate is $K_x k_{ET}$. To provide the reader with examples of applications we refer to the long-term investigations carried out in Davidson's group on ETs between methylamine dehydrogenase, amicyanin and cytochrome c-551i proteins and summarised in various key review papers.[135] Note that other kinetic schemes have been advanced for inter-protein ET, like dynamic docking, and we refer the reader to appropriate references.[136–138]

4.6.3 Electron Transfer beyond the Condon Approximation

4.6.3.1 *Theory of Inelastic Tunnelling*

In this section we consider ET operating under the fast coupling regime ($\tau_{coh} \ll \tau_{FC}$). Several formalisms have been proposed in the literature to deal with this situation.[97,98,139,140] For space reason we will focus our analysis on the one proposed by Troisi *et al.*[97] Most non-Condon ET theories, however, start by distinguishing two categories of nuclear vibrations. A first category encompasses modes that are displaced or mixed upon ET. As already explained in Section 4.5, they supply the system with the energy to enable ETs or to dissipate excess energy. They are often called accepting modes and will be hereafter noted $|v\rangle$ and $|w\rangle$. Structurally accepting modes involve atoms and molecules that are polarised differently in the two redox states. They do not impact significantly the amplitude of H_{DA} and the Condon approximation still applies for them. The second category involves vibrations that strongly affect the electronic coupling (noted $|a\rangle$ and $|b\rangle$). They are typically bridge vibrations localised on the intervening medium. The Condon approximation cannot be made for these modes. Making the hypothesis that the two categories of modes are separable Troisi *et al.*[97] showed that the rate expression can be written as:

$$k_{ET} = \sum_v P_v k_v = \frac{1}{\hbar^2} \sum_v P_v \iint \langle H_{DA}(t) H_{DA}(0) \rangle_B \times \rho_{FCv} \times e^{(i/\hbar)(E_v - E)t} dt dE \quad (4.77)$$

where $\rho_{FCv} = \sum_w S_{vw} \delta(E - E_w)$ is the Franck–Condon weighted density of final states for initial vibronic state $|v\rangle$. $\langle H_{DA}(t) H_{DA}(0) \rangle_B$ denotes the autocorrelation function of H_{DA} over bridge fluctuations. Note that compared to eqn (4.49) the electronic coupling is not factored out of the integral. If one makes the approximation of constant coupling despite the bridge structure fluctuations $\langle H_{DA}(t) H_{DA}(0) \rangle_B$ would equal $\langle H_{DA}^2 \rangle$ leading back to the standard Condon expression. This is not what we do here. Instead, introducing the spectral density:

$$J(E) = \frac{1}{\hbar^2} \int_{-\infty}^{+\infty} \langle H_{DA}(t) H_{DA}(0) \rangle_B e^{(i/\hbar)Et} dt \quad (4.78)$$

the rate expression reads:

$$k_{\mathrm{ET}} = \sum_{v} P_v \int_{-\infty}^{+\infty} J(E - E_v)\rho_{\mathrm{FC}v}(E)\mathrm{d}E \tag{4.79}$$

The authors made a Taylor expansion of the FC weighted density as a function of the energy:

$$k_{\mathrm{ET}} = \sum_{v} P_v \sum_{n}^{\infty} \frac{1}{n!}\left(\frac{\mathrm{d}^n \rho_{\mathrm{FC}v}}{\mathrm{d}E^n}\right)_{E=E_v} \int_{-\infty}^{+\infty} E^n J(E)\mathrm{d}E \tag{4.80}$$

Three hypotheses may further be introduced to obtain useful expressions. Let's first assume that bridge fluctuations are classical. This simplifies the mathematical form of J, and then the calculation of the integrals. Second we assume that $\langle H_{\mathrm{DA}}(t)H_{\mathrm{DA}}(0)\rangle_{\mathrm{B}}$ as well as $\rho_{\mathrm{FC}v}$ have a Gaussian form. The ET rate can then be written:

$$k_{\mathrm{ET}} = k^{(0)} + k^{(2)} + \cdots \tag{4.81}$$

$$k^{(0)} = \frac{2\pi}{\hbar}\left\langle H_{\mathrm{DA}}^2\right\rangle \frac{1}{\sqrt{4\pi\lambda k_{\mathrm{B}}T}} \exp\left(\frac{-(\Delta G^0 + \lambda)^2}{4\lambda k_{\mathrm{B}}T}\right) \tag{4.82}$$

$$k^{(2)} = k^{(0)} \times 2\frac{\hbar^2}{\tau_{\mathrm{coh}}^2}\left[\frac{(\Delta G^0 + \lambda)^2 - 2\lambda k_{\mathrm{B}}T}{(4\lambda k_{\mathrm{B}}T)^2}\right](1 - C) \tag{4.83}$$

Note that the first-order term equals zero by symmetry. The zeroth-order term is the standard non-adiabatic Marcus rate expression with fluctuating electronic coupling. It describes an elastic channel because no energy is exchanged between the tunnelling electron and the inducing modes. The second-order term describes the correction for inelastic channels. Higher-order terms are generally not needed.

The model involves two parameters that we have already met: τ_{coh} and C. As anticipated, non-Condon corrections will be significant only for short τ_{coh}. Interestingly, the coherence parameter C, which is a static property, also puts a condition on the appearance of inelastic effects. For ET ruled by single-pathway mechanisms, for which $C \to 1$ (see above), $k_{\mathrm{ET}} \to k^{(0)}$. On the contrary, non-Condon effects are likely to arise when the mean coupling approaches zero (provided small τ_{coh}). This may happen for example for symmetry forbidden ET for which $H_{\mathrm{DA}} \approx 0$, but also for ET governed by electronic coupling fluctuations. Using a simpler model that contains the same physics as the model of Troisi *et al.*,[97] Medvedev and Stuchebrukhov[139] arrived at the conclusion that non-Condon effects are more likely to be manifest for very long-range ET, because, as we saw in Section 4.6.2.2, electronic coupling fluctuations tend to dominate over the average structure when the D–A distance increases. Eqn (4.81) to (4.83) have been tested on ET in C-clamp molecules,[141] in Ru-modified azurins[122,123] and in cryptochromes,[90] revealing only very moderate corrections to the standard MT (below 1%). A similar, although more sophisticated model, for non-Condon kinetics has been developed by Kakitani and coworkers and was applied to the photoreactive centre.[98] The authors concluded that noticeable non-Condon effects could be observed for ET in the deeply inverted regime, and to some extent in the normal region, a conclusion reminiscent of the older analysis of Medvedev and Stuchebrukhov.[139] Natural biological systems for which ET kinetics would be dominated by non-Condon effects remain to be revealed experimentally.

4.6.3.2 *Manipulation of ET Rates*

The formalism of non-Condon ET has been further developed by Beratan, Skourtis and coworkers.[140,142,143] These authors showed, for example, that inelastic interactions along specific ET pathways destroy interferences with other pathways, in turn affecting ET rates. The modulation of ET rates resulting from the interaction of the transferred electron with inducing modes is interesting because these modes do not affect the thermodynamics of ET reactions. In principle it should be possible to manipulate ET rates by infrared laser pulses targeting specific bridge vibrations. In 2009 Lin *et al.*[144] actually reported a proof-of-concept of these ideas. They were able to conduct experiments that can be seen as analogues to the which-way interferometer experiment but for the case of ET along two parallel pathways. In their study IR pulses were used in the course of ET to probe which pathway the electron was passing through, causing collapse of the electronic wavefunction, and affecting the overall rate constant.

4.7 Beyond the Two-state Approximation

In previous sections we have described ET between single D–A pairs separated by a bridge whose role was simply to extend the range of the electronic wavefunction into the forbidden region. ET was governed by the so-called super-exchange mechanism. We now consider situations where bridge units (B_1,B_2,...,B_N) take a more active role, with the possibility of being transiently oxidised/reduced. This may happen in the case of energy proximity between the [$B_i^{-/+}$] and the donor and acceptor levels. Multisite ET is typically encountered in DNAs or RNAs.[145] For proteins, we mention the photoreactive centre (SP–B_L–H_L–Q chain of cofactors) and the tryptophan triad/tetrad involved in flavin photoreduction in photolyases and cryptochromes.[146] Modified proteins and peptides were also produced in the 2000s[147–149] in which multisite ET is possible. We now portray two limit mechanisms that have been advanced to account for multiple-site transfers: the incoherent hopping model[150] and the multiple-sites coherent tunnelling model caused by flickering resonances.[151]

4.7.1 Incoherent Hopping Model

One way to model electron transport over multiple sites is to assume a succession of elementary ET steps along the chain of cofactors. For an N-site system the overall rate may be obtained by solving the following kinetic scheme:

$$\mathrm{P_d}(t) \xrightarrow{k_{1N}} \mathrm{P_1}(t) \underset{k_{2,1}}{\overset{k_{1,2}}{\rightleftarrows}} \mathrm{P_2}(t) \underset{k_{3,2}}{\overset{k_{2,3}}{\rightleftarrows}} \mathrm{P_3}(t) \cdots$$

$$\mathrm{P_N}(t) \xrightarrow{k_{\mathrm{tr}}} \mathrm{P_a}(t)$$

$\mathrm{P_d}$ and $\mathrm{P_a}$ are initial and final states. k_{1N} and k_{tr} are the rate of production of the first reactive state and the final trapping event. The intermediate $k_{n,m}$ are ET rates between sites P_n and P_m. The incoherent hopping model assumes that electronic dephasing is faster than the time required for two successive steps along the chain. Thus the intermediate states refer to actual transient chemical intermediates. The overall ET chain can be described as a succession of two-level ET steps. Depending on the system of interest standard classical Marcus rate expressions (eqn (4.10)), mixed quantum classical versions (eqn (4.54)) of it or quantum expressions (eqn (4.49)) may be chosen to compute $k_{n,m}$ values. If needed, non-ergodic or non-linear effects may be included in the treatment. A take-home message of the incoherent hopping model is, assuming all $k_{n,m}$ equal k, that the rate decreases with the number of steps (N) as kN^{η} with η comprised between 1 and 2, while the rate of direct ET *via* super-exchange decays exponentially.

4.7.2 Flickering Resonance Model

In 2014, Zhang *et al.*[151] elaborated an original scheme to account for coherent electron tunnelling over more than two sites. The underlying idea is that for a system comprised of N sites, it may be possible to bring more than two energy levels into resonance at once. In such resonant situations, the electron may tunnel over more than two sites before resonance is lost. To formalise the model, the authors proposed to extend the two-state

Hopfield theory of ET[19] to the multiple-level case. Hopfield theory can be seen as an alternative derivation of the Marcus ET rate by analogy with resonance energy transfer processes that are described by the Förster theory (known as FRET for Förster resonant energy transfer). Excitation transfer from a donor to an acceptor is caused by the coupling (U_{DA}) of their transition dipoles. U_{DA} is proportional to the dipoles and to the inverse cube of the D–A distance. Rates are evaluated from the overlap between the emission spectral shape (S_a) of the donor and the absorption spectral shape (S'_b) of the acceptor, integrated over all energies, and multiplied by $|U_{DA}|^2$.

$$k_{FRET} = \frac{2\pi}{\hbar}|U_{DA}|^2 \int_{-\infty}^{+\infty} S_a(E)S'_b(E)\mathrm{d}E \tag{4.84}$$

It is important to notice that S_a and S'_b include all the effects resulting from the interaction of the excitation with its environment (electrons and nuclei). Hopfield proposed to express the rate of an ET by analogy with eqn (4.84). To this end U_{DA} has first to be replaced by H_{DA}. Then the optical line shapes are replaced by the removal (insertion) spectral distributions D_a (D'_b) that are the distribution of vertical energies associated with oxidation (reduction) of the donor (acceptor).

$$k_{ET} = \frac{2\pi}{\hbar}|H_{DA}|^2 \int_{-\infty}^{+\infty} D_a(E)D'_b(E)\mathrm{d}E \tag{4.85}$$

In the high-temperature limit a Gaussian form for the spectral distribution is reasonable and one recovers the standard non-adiabatic Marcus rate expression. However, eqn (4.85) does not assume any specific form for D_a and D'_b functions and non-linear effects, for example, can naturally be accounted for by the model. Nuclear quantum effects can be included as well. The extension to tunnelling over N sites is given by:

$$k_{FR}(N) = \frac{1}{\tau} \times P_{match}(N) \tag{4.86}$$

P_{match} is the probability that N energies are within $\pm H_{DA}$. τ is the characteristic timescale of the overall process. It may be the time taken to bring the levels into degeneracy, the electron transport time or the trapping time. For non-correlated site energy (D_x) fluctuations, P_{match} can be estimated as:

$$P_{match}(2) = \int_{-\infty}^{-\infty} \mathrm{d}H_{DA}\rho_H(H_{DA}) \int_{-\infty}^{-\infty} \mathrm{d}E_1 D_1(E_1) \int_{E_1-H_{DA}}^{E_1+H_{DA}} \mathrm{d}E_2 D_2(E_2) \tag{4.87}$$

$$P_{\text{match}}(N) = \int_{-\infty}^{-\infty} \mathrm{d}H_{\text{DA}}\rho_{\text{H}}(H_{\text{DA}}) \int_{-\infty}^{-\infty} \mathrm{d}E_1 D(E_1) \int_{E_1-H_{\text{DA}}}^{E_1+H_{\text{DA}}} \mathrm{d}E_2 D(E_2) \ldots \int_{\max[E_1,\ldots,E_{N-1}]-H_{\text{DA}}}^{\min[E_1,\ldots,E_{N-1}]+H_{\text{DA}}} \mathrm{d}E_N D_N(E_N) \quad (4.88)$$

In this equation H_{DA} is the nearest-neighbour electronic coupling. For $N=2$, eqn (4.87) is similar to the non-adiabatic Marcus rate if we set in the non-adiabatic limit $\tau = \hbar/\pi H_{DA}$.[151] For higher N values P_{match} may be either evaluated numerically or by an analytical approximate expression relating the FR rate to the MT parameters. The FR theory predicts an exponential decay with the D–A distance, as with the super-exchange mechanism, however with a characteristic decay $1/\Delta R \ln[\sqrt{\pi/2}(\sigma_E/H_{nm})]$ where ΔR and H_{nm} are the inter-site distance and electronic coupling respectively, and σ_E is the standard deviation of the bridge site energies.[151] A critical parameter for this reactive channel to be competitive over incoherent hopping for example is, besides the proximity of the bridge levels, the strength of the electronic coupling. Indeed as can be seen from eqn (4.88), H_{DA} imposes the window widths where electronic population exchange can take place (see Figure 4.17). Flickering resonance has been tested for multistep CTs in

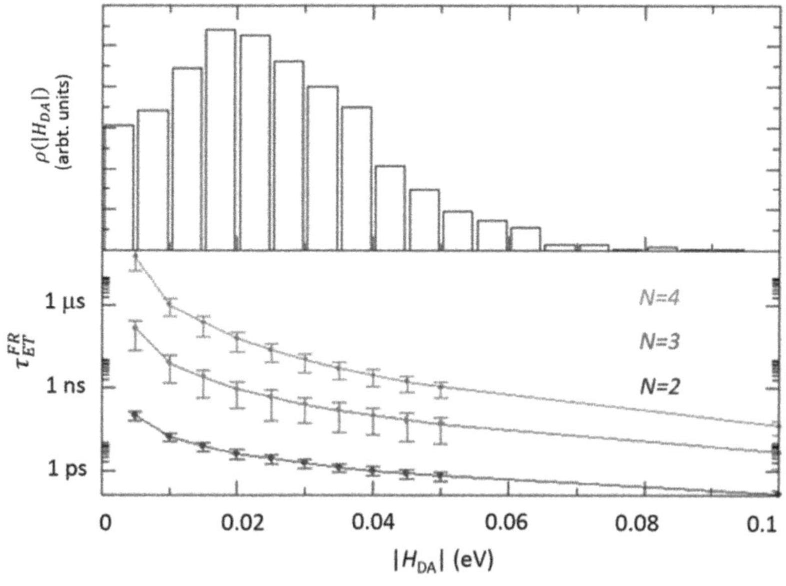

Figure 4.17 FR model for ET in (6–4) photolyase. Top: H_{DA} distribution. Bottom: characteristic rate as a function of the electronic coupling. N is the number of redox cofactors (tryptophan residues in the present case) participating in the charge migration process.
Reprinted with permission from F. Cailliez, P. Müller, T. Firmino, P. Pernot and A. de la Lande, *J. Am. Chem. Soc.*, 2016, **138**, 1904–1915.[152] Copyright (2016) American Chemical Society.

DNA,[151] in photolyases[152] and in multihaem proteins.[40] The average H_{DA} between adjacent bridge units was evaluated at 0.1, 0.03 and 0.001 eV, respectively. Only in the former case was FR shown to clearly compete with incoherent hopping, while for the latter it could be ruled out, photolyase representing an intermediate situation. Recently Liu *et al.*[153] examined the influence of correlations among site energies on the FR rates. The analysis of various types of noise shed new light on the mechanisms favouring FR over super-exchange and incoherent hopping in DNA. More elaborate theoretical models and fresh experimental data will certainly emerge in the coming years to definitely assess the multisite coherent tunnelling processes in proteins.

4.8 Summary and Perspectives

The MT is now more than 60 years old but is still the major paradigm to treat ETs. While the theory was designed for ET between ions in solution, *i.e.* in homogeneous media characterised by fast relaxation, it served as a basis on which theories for describing ET in proteins were developed. ET is indeed one of the most fundamental processes of physical chemistry that is at the heart of many biochemical processes. As such, it has attracted the attention of some of the greatest researchers of our field, as can be seen in this chapter. From their work new theories and new methodologies emerged that are used by many of us. Due to space limitations our choices for this chapter were necessarily restricted to some of the most common theories for ET in proteins. The appendix depicts a graphical historical summary of the main concepts of ET theories, including various excellent review articles for interested readers.

Most researchers will agree that the field of ET in biomolecules is still in its infancy and much remains to be discovered. While we conclude this chapter in the beginning of 2016, we may foresee various conceptual breakthroughs that could emerge in the coming years. At the bottom of the temporal scale, the sub-femtosecond electronic dynamics associated with tunnelling electrons may well hold surprises to be revealed. The advent of attosecond spectroscopies, for the time being reserved to small molecular systems,[154] will certainly be applied to large biomolecules and may bring unexpected results about the intimate nature of the coupling between electrons and nuclei. We have already mentioned in the text the possibility of manipulating ET by IR laser pulses. On the other end of the temporal scale, much remains to be understood concerning the insertion of ET events in complex bio-machineries. The latter involve multiple physicochemical events, such as chemical reactions, protein conformational changes, ligand diffusion, long-range proton hops, *etc.* that may be characterised by space and temporal scales largely overlapping that of ET. There is certainly much to be understood about how these 'elementary steps' are coupled and confer specific functions to biomolecules. The road towards the full understanding of ET in proteins is still a long and winding one.

Appendix: Chronology of Contributions to ET Theory

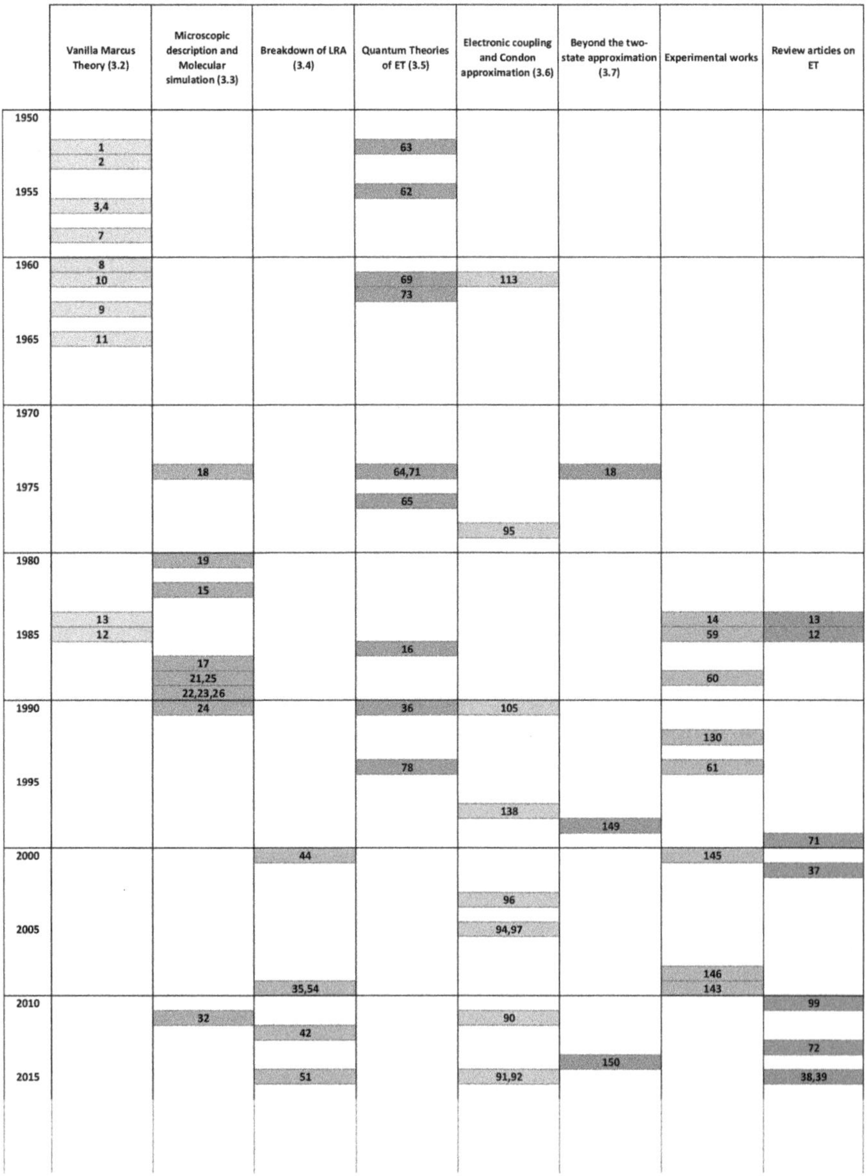

The numbers refer to references used in Chapter 4.

Acknowledgements

D. R. S. is grateful to NSERC-Canada for ongoing Discovery Grants.

References

1. W. F. Libby, *J. Phys. Chem.*, 1952, **56**, 863–868.
2. N. S. Hush, *J. Polym. Sci.*, 1953, **11**, 289–298.
3. R. A. Marcus, *J. Chem. Phys.*, 1956, **24**, 979–989.
4. R. A. Marcus, *J. Chem. Phys.*, 1956, **24**, 966–978.
5. R. A. Marcus, *J. Chem. Phys.*, 1957, **26**, 867–871.
6. R. A. Marcus, *J. Chem. Phys.*, 1957, **26**, 872–877.
7. N. S. Hush, *J. Chem. Phys.*, 1958, **28**, 962–972.
8. R. A. Marcus, *Discuss. Faraday Soc.*, 1960, 21–31.
9. A. de la Lande, N. S. Babcock, J. Rezac, B. Levy, B. C. Sanders and D. R. Salahub, *Phys. Chem. Chem. Phys.*, 2012, **14**, 5902–5918.
10. R. A. Marcus, *J. Phys. Chem.*, 1963, **67**, 853–857.
11. N. S. Hush, *Trans. Faraday Soc.*, 1961, **57**, 557–580.
12. R. A. Marcus, *J. Chem. Phys.*, 1965, **43**, 679–701.
13. R. A. Marcus and N. Sutin, *Biochim. Biophys. Acta*, 1985, **811**, 265–322.
14. M. D. Newton and N. Sutin, *Annu. Rev. Phys. Chem.*, 1984, **35**, 437–480.
15. J. R. Miller, L. T. Calcaterra and G. L. Closs, *J. Am. Chem. Soc.*, 1984, **106**, 3047–3049.
16. A. Warshel, *J. Phys. Chem.*, 1982, **86**, 2218–2224.
17. A. Warshel and J. K. Hwang, *J. Chem. Phys.*, 1986, **84**, 4938–4957.
18. J. K. Hwang and A. Warshel, *J. Am. Chem. Soc.*, 1987, **109**, 715–720.
19. J. J. Hopfield, *Proc. Natl. Acad. Sci. U. S. A.*, 1974, **71**, 3640–3644.
20. L. D. Zusman, *Chem. Phys.*, 1980, **49**, 295–304.
21. R. A. Kuharski, J. S. Bader, D. Chandler, M. Sprik, M. L. Klein and R. W. Impey, *J. Chem. Phys.*, 1988, **89**, 3248–3257.
22. D. Chandler and R. A. Kuharski, *Faraday Discuss.*, 1988, **85**, 329–340.
23. A. Warshel, Z. T. Chu and W. W. Parson, *Science*, 1989, **246**, 112–116.
24. E. A. Carter and J. T. Hynes, *J. Phys. Chem.*, 1989, **93**, 2184–2187.
25. G. King and A. Warshel, *J. Chem. Phys.*, 1990, **93**, 8682–8692.
26. T. Kakitani and N. Mataga, *J. Phys. Chem.*, 1988, **92**, 5059–5068.
27. M. Tachiya, *J. Phys. Chem.*, 1989, **93**, 7050–7052.
28. W. W. Parson, Z. T. Chu and A. Warshel, *Biophys. J.*, 1998, **74**, 182–191.
29. L. H. Hu, M. Farrokhnia, J. Heimdal, S. Shleev, L. Rulisek and U. Ryde, *J. Phys. Chem. B*, 2011, **115**, 13111–13126.
30. M. Cascella, A. Magistrato, I. Tavernelli, P. Carloni and U. Rothlisberger, *Proc. Natl. Acad. Sci. U. S. A.*, 2006, **103**, 19641–19646.
31. J. Blumberger, *Phys. Chem. Chem. Phys.*, 2008, **10**, 5651–5667.
32. W. W. Parson and A. Warshel, *Chem. Phys.*, 2004, **296**, 201–216.
33. P. B. Woiczikowski, T. Steinbrecher, T. Kubar and M. Elstner, *J. Phys. Chem. B*, 2011, **115**, 9846–9863.
34. V. Tipmanee, H. Oberhofer, M. Park, K. S. Kim and J. Blumberger, *J. Am. Chem. Soc.*, 2010, **132**, 17032–17040.
35. N. Gillet, B. Levy, V. Moliner, I. Demachy and A. de la Lande, *J. Chem. Theory Comput.*, 2014, **10**, 5036–5046.

36. D. N. LeBard and D. V. Matyushov, *J. Phys. Chem. B*, 2009, **113**, 12424–12437.
37. J. S. Bader, R. A. Kuharski and D. Chandler, *J. Chem. Phys.*, 1990, **93**, 230–236.
38. A. Warshel and W. W. Parson, *Q. Rev. Biophys.*, 2001, **34**, 563–679.
39. A. de la Lande, N. Gillet, S. F. Chen and D. R. Salahub, *Arch. Biochem. Biophys.*, 2015, **582**, 28–41.
40. J. Blumberger, *Chem. Rev.*, 2015, **115**, 11191–11238.
41. D. W. Small, D. V. Matyushov and G. A. Voth, *J. Am. Chem. Soc.*, 2003, **125**, 7470–7478.
42. J. Blumberger, *J. Am. Chem. Soc.*, 2008, **130**, 16065–16068.
43. R. Vuilleumier, K. A. Tay, G. Jeanmairet, D. Borgis and A. Boutin, *J. Am. Chem. Soc.*, 2012, **134**, 2067–2074.
44. D. N. LeBard and D. V. Matyushov, *Phys. Chem. Chem. Phys.*, 2010, **12**, 15335–15348.
45. D. V. Matyushov and G. A. Voth, *J. Chem. Phys.*, 2000, **113**, 5413–5424.
46. S. Hammes-Schiffer, *Chem. Rev.*, 2010, **110**, 6937–6938.
47. J. M. Saveant, *J. Am. Chem. Soc.*, 1987, **109**, 6788–6795.
48. A. Soudackov and S. Hammes-Schiffer, *J. Chem. Phys.*, 1999, **111**, 4672–4687.
49. R. G. Palmer, *Adv. Phys.*, 1982, **31**, 669–735.
50. D. N. LeBard and D. V. Matyushov, *J. Phys. Chem. B*, 2010, **114**, 9246–9258.
51. D. V. Matyushov, *J. Phys. Chem. B*, 2011, **115**, 10715–10724.
52. D. V. Matyushov, *J. Phys.: Condens. Matter*, 2015, **27**, 473001–473025.
53. D. V. Matyushov, *J. Chem. Phys.*, 2013, **139**, 025102–025114.
54. M. Drobizhev, N. S. Makarov, S. E. Tillo, T. E. Hughes and A. Rebane, *J. Phys. Chem. B*, 2012, **116**, 1736–1744.
55. D. V. Matyushov, *J. Chem. Phys.*, 2009, **130**, 164522–164530.
56. D. N. LeBard, D. R. Martin, S. Lin, N. W. Woodbury and D. V. Matyushov, *Chem. Sci.*, 2013, **4**, 4127–4136.
57. C. Narth, N. Gillet, F. Cailliez, B. Levy and A. de la Lande, *Acc. Chem. Res.*, 2015, **48**, 1090–1097.
58. D. N. LeBard, V. Kapko and D. V. Matyushov, *J. Phys. Chem. B*, 2008, **112**, 10322–10342.
59. C. C. Moser, J. M. Keske, K. Warncke, R. S. Farid and P. L. Dutton, *Nature*, 1992, **355**, 796–802.
60. C. Kirmaier, D. Holten and W. W. Parson, *Biochim. Biophys. Acta*, 1985, **810**, 33–48.
61. G. R. Fleming, J. L. Martin and J. Breton, *Nature*, 1988, **333**, 190–192.
62. M. H. Vos, M. R. Jones, C. N. Hunter, J. Breton and J. L. Martin, *Proc. Natl. Acad. Sci. U. S. A.*, 1994, **91**, 12701–12705.
63. R. Kubo and Y. Toyozawa, *Prog. Theor. Phys.*, 1955, **13**, 160–182.
64. M. Lax, *J. Chem. Phys.*, 1952, **20**, 1752–1760.
65. N. R. Kestner, J. Logan and J. Jortner, *J. Phys. Chem.*, 1974, **78**, 2148–2166.

66. S. Efrima and M. Bixon, *Chem. Phys.*, 1976, **13**, 447–460.
67. R. Borrelli and A. Peluso, *J. Chem. Phys.*, 2008, **128**, 044303–044310.
68. H. C. Jankowiak, J. L. Stuber and R. Berger, *J. Chem. Phys.*, 2007, **127**, 234101–234123.
69. R. R. Dogonadze, V. G. Levich and Y. A. Chizmadzhev, *Zh. Fiz. Khim.*, 1960, **34**, 2320–2327.
70. V. G. Levich and R. R. Dogonadze, *Dokl. Akad. Nauk SSSR*, 1960, **133**, 158–161.
71. R. R. Dogonadze, A. M. Kuznetsov and A. A. Chernenko, *Russ. Chem. Rev.*, 1965, **34**, 759–775.
72. M. Bixon and J. Jortner, *Adv. Chem. Phys.*, 1999, **106**, 35–202.
73. R. Borrelli and A. Peluso, *Wiley Interdiscip. Rev.: Comput. Mol. Sci.*, 2013, **3**, 542–559.
74. N. Sutin, *Annu. Rev. Nucl. Sci.*, 1962, **12**, 285–328.
75. S. Efrima and M. Bixon, *Chem. Phys. Lett.*, 1974, **25**, 34–37.
76. J. Ulstrup and J. Jortner, *J. Chem. Phys.*, 1975, **63**, 4358–4368.
77. X. Y. Song and R. A. Marcus, *J. Chem. Phys.*, 1993, **99**, 7768–7773.
78. K. Ando, *J. Chem. Phys.*, 1997, **106**, 116–126.
79. D. Xu and K. Schulten, *Chem. Phys.*, 1994, **182**, 91–117.
80. S. Tanaka and Y. Sengoku, *Phys. Rev. E: Stat., Nonlinear, Soft Matter Phys.*, 2003, **68**, 031905–031910.
81. N. Makri, E. J. Sim, D. E. Makarov and M. Topaler, *Proc. Natl. Acad. Sci. U. S. A.*, 1996, **93**, 3926–3931.
82. H. D. Meyer, U. Manthe and L. S. Cederbaum, *Chem. Phys. Lett.*, 1990, **165**, 73–78.
83. H. B. Wang and M. Thoss, *J. Chem. Phys.*, 2003, **119**, 1289–1299.
84. F. Shibta, Y. Takahashi and N. Hashitome, *J. Stat. Phys.*, 1977, **17**, 171–187.
85. S. Jurinovich, L. Viani, I. G. Prandi, T. Renger and B. Mennucci, *Phys. Chem. Chem. Phys.*, 2015, **17**, 14405–14416.
86. S. Jurinovich, L. Viani, C. Curutchet and B. Mennucci, *Phys. Chem. Chem. Phys.*, 2015, **17**, 30783–30792.
87. A. Ishizaki and G. R. Fleming, *Annu. Rev. Condens. Matter Phys.*, 2012, **3**, 333–361.
88. P. F. Huo and T. F. Miller, *Phys. Chem. Chem. Phys.*, 2015, **17**, 30914–30924.
89. E. Sim and N. Makri, *J. Phys. Chem. B*, 1997, **101**, 5446–5458.
90. T. Firmino, E. Mangaud, F. Cailliez, A. Devolver, D. Mendive-Tapia, F. Gatti, C. Meier, M. Desouter-Lecomte and A. de la Lande, *Phys. Chem. Chem. Phys.*, 2016, **18**, 21442–21457.
91. L. K. McKemmish, R. H. McKenzie, N. S. Hush and J. R. Reimers, *J. Chem. Phys.*, 2011, **135**, 244110–244121.
92. L. K. McKemmish, R. H. McKenzie, N. S. Hush and J. R. Reimers, *Phys. Chem. Chem. Phys.*, 2015, **17**, 24666–24682.
93. J. R. Reimers, L. K. McKemmish, R. H. McKenzie and N. S. Hush, *Phys. Chem. Chem. Phys.*, 2015, **17**, 24641–24665.

94. C. C. Page, C. C. Moser and P. L. Dutton, *Curr. Opin. Chem. Biol.*, 2003, 7, 551–556.
95. H. B. Gray and J. R. Winkler, *Proc. Natl. Acad. Sci. U. S. A.*, 2005, **102**, 3534–3539.
96. M. A. Ratner and A. Madhukar, *Chem. Phys.*, 1978, **30**, 201–215.
97. A. Troisi, A. Nitzan and M. A. Ratner, *J. Chem. Phys.*, 2003, **119**, 5782–5788.
98. H. Nishioka, A. Kimura, T. Yamato, T. Kawatsu and T. Kakitani, *J. Phys. Chem. B*, 2005, **109**, 15621–15635.
99. D. N. Beratan, C. Liu, A. Migliore, N. F. Polizzi, S. S. Skourtis, P. Zhang and Y. Zhang, *Acc. Chem. Res.*, 2015, **48**, 474–481.
100. S. S. Skourtis, D. H. Waldeck and D. N. Beratan, *Annu. Rev. Phys. Chem.*, 2010, **61**, 461–485.
101. C. Cohen-Tannoudji, B. Diu and F. Laloe, *Quantum Mechanics*, Wiley, New York, 1982, vol. 2.
102. A. A. Stuchebrukhov, *Theor. Chem. Acc.*, 2003, **110**, 291–306.
103. C. P. Hsu, *Acc. Chem. Res.*, 2009, **42**, 509–518.
104. D. N. Beratan, J. N. Onuchic and J. J. Hopfield, *J. Chem. Phys.*, 1987, **86**, 4488–4498.
105. D. N. Beratan and J. N. Onuchic, *Photosynth. Res.*, 1989, **22**, 173–186.
106. D. N. Beratan, J. N. Onuchic, J. N. Betts, B. E. Bowler and H. B. Gray, *J. Am. Chem. Soc.*, 1990, **112**, 7915–7921.
107. W. Bialek, W. J. Bruno, J. Joseph and J. N. Onuchic, *Photosynth. Res.*, 1989, **22**, 15–27.
108. A. de la Lande, N. S. Babcock, J. Rezac, B. C. Sanders and D. R. Salahub, *Proc. Natl. Acad. Sci. U. S. A.*, 2010, **107**, 11799–11804.
109. I. V. Kurnikov, Journal, HARLEM – Molecular Modeling Package.
110. I. A. Balabin, X. Hu and D. N. Beratan, *J. Comput. Chem.*, 2012, **33**, 906–910.
111. J. Řezáč, *J. Comput. Chem.*, 2016, **37**, 1230–1237.
112. R. L. Osborne and J. P. Klinman, in *Copper Oxygen Chemistry*, ed. K. D. Karlin and S. Itoh, J. Wiley &Sons, New York, 2011.
113. A. de la Lande, S. Mart, O. Parisel and V. Moliner, *J. Am. Chem. Soc.*, 2007, **129**, 11700–11707.
114. J. Bardeen, *Phys. Rev. Lett.*, 1961, **6**, 57–59.
115. M. Redi and J. J. Hopfield, *J. Chem. Phys.*, 1980, **72**, 6651–6660.
116. A. A. Stuchebrukhov, *J. Chem. Phys.*, 1996, **104**, 8424–8432.
117. A. A. Stuchebrukhov, *J. Chem. Phys.*, 2003, **118**, 7898–7906.
118. H. Nishioka, A. Kimura, T. Yamato, T. Kawatsu and T. Kakitani, *J. Phys. Chem. B*, 2005, **109**, 1978–1987.
119. H. Nishioka and T. Kakitani, *J. Phys. Chem. B*, 2008, **112**, 9948–9958.
120. T. Kawatsu, T. Kakitani and T. Yamato, *J. Phys. Chem. B*, 2002, **106**, 11356–11366.
121. T. Hayashi and A. A. Stuchebrukhov, *Proc. Natl. Acad. Sci. U. S. A.*, 2010, **107**, 19157–19162.

122. S. S. Skourtis, I. A. Balabin, T. Kawatsu and D. N. Beratan, *Proc. Natl. Acad. Sci. U. S. A.*, 2005, **102**, 3552–3557.
123. I. Daizadeh, E. S. Medvedev and A. A. Stuchebrukhov, *Proc. Natl. Acad. Sci. U. S. A.*, 1997, **94**, 3703–3708.
124. E. Neria and A. Nitzan, *J. Chem. Phys.*, 1993, **99**, 1109–1123.
125. O. V. Prezhdo, *J. Chem. Phys.*, 1999, **111**, 8366–8377.
126. D. M. Lockwood, H. Hwang and P. J. Rossky, *Chem. Phys.*, 2001, **268**, 285–293.
127. H. Hwang and P. J. Rossky, *J. Phys. Chem. B*, 2004, **108**, 6723–6732.
128. D. M. Lockwood, Y. K. Cheng and P. J. Rossky, *Chem. Phys. Lett.*, 2001, **345**, 159–165.
129. C. Narth, N. Gillet, B. Levy, I. Demachy and A. de la Lande, *Can. J. Chem.*, 2013, **91**, 628–636.
130. I. A. Balabin and J. N. Onuchic, *Science*, 2000, **290**, 114–117.
131. J. R. Winkler and H. B. Gray, *Chem. Rev.*, 1992, **92**, 369–379.
132. J. R. Winkler, A. J. Di Bilio, N. A. Farrow, J. H. Richards and H. B. Gray, *Pure Appl. Chem.*, 1999, **71**, 1753–1764.
133. T. R. Prytkova, I. V. Kurnikov and D. N. Beratan, *Science*, 2007, **315**, 622–625.
134. I. A. Balabin, D. N. Beratan and S. S. Skourtis, *Phys. Rev. Lett.*, 2008, 101.
135. V. L. Davidson, *Acc. Chem. Res.*, 2008, **41**, 730–738.
136. Z. X. Liang, J. M. Nocek, K. Huang, R. T. Hayes, I. V. Kurnikov, D. N. Beratan and B. M. Hoffman, *J. Am. Chem. Soc.*, 2002, **124**, 6849–6859.
137. Z. X. Liang, I. V. Kurnikov, J. M. Nocek, A. G. Mauk, D. N. Beratan and B. M. Hoffman, *J. Am. Chem. Soc.*, 2004, **126**, 2785–2798.
138. P. Xiong, J. M. Nocek, J. Vura-Weis, J. V. Lockard, M. R. Wasielewski and B. M. Hoffman, *Science*, 2010, **330**, 1075–1078.
139. E. S. Medvedev and A. A. Stuchebrukhov, *J. Chem. Phys.*, 1997, **107**, 3821–3831.
140. S. S. Skourtis, D. H. Waldeck and D. N. Beratan, *J. Phys. Chem. B*, 2004, **108**, 15511–15518.
141. A. Troisi, M. A. Ratner and M. B. Zimmt, *J. Am. Chem. Soc.*, 2004, **126**, 2215–2224.
142. H. Carias, D. N. Beratan and S. S. Skourtis, *J. Phys. Chem. B*, 2011, **115**, 5510–5518.
143. P. Antoniou, Z. Ma, P. Zhang, D. N. Beratan and S. S. Skourtis, *Phys. Chem. Chem. Phys.*, 2015, **17**, 30854–30866.
144. Z. W. Lin, C. M. Lawrence, D. Q. Xiao, V. V. Kireev, S. S. Skourtis, J. L. Sessler, D. N. Beratan and I. V. Rubtsov, *J. Am. Chem. Soc.*, 2009, **131**, 18060–18062.
145. P. A. Sontz, N. B. Muren and J. K. Barton, *Acc. Chem. Res.*, 2012, **45**, 1792–1800.
146. C. Aubert, M. H. Vos, P. Mathis, A. P. M. Eker and K. Brettel, *Nature*, 2000, **405**, 586–590.

147. C. Shih, A. K. Museth, M. Abrahamsson, A. M. Blanco-Rodriguez, A. J. Di Bilio, J. Sudhamsu, B. R. Crane, K. L. Ronayne, M. Towrie, A. Vlcek, J. H. Richards, J. R. Winkler and H. B. Gray, *Science*, 2008, **320**, 1760–1762.
148. J. J. Warren, M. E. Ener, A. Vlcek, J. R. Winkler and H. B. Gray, *Coord. Chem. Rev.*, 2012, **256**, 2478–2487.
149. B. Giese and S. Eckhardt, *Chimia*, 2013, **67**, 200–203.
150. J. Jortner, M. Bixon, T. Langenbacher and M. E. Michel-Beyerle, *Proc Natl Acad Sci*, 1998, **95**, 12759–12765.
151. Y. Q. Zhang, C. R. Liu, A. Balaeff, S. S. Skourtis and D. N. Beratan, *Proc Natl Acad Sci*, 2014, **111**, 10049–10054.
152. F. Cailliez, P. Müller, T. Firmino, P. Pernot and A. de la Lande, *J. Am. Chem. Soc.*, 2016, **138**, 1904–1915.
153. C. Liu, D. N. Beratan and P. Zhang, *J. Phys. Chem. B*, 2016, **120**, 3624–3633.
154. F. Calegari, D. Ayuso, A. Trabattoni, L. Belshaw, S. De Camillis, S. Anumula, F. Frassetto, L. Poletto, A. Palacios, P. Decleva, J. B. Greenwood, F. Martin and M. Nisoli, *Science*, 2014, **346**, 336–339.

CHAPTER 5

Kinetic Isotope Effects

I. H. WILLIAMS* AND P. B. WILSON

Department of Chemistry, University of Bath, Bath BA2 7AY, UK
*Email: i.h.williams@bath.ac.uk

5.1 Introduction

A kinetic isotope effect (KIE) is a ratio of rate constants deviating from unity for a pair of reactions under identical conditions except for the isotopic composition of the reacting species, and is a consequence of the quantum-mechanical nature of nuclear motion. Historically, the theory of KIEs for reactions of gas-phase molecules was developed within the framework of transition state theory (TST) by the methods of statistical mechanics in terms of partition functions evaluated within the Born–Oppenheimer, harmonic-oscillator, rigid-rotor and ideal-gas approximations.[1,2] Neglecting tunnelling and transmission-factor contributions, the ratio of rate constants k/k' (where the prime denotes the heavier isotopologue) may be expressed as the product of three factors (eqn (5.1)) involving translational, rotational and vibrational partition functions for the isotopologous reactant state (RS) and transition state (TS) species.[3,4] The first is the mass and moments of inertia factor (MMI), where $|I|$ is the determinant of the moment-of-inertia tensor (*i.e.* the product of the three principal moments of inertia) of a non-linear N-atomic molecule of molecular mass M; rotational symmetry factors are omitted here because they are generally equal to unity for applications to enzymatic reactions. The second is the Boltzmann excitational (EXC) factor corresponding to the relative populations of the higher vibrational quantum states, and the third is the zero-point energy (ZPE). EXC involves a product, and ZPE a summation, over $3N-6$ and $3N-7$ real vibrational frequencies of RS and

RSC Theoretical and Computational Chemistry Series No. 9
Simulating Enzyme Reactivity: Computational Methods in Enzyme Catalysis
Edited by Iñaki Tuñón and Vicent Moliner

Published by the Royal Society of Chemistry, www.rsc.org

TS, respectively, where $u = hc\nu/k_BT$ with frequency (s^{-1}) replaced by wavenumber ν (cm^{-1}) and h, k_B, c and T are the Planck and Boltzmann constants, the velocity of light and the absolute temperature.

$$k/k' = \underbrace{\frac{\left[\frac{|I'|}{|I|}\right]_{RS}^{1/2}\left[\frac{M'}{M}\right]_{RS}^{3/2}}{\left[\frac{|I'|}{|I|}\right]_{TS}^{1/2}\left[\frac{M'}{M}\right]_{TS}^{3/2}}}_{\text{MMI}} \; \frac{\prod_i^{3N-6}\left[\frac{[1-\exp(-u_i)]}{[1-\exp(-u_i')]}\right]_{RS}\left[\frac{\exp(u_i/2)}{\exp(u_i'/2)}\right]_{RS}}{\prod_i^{3N-7}\underbrace{\left[\frac{[1-\exp(-u_i)]}{[1-\exp(-u_i')]}\right]_{TS}}_{\text{EXC}}\underbrace{\left[\frac{\exp(u_i/2)}{\exp(u_i'/2)}\right]_{TS}}_{\text{ZPE}}} \tag{5.1}$$

The MMI, EXC and ZPE factors in eqn (5.1) are derived from the partition functions for the molecular species involved. Cancellation of physical constants, volume and temperature common to the translational and rotational partition functions for all four species leads to the MMI factor as a ratio of 'reduced' partition functions for isotopologous species.[1,2] EXC describes the varying population statistics of harmonic vibrational energy levels, by means of vibrational partition functions evaluated with respect to the lowest vibrational energy level, and ZPE is, of course, obtained simply from the differences between that quantised energy level and the potential energy at a classical stationary point. It is common to replace MMI in eqn (5.1) by an equivalent factor containing only vibrational frequencies. According to the Teller–Redlich product rule,[5] the masses and moments of inertia for a pair of isotopologues are related to the vibrational frequencies: the equality expressed by eqn (5.2) assumes separability of translational and rotational motions from vibrational motions within the harmonic approximation. Substituting the vibrational product VP for each of RS and for TS in eqn (5.1), the KIE may be written as eqn (5.3), the Bigeleisen equation, where VPR is the vibrational product ratio.[1,6,7] Two points should be noted: (i) the product of ratios of atomic masses m that appears in eqn (5.2) vanishes from the KIE because it is identical for both RS and TS; (ii) it is conventional to consider the ratio of imaginary transition frequencies for the TS as a separate factor, so that (in eqn (5.3) and (5.4)) the VPR and EXC products and the ZPE summation are all taken over $3N-7$ real TS frequencies.

$$\prod_i^{3N-6}\frac{\nu_i'}{\nu_i} = \text{VP} = \text{MI} = \left[\frac{|I'|}{|I|}\right]^{1/2}\left[\frac{M'}{M}\prod_j^{N}\frac{m_j}{m_j'}\right]^{3/2} \tag{5.2}$$

$$k/k' = (\nu_{\ddagger}/\nu_{\ddagger}')\times\text{VPR}\times\text{EXC}\times\text{ZPE} \tag{5.3}$$

$$\text{VPR} = \prod_i^{3N-6}\left[\frac{\nu_i'}{\nu_i}\right]_{RS}\Big/\prod_i^{3N-7}\left[\frac{\nu_i'}{\nu_i}\right]_{TS} \tag{5.4}$$

Use of the Bigeleisen equation (written more compactly as eqn (5.5)) enables the determination of KIEs from normal-mode frequencies without the need to consider the MMI factor explicitly; this fact has been exploited in several computer codes for KIE calculations.[8–10]

$$\mathrm{KIE} = \frac{\prod_i^{3N-6}\left[\dfrac{\nu_i \sinh\left(\frac{1}{2}u_i\right)}{\nu_i' \sinh\left(\frac{1}{2}u_i'\right)}\right]_{\mathrm{RS}}}{\prod_i^{3N-7}\left[\dfrac{\nu_i \sinh\left(\frac{1}{2}u_i\right)}{\nu_i' \sinh\left(\frac{1}{2}u_i'\right)}\right]_{\mathrm{TS}}\left[\dfrac{\nu_\ddagger'}{\nu_\ddagger}\right]_{\mathrm{TS}}} \tag{5.5}$$

The ratio of imaginary transition frequencies in eqn (5.3) and (5.5) is a temperature-independent factor that arises because motion in the direction of the transition vector is considered to be a classical translation in TS theory. The other factors are purely quantal and their values tend to unity in the limit of very high temperature. The products over real vibrational frequencies in eqn (5.1), which determine the EXC and ZPE factors, correspond to ratios of quantum corrections to the classical vibrational partition functions. It is perfectly logical and consistent to include also the quantum correction to the classical partition function for motion in the TS vibrational mode with the imaginary frequency. Indeed, Bigeleisen did this, following Wigner's approximate treatment,[11] in his original formulation of KIE theory.[1] Application of a quantum correction to every degree of freedom of both RS and TS is equivalent to consideration of both ZPE and tunnelling. Assuming that motion along the reaction coordinate in a TS is separable from the other vibrational degrees of freedom, then use of the simplified form of Bell's treatment for tunnelling through a one-dimensional parabolic barrier[12] gives a particularly elegant expression for the KIE (eqn (5.6)) which involves $3N-6$ harmonic vibrational frequencies for both the RS and TS.

$$\mathrm{KIE} = \frac{\prod_i^{3N-6}\left[\dfrac{\nu_i \sinh\left(\frac{1}{2}u_i\right)}{\nu_i' \sinh\left(\frac{1}{2}u_i'\right)}\right]_{\mathrm{RS}}}{\prod_i^{3N-7}\left[\dfrac{\nu_i \sinh\left(\frac{1}{2}u_i\right)}{\nu_i' \sinh\left(\frac{1}{2}u_i'\right)}\right]_{\mathrm{TS}}\left[\dfrac{\nu_\ddagger' \sin\left(\frac{1}{2}u_i\right)}{\nu_\ddagger \sin\left(\frac{1}{2}u_i'\right)}\right]_{\mathrm{TS}}} \tag{5.6}$$

These equations provide the starting point for discussion of computational simulation of KIEs for enzymic reactions in the rest of this chapter. A completely different formulation of nuclear quantum effects, and thereby of KIEs, is provided by path integral methods, which are the subject of Chapter 10 in this volume and therefore are not discussed here.

5.2 The Cut-off Approximation

Use of the equations above requires separate Hessian matrices for the RS and for the TS in order to obtain harmonic frequencies of vibration, but the same Hessian may be employed with different atomic masses for each member of an isotopologous pair. However, to describe an entire system of enzyme and substrate(s), the large number of atoms necessarily implies very large Hessians for RS and TS. Early applications of the Bigeleisen equation, even to reactions of simple organic molecules, were restricted by the practical limitations of computer memory and processing power available in those times so that it was often necessary to employ truncated molecular models containing fewer atoms.

Furthermore, there existed another very serious problem that, in effect, also limited the size of system that could be treated in KIE computations. The advantage of using the Bigeleisen equation involving only vibrational frequencies is that, in principle, these could be obtained from experimental observations for isotopologous species. In practice, for most molecules of interest, there were insufficient data to allow solution of the problem. Even if observed fundamentals were available, the Bigeleisen equation required harmonic frequencies which typically were not available. Moreover, no experimental frequencies were available at all for TSs!

Much effort in the 1960s (and later) went into vibrational analysis to extract force constants from spectra for stable molecules and, with the assumption of transferability, to construct Hessians for analogous species. The computer programs in use then[13] were implementations of the FG matrix method[14] employing a valence-coordinate representation of the molecule (in terms of bond stretching, angle bending, *etc.*) involving $3N-6$ non-redundant internal coordinates for a non-linear N-atomic molecule. The matrix of force constants was usually constructed to contain only diagonal elements, each corresponding to an individual bond or angle for which a value could be chosen by analogy with similar coordinates in other molecules, with the exception that a small number of off-diagonal elements were included for a TS with values carefully engineered in order to obtain a transition vector with an imaginary frequency.[4]

Wolfsberg and Stern[7] noted that it was possible to leave out parts of a 'large' molecule without significantly affecting the value of a calculated KIE. A 'cut-off' that excluded all atoms more distant from a site of isotopic substitution than three bonds was found to yield isotope effects negligibly different from those obtained from a 'full' model system, and they concluded that a three-bond cut-off was always a justifiable procedure at room temperature and above. Using models comprising only diagonal elements in the valence force constant matrix, non-negligible KIEs were found only in cases with significant changes in force constants for valence coordinates directly involving an isotopically substituted atom as between RS and TS. A two-bond cut-off was usually justifiable, as this would include any force constant changes in both stretching and angle bending coordinates, despite excluding torsional coordinates involving the site of isotopic substitution as a terminal atom defining a dihedral angle. The other requirement for the cut-off procedure to work was that values of force constants retained in the cut-off model must be the same as those for corresponding valence coordinates in the full model. Melander and Saunders discussed the practical usefulness of this simplification, as a means by which to reduce the computational cost of isotope-effect calculations, but added a further wise precautionary note that the cut-off procedure would be valid only if it were justifiable to ignore the influence of medium effects on the isotope effect.[4] In addition to making computations more tractable by reducing the dimension of the force constant matrix **F** in valence coordinates (or of a matrix of non-redundant symmetry force constants constructed from redundant valence force constants), the cut-off procedure reduced the number of coordinates for which force constant values needed to be assigned. In view of the paucity of reliable data, this was a definite benefit.

5.3 The Bebo Vibrational Analysis Method for KIE Calculations

An alternative method for computational solution of the vibrational secular equations was proposed by Gwinn,[15] which simply involved diagonalisation of the mass-weighted Hessian matrix **V** in Cartesian coordinates. The Hessian could still be constructed from valence force constants, as before, but any (immediate) problem with redundant valence coordinates was removed, and the $3N$ eigenvalues of **V** corresponded (in general) to $3N-6$ vibrations and six zero-valued translations and rotations. Although the dimension of the matrix to be diagonalised was larger (by 6) than in the FG method, increases in computational power soon made this an irrelevant point. Essentially all harmonic vibrational frequency calculations are now performed by this method, but in 1978 one eminent scientist still seemed to remain unconvinced of its advantages for large unsymmetrical systems after a robust discussion during the PhD thesis defence of one of the present authors!

Gwinn's method was incorporated into a computer program (QREL)[16] for calculating relative rate constants, including (but not limited to) KIEs, written by Sims in 1975 during a sabbatical in Sheffield, which was the precursor to the well-known BEBOVIB-IV program[17] for KIE calculations (as well as to the present author's CAMVIB and CAMISO programs).[18,19] BEBOVIB utilised Johnston's bond-energy–bond-order method,[20] together with empirical relationships between bond order, bond length and force constants, in order to provide estimates for otherwise unknown geometrical and force constant parameters for TSs.[21,22] The details of the method as applied to calculations of KIEs for enzymic reactions have been described very thoroughly by Berti.[23]

In what appears to have been the first example of KIE calculations for an enzymic reaction, Schowen and coworkers applied the BEBOVIB program to modelling of KIEs for the reaction catalysed by the enzyme catechol *O*-methyltransferase (COMT).[24] Their study was initiated in 1978 and, in view of its seminal importance, it is worth reviewing its key features here. A two-bond cut-off was employed to construct a 17-atom truncated structure

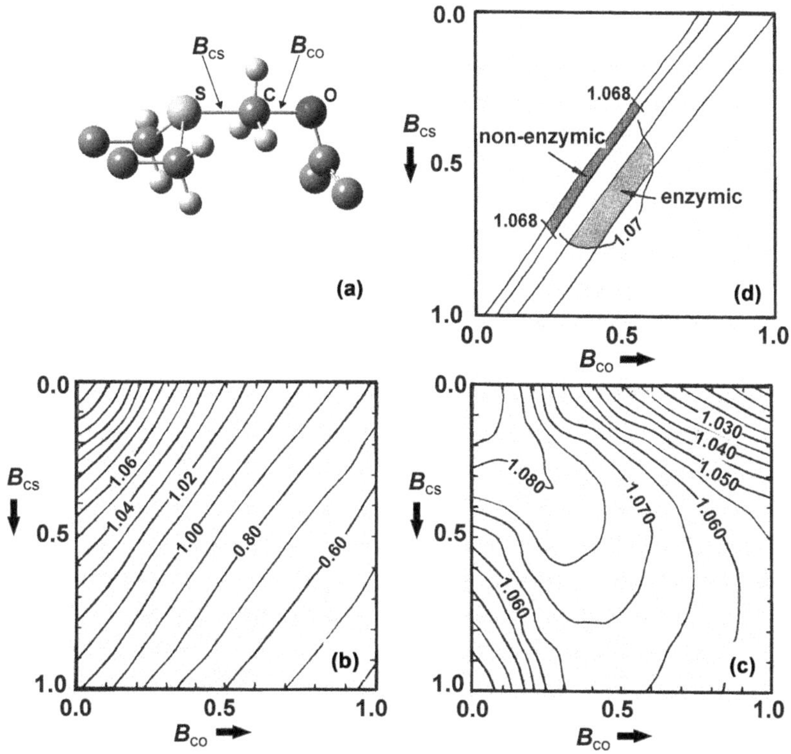

Figure 5.1 (a) Two-bond cut-off model for transition state of methyl transfer between sulfur of AdoMet and oxygen of catecholate used in the BEBOVIB modelling of Rodgers *et al.*[24] (b) Map of calculated 2° $^2H_3$ KIEs as a function of bond orders B_{CO} and B_{CS} to nucleophile and nucleofuge. (c) Similar map of calculated 1° ^{13}C KIEs. (d) Superposition of (b) and (c) showing regions consistent with experimental KIEs for non-enzymic (dark shading) and enzymic (light shading) methyl transfer. Adapted with permission from J. Rodgers, D. A. Femec and R. L. Schowen, *J. Am. Chem. Soc.*, 1982, **104**, 3263–3268.[24] Copyright (1982) American Chemical Society.

(Figure 5.1a) for which RS and TS geometries and valence force constants were derived from literature values for analogous coordinates together with simple relationships based upon the Pauling bond orders B_{CO} and B_{CS} of the making and breaking bonds between the carbon atom of the transferring methyl group and the oxygen and sulfur atoms of the nucleophile and nucleofuge, respectively. In principle the structural space containing the TSs was continuously variable. In practice, using these relations, a 2D grid of TSs was constructed and the primary (1°) ^{13}C, ^{18}O and ^{34}S and secondary (2°) $^2H_3$ KIE was computed as a function of the parameters B_{CO} and B_{CS} at each point on the grid, and a map of contours of constant KIE was plotted for each isotopic substitution (Figure 5.1). Experimental values for the ^{13}C and $^2H_3$ KIEs had been measured and could be compared with the computed maps: consideration of the values of the mean of the experimental KIE, plus and minus one standard deviation, enabled a range of TS structures to be delineated that were consistent with each KIE. Superimposition of the two maps generated smaller ranges of allowed TS structures that were consistent with both the ^{13}C KIE and the $^{2}H_3$ KIE observed for the COMT-catalysed reaction and separately for a model uncatalysed reaction in solution. These results indicated that the TS for the enzyme-catalysed reaction might have partial bond orders B_{CO} and B_{CS} significantly larger than those in the TS for the uncatalysed reaction, which might correspond to these bonds each being about 0.15 Å shorter in the enzymic TS, consistent with the idea that the enzyme's catalytic power might arise from compression of the S_N2 TS.

Amongst enzymologists, the BEBOVIB approach was subsequently taken up by Hermes and Cleland to model the glucose-6-phosphate hydrogenase reaction,[25] and then most notably by Schramm and coworkers in a series of papers over many years, beginning with their study of adenosine monophosphate (AMP) nucleosidase[26,27] and continuing with applications to nucleoside hydrolysis,[28] AMP deaminase,[29] purine nucleoside phosphorylase,[30] NAD^+ hydrolysis[31–33] and ADP ribosylation.[34,35] Uses of BEBOVIB by others include applications to glucoside hydrolysis,[36] orotate phosphoribosyltransferase,[37] oxalate decarboxylase[38] and adenosine nucleosidase.[39]

The isotopic mapping approach of using BEBOVIB in combination with multiple experimental KIE measurements for different isotopic substitutions, pioneered by Schowen and coworkers,[24] was formalised by Berti and Schramm[40] as the 'structure interpolation method'. Manual adjustment of bond order parameters to give a TS structure yielding calculated KIEs that agree with experimental values is not only laborious but also error-prone and unable to predict possible alternative TS structures. A systematic procedure to use BEBOVIB calculations to explore many TS structures within the reaction space was first demonstrated by Sims and coworkers for a non-enzymic reaction,[41] but Berti and Schramm's protocol[40] used either X-ray crystal structures or QM optimised structures for RS and TS, and interpolations in all the internal coordinates are made without recourse to the empirical relationships used in the early work. An alternative for BEBO-style vibrational analysis and isotope effect calculations was Huskey's VIBIE program.[42]

Although when used carefully within a systematic procedure, the BEBOVIB method was capable of providing valuable insight concerning TS structure leading, for example, to successful design of TS-analogue enzyme inhibitors as potential drugs,[43,44] the fundamental lack of data for TS geometries and force constants was always a concern, to some at least.

5.4 QM Cluster Calculations of KIEs

5.4.1 Early Examples

As developments in semi-empirical molecular orbital (MO) theory began to allow for calculations of vibrational frequencies for TSs, and thereby of

KIEs,[45,46] so applications to enzymic reactions began to appear. At first, even for modelling of non-enzymic reactions, it was often necessary to use a truncated version of the molecular system in order to reduce the number of atoms (and electrons) in the calculations. Certainly it was not possible to include any atoms of the protein environment in a model KIE calculation for an enzymic reaction, meaning that the cut-off approximation was being used implicitly. This fact also made the distinction between calculations for enzymic or non-enzymic reactions rather meaningless! Nonetheless, arguably the first paper to report KIE calculations for a QM model of an enzymic reaction would appear to be the MNDO study by Ford and coworkers on cytochrome P450 oxidations of aliphatic substrates,[47] which was followed by similar studies of H-atom abstraction by Loew and coworkers for other substrates.[48]

These calculations used the Bigeleisen equation (eqn (5.5)) with the rigid-rotor-harmonic-oscillator approximation and neglected tunnelling effects. However, some time ago one of us pointed out[19] that there was some practical merit in separately determining the VP and MI terms for a pair of isotopologues (eqn (5.2)), rather than relying upon their equality, particularly when the vibrational frequencies had been obtained using Hessians in Cartesian coordinates from standard packages for *ab initio* or semi-empirical MO calculations. Older implementations involving numerical second derivatives tended to produce Hessians which, upon mass weighting and diagonalisation,[15] yielded eigenvectors corresponding to translation and rotation of the whole molecule which did not have zero eigenvalues: this meant that the Teller–Redlich product rule was not exactly satisfied, and consequently that application of the Bigeleisen equation incurred some error. Model calculations of KIEs for enzymic methyl transfer[49] (*ab initio* Hartree–Fock, HF) and carbonyl reduction by lactate dehydrogenase[50] (semi-empirical AM1) both used explicit evaluation of the partition functions for translation, rotation and vibration after a projection to remove spurious translational and rotational components from the Hessian and to obtain six properly zero eigenvalues.[19]

5.4.2 Dehydrogenases

An interest in KIE calculations for enzymic hydride-transfer reactions was initiated by the work of Jones and Urbauer on the identity hydride transfer from formate anion to carbon dioxide using a range of semi-empirical and *ab initio* HF methods.[51] No solvent was included because it could not be evaluated at the time, and the authors commented that "since formate dehydrogenase is believed to act by desolvation of the substrate the gas phase should be a good model of the reaction in the enzyme active site provided specific amino acid interactions in the active site are small". In making comparison between experimental (V_{max}/K_m) and calculated 1° ^{2}H, ^{13}C and ^{18}O KIEs (including a Wigner tunnelling correction) for the formate moiety, they did not include the mass component of the MMI factor, but noted that both this and the inertial component (*i.e.* the translational and the rotational partition function contributions) might be affected by binding in a Michaelis complex. This was a prescient observation but it was not possible then to provide better estimates for the isotopic sensitivities of entropy changes in condensed-phase systems. Finally, Jones and Urbauer noted that a QM calculation yielded only a single KIE for isotopic substitution in a reaction whereas BEBOVIB was capable of allowing TS variations to be explored. Nonetheless they did suggest that the QM approach "has the advantage of readily providing quality force fields, especially for larger systems, which otherwise would have to be fabricated from standard values or guesswork and trial and error".[51]

The QM cluster approach was used by Andrés *et al.*[52] to estimate 1° and 2° ^{2}H KIEs for the hydride-transfer step in lactate dehydrogenase (LDH), which catalyses the reduction of pyruvate to lactate. Their model included the pyruvate substrate, the nicotinamide ring of the hydride-donor NADH cofactor, the imidazolium ring of the proton-donor histidine residue, and the guanidinium moiety of the carboxylate-binding arginine residue in the active site, a total of 44 atoms. Despite the lack of any further components of the protein or solvent environment, the TS structures located using several semi-empirical MO methods were consistent with what was known about other dehydrogenases. The RS structures were located by following the intrinsic reaction coordinate down from each TS to the adjacent energy minimum. The semiclassical KIEs (excluding tunnelling) for substitution of either the transferring (1°) or the non-transferring (2°) hydrogen at the C4 position of the nicotinamide ring were of plausible magnitudes as compared with experimental KIEs for another dehydrogenase (glutamate dehydrogenase). Note that since the calculated RS for the reaction was a model for the Michaelis complex within the enzyme active site, the calculated values presumably corresponded to KIEs on V_{max}. Hydride transfer is not the rate-determining step for the reaction catalysed by wild-type LDH, so the intrinsic KIEs are not immediately accessible experimentally.

Andrés and coworkers[53] applied a similar approach to calculate 1° ^{2}H and ^{3}H KIEs for hydride transfer from NADH to flavin adenine dinucleotide (FAD), using AM1 with a 45-atom truncated model and HF/6-31G for a 23-atom model. The equation they seem to have used to calculate the semiclassical KIE appears to include only the ZPE factor and neglects the ratio of TS imaginary frequencies. Reasonable agreement with experimental isotope effects on V_{max}/K_m was apparently obtained but, as before, the RS in the calculations refers to the active-site complex immediately preceding the TS.

5.4.3 Binding Isotope Effects and Software

Oxamate is an inhibitor of LDH, and its ^{18}O equilibrium isotope effect (EIE) for binding to LDH from aqueous solution was measured experimentally by Paneth and coworkers and computed by a variety of QM cluster methods.[54]

This very small inverse EIE (0.9840 ± 0.0027) reflects formation of a hydrogen bond to the carboxylate O atom in the enzyme active site. Using force constants computed for 'full' models of oxamate in explicit water (22 molecules within a 5 Å radius) and oxamate in LDH (132 atoms including the side chains of four amino acid residues and the nicotinamide moiety of NADH besides oxamate) with the AM1 semi-empirical method gave a calculated EIE of 0.9860, in very good agreement with experiment. However, using force constants computed for 'cut' models of oxamate only within the frozen environment of the active site or solvent resulted in the calculated EIE being very much closer to unity (0.9988). This indicated the necessity of including more atoms in the Hessian evaluation than those of the inhibitor (or substrate) alone. Using some continuum solvation models to describe the aqueous solvation of oxamate gave good results in conjunction with the full model for the enzyme active site, but others did not, and the PM3 semi-empirical method performed badly by predicting the wrong direction for the EIE.

These calculations employed a version of Paneth's ISOEFF program[54] which enabled structures and Hessians to be input directly from several quantum chemistry packages (MOPAC/AMPAC, AMSOL, SIBIQ, GAMESS and GAUSSIAN) and computed KIEs by means of the Bigeleisen equation (eqn (5.5)) and EIEs by a simple modification involving replacement of the TS terms by 'PS' (product state) terms involving $3N-6$ frequencies, just as for RS. (Our CAMVIB/CAMISO programs[18,19] performed essentially the same function.) An upgraded program, ISOEFF98,[55] served to fill the gap between BEBOVIB and QM cluster calculations such that the user could modify selected QM force constants to see how these modifications affected the value of an isotope effect. First, the Hessian in Cartesian coordinates was transformed into (non-redundant) internal coordinates, then diagonal elements of the resulting force constant matrix were changed by hand while leaving the off-diagonal elements unchanged, and finally the back-transformation to Cartesians was performed before obtaining new vibrational frequencies from the modified Hessian. This new functionality was described by Anisimov and Paneth as 'quantum-BEBOVIB'.[55] The other widely used program for KIE calculations, QUIVER, described by Saunders *et al.*,[56] took Cartesian force constant input from GAUSSIAN and computed the reduced isotopic partition function ratio for a pair of isotopologues by means of the Bigeleisen equation.

5.4.4 Glycosyl Transfer

A QM cluster calculation for a truncated model for adenosine monophosphate hydrolysis by either the S_N1 or the S_N2 mechanisms in order was performed to compare the AM1 method against BEBOVIB for 2° α-^{3}H and β-^{2}H and 1° ^{15}N and ^{13}C KIEs, and concluded that BEBOVIB-derived TS structures were "certainly no worse" than those directly calculated by AM1.[57]

EIEs for a truncated model of cytidine monophosphate *N*-acetylneuraminate (CMP-NeuAc) hydrolysis were calculated by Horenstein using QUIVER using HF/6-31G* structures and force constants in vacuum and in continuum solvent (self-consistent isodensity polarised continuum model) were compared with experimental values for the 1° α-^{14}C and 2° carboxylate ^{14}C and β-$^{2}H_2$ KIEs.[58] Good agreement was found for formation of a planar sialyl oxocarbenium ion. Bruner and Horenstein also used the same method to calculate EIEs for conversion of a CMP-NeuAc model to species representing different active site interactions involving the carboxylate group in the sialyl glycoside to aid interpretation of experimental KIEs for the reaction catalysed by α-(2→6)-sialyltransferase.[59]

The principle of using calculated EIEs to aid interpretation of experimental KIEs had been proposed in earlier studies of glycoside hydrolysis.[57,60] It was also employed by Chen *et al.*[61] in a study of KIEs for hydrolysis of an adenosine group (adenine depurination) within stem-loop RNA catalysed by the A-chain of ricin, in which EIEs were calculated from fractionation factors (*i.e.* isotopic partition function ratios) for RS and PS determined with a density-functional method by means of QUIVER and TS structures were interpolated between these using VIBIE. A similar method was used by the same authors for the analogous depurination of DNA,[62] except that TS structures were directly determined on the energy surface for a 20-atom truncated model in the gas phase. Comparison between experimental KIEs (1° ^{14}C and ^{15}N, 2° α-^{3}H, β-$^{3}H_{(S)}$ and β-$^{3}H_{(R)}$) and values calculated for alternative stepwise mechanisms, in which either formation or breakdown of an enzyme•deoxy-ribooxocarbenium ion complex as a discrete intermediate was rate-determining, was unable to distinguish the two possibilities in view of some doubt about the reliability of the 2° β-^{3}H calculated values. The calculated 2° α-^{3}H KIE for each alternative TS was larger than any experimental KIE for this reaction. Although the source of the computational discrepancy was not known, it was suggested that it did not influence the TS determination for KIEs at other positions "because it is possible to adjust the bending force constants" involving this α-H by a BEBO-style method in order to bring the errant KIEs into line without changing other features of the TS structure.[62] A gas-phase DFT study for a 42-atom truncated model for adenosine diphosphate ribosylation calculated KIEs by means of ISOEFF98, involving a different BEBO-style method for force constant interpolation between directly calculated TS structures.[63] Similarly, TSs were determined for (gas-phase) reaction of thymidine with dianionic phosphate by means of saddle point optimisations with a pair of constrained valence coordinates (bond lengths from the anomeric C atom to the nucleofuge N atom and to the nucleophile O atom) and calculated KIEs compared with experimental values for eight different isotopic substitutions in the reaction catalysed by thymidine phosphorylase.[64,65] The TS obtained by matching calculated KIEs with experiment by means of force constant modifications with ISOEFF98 possessed three imaginary frequencies (instead of one), but how the extra imaginaries were treated in the KIE calculations was not explained.[64] Similar methods have been used by Lewandowicz and Schramm[66] and Luo *et al.*[67] to study purine nucleoside phosphorylase and adenosine deaminase.[68] Their 2008 paper makes it clear that KIEs are calculated for TSs whose ISOEFF98-modified Hessians yielded an imaginary frequency $\geq 50i\,\text{cm}^{-1}$, but EIEs were calculated for analogous 'TS' structures with no or $<50i\,\text{cm}^{-1}$ imaginary frequency.[67]

The interpolation procedures used so successfully by Schramm and his collaborators, first with BEBOVIB and later by means of ISOEFF98-modified QM force constants, involves calculation of KIEs for structures that do not correspond to stationary points on a potential energy surface. This lays the theoretical rigour of the method open to question, because frequencies are not reliable at non-stationary points. Hirschi *et al.*[69] performed a critical assessment of Schramm's methodology by comparing C–O bond distances in TSs for oxaziridine epoxidation (gas phase, B3LYP/6-31G*) as predicted by interpolation from calculated non-stationary structures with those directly calculated at stationary points. The deviations between the two sets of bond lengths provided an estimate of the error induced by the Schramm process. In view of the small average deviations, it was judged that "the Schramm process is strikingly successful".

Rios-Font *et al.*[70] applied a gas-phase DFT method to compute KIEs for hydrolysis of 2′-deoxyguanosine, yielding 1° α-^{14}C and ^{15}N and 2° β-^{2}H values in the expected ranges, but no 2° α-^{3}H KIE was reported. McCann and Berti[71] performed DFT calculations for hydrolysis of deoxyadenosine monophosphate (dAMP) proceeding by A_ND_N, $D_N^{\ddagger}*A_N$, $D_N*A_N^{\ddagger}$ and D_N+A_N mechanisms (Figure 5.2). After taking into account 20% of D_N+A_N, there was an excellent match of calculated to experimental KIEs (1° 1′-^{14}C and 9-^{15}N; 2° 6-^{15}N, 7-^{15}N, α-1′-^{3}H, β-2′*S*-^{2}H, β-2′*R*-^{2}H and δ-5′,5′-$^{3}H_2$) for 80% of the reaction having a $D_N*A_N^{\ddagger}$ mechanism; C–N bond cleavage is reversible, with dAMP and the adenine oxacarbenium ion complex in equilibrium, and the first irreversible step being water attack on the oxacarbenium ion. They pointed out that stepwise D_N*A_N reactions have two TSs, and the observable KIEs are a function of partitioning of the oxacarbenium ion intermediate in the forward ($k5$) and reverse ($k4$) directions, as given by eqn (5.7)–(5.9) where kie_n is the intrinsic KIE on step n.[62,71]

$$\text{KIE} = \frac{\dfrac{\text{kie1 kie3}}{\text{kie2}}\left(\dfrac{\text{kie5}}{\text{kie4}} + \dfrac{k5}{k4}\right)}{1+\dfrac{k5}{k4}} \tag{5.7}$$

$$k_5 \gg k_4 \Rightarrow \text{KIE} \approx \frac{\text{kie}_1\text{kie}_3}{\text{kie}_2} \tag{5.8}$$

$$k_5 \ll k_4 \Rightarrow \text{KIE} \approx \frac{\text{kie}_1\text{kie}_3\text{kie}_5}{\text{kie}_2\text{kie}_4} \tag{5.9}$$

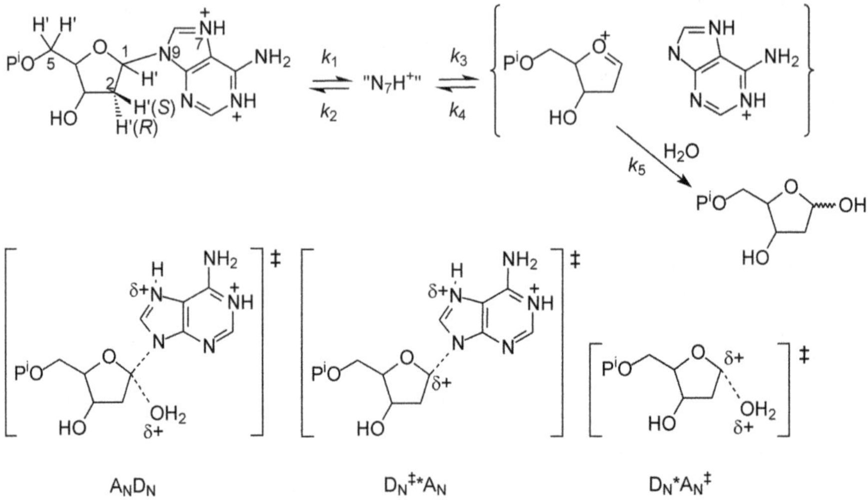

Figure 5.2 Scheme for hydrolysis of deoxyadenosine monophosphate showing TSs for alternative concerted and stepwise mechanisms discussed by McCann and Berti,[71] and rate constants involved in the relationship between observed KIE and intrinsic KIEs on individual steps.
Adapted with permission from J. A. B. McCann and P. J. Berti, *J. Am. Chem. Soc.*, 2007, **129**, 7055–7064.[71] Copyright (2007) American Chemical Society.

McCann and Berti[71] also noted that the calculated α-^{3}H KIEs were somewhat higher than the experimental value and were sensitive to the position of the water nucleophile, commenting that this sensitivity to the environment was consistent with that previously encountered, causing difficulties in the interpretation of 2° α-hydron KIEs. Using the same methods to calculate KIEs for the corresponding reaction in DNA,[72] these authors remarked, with characteristic lucidity, that there was no guarantee that TS structures directly determined by QM computation on a model reaction that did not include the protein environment would match the TS of the real enzymic reaction: "However, because vibrational frequencies from hybrid DFT calculations accurately reflect molecular structure, incorrect computational TS structures would have yielded computational KIEs that did not match experimental values. In that case, bond order vibrational analysis would have been needed to interpret the KIEs." The fact that calculated and experimental KIEs actually did agree for this reaction was good evidence that the computed TS did match the true TS.[72]

Bennet and coworkers used a DFT method to calculate KIEs for comparison with experimental V_{max}/K_m values for isotopic substitutions in the 1° (anomeric) α-^{13}C and nucleofuge ^{18}O, and 2° ring ^{18}O and β-^{2}H$_{(S)}$ and β-^{2}H$_{(R)}$ positions of a truncated model for a sialyl glycoside substrate of *Vibrio cholerae* sialidase.[73] The RS corresponded to the free substrate in solution whereas the TS included the truncated side chains of active-site aspartate, glutamate and tyrosine residues. By means of constraints imposed on the bond lengths between the anomeric C atom and the nucleofuge and nucleophile O atoms, a series of TS structures were generated. Best agreement between calculated and experimental KIEs was obtained for a TS corresponding to a dissociative $D_N^{\ddagger}{*}A_N$ mechanism. However, good agreement could not be found for the β-^{2}H KIEs, but it was decided to give priority to matching the heavy-atom KIEs until it became computationally feasible to include more of the critical active-site residues, together with the enzyme framework that mediates catalysis, within a QM/MM calculation.

Roitberg and coworkers performed B3LYP/6-31G(d) KIE calculations with ISOEFF98 for a sialyllactose substrate together with either hydronium ion in PCM water or else the side chains of catalytically active aspartate, glutamate and tyrosine residues in the active site of the *trans*-sialidase from *Trypanosoma cruzi* to model, respectively, the acid-catalysed and enzymic reactions.[74] The 1° (anomeric) α-^{13}C and 2° β-$^2H_2$ KIEs computed for a mechanism involving formation of a covalent intermediate to the tyrosine residue were found to be in accord with experimental results.

5.4.5 Other Enzymes

Götze and Saalfrank[75] considered a 77-atom cluster model for the substrate and active site of ribulose-1,5-bisphosphate carboxylase/oxygenase (RuBisCO), using the B3LYP/6-31G(d,p) method with the polarisable continuum model (PCM) and dielectric constant $\varepsilon = 4.71$ to roughly resemble the protein environment beyond those amino acid residues which were explicitly included in the QM model. A Wigner tunnelling correction was included in their calculation of the 1° ^{13}C KIE for the C atom of CO_2 which yielded quite good agreement with experiment.

Manta *et al.*[76] used a 145-atom cluster model for a zinc-dependent cytosine deaminase to perform B3LYP calculations with the conductor-like polarisable continuum model (CPCM) and dielectric constant $\varepsilon = 4$. They estimated the solvent deuterium KIE by consideration of only ZPE changes between an intermediate structure and its adjacent TS each containing two water molecules: the calculated KIE of 0.74 was considered to be in good agreement with the experimental isotope effect on k_{cat} of 0.62.

Hydrolysis of the enol ether moiety of phosphoenolpyruvate catalysed by acid and by enolpyruvylshikimate-3-phosphate synthase ('AroA') has been treated at the B3PW91/6-31+G** level of theory with CPCM and the dielectric constant of water to obtain constrained TS structures, EIEs and KIEs (using partition functions derived from QUIVER).[77] The cluster contained 147 atoms from the seven residue within 4 Å of the carboxyvinyl moiety of EPSP plus six other key residues, although additional measures were taken to reduce the size of the system. Although none of the computed TS structures gave a clear match with the experimental KIEs for seven isotopic substitutions, bond order-based relationships did allow their interpretation, and the cluster model indicated that of the AroA active-site environment promotes an earlier TS than for the non-enzymic reaction.

York and coworkers found that the B3LYP/6-31++G(d,p) method with PCM/UAKS solvation (dielectric constant for water) gave the most consistent agreement with experimental ^{18}O KIEs for a benchmark series of phosphate ester hydrolyses, and this method was then applied to predict KIEs in a model for transphosphorylation catalysed by ribonuclease A.[78]

5.5 QM/MM Calculations of KIEs

5.5.1 Early Examples

Possibly the first example of a hybrid QM/MM method being applied to calculate KIEs was the study of alternative stepwise D_N*A_N pre-association and concerted A_ND_N mechanisms for acid-catalysed AMP hydrolysis.[79] A 23-atom model for AMP and a nucleophilic water were described by AM1, and were surrounded by a sphere of about 500 TIP3P water molecules. A 2D grid search with respect to the bond distances from the anomeric C atom to the nucleofuge N and nucleophile O atoms generated an energy surface showing both pathways. Using a Hessian computed for the QM atoms only, the calculated 1° (anomeric) α-^{14}C and nucleofuge ^{15}N, and 2° β-^{2}H KIEs obtained from TS structures for both mechanisms agreed with each other and with the experimental values and could not serve to discriminate between the mechanisms. However, the calculated value for α-^{3}H was in excellent agreement with experiment for the stepwise TS but not the concerted TS. This result was in accord with Schramm's BEBOVIB-based interpretation.

The first application to an enzymic reaction was probably the AM1/CHARMM study of *p*-nitrophenyl-*N*-acetylneuraminide hydrolysis catalysed by influenza B and *Salmonella typhimurium* sialidases.[80] This work employed a QM region of 65 atoms (including a nucleophilic water, truncated substrate and side chains of active-site aspartate, glutamate and tyrosine residues) and an MM region of about 1500 surrounding atoms within a stochastic boundary. Approximate RS and TS structures were located by a 2D grid search (as for AMP hydrolysis) and 1° nucleofuge ^{18}O and 2° β-$^{2}H_{(R)}$, β-$^{2}H_{(S)}$ and β-$^{2}H_2$ KIEs were computed. It was commented that comparison of calculated KIEs with experiment provided a very stern test, and their use provided "a very strong anchor to prevent theoretical modelling from drifting into unreality". The less-than-wholly-impressive agreement of calculated and experimental KIEs (for influenza A sialidase) led Barnes and Williams to suggest that their QM/MM modelling "had not yet passed this test" and to speculate that the procedure employed for optimising and characterising stationary points might have been at fault.[80]

5.5.2 Hydride and Hydron Transfer

A more rigorous procedure was developed as a suite of utilities (GRACE) for optimisation of minima and saddle points, IRC paths and vibrational frequencies in large flexible QM/MM systems, which enabled TSs to be refined using the same criteria as were normally used in QM applications with small molecules, and this methodology was applied to LDH.[81] A Hessian was computed for a 'core' of 147 atoms comprising 35 QM (AM1) atoms and 112 MM (CHARMM) atoms within the active site; the QM region was embedded mechanically and electronically within the MM 'environment' of about 1900 atoms, and was a subset of the core atoms included in the Hessian. Although the residual force on every core atom was tolerably close to zero within the full QM/MM system, the core was not itself a stationary structure in isolation. Transformation from Cartesians to non-redundant valence coordinates, followed by back-transformation, served to project out the components of translation and rotation of the core as a rigid entity with respect to its environment,[18,19] equivalent to treating the core atoms as gas phase cluster with force constants determined within the protein environment. The average value of the 1° ^{2}H for hydride transfer from NADH to pyruvate, taken over six different TS configurations within the flexible enzyme active site, was 2.73 ± 0.06 at 298 K, which compared well with experimental values for mutant LDHs with rate-limiting chemistry as opposed to a conformational change as in the wild type.[81] Castillo *et al.*[82] subsequently applied the same method to the hydride-transfer step in dihydrofolate reductase.

The proton transfer reaction involved the conversion of 2-phospho-D-glycerate to phosphoenolpyruvate catalysed by yeast enolase was studied by Alhambra *et al.*[83] using a AM1/CHARMM hybrid method with a 25-atom QM region surrounded by 8863 MM atoms. Their calculation of the semiclassical 1° ^{2}H KIE was very similar to that used for LDH,[81] except that the generalised hybrid orbital method was used to define the QM/MM boundary instead of the link-atom method; the (conventional) TS Hessian was determined for the $N_1 = 25$ QM atoms at the saddle point in what they termed as the primary zone, with vibrational partition functions being evaluated for $3N_1 - 1$ degrees of freedom. However, quantum effects along the reaction coordinate were treated by the microcanonical optimised multidimensional tunnelling approximation with small-curvature tunnelling (SCT), which also includes non-classical reflection by the reaction barrier: this gave a KIE of 3.5 at 300 K as compared to the experimental value of 3.3. A separate determination of the quantal corrections within the primary zone was required for each isotopic substitution.

A similar treatment of hydride transfer from a benzyl alcoholate substrate to NAD$^+$ catalysed by liver alcohol dehydrogenase (LADH) was also reported by Alhambra *et al.*,[84] but ^{2}H and ^{3}H 1° and 2° KIEs were determined as averages over 20 configurations of the 21-atom (mobile) primary zone within the 5539-atom (frozen) secondary zone. The merits of the multidimensional treatment of tunnelling were clearly stated in comparison with the one-dimensional Bell correction applied by Rucker and Klinman in a BEBOVIB analysis of a truncated model for LADH.[85] In order to obtain satisfactory agreement between calculation and experiment for the 2° ^{3}H KIE in the 1D parabolic method, the tunnelling correction was overestimated, whereas the

63-dimensional non-parabolic method reproduces this KIE and the Swain–Schaad exponent in agreement with experiment. Another paper on LADH by the same group made further refinements to the method and noted that the quantum effects in the primary zone were included "only on the $3N_1 - 7$ highest-frequency vibrational modes".[86] The transmission coefficient was now evaluated within a relaxed secondary zone, and the procedure allowed for Boltzmann averaging over an ensemble of RS and TS conformations.

Cui *et al.*[87] used related methodology, based upon a QM/MM method using the self-consistent-charge-density-functional-tight-binding (SCC-DFTB) method with CHARMM, with a much larger number of mobile atoms (278) in the secondary zone but with no averaging over different enzyme structures; these authors obtained similar KIE results to those of Alhambra *et al.*[84] In contrast, Ferrer *et al.*[88] noted that reactions taking place within separate valleys, with distinctly different reactant conformations with the active site of LDH, gave different 1° ^{2}H KIEs as evaluated by a slightly modified version of the ensemble-averaged variational TS theory (EA-VTST) with multidimensional tunnelling contributions. Tresadern *et al.*[89] also used a similar approach to calculate 1° ^{2}H KIEs for proton transfer from a methylamine substrate catalysed by methylamine dehydrogenase (MADH) for a range of enzyme conformations; large tunnelling corrections were predicted for proton transfer with both semi-empirical and *ab initio* QM/MM methods. Ranaghan *et al.*[90] performed similar calculations for MADH, using PM3 (with specific reaction parameters) and CHARMM together with EA-VTST/SCT, and averaged their 1° ^{2}H KIEs over different minimum-energy paths. They found that the protein configuration used had a significant effect on the computed KIE, with average values of 14 or 41 being obtained for proton transfer to one of the O atoms of an aspartate carboxylate side chain for two ensembles of structures taken from different regions of the reaction coordinate. These authors also applied the same methodology to calculate KIEs for oxidation of tryptamine by aromatic amine dehydrogenase.[91]

The hydrogen-atom transfer catalysed by coenzyme B_{12}-dependent methylmalonyl-CoA mutase was studied by Dybala-Defratyka *et al.*[92] using AM1/CHARMM and EA-VTST with averaging over eight reaction valleys descending from points selected from the TS ensemble. The importance of corner-cutting tunnelling paths in order to obtain a 1° ^{2}H KIE that agreed with experiment within experimental error was demonstrated. The 45-atom QM region was described by an unrestricted Hartree–Fock implementation of the semi-empirical MO method within the CHARMM program, and the dynamics calculations employed CHARMMRATE[93] with a 45-atom primary zone. Agreement with experiment could not be obtained from smaller QM cluster models that included part or all of the substrate and cofactor moieties but excluded active-site protein residues.[94] Without the enzyme, a larger KIE of about 100 suggested the existence of a higher commitment to catalysis, and consequently led to the conclusion that the intrinsic KIE was masked by the kinetic complexity, whereas with enzyme residues included in the QM/MM model, the good agreement between calculation and experiment was consistent with a minimal commitment.

Note that alternative theoretical treatments of nuclear quantum effects in LADH and related enzymes are discussed thoroughly in Chapter 10 of this volume.

5.5.3 Chorismate Mutase

Chorismate mutase has been to QM/MM studies of enzyme catalysis in the 2000s what the water molecule was to electronic structure theory in the 1960s: a very convenient benchmark. Since it catalyses the pericyclic rearrangement of chorismate to prephenate, including these substrate atoms in the QM region, while treating the whole protein within the MM region, means that the QM/MM boundary does not cut through any covalent bonds. A 2003 study by Martí *et al.*[95] used an AM1/CHARMM24/TIP3P hybrid method to locate any characteristic RS and TS structures with full gradient relaxation of >5200 atoms (out of a total of >17 000 MM atoms) for the enzymic simulations, and with a box containing 711 water molecules for the corresponding reaction in aqueous solution. Inclusion of the chorismate conformational pre-equilibrium step in the modelled kinetic scheme led to better agreement between calculated and experimental 1° ^{13}C and ^{18}O KIEs. Good agreement for the 2° 5-^{3}H and 9-$^{3}H_2$ KIEs in aqueous solution was obtained provided that the Hessian water molecules hydrogen-bonded to the substrate, but poorer agreement for the enzymic reaction may have been due to the available experimental V_{max}/K_m KIEs being for chorismate mutase from a different organism to that for the available X-ray structure used as the starting point for the calculations. Worthington *et al.*[96] calculated KIEs for the rearrangement step, but not the conformational pre-equilibrium, using an *ab initio* HF treatment of the substrate with effective fragment potentials to represent the protein environment. The same system was later studied using 'high-level' (DFT) corrections that were applied to the 'low-level' (AM1) QM subsystem during QM/MM evaluations of gradients and Hessians.[97] This procedure arguably gave more reliable estimates of the intrinsic KIEs for the rearrangement, and discrepancies with experiment may have been due to the rearrangement occurring after the rate-limiting step for the enzymic reaction.

Ruggiero *et al.*[98] used their original AM1/CHARMM24/TIP3P hybrid method to investigate the sensitivity of calculated KIEs to the size of the Hessian for the chorismate rearrangement in water and in the active site of chorismate mutase. As the system was expanded to include not only the atoms of chorismate/prephenate itself but also an increasing number of surrounding water molecules (up to 99) or active-site residues (up to 225 atoms), 1° ^{13}C and ^{18}O KIEs were not sensitive to the size of the Hessian, but 2° 5-^{3}H and 9-$^{3}H_2$ KIEs required the inclusion of at least those atoms directly involved in hydrogen bonds to the substrate or, better, a complete first solvation shell or cage of active-site amino acid residues.

5.5.4 Methyl Transfer

The compression hypothesis for the origin of catalysis in COMT (and other group-transfer enzymes) has had both its supporters and detractors over the years since the pioneering BEBOVIB studies of Schowen and coworkers. A hybrid AM1/TIP3P/CHARMM calculation of 2° α-$^2H_3$ KIEs for the reaction of *S*-adenosylmethionine (AdoMet) with catecholate anion in water and catalysed by COMT gave calculated KIEs in agreement with experiment, with a significantly more inverse KIE for the enzymic reaction than for the one in aqueous solution.[99] However, this result did not support the compression hypothesis because the sum of the bond lengths from the α-C atom to the S atom of the nucleofuge and to the O atom of the nucleophile was essentially the same in the COMT active site as it was in water; in other words, there was no evidence for geometrical compression. The point was further reinforced by a subsequent study in which the KIEs were evaluated as averages over RS and TS structures taken from QM/MM molecular dynamics (MD) trajectories.[100] The QM region comprised 63 atoms (catecholate + AdoMet) and the full QM/MM system for the MD simulations contained 17 162 atoms including 4614 water molecules; the flexible region for local relaxation of 'snapshot' structures (to either a local minimum or saddle point) contained 2610 atoms; and the Hessian included the QM atoms and the active-site Mg^{2+} cation (64 atoms). The uncatalysed methyl transfer in solution was the same as that studied experimentally in the original work by Schowen's group, namely reaction of methoxide anion with *S*-methyldibenzothiophenium cation in methanol; the QM region comprised 30 atoms and the Hessian included at least a complete first solvation shell of solvent molecules around the solute. MD trajectories of 1 ns duration were performed in RS and TS windows from a potential of mean force (PMF) along the reaction coordinate and 100 snapshots were taken at 10 ps intervals in order to ensure that all structures were independent, and all simulations were performed at the temperatures of the experimental studies. The mean value of the calculated KIE for the enzymic reaction, averaged over 10 000 combinations of RS and TS (0.82 ± 0.05) agreed well with the experimental value (0.83 ± 0.05), but it was noted that the standard deviation for the calculated KIE reflected the uncertainty of averaging over precisely computed individual TS-theoretical rate constant ratios for a limited sample of 100 RS and 100 TS configurations spanning a distribution of structures determined by the model potential, whereas the experimental errors arose from uncertainties in rate constant determinations for (effectively) complete sampling of all thermally accessible RS and TS configurations that may also span a distribution of structures. It was suggested that averaging over thermal fluctuations of the environment allowed for meaningful comparison of computed with experimental KIE values. Furthermore, the authors commented that apparent differences between KIEs calculated by different QM methods were often smaller than the fluctuations found between different configurations within the same method, with the implication that computational studies based upon a single conformation of an enzyme might be unrepresentative.[100]

Analysis of valence force constants computed in the 2004 QM/MM study for the COMT-catalysed methyl transfer showed that the methyl CH stretching force constants were larger in TS than in RS, as expected.[99] What was not expected, however, was that the HCS and HCO bending force constants, involving these bonds and either the nucleofuge S atom or the nucleophile O atom, were also significantly larger in TS than in RS, both in the COMT active site and in solution. These comparisons were made using relaxed force constants (reciprocals of diagonal elements of the compliance matrix in valence coordinates), which have the property of being independent of the choice of redundant internal coordinates.[18] This finding shed additional light on the compression hypothesis. The TS bending force constants used in the original BEBOVIB modelling were obtained by use of a rule that took the product of the RS force constant (for HCS) or the PS force constant (for HCO) and the appropriate TS bond order (B_{CS} or B_{CO}) which always had a value <1; the TS bending force constants were, therefore, always less than the corresponding RS (or PS) value. In order to calculate an inverse 2° α-$^2H_3$ KIE to match that of the enzymic reaction, it was necessary to use larger values for B_{CS} or B_{CO} than for the uncatalysed reaction. The apparently shorter bond lengths for the C...S and C...O partial bonds in the TS were probably an artefact of the arbitrary rule used to generate TS bending force constants within BEBOVIB.

5.5.5 Other Enzymes

Kwiecień *et al.*[101] studied the cytochrome-P450 2A6-catalysed hydroxylation of nicotine, at either of two positions, using B3LYP with the 6-31G(d) basis (LACVP for Fe) in combination with the AMBER force field in an ONIOM method with mechanical embedding. The approximately 70-atom QM region consisted of the truncated haem, cysteine and nicotine, and about 1600 MM atoms of the active site were included. KIEs for six different isotopic substitutions were computed for both low-spin and high-spin states of iron using ISOEFF98. Comparison of the KIEs predicted for reaction at each of the two sites gave mechanistic insight.

A QM/QM ONIOM method, with 30–40 atoms in the higher layer (B3LYP/6-31G(d)) and 800–900 atoms in the lower layer (PM6), was applied by Siwek *et al.*[102] to calculate 1° ^{37}Cl KIEs on the nucleofuge in hydrolysis of both enantiomers of 2-chloropropionate by DL-2-haloacid dehalogenase. Excellent agreement with experiment was obtained provided that account was taken not only of the KIE on the S_N2 step but also of the EIEs for binding of the substrate (first) from water to an initial binding site and (second) from there to the active site. Krzemińska *et al.*[103] performed QM/MM calculations of heavy atoms (^{13}C, ^{15}N and ^{18}O) EIEs for binding sites of different inhibitors to HIV-1 reverse transcriptase. The QM region (either AM1 or B3LYP/6-31 + G(d,p)) consisted of the inhibitor molecule alone and the MM region

(AMBER for protein or TIP3P for aqueous solution) comprised the rest of the system. The subset Hessian was projected to remove the translational and rotational contributions, and the binding isotope effects (BIEs) were obtained as the average over 121 pairwise combinations of 11 RS (solution) and 11 PS (enzyme-bound) structures. Significant BIEs (especially ^{18}O) were found for binding of nucleoside HIV-1 RT inhibitors by means of specific interactions in a hydrophilic binding site, whereas no BIEs were found for non-nucleoside HIV-1 RT inhibitors binding within a hydrophobic cavity on the protein. Essentially the same method was used by Świderek *et al.*[104] to compute KIEs for hydride transfer from tetrahydrofolate to the exocyclic methylene–dUMP intermediate in the reaction catalysed by thymidylate synthase. QM/MM free energy surfaces were determined by means of both 1D and 2D PMFs, which allowed concerted and stepwise reaction paths to be characterised. The 54-atom QM region was described by AM1 and by a dual-level method involving M06-2X corrections; the rest of the system was described using the OPLS-AA and TIP3P force fields. The AM1/OPLS-AA KIEs were averaged over all combinations of 10 RS and 10 TS structures. The 2° ^{3}H KIE for substitution at C6 of tetrahydrofolate (from which hydride transfer occurs) was revealing: a small normal effect, consistent with experiment, was calculated for the concerted mechanism, but an inverse effect was predicted for the alternative stepwise mechanism. The calculated KIEs served to discriminate between the two mechanisms.

5.6 KIE Calculations in the Supramolecular Age

5.6.1 KIEs and Isotopic Partition Function Ratios (IPFRs) for Subsets

The traditional theory of KIEs based on molecular partition functions and the Bigeleisen equation, as described in Section 5.1, assume that each molecular system, either RS or TS, corresponds to a zero-gradient stationary point on a potential energy surface. Reliable optimised geometries and Hessians are readily available from modern quantum-chemical packages employing second-derivative methods, provided that convergence thresholds are set suitably tightly. However, since the turn of the millennium, it has become common to perform simulations for very large molecular systems (either molecules in solution or within enzyme active sites), but to compute explicit Hessians for only a subset of the total number of atoms in the system. For example, relaxation of a specified subset of atoms to a local minimum (or saddle point) may be performed within a frozen environment of the remaining atoms. In the context of QM/MM methods (as reviewed in Section 5.5), the mobile subset and the frozen environment may be the same as the QM and MM regions, although different selections may also be made. In these circumstances the N_s subset atoms do not in themselves constitute a stationary structure in which vibrational degrees of freedom are separable from translations and rotation. The constraining influence of the environment means that diagonalisation of the mass-weighted $3N_s \times 3N_s$ Hessian in Cartesian coordinates generally yields $3N_s$ non-zero eigenvalues which include six corresponding to libration of the whole subset with respect to its environment: 'translation' and 'rotation' of the subset as a whole are not free or separable motions but are coupled with the internal vibrational degrees of freedom.

An approximate way to calculate a KIE for a subset of N_s atoms (within a larger environment of $N - N_s$ atoms) is project out the six librational degrees of freedom and to treat them instead as translations and rotations, as mentioned previously.[18,19] In practice, the $3N_s \times 3N_s$ Hessian in Cartesian coordinates is first transformed into a $(3N_s - 6) \times (3N_s - 6)$ Hessian in a set of internal coordinates, and then back-transformed, whereupon $3N_s - 6$ pure vibrational eigenvalues and six zero eigenvalues for translational and rotational motion are obtained. This procedure is equivalent to treating the subset as if it were a freely translating and freely rotating cluster isolated from its environment but whose internal vibrations are governed by force constants whose values are determined by the influence of the environment. The subset KIEs may then be determined by replacing N by N_s in eqn (5.1) to (5.6). However, is there a form, analogous to the Bigeleisen equation, which expresses the KIE in terms only of vibrational frequencies for a subset? The product of the $3N$ eigenvalues of the vibrational secular matrix is equal (eqn (5.8)) to the product of determinants of the Wilson **F** and **G** matrices for potential energy and kinetic energy,[14] respectively, which in turn is equal to the product of determinants of the potential energy matrix **V** and kinetic energy matrix **T** in mass-weighted Cartesian coordinates.[15] There is a problem when there are zero eigenvalues, as for translation and rotation, because then eqn (5.10) is not valid; the solution is the Teller–Redlich product rule for molecules whose $3N$ degrees of freedom are separable into three translations, 3 rotations and $3N - 6$ vibrations.

$$\prod_i^{3N} \lambda_i = |\mathbf{F}|\,|\mathbf{G}| = |\mathbf{V}|\,|\mathbf{T}| \tag{5.10}$$

However, using the Cartesian formulation of the vibrational problem, it is easy to note that since the frequencies are proportional to the square roots of the vibrational eigenvalues, and **T** is the unit matrix, the product of the ratios of frequencies for isotopologues with the same Hessian (within the Born–Oppenheimer approximation) reduces simply (eqn (5.11)) to the ratio of products of atomic masses used to mass-weight the Hessian prior to diagonalisation. Furthermore, the atomic masses are the same for both isotopologues, except for the isotopically substituted atom(s) alone.

$$\prod_i^{3N} \frac{\nu_i'}{\nu_i} = \left[\frac{|\mathbf{V}'|}{|\mathbf{V}|}\right]^{1/2} = \left[\prod_j^{N} \frac{m_j}{m_j'}\right]^{3/2} \tag{5.11}$$

With a subset Hessian there are, in general, $3N_s$ non-zero vibrational frequencies and so there is no difficulty in evaluating the product on the left-hand side of eqn (5.11) over $3N_s$ ratios ν_i'/ν_i for an isotopologous pair. Thus there is no need to invoke the Teller–Redlich product rule when a subset Hessian is used. The product of the isotopic ratios for all $3N_s$ frequencies is equal to the three-halves power of the product of the ratios of atomic masses for the specific isotopic substitution. In the context of a typical calculation, satisfaction of this rule requires nothing more than that the matrix diagonalisation is performed correctly.[105]

Inspection of either eqn (5.5) or (5.6) shows that a KIE consists of a product of RS factors divided by a product of TS factors, which may be rewritten in the form of eqn (5.12):

$$k/k' = f_{RS}/f_{TS} \tag{5.12}$$

The terms f_{RS} and f_{TS} are IPFRs, as first introduced by Bigeleisen.[1] Whereas an isotopic fractionation factor is defined as the ratio of a heavier isotope relative to a lighter isotope at a particular site with respect to the corresponding ratio at a site in a reference molecule, the definition of an IPFR does not involve any external reference. The logarithm of an IPFR is proportional to the free energy change due to isotopic substitution at a particular site; an IPFR defined as 'heavy'/'light' always has a value >1, becoming larger at lower temperatures but tending towards unity as the temperature increases.

In the light of these considerations, an alternative way to calculate KIEs and IPFRs for subsets of atoms is simply to replace the MMI factor in eqn (5.1) by extending the products for EXC and the summations for ZPE over $3N_s$ (for RS) or $3N_s - 1$ (for TS) vibrations and librations, as in eqn (5.13) and (5.14), in which the contributing factors VP, EX and ZP are identified for clarity. Note that VP may be replaced by the MI factor, which differs from unity only for isotopically substituted atoms within the subset N_s; however, these factors cancel from the KIE (eqn (5.13)) because the same isotopic substitution applies to both RS and TS. Note also that the one-dimensional Bell quantum correction (QC) for an inverted parabola is included in the TS IPFR. Within the approximation of separability, this also corresponds to a free energy change due to isotopic substitution in the TS: for a vibrational mode with a real frequency, a heavier isotope lowers the free energy, but for a vibrational mode with an imaginary frequency, a heavier isotope raises the free energy.

$$\begin{aligned} f_{RS} &= \left[\prod_i^{3N_s} \underbrace{\frac{\nu_i'}{\nu_i}}_{VP} \underbrace{\frac{[1-\exp(-u_i)]}{[1-\exp(-u_i')]}}_{EX} \underbrace{\frac{\exp(u_i/2)}{\exp(u_i'/2)}}_{ZP} \right]_{RS} \\ &= \left[\prod_j^{N_s} \frac{m_j}{m_j'} \right]^{3/2} \prod_i^{3N_s} \left[\frac{\sinh(u_i/2)}{\sinh(u_i'/2)} \right]_{RS} \end{aligned} \tag{5.13}$$

$$f_{\mathrm{TS}} = \frac{\nu'_{\ddagger}}{\nu_{\ddagger}} \left[\prod_{i}^{3N_s-1} \underbrace{\frac{\nu'_i}{\nu_i}}_{\mathrm{VP}} \underbrace{\frac{[1-\exp(-u_i)]}{[1-\exp(-u'_i)]}}_{\mathrm{EX}} \underbrace{\frac{\exp(u_i/2)}{\exp(u'_i/2)}}_{\mathrm{ZP}} \right]_{\mathrm{TS}} \underbrace{\frac{\nu'_{\ddagger} \sin(u_{\ddagger}/2)}{\nu_{\ddagger} \sin(u'_{\ddagger}/2)}}_{\mathrm{QC}}$$
$$= \left[\prod_{j}^{N_s} \frac{m_j}{m'_j} \right]^{3/2} \prod_{i}^{3N_s-1} \left[\frac{\sinh(u_i/2)}{\sinh(u'_i/2)} \right]_{\mathrm{TS}} \frac{\nu'_{\ddagger} \sin(u_{\ddagger}/2)}{\nu_{\ddagger} \sin(u'_{\ddagger}/2)} \quad (5.14)$$

$$k/k' = \frac{\prod_{i}^{3N_s} \left[\frac{\sinh(u_i/2)}{\sinh(u'_i/2)} \right]_{\mathrm{RS}}}{\prod_{i}^{3N_s-1} \left[\frac{\sinh(u_i/2)}{\sinh(u'_i/2)} \right]_{\mathrm{TS}} \frac{\nu'_{\ddagger} \sin(u_{\ddagger}/2)}{\nu_{\ddagger} \sin(u'_{\ddagger}/2)}} \quad (5.15)$$

The use of eqn (5.15) to calculate a KIE for a subset of atoms within a supramolecular system means that the isotopic sensitivity of all $3N_s$ degrees of freedom is considered, including that of the librational modes that arise from coupling between the subset and its environment. As mentioned above, to project out these contributions to the subset Hessian is equivalent to treating the subset atoms as a cluster in a vacuum. Of course, the supramolecular environment (protein or solvent) affects the values of the Hessian elements corresponding to the $3N_s - 6$ degrees of freedom, which are not the same as they would be from a strictly gas-phase calculation for the subset atoms in isolation, but exclusion of the six external degrees of freedom implies a loss of information in regard to coupling between the subset and its environment. The lowest six of the $3N_s$ frequencies from a subset Hessian calculation for either a RS or TS species typically have values <50 cm^{-1} (a threshold value used by some authors to exclude frequencies from KIE calculations in ISOEFF98), but it is not possible to associate these with the six librations without inspection of the corresponding normal modes, because there is coupling between the internal and external degrees of freedom. Simply to exclude the lowest six (real) frequencies from a subset KIE calculation would appear to be an arbitrary and risky action for the sake of keeping $3N_s - 6$ frequencies for use with the Bigeleisen equation when this is both invalid and unnecessary for a subset embedded within a supramolecular environment.[105] Pragmatically it may be justifiable provided that these excluded frequencies are isotopically insensitive, but it should not be assumed that this is always the case.

If a Hessian is computed for an atomic subset whose geometry has been relaxed to a local energy minimum (for RS) or saddle point (for TS) within a frozen environment, there may remain a small number of very low-valued imaginary frequencies. The question then arises as to how these should be treated in an IPFR or KIE calculation. Should they simply be excluded, or should they be treated as if they were real frequencies? An argument from experience in favour of adopting the latter option is that it leads to smaller standard deviations from the mean in averages taken over samples of different thermally accessible conformations. Also, exclusion of low-valued but nonetheless isotopically sensitive imaginary frequencies can lead to physically unrealistic IPFRs with values <1.[106]

From an operational point of view, it is easier to retain all $3N_s$ frequencies in a subset Hessian KIE calculation than it is to perform the projection required to obtain $3N_s - 6$ pure vibrations and six translations and rotations. Moreover, in a study of QM/MM subset Hessians for explicitly solvated AdoMet, it was found that inclusion of all frequencies (in eqn (5.13)) led to significantly smaller errors in KIEs calculated within models with increasingly severe cut-offs than were involved in the more traditional method (eqn (5.5)) which relied on the Teller–Redlich product rule.[105] Our recommendation is to include all $3N_s$ frequencies and to treat small imaginary frequencies as if they were real. Furthermore, for reactions not involving making or breaking of bonds to hydrogen atoms, we recommend routine application of the one-dimensional approximate QC to the reaction coordinate frequency in the IPFR expression for a TS (eqn (5.12)) and in evaluation of the KIE (eqn (5.13)); in this way the isotopic sensitivity of all $3N_s$ frequencies is considered for both RS and TS.

Garcia-Viloca *et al.*[107] described a method for inclusion of QM vibrational energy into a PMF, which is a necessary part of their EA-VTST procedure.[83,84,86] At any point along a reaction path, the reaction coordinate is projected out (to first order) of the Hessian leading to $3N_s - 1$ generalised normal modes and frequencies. They noted that, for a gas-phase molecule, translations and rotations would also be projected out, but that in a condensed phase these are converted into librations, and so the projection does not include these degrees of freedom. However, they also commented that the six lowest-frequency modes of the generalised TS corresponded to very low-frequency librational motion, for which the harmonic approximation was inapplicable and, noting that the QC to low-frequency modes was expected to be small, stated that these modes were omitted. Thus the summation of (harmonic) ZPE and vibrational partition function contributions to the free energy was "calculated over the $3N_s - 7$ highest-frequency modes, not the $3N_s - 1$ modes".[107] While it is true that the librational modes are likely to be anharmonic in nature, and that the harmonic expressions for ZPE and the vibrational partition function are probably inaccurate, our view, however, is that it is still better to include the isotopic sensitivity of these modes, notwithstanding the limitations of the harmonic approximation, than to exclude them altogether.

5.6.2 Conformational Averaging of KIEs and IPFRs

The early QM/MM studies on the coupled hydride/hydron transfer reaction catalysed by LDH that showed different KIEs for different TS structures, each connected to a different RS by means of an IRC path, raised the question as to how these values should be averaged.[81] Possibilities might include the following procedures. The individual KIE could be evaluated for each RS/TS pair, connected by an IRC path, and a simple arithmetic mean could be taken over all the individual values, as indeed was originally done.[81] However, each RS/TS pair was associated with a different energy barrier height, so perhaps the contribution of each individual KIE should be Boltzmann-weighted accordingly? But each RS structure has a different energy, so should the relative populations of each at thermal equilibrium be taken into account? And if interconversion of different RS species is rapid as compared with the chemical reaction, then is the overall rate constant not determined solely by consideration of the relative TS energies?

Truhlar's EA-VTST method with multidimensional tunnelling has been highly successful in application to KIEs for enzymic reactions involving transfer of hydride, hydrogen or hydron.[108] The fundamental theory is, of course, applicable to any reaction, even those not involving C–H bond making or breaking in a rate-determining TS, and those in which the contribution of tunnelling is small. Averaging over RS configurations occurs, in the first instance, by means of the umbrella sampling during the determination of the classical PMF. Secondly, instantaneous normal modes and frequencies are determined for each configuration frequencies sampled and average frequencies are obtained and used in the evaluation of ZPE and the vibrational partition function contribution to the free energy. The same is done for TS configurations. Subsequently, the re-crossing and tunnelling transmission coefficients computed for a further sample from the 'quasiclassical TS ensemble' are averaged. At each stage it seems that the arithmetic mean is taken.[109,110]

The approach developed by Ruiz-Pernía and Williams[106] bears similarity to that of Truhlar and Gao, but is simply an extension of the subset Hessian approach to evaluation of IPFRs and KIEs as described in Section 5.6.1. Partly in view of slightly different terminology that is used, it is worth outlining here for the sake of clarity. First, a one-dimensional PMF is computed with respect to a distinguished coordinate, and then at the free energy minimum (RS) and maximum (TS), extended QM/MM MD trajectories are followed; the TS is subject to a constraint to prevent it collapsing to the RS or product. From the RS and TS trajectories, numerous independent configurations are extracted at regular time intervals; provided that the MD simulation is adequately equilibrated, this should be a sample from a Boltzmann distribution. For each 'snapshot', QM/MM geometry optimisation is performed to either a local minimum or saddle point within a frozen environment. The QM/MM Hessian is computed for a subset of selected atoms within the relaxed region and thence the IPFR for each isotopic substitution of interest, for either RS or TS, by means of eqn (5.11) and (5.12) involving all $3N_s$ or $3N_s - 1$ frequencies (and the imaginary reaction coordinate frequency if the one-dimensional tunnelling correction is included). Finally, the arithmetic mean is taken of the IPFRs for the individual RS and TS configurations, and the average

(Boltzmann-weighted) KIE is obtained as the ratio of the two values (eqn (5.14)).

$$\langle k/k' \rangle = \langle f_{RS} \rangle / \langle f_{TS} \rangle \tag{5.16}$$

This method considers the IPFRs as thermodynamic (or quasi-thermodynamic) functions of state; the value of the KIE is independent of the actual path taken from the RS state to the TS state. Alternatively, $n_{RS} \times n_{TS}$ individual KIEs may be calculated for each possible combination of n_{RS} RS and n_{TS} TS configurations, and the arithmetic mean taken over these $n_{RS} \times n_{TS}$. This method also assumes that there is an equal probability of each and every individual RS structure going to each and every individual TS structure. This assumption is entirely consistent with conventional TST but could be inappropriate if relaxation of the protein or solvent environment were slow compared with the chemical reaction step. The individual RS or TS configurations may not necessarily all contain the same number of atoms; in this case it would not be meaningful to perform the averaging upon the individual partition functions, but it is meaningful to take the averages of IPFRs each obtained from isotopologous species with the same number of atoms sharing and the same Hessian. When applied to QM/MM KIEs for six isotopic substitutions for the (non-enzymic) reaction of cyanide anion with chloroethane in explicit DMSO solvent, the means and standard deviations differed very slightly between the two methods of averaging (if enough decimal places were inspected) but were indistinguishable if the number of significant figures was restricted to what was justified by the magnitudes of the standard deviations.[106] The 'all possible combinations' method was used in our study of COMT[100] and subsequently by others;[103,104] it is perhaps less elegant than the quotient of two average IPFRs (eqn (5.14)), but it does allow the distribution of individual KIE values to be inspected. However, when eqn (5.14) is used, consideration of the distribution about the mean IPFR value should provide a way to assess whether the limited sample of configurations is adequately representative of the population from which it is taken.

In view of the magnitude of fluctuation in the values of IPFRs for individual RS or TS structures usually found, averaging of IPFRs and KIEs in simulations of enzyme reactivity should be an obligatory requirement. It should no longer be considered acceptable to publish results based upon a single RS structure and a single TS structure unless it can be argued convincingly that these are truly representative of the RS and TS ensembles. Otherwise, when comparison is attempted between calculated and experimental KIEs, there is no way of knowing that the computational result is not an outlier from a distribution about a mean with a significantly different value.

5.6.3 Does TS Theory Still Work for KIEs?

The usefulness and validity of modern TST in application to enzyme simulations, including KIE calculations, has been argued with clarity and cogency by Truhlar in a recent review.[111] Despite suggestions that TST was no longer adequate as a basis for understanding KIEs (and their temperature dependence) for enzymic reactions involving significant tunnelling of a transferring hydride/hydrogen/hydron,[112] TST with multidimensional tunnelling and ensemble averaging has been shown to be generally fit for purpose.[113] In particular, the role of non-Boltzmann distributions of vibrational energy (promoting vibrations) has been discussed,[114] with the implication that something beyond TST is required for their full description. However, it has been demonstrated that the complex temperature dependences of KIES for enzyme-catalysed reactions involving significant quantum tunnelling can be reproduced using a TST framework with inclusion of multiple conformations with different reactivity.[115] QM/MM simulations have been employed with EA-VTST/MT to calculate the 'enzyme' KIE on the hydride-transfer reaction catalysed by DHFR, which is the ratio of k_{cat} rate constants for the natural 'light' enzyme and the 'heavy' enzyme, in which ^{12}C, ^{14}N and non-exchangeable ^{1}H have been replaced by ^{13}C, ^{15}N and ^{2}H.[116–118] Good agreement between experiment and calculation was obtained with both wild-type and mutant DHFR for this unusual 2° KIE: the small normal values were attributed to the transmission coefficient factor, reflecting differences in environmental coupling, with no effect on the barrier and the QM tunnelling contribution. The influence of protein dynamics was small and could be reproduced accurately by TST.

Recently, it has been reported that common TST-based methods for enzyme simulation lead to TS ensembles that are considerably different from those obtained from a dynamics-based (transition path sampling) method, thereby raising fresh doubt about the validity of the TST approach.[119] On the other hand, the necessity of having a good reaction coordinate in applications of TST to enzymic simulations is recognised, and new methods have been proposed for TS ensemble optimisation.[120]

5.6.4 Cut-off Rules Revisited

The motivation underlying the original cut-off rule proposed by Stern and Wolfsberg[7] was to make KIE calculations tractable in regard both to computational resource and availability of force constant data for RS and TS structures. In view of the fact that they considered only covalently-bound molecules, the rule was stated in terms of how many covalent bonds a particular atom is removed from the site of isotopic substitution.[7] Simulations of chemical reactivity in enzyme active sites or in solution (or other condensed media) inevitably involve consideration of non-covalent interactions. The range of QM and QM/MM applications surveyed above testifies that neither lack of computer power nor the ability to compute a Hessian is now a limiting factor. But a question remains: what size does a Hessian need to be in order to provide reliable estimates for IPFRs, and thence KIEs? How should the cut-off rule be reframed for KIE calculations in the supramolecular age?

In a QM/MM study of cut-offs on calculated KIEs for methyl transfer AdoMet to catecholate in water,[105] it was found that, for more severe cut-offs, significantly smaller errors were obtained when the six librational degrees of freedom were treated as harmonic vibrations than as being projected out as translations and rotations. Errors were evaluated by comparison of the natural logarithm of the KIE for a cut-off Hessian against that obtained from a large Hessian, in this case for 324 atoms of AdoMet and 87 water molecules. The QM region (AM1) contained 63 atoms of AdoMet and the MM region contained about 500 waters (TIP3P). Cut-off Hessians were obtained by deletion of appropriate rows and columns from the full Hessian; the values of retained elements in each Hessian remained constant throughout. Exclusion of all solvent water molecules from the Hessian yielded α-$^{2}H_3$, α-^{13}C and nucleofuge ^{34}S KIEs with equally small errors using either the 'all frequencies' method or the projection method, but the latter method (using only $3N_s - 6$ frequencies) gave a much larger error for the nucleophile ^{18}O KIE. The standard two-bond cut-off rule performed significantly better with the all $3N_s$ frequencies method for all isotopic substitutions except α-^{13}C, for which the error was the same with both methods. Larger errors were obtained with both methods when the one-dimensional Bell tunnelling correction was included, because the imaginary TS reaction coordinate frequencies were surprisingly sensitive to the size of the subset Hessian. It was concluded that, to ensure that the error in ln(KIE) remained less than 1% (or 2% for the quantum-corrected KIE) for all four isotopic substitutions considered, it was necessary to use a less restrictive cut-off procedure which retained all covalently-bonded atoms to a distance of at least three-bonds together with solvent water molecules whose O atom was less than 4.5 Å from any isotopically substituted atom. To the extent that KIEs may be treated to a satisfactory degree of approximation by means of a cut-off procedure, it implies that these ratios of rate constants are essentially local properties of the system, reflecting changes in curvature of the potential energy surface in the immediate vicinity of the position(s) of isotopic substitution. However, the surprising degree of sensitivity of KIE values to the extent of cut-off implies that these are not entirely local properties of the system. In particular, it seems that a cut-off should not cause significant change in the nature of an isotopically sensitive reaction coordinate vibrational mode: exclusions of atoms whose motions are coupled to the reaction coordinate may lead to error. These findings have important practical implications for reliable computational simulation of KIEs for large systems in solution or within an enzyme active site. To obtain reliable KIEs requires not only an appropriate choice of QM or QM/MM method but also an adequate description of the environment in regard to both its potential energy and kinetic energy influence upon the positions of isotopic substitution.[105]

In QM/MM applications, the atoms included in the subset Hessian have commonly been the same as those in the QM region. However, the selection criteria for inclusion of an atom in the QM region are not necessarily the sameas those for inclusion in the Hessian. A QM region may be chosen so as to avoid the need for the QM/MM boundary to cut through covalent bonds. However, this issue does not affect the Hessian: there is no reason why even just a single atom of a group connected, either covalently or non-covalently, within two bonds distance from a site of isotopic substitution cannot be included in a subset Hessian. The size and shape of the atomic subsets included in the Hessian and in the QM region do not need to be the same; better KIE results may be obtained using QM/MM Hessians computed for roughly spherical subsets, centred on the site of isotopic substitution, whose size is determined by a 'cut-off' radius. The quality of the data contained within a computed subset Hessian is as important as the quantity: reliable KIE calculations for enzyme-catalysed reactions require an adequate description of the influence of the (generally inhomogeneous) environment surrounding the subset, which in turn may require a larger QM region than has typically been used in the past.

References

1. J. Bigeleisen, *J. Chem. Phys.*, 1949, **17**, 675–678.
2. J. Bigeleisen and M. Wolfsberg, *Adv. Chem. Phys.*, 1958, **1**, 15–76.
3. E. K. Thornton and E. R. Thornton, in *Isotope Effects in Chemical Reactions*, ed. C. J. Collins and N. S. Bowman, Van Nostrand Reinhold, New York, 1970.
4. L. Melander and W. H. Saunders, *Reaction Rates of Isotopic Molecules*, Wiley, New York, 1980.
5. (a) E. Teller, as attributed by W. R. Angus, C. R. Bailey, J. B. Hale, C. K. Ingold, A. H. Leckie, C. G. Raisin, J. W. Thompson and C. L. Wilson, *J. Chem. Soc.*, 1936, 971–987; (b) O. Redlich, *Z. Physik. Chem. B*, 1935, **28**, 371–382.
6. J. Bigeleisen and M. G. Mayer, *J. Chem. Phys.*, 1947, **15**, 261–267.
7. (a) M. Wolfsberg and M. J. Stern, *Pure Appl. Chem.*, 1964, **8**, 225–242; (b) M. J. Stern and M. Wolfsberg, *J. Chem. Phys.*, 1966, **45**, 4105–4124.
8. Quiver, M. Saunders, K. E. Laidig and M. Wolfsberg, *J. Am. Chem. Soc.*, 1989, **111**, 8989–8994.
9. ISOEFF98, V. Anisimov and P. Paneth, *J. Math. Chem.*, 1999, **26**, 75–86.
10. S. Wolfe, S. Hoz, C.-K. Kim and K. Yang, *J. Am. Chem. Soc.*, 1990, **112**, 4186–4191.
11. E. Wigner, *Z. Physik. Chem. B*, 1932, **19**, 203–216.
12. R. P. Bell, *Trans. Faraday Soc.*, 1959, **55**, 1–4.
13. J. H. Schachtschneider and R. G. Snyder, *Spectrochim. Acta*, 1963, **19**, 117–168.
14. E. B. Wilson, J. C. Decius and P. C. Cross, *Molecular Vibrations*, McGraw-Hill, New York, 1955.
15. W. D. Gwinn, *J. Chem. Phys.*, 1971, **55**, 477–481.
16. C. D. Chalk, B. G. Hutley, J. McKenna, L. B. Sims and I. H. Williams, *J. Am. Chem. Soc.*, 1981, **103**, 260–268.

17. L. B. Sims, G. W. Burton and D. E. Lewis, BEBOVIB-IV, Program 337, Quantum Chemistry Program Exchange, Department of Chemistry, University of Indiana, Bloomington, IN, 1977.
18. I. H. Williams, *Chem. Phys. Lett.*, 1982, **88**, 462–466.
19. I. H. Williams, *J. Mol. Struct.: Theochem*, 1983, **94**, 275–284.
20. H. S. Johnston, *Gas Phase Reaction Rate Theory*, Ronald Press, New York, 1966.
21. G. W. Burton, L. B. Sims, J. C. Wilson and A. Fry, *J. Am. Chem. Soc.*, 1977, **99**, 3371–3379.
22. L. B. Sims and D. E. Lewis, in *Isotope Effects in Organic Chemistry*, ed. E. Buncel and C. C. Lee, Elsevier, Amsterdam, 1984, vol. 6, pp. 161–259.
23. P. J. Berti, *Methods Enzymol.*, 1999, **308**, 355–397.
24. J. Rodgers, D. A. Femec and R. L. Schowen, *J. Am. Chem. Soc.*, 1982, **104**, 3263–3268.
25. J. D. Hermes and W. W. Cleland, *J. Am. Chem. Soc.*, 1984, **106**, 7263–7264.
26. F. Mentch, D. W. Parkin and V. L. Schramm, *Biochemistry*, 1987, **26**, 921–930.
27. D. W. Parkin, F. Mentch, G. A. Banks, B. A. Horenstein and V. L. Schramm, *Biochemistry*, 1991, **30**, 4586–4594.
28. B. A. Horenstein, D. W. Parkin, B. Estupinan and V. L. Schramm, *Biochemistry*, 1991, **30**, 10788–10795.
29. D. J. Merkler, P. C. Kline, P. Weiss and V. L. Schramm, *Biochemistry*, 1993, **32**, 12993–13001.
30. (a) P. C. Kline and V. L. Schramm, *Biochemistry*, 1993, **32**, 13212–13219; (b) P. C. Kline and V. L. Schramm, *Biochemistry*, 1995, **34**, 1153–1162.
31. J. Scheuring and V. L. Schramm, *Biochemistry*, 1997, **36**, 4526–4534.
32. K. A. Rising and V. L. Schramm, *J. Am. Chem. Soc.*, 1997, **119**, 27–37.
33. P. J. Berti, S. R. Blanke and V. L. Schramm, *J. Am. Chem. Soc.*, 1997, **119**, 12079–12088.
34. J. Scheuring and V. L. Schramm, *Biochemistry*, 1997, **36**, 8215–8223.
35. J. Scheuring, P. J. Berti and V. L. Schramm, *Biochemistry*, 1998, **37**, 2748–2758.
36. Y. Tanaka, W. Tao, J. S. Blanchard and E. J. Hehre, *J. Biol. Chem.*, 1994, **269**, 32306–32312.
37. W. Tao, C. Grubmeyer and J. S. Blanchard, *Biochemistry*, 1996, **35**, 13–21.
38. L. A. Reinhardt, D. Svedruzic, C. H. Chang, W. W. Cleland and N. G. J. Richards, *J. Am. Chem. Soc.*, 2003, **125**, 1244–1252.
39. C. Bates, Z. Kendrick, N. McDonald and P. C. Kline, *Phytochemistry*, 2006, **67**, 5–12.
40. P. J. Berti and V. L. Schramm, *J. Am. Chem. Soc.*, 1997, **119**, 12069–12078.
41. D. E. Lewis, L. B. Sims, H. Yamataka and J. McKenna, *J. Am. Chem. Soc.*, 1980, **102**, 7411–7419.
42. W. P. Huskey, *J. Am. Chem. Soc.*, 1996, **118**, 1663–1668.

43. V. L. Schramm, B. A. Horenstein and P. C. Kline, *J. Biol. Chem.*, 1994, **269**, 18259–18262.
44. (a) V. L. Schramm, *Acc. Chem. Res.*, 2003, **36**, 588–596; (b) V. L. Schramm, *J. Biol. Chem.*, 2007, **282**, 28297–28300; (c) V. L. Schramm, *J. Biol. Chem.*, 2008, **284**, 33201–32208.
45. M. J. S. Dewar and G. P. Ford, *J. Am. Chem. Soc.*, 1977, **99**, 8343–8344.
46. S. B. Brown, M. J. S. Dewar, G. P. Ford, D. J. Nelson and H. S. Rzepa, *J. Am. Chem. Soc.*, 1978, **100**, 7832–7836.
47. J. P. Shea, S. D. Nelson and G. P. Ford, *J. Am. Chem. Soc.*, 1983, **105**, 5451–5454.
48. K. Korzekwa, W. Trager, M. Gouterman, D. Spangler and G. H. Loew, *J. Am. Chem. Soc.*, 1985, **107**, 4273–4279.
49. I. H. Williams, *J. Am. Chem. Soc.*, 1984, **106**, 7206–7212.
50. J. Wilkie and I. H. Williams, *J. Am. Chem. Soc.*, 1992, **114**, 5423–5425.
51. J. P. Jones and J. L. Urbauer, *J. Comput. Chem.*, 1991, **12**, 1134–1141.
52. J. Andrés, V. Moliner and V. S. Safont, *J. Chem. Soc. Faraday Trans.*, 1994, **90**, 1703–1707.
53. J. Andrés, V. Moliner, V. S. Safont, L. R. Domingo and M. T. Picher, *J. Org. Chem.*, 1996, **61**, 7777–7783.
54. E. Gawlita, V. E. Anderson and P. Paneth, *Eur. Biophys. J.*, 1994, **23**, 353–360.
55. V. Anisimov and P. Paneth, *J. Math. Chem.*, 1999, **26**, 75–86.
56. M. Saunders, K. E. Laidig and M. Wolfsberg, *J. Am. Chem. Soc.*, 1989, **111**, 8989–8994.
57. J. A. Barnes and I. H. Williams, in *Organic Reactivity: Physical and Biological Aspects (Proc. ESOR-IV/MMBP-II, Newcastle, 1993)*, ed. B. T. Golding, R. J. Griffin and H. Maskill, Royal Society of Chemistry, 1995, pp. 437–443.
58. B. A. Horenstein, *J. Am. Chem. Soc.*, 1997, **119**, 1101–1107.
59. M. Bruner and B. A. Horenstein, *Biochemistry*, 1998, **37**, 289–297.
60. A. J. Bennett and M. L. Sinnott, *J. Am. Chem. Soc.*, 1986, **108**, 7287–7294.
61. X.-Y. Chen, P. J. Berti and V. L. Schramm, *J. Am. Chem. Soc.*, 2000, **122**, 1609–1617.
62. X.-Y. Chen, P. J. Berti and V. L. Schramm, *J. Am. Chem. Soc.*, 2000, **122**, 6527–6534.
63. S. L. Parikh and V. L. Schramm, *Biochemistry*, 2004, **43**, 1204–1212.
64. M. R. Birck and V. L. Schramm, *J. Am. Chem. Soc.*, 2004, **126**, 2447–2453.
65. P. A. Schwartz, M. J. Vetticatt and V. L. Schramm, *J. Am. Chem. Soc.*, 2010, **132**, 13425–13433.
66. A. Lewandowicz and V. L. Schramm, *Biochemistry*, 2004, **43**, 1458–1468.
67. M. Luo, L. Li and V. L. Schramm, *Biochemistry*, 2008, **47**, 2565–2576.
68. M. Luo, V. Singh, E. A. Taylor and V. L. Schramm, *J. Am. Chem. Soc.*, 2007, **129**, 8008–8017.
69. J. S. Hirschi, T. Takeya, C. Hang and D. A. Singleton, *J. Am. Chem. Soc.*, 2009, **131**, 2397–2403.

70. R. Rios-Font, L. Rodríguez-Santiago, J. Bertran and M. Sodupe, *J. Phys. Chem. B*, 2007, **111**, 6071–6077.
71. J. A. B. McCann and P. J. Berti, *J. Am. Chem. Soc.*, 2007, **129**, 7055–7064.
72. J. A. B. McCann and P. J. Berti, *J. Am. Chem. Soc.*, 2008, **130**, 5789–5797.
73. J. Chan, A. R. Lewis, D. Indurugalla, M. Schur, W. Wakarchuk and A. J. Bennet, *J. Am. Chem. Soc.*, 2012, **134**, 3748–3757.
74. G. Pierdominici-Sottile, N. A. Horenstein and A. E. Roitberg, *Biochemistry*, 2011, **50**, 10150–10158.
75. J. P. Götze and P. Saalfrank, *J. Mol. Model.*, 2012, **18**, 1877–1883.
76. B. Manta, F. M. Raushel and F. Himo, *J. Phys. Chem. B*, 2014, **118**, 5644–5652.
77. (a) M. Lou, M. E. Gilpin, S. K. Burger, A. M. Malik, V. Gawuga, V. Popović, A. Capretta and P. J. Berti, *J. Am. Chem. Soc.*, 2012, **134**, 12958–12969; (b) M. Lou, S. K. Burger, M. E. Gilpin, V. Gawuga, A. Capretta and P. J. Berti, *J. Am. Chem. Soc.*, 2012, **134**, 12958–12969.
78. (a) H. Gu, S. Zhang, K.-Y. Wong, B. K. Radak, T. Dissanayake, D. L. Kellermana, Q. Dai, M. Miyagi, V. E. Anderson, D. M. York, J. A. Piccirilli and M. E. Harris, *Proc. Natl. Acad. Sci. U. S. A.*, 2013, **110**, 13002–13007; (b) H. Chen, T. J. Giese, M. Huang, K.-Y. Wong, M. E. Harris and D. M. York, *Chem. – Eur. J.*, 2014, **20**, 14336–14343; (c) H. Chen, J. A. Piccirilli, M. E. Harris and D. M. York, *Biochim. Biophys. Acta*, 2015, **1854**, 1795–1800.
79. J. A. Barnes and I. H. Williams, *Chem. Commun.*, 1996, 193–194.
80. J. A. Barnes and I. H. Williams, *Biochem. Soc. Trans.*, 1996, **24**, 263–268.
81. (a) V. Moliner, A. J. Turner and I. H. Williams, *Chem. Commun.*, 1997, 1271–1272; (b) A. J. Turner, V. Moliner and I. H. Williams, *Phys. Chem. Chem. Phys.*, 1999, **1**, 1323–1331.
82. R. Castillo, J. Andrés and V. Moliner, *J. Am. Chem. Soc.*, 1999, **121**, 12140–12147.
83. C. Alhambra, J. Gao, J. C. Corchado, J. Villa and D. G. Truhlar, *J. Am. Chem. Soc.*, 1999, **121**, 2253–2258.
84. C. Alhambra, J. C. Corchado, M. L. Sanchez, J. Gao and D. G. Truhlar, *J. Am. Chem. Soc.*, 2000, **122**, 8197–8203.
85. J. Rucker and J. P. Klinman, *J. Am. Chem. Soc.*, 1999, **121**, 1997–2006.
86. C. Alhambra, J. C. Corchado, M. L. Sanchez, M. Garcia-Viloca, J. Gao and D. G. Truhlar, *J. Phys. Chem. B*, 2001, **105**, 11326–11340.
87. Q. Cui, M. Elstner and M. Karplus, *J. Phys. Chem. B*, 2002, **106**, 2721–2740.
88. S. Ferrer, I. Tuñón, S. Martí, V. Moliner, M. Garcia-Viloca, A. González-Lafont and J. M. Lluch, *J. Am. Chem. Soc.*, 2006, **128**, 16851–16863.
89. G. Tresadern, S. Nunez, P. F. Faulder, H. Wang, I. H. Hillier and N. A. Burton, *Faraday Discuss.*, 2003, **122**, 223–242.
90. K. E. Ranaghan, L. Masgrau, N. S. Scrutton, M. J. Sutcliffe and A. J. Mulholland, *ChemPhysChem*, 2007, **8**, 1816–1835.
91. L. Masgrau, K. E. Ranaghan, N. S. Scrutton, A. J. Mulholland and M. J. Sutcliffe, *J. Phys. Chem. B*, 2007, **111**, 3032–3047.

92. A. Dybala-Defratyka, P. Paneth, R. Banerjee and D. G. Truhlar, *Proc. Natl. Acad. Sci. U. S. A.*, 2007, **104**, 10774–10779.
93. C. Alhambra, J. C. Corchado, M. L. Sanchez, J. Villa, J. Gao and D. G. Truhlar, CHARMMRATE (Univ. Minnesota, Minneapolis), Version 1.0, 1999.
94. A. Dybala-Defratyka, M. Rostkowski and P. Paneth, *Arch. Biochem. Biophys.*, 2008, **474**, 274–282.
95. S. Martí, V. Moliner, I. Tuñón and I. H. Williams, *Org. Biomol. Chem.*, 2003, **1**, 483–487.
96. S. E. Worthington, A. E. Roitberg and M. Krauss, *Int. J. Quantum Chem.*, 2003, **94**, 287–292.
97. S. Martí, V. Moliner, I. Tuñón and I. H. Williams, *J. Phys. Chem. B*, 2005, **109**, 3707–3710.
98. G. D. Ruggiero, S. J. Guy, S. Martí, V. Moliner and I. H. Williams, *J. Phys. Org. Chem.*, 2004, **17**, 592–601.
99. G. D. Ruggiero, I. H. Williams, M. Roca, V. Moliner and I. Tuñón, *J. Am. Chem. Soc.*, 2004, **126**, 8634–8635.
100. N. Kanaan, J. J. Ruiz-Pernía and I. H. Williams, *Chem. Commun.*, 2008, 6114–6116.
101. R. A. Kwiecień, J.-Y. Le Questel, J. Lebreton, M. Delaforge, F. Andre, E. Pihan, A. Roussel, A. Fournial, P. Paneth and R. J. Robins, *J. Phys. Chem. B*, 2012, **116**, 7827–7840.
102. A. Siwek, R. Omi, K. Hirotsu, K. Jitsumori, N. Esaki, T. Kurihara and P. Paneth, *Arch. Biochem. Biophys.*, 2013, **540**, 26–32.
103. A. Krzemińska, P. Paneth, V. Moliner and K. Świderek, *J. Phys. Chem. B*, 2015, **119**, 917–927.
104. K. Świderek, A. Kohen and V. Moliner, *Phys. Chem. Chem. Phys.*, 2015, **17**, 30793–30804.
105. I. H. Williams, *J. Chem. Theor. Comput.*, 2012, **8**, 542–553.
106. J. J. Ruiz-Pernía and I. H. Williams, *Chem. – Eur. J.*, 2012, **18**, 9405–9414.
107. M. Garcia-Viloca, C. Alhambra, D. G. Truhlar and J. Gao, *J. Chem. Phys.*, 2001, **114**, 9953–9958.
108. D. G. Truhlar, in *Isotope Effects in Chemistry and Biology*, ed. A. Kohen and H.-H. Limbach, CRC Press, Boca Raton, FL, 2006, pp. 579–619.
109. T. D. Poulsen, M. Garcia-Viloca, J. Gao and D. G. Truhlar, *J. Phys. Chem. B*, 2003, **107**, 9567–9578.
110. D. G. Truhlar, *Acc. Chem. Res*, 2015, **48**, 431–438.
111. D. G. Truhlar, *Arch. Biochem. Biophys.*, 2015, **582**, 10–17.
112. Z. D. Nagel and J. P. Klinman, *Nat. Chem. Biol.*, 2009, **5**, 543–550.
113. K. E. Ranaghan and A. J. Mulholland, *Interdiscip. Sci.: Comput. Life Sci.*, 2010, **2**, 78–97.
114. S. Hay and N. S. Scrutton, *Nat. Chem.*, 2012, **4**, 161–168.
115. D. R. Glowacki, J. N. Harvey and A. J. Mulholland, *Nat. Chem.*, 2012, **4**, 169–176.

116. L. Y. P. Luk, J. J. Ruiz-Pernía, W. M. Dawson, M. Roca, E. J. Loveridge, D. R. Glowacki, J. N. Harvey, A. J. Mulholland, I. Tuñón, V. Moliner and R. K. Allemann, *Proc. Natl. Acad. Sci. U. S. A.*, 2013, **110**, 16344–16349.
117. J. J. Ruiz-Pernía, L. Y. P. Luk, R. García-Meseguer, S. Martí, E. J. Loveridge, I. Tuñón, V. Moliner and R. K. Allemann, *J. Am. Chem. Soc.*, 2013, **135**, 18689–18696.
118. K. Świderek, J. J. Ruiz-Pernía, V. Moliner and I. Tuñón, *Curr. Opin. Chem. Biol.*, 2014, **21**, 11–18.
119. D. Doron, A. Kohen, K. Nam and D. T. Major, *J. Chem. Theory Comput.*, 2014, **10**, 1863–1871.
120. K. Zinovjev and I. Tuñón, *J. Chem. Phys.*, 2015, **143**, 134111.

CHAPTER 6

Free Energy Calculation Methods and Rare Event Sampling Techniques for Biomolecular Simulations

JENS SMIATEK,*[a] NIELS HANSEN*[b] AND JOHANNES KÄSTNER*[c]

[a] Institute for Computational Physics, University of Stuttgart, Allmandring 3, 70569 Stuttgart, Germany; [b] Institute of Thermodynamics and Thermal Process Engineering, University of Stuttgart, Pfaffenwaldring 9, 70569 Stuttgart, Germany; [c] Institute for Theoretical Chemistry, University of Stuttgart, Pfaffenwaldring 55, 70569 Stuttgart, Germany
*Email: smiatek@icp.uni-stuttgart.de; hansen@itt.uni-stuttgart.de; kaestner@theochem.uni-stuttgart.de

6.1 Introduction

Over the last three decades, the development of free energy methods and rare event sampling techniques for biomolecular simulations has grown enormously. This can be mostly rationalised by the significant increase of computational power and the development of meaningful simulation approaches which allows a reasonable analysis of protein and enzyme properties. Nowadays, the detection of stable protein or complex conformations, reaction pathways and important intermediate structures by the study of free energy landscapes and rare event sampling techniques has become a

RSC Theoretical and Computational Chemistry Series No. 9
Simulating Enzyme Reactivity: Computational Methods in Enzyme Catalysis
Edited by Iñaki Tuñón and Vicent Moliner

Published by the Royal Society of Chemistry, www.rsc.org

standard approach in computational biosciences. Combined with experimental structure data, *ab initio* or atomistic molecular dynamics (MD) simulations often provide new insights into conformational transitions, dynamic behaviour as well as enzymatic reactions. Next-generation force fields allow accurate simulations of DNA and RNA sequences[1] and proteins,[2] including enzymes.[3] In addition, the development of polarisable force fields for even more accurate predictions is an ongoing field of research.[4] Also enzymatic reactions, which usually include the cleavage of covalent bonds, can be efficiently modelled either by sophisticated QM/MM methods[5] or classical reactive force fields.[6] In summary, with the available methods and approaches, systematic studies of protein and enzyme properties are possible, which is important for a deeper understanding of diseases as well as medical and bioengineering applications.

The development of distributed computing algorithms,[7] high-performance super computers[8] or GPU-based software packages[9] significantly increased the simulation timescale for small proteins up to the millisecond scale. Novel analysis methods such as Markov state models[10] rely on trajectories which were obtained by long-time scale MD simulations. With regard to these developments, the question remains whether it is still beneficial to use sophisticated sampling and free energy methods? The answer is two-fold: on the one hand it is desirable to avoid sampling techniques which are often based on order parameters or reaction coordinates that introduce an inherent simplification of the dynamics by a significant reduction of the degrees of freedom. On the other hand, sophisticated sampling techniques or free energy methods provide detailed access to thermodynamic data or reaction pathways that are otherwise inaccessible. Hence, it is advantageous to combine long timescale MD simulations with sophisticated free energy and sampling techniques such that even more detailed insights into the system behaviour can be gained. As an example, the knowledge of free energy differences can be used to distinguish between enthalpic and entropic contributions.[11] This is specifically of importance for the study of solvation effects and intramolecular interactions.[12] Moreover, the famous folding funnel approach for proteins[13–16] heavily relies on a description in terms of a free energy landscape. In most cases, the quality of the calculated free energy landscape relies on the considered reaction coordinates. It has to be mentioned that all low-dimensional free or potential energy landscapes represent a simplified description of the full phase space free energy where unimportant degrees of freedom were neglected and reaction coordinates were introduced to monitor the essential dynamics of the system or to distinguish between different conformations. Poorly chosen reaction coordinates might introduce artifacts such as hidden complexities or hysteresis effects.[16,17] Thus, the reliability of the free energy landscape has to be carefully proven.[18] The fundamental dependence on the reaction coordinate can be interpreted as the most important drawback for so-called biasing free energy methods like metadynamics[19] or the adaptive biasing force method.[20] In fact, these methods introduce an artificial biasing energy to the system that is intended to accelerate the dynamics or to increase the sampling accuracy at some regions in the reaction coordinate phase space. In contrast, methods like replica exchange,[21] transition path sampling[22] or forward flux sampling[23] avoid the use of a biasing energy.

Despite the chosen method, it is obvious that estimating free energy barriers and reaction free energies is of paramount importance for the understanding of enzymatic reactivity. With regard to thermodynamics or the concepts of statistical mechanics, it can be stated that the free energy is *the* essential main driving force of enzymes in the thermodynamic limit. A negative reaction free energy means that the forward reaction is more beneficial, a positive reaction free energy favours the backward reaction. The reaction rate constant k is given in transition state theory[24] by:

$$k = \frac{k_B T}{h} \exp[-\Delta^{\ddagger} A / k_B T] \tag{6.1}$$

with T being the absolute temperature, k_B being Boltzmann's constant, h Planck's constant, and $\Delta^{\ddagger}A$ the free energy of activation. The concept of thermodynamic state functions like the free energy is based on equilibrium quantities. Dynamic effects on the reaction rate and the reactivity are neglected or integrated out. Thermal fluctuations of the enzyme and the active centre are accounted for, neglected effects include inertia along the reaction coordinate without the fast dissipation of energy into other degrees of freedom. While the potential energy $E(\boldsymbol{r})$ depends on one specific atomic geometry $\boldsymbol{r}$, the free energy A for the canonical ensemble or the Gibbs energy G for the constant-pressure ensemble depend on the state of a system. Both can be obtained from the potential energy *via* a phase space integral in terms of a simple expression by:

$$A = -\frac{1}{\beta} \ln \int \exp[-\beta E(\boldsymbol{r})] \mathrm{d}^N r \tag{6.2}$$

with $\beta = 1/k_B T$. The integral is taken over the whole configurational space, *i.e.* all accessible coordinates of all atoms. In principle the integral also covers momentum space. However, for a potential energy that is independent of momenta, that contribution is constant and can be integrated out. Because A is only defined up to an additive constant in any case, the absolute value of A is unimportant and cannot be calculated explicitly. The more important free energy difference between two states as defined by an arbitrarily chosen reaction coordinate $\xi(\boldsymbol{r})$ as a function of the spatial coordinates $\boldsymbol{r}$ is given by:

$$\Delta A(\xi) = -\frac{1}{\beta} \ln \left(\frac{P(\xi)}{P(\xi_0)} \right) \tag{6.3}$$

with the probabilities $P(\xi)$ of finding the system at a position ξ in reaction coordinate space where ξ_0 can be interpreted as the reference state. It is worth noting that most methods which avoid the introduction of biasing potential energy rely on a high statistical sampling accuracy for the probabilities $P(\xi)$ such that the resulting free energy difference can be calculated in accordance to eqn (6.3) in a straightforward way. For a given potential energy, the probability distribution along ξ can be obtained by integrating over all spatial coordinates but ξ:

$$P(\xi) = \frac{\int \exp[-\beta E(\boldsymbol{r})]\delta[\xi(\boldsymbol{r}) - \xi]\mathrm{d}^N\boldsymbol{r}}{\int \exp[-\beta E(\boldsymbol{r})]\mathrm{d}^N\boldsymbol{r}}. \tag{6.4}$$

Here, ξ is a specific value of the spatial coordinates which is constant in the calculation of the configuration space integral. Such a probability distribution can be used to define a free energy depending on a reaction coordinate:

$$A(\xi) = -\frac{1}{\beta}\ln P(\xi) \tag{6.5}$$

which exactly results in eqn (6.3) for the free energy differences.

In general, the probability $P(\boldsymbol{r})$ of finding the system at a point in configuration space is proportional to the Boltzmann factor $\exp[-\beta E(\boldsymbol{r})]$ and fulfils the normalisation condition $\int P(\boldsymbol{r})d^N\boldsymbol{r} = 1$. Using eqn (6.2) one can see that the proportionality factor is given by $\exp(\beta A)$ resulting in:

$$P(\boldsymbol{r}) = \exp[\beta(A - E(\boldsymbol{r}))], \tag{6.6}$$

which demonstrates the inclusion of the free energy A in terms of a probabilistic approach in accordance with eqn (6.3) and (6.5). An ensemble average $\langle F \rangle$ of any quantity $F(\boldsymbol{r})$ can be calculated as:

$$\langle F \rangle = \frac{\int F(\boldsymbol{r})\exp[-\beta E(\boldsymbol{r})]\mathrm{d}^N\boldsymbol{r}}{\int \exp[-\beta E(\boldsymbol{r})]\mathrm{d}^N\boldsymbol{r}} = \int P(\boldsymbol{r})F(\boldsymbol{r})\mathrm{d}^N\boldsymbol{r} \tag{6.7}$$

which is also valid in the reaction coordinate space spanned by ξ. With regard to these considerations, it becomes clear that the knowledge of the free energy is of fundamental importance for a detailed understanding of the system properties, which also explains the broad variety of free energy methods and sampling techniques that have been used in computer simulations.

In the present chapter, we will describe different methods that have proven their success in terms of biomolecular simulations. We will mostly describe flexible free energy methods as well as rare event sampling techniques that are applicable for a large number of systems and can also be used at different length scales. The choice of the presented methods reflects the personal experience of the authors and does not convey their success over methods not discussed here. Our intention was mostly driven by the fact that we want to give a short overview on advanced sampling techniques and their inherent principles instead of a complete introduction into free energy methods, which is beyond the scope of this chapter. We refer the reader to some excellent textbooks[25,26] as well as further review papers[27,28] for a more complete introduction. In the next section, we will give a short introduction into the usage of reaction coordinates, which is basically a prerequisite for the usage of biasing potential energies in some approaches. The following sections will be devoted to the discussion of free energy methods as well as rare event sampling techniques. At the end, we will discuss averaging techniques for QM/MM simulations, which are mostly important for the efficient description of enzymes. We briefly conclude in the last section.

6.2 Reaction Coordinates

For all systems a reaction coordinate, sometimes also called a collective variable ξ can be defined, which can be used to monitor the dynamical behaviour of the system by an effective reduction of the degrees of freedom.[29] In many cases, a function of spatial coordinates like atomic positions are used to define the reaction coordinates and the corresponding configurations. The resulting configuration phase space is low in dimensionality and is usually formed by one or two reaction coordinates. For historic reasons, the corresponding free energy depending on a reaction coordinate is sometimes also called potential of mean force. In principle, the corresponding configuration space integrals according to eqn (6.4) and (6.5) can be approximated by MD calculations. The ergodicity theorem states that with infinite sampling time the distribution of the system along a MD trajectory will approach $P(\xi)$ as defined in eqn (6.4). However, in reality, the sampling times are always limited. Several techniques have been developed to improve the sampling accuracy. A selection of those will be discussed in this chapter.

Specifically for methods which use biasing energies that are evaluated in the reaction coordinate configurational space, the corresponding reaction coordinate has to be carefully chosen. In fact, it is not obvious that the dynamics of a system, *e.g.* the unfolding of a protein, is properly described by a set of *a priori* chosen reaction coordinates. The usage of reaction coordinates is usually intended to reduce the problem of high dimensionality to a low-dimensional problem. It becomes immediately clear that this crucial reduction might also result in the loss of important information. Nevertheless, specific reaction coordinates might provide a proper description of the most important system dynamics. Often used examples for biomacromolecules are given by the root mean square deviation with regard to reference configurations, torsional angles like $\psi - \phi$ backbone angles, number of internal hydrogen bonds, eigenvectors or end-to-end distances. Although these reaction coordinates often provide meaningful results, it is not *a priori* clear whether a reasonable projection is achieved. Thus, one has to carefully prove whether the chosen reaction coordinate adequately describes the dynamics and allows us to distinguish between different stable configurations.[30]

Specific methods to analyse the applicability of the chosen reaction coordinates were developed in a series of publications.[16,31–33] Most of the proposed methods focus on the dynamics of the system. Specifically with regard to essential dynamics,[34] where the eigenvectors of the system are taken into consideration, a 'non-Gaussian parameter method'[16] has been proposed which provides an estimate for the minimal number of reaction coordinates. The approach relies on a critical evaluation of stochastic and deterministic motion as induced by the presence or absence of free energy minima and is therefore broadly applicable.

In addition, unbiased methods like transition path sampling and transition interface sampling or forward flux sampling (FFS) rely on an optimal choice of the order parameter. In this chapter, we ignore the slight differences in the definition of an order parameter and a reaction coordinate and use the terms interchangeably. However, for all methods it should be noted that a projection into a different reaction coordinate space after the simulation is often not acceptable.[35] The missing information and the neglected degrees of freedom often prohibit a reasonable interpretation *a posteriori* such that the reaction coordinates have to be chosen before the simulation starts. Therefore, high-temperature simulations might represent a computationally efficient method to decide about the quality of the reaction coordinates *a priori*.[36]

6.3 Methods

6.3.1 Thermodynamic Integration

As a general principle, integration methods determine the change in free energy between two states of a system from the integral of the work required to go from an initial state to a final state *via* a reversible path. The path itself may be physical or non-physical. Here we focus on non-physical paths in the context of so-called alchemical free energy differences, in which the two states differ in their Hamiltonian. The transformation of a Hamiltonian H_A to a Hamiltonian H_B is performed gradually by introducing $H(\boldsymbol{p}, \boldsymbol{r}, \lambda)$ with the momenta $\boldsymbol{p}$, as a function of a coupling parameter λ which has values between 0 and 1. The difference in free energy $\Delta A = A_B - A_A$ between the two systems is characterised by the Hamiltonians $H(\boldsymbol{p}, \boldsymbol{r}; \lambda_B)$ and $H(\boldsymbol{p}, \boldsymbol{r}; \lambda_A)$ and becomes:

$$A_B - A_A = \int_{\lambda_A}^{\lambda_B} \left\langle \frac{\partial H}{\partial \lambda} \right\rangle_{\lambda} \mathrm{d}\lambda. \tag{6.8}$$

This expression is formally exact. Because the free energy is a state function, the result of the integration should be independent of the path chosen. However in practice the choice of the paths can seriously affect the convergence of the calculation such that it is important to develop criteria for suitable paths.[37–39] The choice of λ-dependence is often determined by the properties of the derivative $\partial H/\partial\lambda$ which needs to be evaluated during the course of the simulation. For force field-based methods, the particular values of λ usually correspond to the gradual increase or decrease of electrostatic and Lennard-Jones interactions. The modifications of these so-called non-bonded potentials in the Hamiltonian can be used to estimate the free energy of binding between two molecules, for example between enzymes and substrates[25,40] or free solvation energies[41,42] of solutes. The main concept behind these fundamental processes corresponds to the use of thermodynamic cycles.[43] The calculation of free energy differences related to computationally challenging processes, *e.g.* the binding of substrates to enzymes, can be easily approximated by *alchemical transformations*. Herewith, slight mutations of the original molecule are introduced, *e.g.* often so-called dummy atoms or modified molecular groups. In fact, the path independence of the free energy difference allows us to use these non-physical paths and to calculate the free energy difference of the reference scheme. An example can be found in Figure 6.1. If one is interested in an enzyme–substrate binding free energy difference between substrates X and Y as defined by $\Delta\Delta A_{\mathrm{YX}} = \Delta A_{\mathrm{bind,Y}} - \Delta A_{\mathrm{bind,X}}$, it is often infeasible to consider the full binding pathway between the unbound and the bound substrate as denoted by the horizontal reactions. In fact, these reactions can be replaced by the vertical reactions that only consider alchemical transformations between the substrates in the bound and the unbound state, respectively. Thus, one does not have to consider the whole reaction pathway but alchemically transforms substrate X to substrate Y. The free energy difference of interest $\Delta\Delta A_{\mathrm{YX}}$ can then be calculated by $\Delta\Delta A_{\mathrm{YX}} = \Delta A_{\mathrm{bound,XY}} - \Delta A_{\mathrm{unbound,XY}}$.

The use of thermodynamic integration also provides the possibility of distinguishing between the entropic and the enthalpic contributions to the free energy. This is of specific interest for free solvation energies as well as hydrophobic hydration where entropic effects are supposed to impose an important contribution.[44,45]

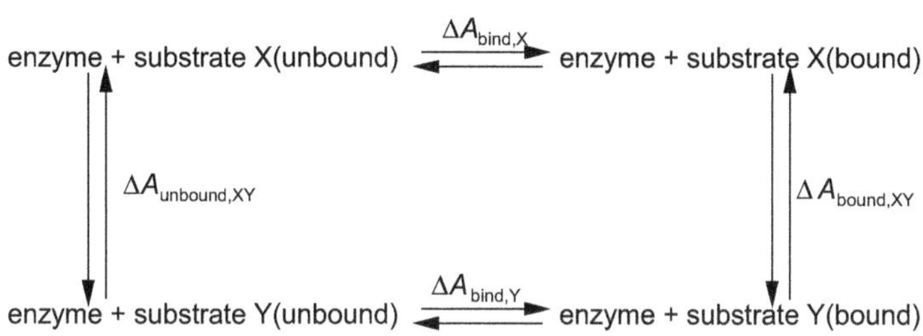

Figure 6.1 A thermodynamic cycle to estimate the relative enzyme–substrate binding free energy difference between substrates X and Y. The horizontal transformations are often not accessible in computer simulations. The alchemical transformations from a substrate X to a substrate Y in free solution (unbound state to the enzyme at the left side of the cycle) or to a bound state can be calculated by thermodynamic integration to yield the final estimate $\Delta\Delta A_{\mathrm{YX}} = \Delta A_{\mathrm{bind,Y}} - \Delta A_{\mathrm{bind,X}} = \Delta A_{\mathrm{bound,XY}}$ which corresponds to the binding free energy difference between substrates X and Y.

The entropy difference between two states can be estimated by using a finite difference expression[46,47] according to:

$$\Delta S \approx -\frac{\Delta A(T+\Delta T) - \Delta A(T-\Delta T)}{2\Delta T} \tag{6.9}$$

where the free energy difference between two states ΔA has to be evaluated at different temperatures T with the corresponding temperature differences ΔT. It should be noted that the temperature differences are usually small and of maximum difference $\Delta T \approx 20$ K to the target temperature. Note that eqn (6.9) can be applied to all free energy methods but often the free energy differences are rather inaccurate such that it is often only applied in the context of thermodynamic integration.[48] Furthermore, it should be noted that the inherent error in the calculated entropy difference is at least one order of magnitude larger than that in the free energy difference.[49]

6.3.1.1 Replica-exchange MD

Replica-exchange MD, also often called parallel tempering, was mainly developed for the study of systems like spin glasses.[50] Later, the method was reformulated[51] and finally applied to biomolecular simulations.[21,52] The basic motivation behind replica exchange is given by the fact that the dynamics of the system, *e.g.* the movement of the internal degrees of freedom, is significantly faster at higher temperatures. In general, a replica-exchange set includes a series of simulation runs which are executed in parallel, where each simulation is conducted at a different temperature. The optimal difference between the temperatures in the simulations hardly affects the efficiency of the replica-exchange algorithm.[53] Each simulation starts with a different protein conformation and therefore represents a possible replica of the system. The main idea now relies on the exchange of the conformations by the use of the Metropolis criterion and the evaluation of the underlying potential energies. The probability for a replica exchange is given by:

$$P = \min\{1, \exp[(\beta_i - \beta_j)(E(\mathbf{r}_i) - E(\mathbf{r}_j))]\} \tag{6.10}$$

with the potential energies $E(\mathbf{r}_i)$ for replicas i and j with the corresponding inverse temperatures denoted by $\beta_i = 1/k_B T_i$. In practice, at certain time steps, so-called swap moves are introduced where the energy of the replicas is calculated and compared. If the Metropolis criterion is fulfilled, an exchange between the replicas is performed which means that replica i originally simulated at T_i now has a temperature T_j and *vice versa*. If the Metropolis criterion is not fulfilled, the exchange is rejected. After a while, the whole phase space is explored, which means that all free energy landscapes at different temperatures according to eqn (6.3) are accessible. The effectiveness of the replica-exchange method has been frequently discussed and criticised.[54] Nevertheless, modern improvements[54,55] and the introduction of the Hamiltonian replica-exchange approach[56] have led to a renaissance of the method. The main drawbacks are surely the slow convergence at lower temperatures as well as the large computational effort to run hundreds of parallel simulations. However, due to the increased computational power available in the last years, these points can be neglected. What remains is a relatively simple method, which relies on statistical mechanics, avoids the introduction of reaction coordinates or additional degrees of freedom and provides free energy differences that are not affected by the application of any biasing force.

The combination of replica exchange and thermodynamic integration results in significant improvement of conventional thermodynamic integration in cases where high barriers have to be overcome along the thermodynamic path.[57,58] More detailed insights into the broad applicability of replica-exchange MD simulations and some variants and extensions in the context of biomolecular simulation can be found in ref. 21, 54, 56, 59 and 60.

6.3.2 Free Energy Perturbation Approaches

6.3.2.1 *Fast Growth*

It was shown by Jarzynski in 1997[61] that the work W required to pull a system over a barrier in a set of non-reversible paths averages to the free energy difference:

$$\Delta A = -\frac{1}{\beta}\ln\langle\exp(-\beta W)\rangle \tag{6.11}$$

under non-equilibrium conditions. The non-reversible paths need to be drawn from a canonical ensemble of, for example, the reactant state. Several computational techniques are based on Jarzynski's identity. A method called fast growth[62,63] takes samples from an equilibrated reactant state's ensemble and moves them with a constraint towards the target state. The constraint can act on a reaction coordinate which depends on atom coordinates. The work along each path is sampled and averaged according to eqn (6.11). The constraint can be moved quite fast, because an intermediate equilibration is not necessary (in contrast to a different method called slow growth which is analysed similarly to thermodynamic integration). However, the faster the movement, the larger the spread in W, such that more trajectories are required to obtain an accurate average for ΔA.

6.3.2.2 *Free Energy Perturbation*

Free energy perturbation (FEP) allows the calculation of the free energy difference between two states without the definition of a reaction coordinate. Let us assume that the potential energy of the full system $E(\boldsymbol{r})$ can be written as a sum of a reference energy $E_{\mathrm{ref}}(\boldsymbol{r})$ and a perturbation, $E = E_{\mathrm{ref}} + E_{\mathrm{pert}}$. The reference energy may be the energy of a reactant state, the perturbation due to the change in bonding during an enzymatic reaction. From eqn (6.2) it follows that the free energy contributions can also be split into $A = A_{\text{ref}} + A_{\text{pert}}$:

$$\begin{aligned} A_{\text{pert}} &= -A_{\text{ref}} + A \\ &= -A_{\text{ref}} - \frac{1}{\beta}\ln\int \exp[-\beta(E_{\text{ref}} + E_{\text{pert}})]\mathrm{d}^N r \\ &= -\frac{1}{\beta}\ln\int \frac{\exp[-\beta E_{\text{ref}}(\boldsymbol{r})]}{\exp[-\beta A_{\text{ref}}]}\exp[-\beta E_{\text{pert}}]\mathrm{d}^N r \end{aligned} \tag{6.12}$$

Using eqn (6.6) one can rewrite that as:

$$A_{\text{pert}} = -\frac{1}{\beta}\ln\int P_{\text{ref}}(\boldsymbol{r})\exp[-\beta E_{\text{pert}}]\mathrm{d}^N r = -\frac{1}{\beta}\ln\langle\exp[-\beta E_{\text{pert}}]\rangle_{\text{ref}} \tag{6.13}$$

where eqn (6.7) has been used in the last step. The derivation is formally exact, no approximations have been introduced. Eqn (6.13) essentially means that the free energy contribution of a change in the potential energy is the exponential ensemble average of that change over the *unperturbed* ensemble. The method works, as mentioned, without the definition of a reaction coordinate. The two states can be distinguished by the energy expression alone.

In practical application, the system is propagated using the unperturbed Hamiltonian. At intervals, the perturbed energy is calculated and finally averaged according to eqn (6.13). The exponential average is dominated by the smallest values of E_{pert}, *i.e.* for configurations where E is smaller than E_{ref}. Because the system mainly resides in areas with small E_{ref}, the method is efficient only if the distributions of reference state and perturbed state overlap to a sufficient degree. To improve sampling, the perturbation can be split up into smaller steps, each sampled individually. A different way to improve sampling efficiency is to additionally exchange reference and perturbed states, *i.e.* to perturb backwards, and then to average the outcome. However, forward and backward perturbation may have very different convergence properties such that appropriate averaging techniques are required.[64–66]

FEP can be used to describe physical transformation between states, alchemical changes (like turning one atom or group of atoms into another one) or changes in the accuracy of the Hamiltonian. An example of the latter is to sample using a force field approach and calculate the free energy influence by an *ab initio* or QM/MM treatment.

The term FEP has historic origin and is nowadays somewhat confusing, because it is an exact derivation and has no relation to perturbation theory in its usual sense. A special case of FEP is the sampling of free energy differences of geometric changes in a QM/MM description. A quantum mechanical description of the interatomic forces is typically required for the chemical step during an enzymatic process, while binding and unbinding and possible structural rearrangements can often be described by force fields with sufficient accuracy. QM/MM with a density functional theory or higher description of the QM part is usually computationally prohibitively expensive for full sampling of free energy differences. However, the main entropic changes during the chemical step of an enzymatic reaction are often caused by the environment of the active centre, which is described by a force field. The QM part is typically responsible for the main enthalpic changes. QM/MM-FEP[67–71] offers a possibility to sample the MM part while keeping the QM part frozen. The algorithm works as follows: first, a series of optimised geometries along the reaction path is found by standard QM/MM geometry optimisations with a constraint along the reaction coordinate. Then the QM part is frozen, and the MM part is subjected to a MD sampling at finite temperature. FEP to the previous and the next QM geometry is performed. In that way, an energy profile is obtained, which includes the entropy of the MM part and the QM/MM interaction. The entropy of the QM part is neglected but can be added in the harmonic approximation. The method has been shown to provide reliable results for enzymatic reactions.[70,71]

6.3.2.3 Bennett's Acceptance Ratio Method

While thermodynamic integration can be viewed as one of the most widely used free energy approaches, indeed many alternatives are available. The Bennett acceptance ratio (BAR) approach[64] combines the distributions of the potential energy difference from a forward and backward FEP calculation in a statistical optimal way leading to:

$$\Delta A_{ij} = -\frac{1}{\beta}\ln\left(\frac{\langle\{1+\exp[\beta(\Delta E_{ij}-C)]\}^{-1}\rangle_i}{\langle\{1+\exp[-\beta(\Delta E_{ij}-C)]\}^{-1}\rangle_j}\right) + C \tag{6.14}$$

in which the constant C is given by:

$$C = \Delta A_{ij} + \frac{1}{\beta}\ln\frac{n_i}{n_j} \tag{6.15}$$

where n_i and n_j represent the number of samples collected from each state. Convergence of this iterative process can only be reached if there is a sufficient overlap between the forward and backward energy differences. However significantly less overlap between the configurational space of each state is required in order to converge results when compared to the FEP method. BAR can also be viewed as a maximum likelihood estimate of the free energy for the samples from two states.[72] An extension of the theory to a multistate framework is available.[73] Comparisons between thermodynamic integration and BAR in practical settings suggest the better efficiency and robustness of BAR.[74,75]

6.3.3 Umbrella Sampling

Umbrella sampling[76–79] is a technique to enhance ergodicity in the simulation, *i.e.* to make sure that all regions along the reaction coordinate are sampled with similar probability. This is achieved by placing a bias, an additional artificial energy term $w(\xi)$, along a predefined reaction coordinate ξ. It can be chosen to increase sampling in high-energy regions and to reduce sampling close to the minima. The bias depends only on the reaction coordinate and on no other degree of freedom. The resulting biased distribution along the reaction coordinate is

$$P^{\mathrm{b}}(\xi) = \frac{\int \exp[-\beta E(\boldsymbol{r}) - \beta w(\xi(\boldsymbol{r}))]\ \delta[\xi(\boldsymbol{r}) - \xi]\mathrm{d}^N r}{\int \exp[-\beta E(\boldsymbol{r}) - \beta w(\xi(\boldsymbol{r}))]\mathrm{d}^N r} \tag{6.16}$$

as a modification of eqn (6.4). The corresponding distribution is obtained from a MD sampling run with a bias. In fact, the main task is to calculate a free energy profile from $P^{\mathrm{b}}(\xi)$. The integration in the numerator is performed over all coordinates but ξ, while the bias depends only on ξ. Therefore, $\exp[-\beta w(\xi(\boldsymbol{r}))]$ can be taken out of the numerator's integral in accordance with:

$$P^{\mathrm{b}}(\xi) = \exp[-\beta w(\xi(\boldsymbol{r}))] \frac{\int \exp[-\beta E(\boldsymbol{r})]\ \delta[\xi(\boldsymbol{r}) - \xi]\mathrm{d}^N r}{\int \exp[-\beta E(\boldsymbol{r}) - \beta w(\xi(\boldsymbol{r}))]\mathrm{d}^N r}. \tag{6.17}$$

which leads with eqn (6.4) to:

$$P(\xi) = P^{\mathrm{b}}(\xi) \times \exp[\beta w(\xi(\boldsymbol{r}))] \times \frac{\int \exp[-\beta E(\boldsymbol{r}) - \beta w(\xi(\boldsymbol{r}))]\mathrm{d}^N r}{\int \exp[-\beta E(\boldsymbol{r})]\mathrm{d}^N r} \tag{6.18}$$

for the true probability $P(\xi)$. In this expression, $P^{\mathrm{b}}(\xi)$ is obtained from a MD sampling run, $\exp[\beta w(\xi(\boldsymbol{r}))]$ can be calculated analytically from the definition of the bias, and the last term can be interpreted as an ensemble average of $\exp[-\beta w(\xi(\boldsymbol{r}))]$ denoted as $\langle \exp[-\beta w(\xi(\boldsymbol{r}))] \rangle$, which is independent of ξ. The free energy depending on ξ can be calculated by inserting eqn (6.18) into eqn (6.5):

$$A(\xi) = -\frac{1}{\beta} \ln P^{\mathrm{b}}(\xi) - w(\xi) + F \tag{6.19}$$

with $F = -\frac{1}{\beta} \ln\langle \exp[-\beta w(\xi(\boldsymbol{r}))] \rangle$. No approximations were introduced so far, therefore eqn (6.19) results in the correct free energy, under the assumption that the sampling of $P^{\mathrm{b}}(\xi)$ is sufficient.

Different flavours of umbrella sampling are in use to ensure a proper sampling. It is obvious from eqn (6.19) that the perfect bias, resulting in a uniform distribution of $P^{\mathrm{b}}(\xi)$ is $w(\xi) = -A(\xi)$, which is, of course, unknown at the start of the simulation. It can be iteratively approximated in adaptive-bias umbrella sampling such that an initial simulation provides the first approximation of $w(\xi) = -A(\xi)$ which is refined in an iterative process. A different approach is to split the reaction coordinate into windows. Different, independent simulations are run for each window i with a different bias w_i in each of them to provide free energy curves $A_i(\xi)$ for each window:

$$A_i(\xi) = -\frac{1}{\beta}\ln P_i^{\mathrm{b}}(\xi) - w_i(\xi) + F_i \tag{6.20}$$

where the values F_i are different. The curves from $A_i(\xi)$ have to be matched in order to achieve a global $A(\xi)$. This matching procedure, which results in appropriate values for F_i, can be performed with the weighted histogram analysis method (WHAM).[80–83] In many cases, second-order polynomials (*i.e.* harmonic potentials) are used as a bias in w_i. In these cases, it is favourable to calculate the mean force (for which the F_i vanish) and numerically integrate it to arrive at $A(\xi)$. This method was termed umbrella integration.[84–87] Contrary to WHAM, umbrella integration can also be applied if there is no overlap in the distribution of two adjacent images. Recently, the availability of mean forces from umbrella integration in more dimensions was used to find minimum free energy pathways,[88,89] *i.e.* the most probable path a reaction is expected to take.

6.3.4 Enveloping Distribution Sampling

Instead of perturbing a system from its state A directly to state B such as in FEP or using various intermediate simulations such as in thermodynamic integration, a non-physical intermediate state, a so-called reference state can be defined, which allows the computation of the free energy difference:

$$\Delta A_{\mathrm{BA}} = A_{\mathrm{B}} - A_{\mathrm{A}} = A_{\mathrm{BR}} - A_{\mathrm{AR}} = -\frac{1}{\beta}\ln\frac{\langle \exp[-\beta(E_{\mathrm{B}}(\boldsymbol{r}) - E_{\mathrm{R}}(\boldsymbol{r}))]\rangle_{\mathrm{R}}}{\langle \exp[-\beta(E_{\mathrm{A}}(\boldsymbol{r}) - E_{\mathrm{R}}(\boldsymbol{r}))]\rangle_{\mathrm{R}}} \tag{6.21}$$

from a single simulation of this reference state R, where $\langle\ldots\rangle_{\mathrm{R}}$ denotes an ensemble average over R. In enveloping distribution sampling (EDS), this reference state is chosen such that its configurational ensemble envelops all important configurations of states A and B. The potential energy term of the Hamiltonian of the reference state reads[90–94]

$$E_{\mathrm{R}}(\boldsymbol{r};\, s, \Delta E_{\mathrm{BA}}^{\mathrm{R}}) = -\frac{1}{\beta s}\ln\{\exp[-\beta s(E_{\mathrm{A}}(\boldsymbol{r}) - E_{\mathrm{A}}^{\mathrm{R}})] + \exp[-\beta s(E_{\mathrm{B}}(\boldsymbol{r}) - E_{\mathrm{B}}^{\mathrm{R}})]\} \tag{6.22}$$

where $E_{\mathrm{X}}(\boldsymbol{r})$ is the potential energy of state X = A,B. The smoothness parameter s with values $0 < s \le 1$ controls the barriers that determine the rate of transitions between the two end states. The energy offsets $E_{\mathrm{X}}^{\mathrm{R}}$ of the different states serve to adjust their energies to the same level for a homogeneous sampling. Because the equations of motion depend on the energy offset difference $\Delta E_{\mathrm{BA}}^{\mathrm{R}} = E_{\mathrm{B}}^{\mathrm{R}} - E_{\mathrm{A}}^{\mathrm{R}}$ and not on individual quantities, $E_{\mathrm{A}}^{\mathrm{R}}$ is set to zero for a two-state system. Therefore, two parameters have to be determined iteratively prior to an EDS production simulation.[95–97] The EDS reference state is not limited to two end states but has also been extended to multiple end states.[95,98] While the method was originally applied to alchemical perturbations, conformational free energy differences can be calculated as well.[99–101]

Recently, EDS has been extended to probe a single-simulation alternative to the thermodynamic cycle that is standardly used for measuring the effect of a modification of a chemical compound, *e.g.* from a given species to a chemical derivative for a ligand or solute molecule, on the free energy change associated with a change in environment, *e.g.* from the unbound state to the bound state for a protein–ligand system according to Figure 6.1 or from one solvent to another one for a solute molecule.[102,103]

The efficiency of EDS depends on the amount of phase space overlap between the end states. For large perturbations, *i.e.* point mutations in proteins,[104] it might be useful to use a limited number of intermediate states, *e.g.* by introducing an additional coupling parameter.[105]

Recently, EDS was applied in constant pH simulations by mixing the energy surface of the protonated and unprotonated states for a small set of preselected ionisable sites. Systems with different values of the EDS smoothing parameter are then connected *via* a Hamiltonian replica exchange procedure to facilitate interconversion between the different protonation states during a hybrid MD–Monte Carlo simulation.[106]

6.3.5 Transition Path Sampling

Transition path sampling (TPS) is one of the oldest rare event sampling methods. It was first introduced by Dellago *et al.* in 1998.[22] A general overview can be found in some older review articles.[107,108] A recent perspective is given by Bolhuis.[109] Although the method has already proven its applicability in recent years, it might become of new importance with regard to the currently available computer hardware. As in forward flux sampling, the computational procedure to perform TPS is quite expensive with regard to the amount of required computation time and necessary data storage capability. The application of TPS mostly focuses on the study of the transition state ensemble, which requires a reliable sampling method to adequately describe the transition state between two regions A and B. In principle, the sampling of the transition state is a challenging process. Often the statistics are so poor that no reliable description is possible. TPS can be regarded as a method to circumvent this problem. The basic idea behind TPS is quite simple. As a prerequisite, it is assumed that all transitions between regions A and B have to be regarded as rare events due to high activation energies. Rare events in computer simulations are usually defined by their transition time τ_{AB}, which is of the same order or even larger than the simulation time τ_s. TPS introduces a population order parameter $h(\xi)$ where ξ denotes a point in phase space spanned by the positions $\boldsymbol{r}^N$ and the momenta $\boldsymbol{p}^N$ of all N particles. When the actual position of the system is in region A, one defines

$h_A(\xi)=1$ and $h_A(\xi)=0$ otherwise. The corresponding population operator $h_B(\xi)$ for region B is defined similarly. Furthermore it is assumed that the transitions between A and B can be described by trajectories $\xi(t)=(\xi_0, \xi_{1,\ldots}, \xi_t)$ as a chronological sequence of phase space points following Liouville's equation. The statistical weight of the trajectories is given by the expression $h_A(\xi_0)\rho[\xi(t)]h_B(\xi_t)$ where $\rho[\xi(t)]$ is defined by:

$$\rho[\xi(t)]=\rho(\xi_0)\prod_{0<t'\leq t}\delta[\xi_{t'}-\xi_{t'}(\xi_0)] \tag{6.23}$$

as an unconstrained functional for initial distributions of phase space points starting at ξ_0. TPS can therefore be interpreted as a random walk in phase space with a focus on the transition state ensemble by an importance sampling of the weighting functional $h_A(\xi_0)\rho[\xi(t)]h_B(\xi_t)$.

In fact, the computational procedure is quite simple. A basic prerequisite is given by a representative trajectory that connects states A and B. The transition state ensemble is now generated by the use of 'shooting moves'. A shooting move initiates a new starting point which is generated from the old trajectory. Therefore, a randomly chosen point in phase space along the trajectory that connects regions A and B is slightly perturbed by changing the momenta of the particles. Furthermore, it is assumed that the system obeys Newton's dynamics. The new trajectory is then integrated forward and backward in time. A detailed description of the shooting move procedure is presented by Dellago *et al.*[110] Recently, improved methods with regard to the conservation of energy and momentum have also been developed.[111] The main prerequisite is given by the fact that the combined forward and backward time integrated trajectory must fulfil the condition that regions A and B are still connected. This requirement is imposed by the constraints of importance sampling[107] which accepts all trajectories that connect regions A and B.

The set of trajectories generated by different initial starting points then provides reasonable statistical accuracy to adequately analyse the transition state. The main interest focuses on the calculation of the committor and the separatrix. The committors p_A and p_B are defined as the probabilities or the fraction of trajectories $\bar{\xi}(t_s)$ that end in region A or B from a specific configuration at ξ. Furthermore, an identical definition for p_B is introduced. Detailed expressions can be found in ref. 22 and 107. The introduction of the committor allows one to define the separatrix at the saddle point. If the values for $p_A=p_B\approx 0.5$, one can directly assume that the position of the so-called transition state or separatrix is known. In fact, this concept is closely related to p-folding analysis for proteins.[112] Moreover, through the good statistics obtained by TPS, one can simply evaluate the free energy landscape according to eqn (6.5). Also rate constants can be easily obtained by an approach first introduced by Chandler and Bennett.[113,114] The rate constant for the transition from A to B can be determined by the relation[115]

$$k_{AB}=\frac{\mathrm{d}}{\mathrm{d}t}C(t) \tag{6.24}$$

with the correlation function

$$C(t)=\frac{h_{\mathrm{A}}(t_0)h_{\mathrm{B}}(t)}{h_{\mathrm{A}}} \tag{6.25}$$

with $h_{\mathrm{A}}=1$ or 0 if the configuration is in A or not.

In summary, it has been shown that TPS has a broad applicability in the context of biomolecular simulations. Although computationally expensive, recent applications include protein folding,[116,117] the study of reaction pathways[118] and enzymatic catalysis.[119,120] Nevertheless, a recent publication studied reaction pathways for enzymatic reactions as obtained by TPS and some other transition state theory methods and found significant differences between the approaches.[121] The authors conclude that the equilibration in TPS might be insufficient due to a rapid climb to the activation energy barrier, which could be a potential issue for most rare event and free energy methods.

6.3.6 Forward Flux Sampling

FFS is a method to simulate rare events with a broad applicability.[23,89,122–124] The main concept of FFS relies on the introduction of the reaction coordinate ξ which continuously connects the states A and B without any further requirements. The reaction coordinate is divided by hypersurfaces λ_i which correspond to specific values of $\xi_i=\xi(\lambda_i)$ with $\xi(\lambda_i)<\xi(\lambda_{i+1})$. The borders of region A and B are represented by the hypersurfaces λ_0 and λ_n. The forward transition probability $p(\lambda_n|\lambda_0)$ to reach B from a starting point in A is given by:

$$p(\lambda_n\,|\,\lambda_0)=\prod_{i=0}^{n-1}p(\lambda_{i+1}\,|\,\lambda_i) \tag{6.26}$$

which yields the forward transition rate

$$k_{\mathrm{AB}}=\phi_{\mathrm{A}}\prod_{i=0}^{n-1}p(\lambda_{i+1}\,|\,\lambda_i) \tag{6.27}$$

where ϕ_{A} denotes the outcoming flux of trajectories from A through hypersurface λ_0. The backward transition rate is defined by

$$k_{\mathrm{BA}}=\phi_{\mathrm{B}}\prod_{i=0}^{n-1}p(\lambda_i\,|\,\lambda_{i+1}) \tag{6.28}$$

with the flux of outcoming trajectories ϕ_{B} from B through hypersurface λ_n.

In accordance with the detailed balance condition, the stationary distribution function is composed of

$$P(\xi)=\Psi_{\mathrm{A}}(\xi)+\Psi_{\mathrm{B}}(\xi) \tag{6.29}$$

with

$$\Psi_{\mathrm{A}}(\xi) = \rho(A)\phi_{\mathrm{A}}\tau_{+}(\xi|\lambda_0) \tag{6.30}$$

and

$$\Psi_{\mathrm{B}}(\xi) = \rho(B)\phi_{\mathrm{B}}\tau_{-}(\xi|\lambda_n) \tag{6.31}$$

where $\tau_+(\xi/\lambda_0)$ and $\tau_-(\xi/\lambda_n)$ are the mean forward and backward time spent at ξ for trajectories that originate either from state A at λ_0 or state B at λ_n.

In fact, the computational procedure of FFS is closely related to TPS,[107,108] transition interface sampling[125] or stochastic path sampling methods.[126] More details about the protocol can be found in ref. 23. As a first step for FFS, one has to carry out simulations which produce a collection of configurations that cross a fixed hypersurface at λ_0. Trial runs are initiated from a few randomly chosen and predefined configurations at λ_0, which are continued as long as they reach either λ_1 or fall back to λ_0. The end points of the trajectories at λ_1, respectively the corresponding configurations, will be stored and used as initiators for new trial runs, which have to reach either λ_2 or completely return to λ_0. The procedure is repeated as long as the final hypersurface λ_n is reached. A schematic representation is shown in Figure 6.2. The corresponding transition probabilities $p(\lambda_{i+1}|\lambda_i)$ can be estimated by the fraction of successful trial runs. Due to the large amount of sufficient sampling data, the free energy and the free energy difference between any arbitrarily chosen states ξ and ξ_0 can be evaluated by eqn (6.3) and (6.5).

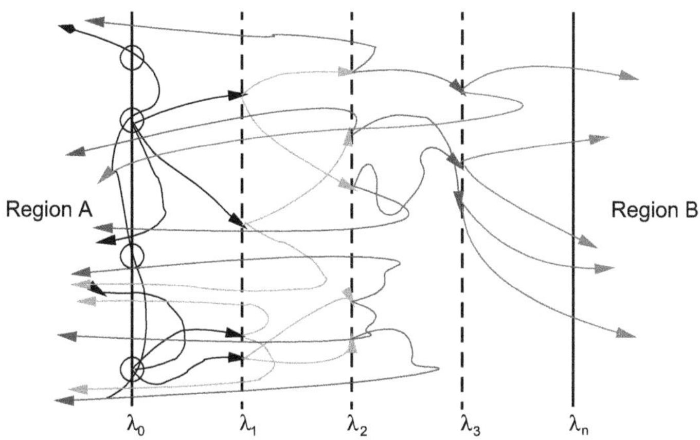

Figure 6.2 Schematic representation of the computational procedure of FFS. The black curved line indicates the initial simulation that is used to generate starting points at λ_0 as denoted by the black circles. From these points, new trajectories are produced that either approach λ_1 or fall back to λ_1. The configurations at λ_1 are used as new starting points for trajectories until finally region B is reached.

In the context of biomolecular simulations, FFS has been used to study polymer translocation,[122,127,128] conformational behaviour of DNA[129] or proteins and peptides[89,130] as well as protein association.[131] A more detailed overview can be found in ref. 132.

6.3.7 Metadynamics

The metadynamics method was first introduced by Laio and Parrinello in 2002.[19] The original idea behind metadynamics is closely related to previously developed approaches like the local elevation method[133] and conformational flooding.[134] All these methods have in common that the introduction of a bias potential along predefined reaction coordinates is essential. The main idea relies on a successive flattening of the free energy landscape by an additional history-dependent potential energy in the form of small Gaussian hills which are deposited after a fixed time at the actual position of the system in phase space. It has been shown that this method is in principle applicable for all reasonable reaction coordinates and offers a broad range of applicability.[17]

The mathematical description is straightforward such that we consider a canonical system which is described by a set of atomic coordinates $\boldsymbol{r}$ evolving under the action of dynamics following the trajectory $\boldsymbol{r}(t)$ at a fixed temperature T. The set of coordinates $\boldsymbol{r}$ are projected onto reaction coordinates $\xi(\boldsymbol{r}(t))$ which might reflect functions of atomic positions or angles as well as any other auxiliary collective variable representing the characteristics of the system. If rare event properties are inherent, for example by the fact that usually some regions of the phase space are separated by large energy barriers, metadynamics can be regarded as a tool to simulate the transition in a reasonable simulation time. The bias potential which allows the system to transit the free energy minima is given by

$$V_{\mathrm{B}}(\xi,t)=\omega\sum_{t'=\tau_{\mathrm{G}},2\tau_{\mathrm{G}},3\tau_{\mathrm{G}},\ldots}\exp\left(-\frac{(\xi(t)-\xi(t'))^2}{2\delta\xi^2}\right) \tag{6.32}$$

which can be interpreted as a sum of Gaussian hills. The individual Gaussian hills with the energy height ω and the width $\delta\xi$ in the reaction coordinate space are deposited after fixed time intervals τ_{G} such that the resulting actual force at t' is

$$\boldsymbol{F}(\xi,t')=-\frac{\mathrm{d}}{\mathrm{d}\xi}V_{\mathrm{B}}(\xi,t') \tag{6.33}$$

which depends on the history of the visited phase space. One of the main properties of metadynamics is reflected by its quasi-equilibrium approach. Therefore, it has to be guaranteed that the depositing times τ_{G} exceed the internal relaxation times such that the system is able to return into local equilibrium before a new hill is deposited. If this condition is fulfilled, it can be rigorously shown by a Liouville type description,[135] that

$$A(\xi) = -\lim_{t_s \to \infty} V_B(\xi, t_s) \tag{6.34}$$

which indicates that the resulting bias potential exactly becomes the negative mirror of the free energy landscape $A(\xi)$ in the limit of infinite simulation times t_s.

Although the validity of eqn (6.34) has been rigorously shown, it has to be mentioned that the original method strongly depends on errors for finite simulation times that are introduced by the values of system parameters like the height and the width of the hills.[135,136] Intuitively, it becomes evident that the errors can be mostly related to the absence of a fully converged landscape which is specifically crucial for large hill heights or broad hill widths.

Several improvements were introduced to overcome this inherent drawback of metadynamics. The most prominent and often used method is *well-tempered metadynamics*,[137] which provides the possibility of adaptively adjusting the hill height and to effectively sample the regions of interest by varying hill heights depending on the frequency of the visited phase space points. The hill height at the actual position ξ is given by:

$$\omega_{\text{WTM}}(\xi, t) = \omega \exp\left(-\frac{V_B(\xi, t)}{k_B \Delta T}\right) \tau_G \tag{6.35}$$

where ΔT is a tunable temperature-like biasing factor which controls how quickly $\omega_{\text{WTM}}(\xi, t)$ decreases. The resulting free energy landscape can be calculated by

$$A(\xi) = -\frac{T + \Delta T}{\Delta T} V_{\text{WTM}}(\xi, t_s) \tag{6.36}$$

with biasing factors between ΔT = 15 and 20 K and an expression for a modified form of the biasing energy V_{WTM} as given in ref. 137.

An improved and thermodynamically consistent convergence scheme has also been introduced in ref. 35. Herewith, the outcomes of the rough free energy landscape estimated by a modification of the metadynamics method are combined with WHAM, which provides exact results within a global error tolerance.[80–83] Therefore, the biased free energy landscape is used as a rough estimate to conduct multiple and short simulation runs within a constant potential to refine the free energy surface in the spirit of umbrella sampling. The overlapping distribution functions are then combined by the WHAM scheme.

Metadynamics, and specifically well-tempered metadynamics, can also be used to determine the rate constants of rare event processes[89,138–140] by an approach originally presented in ref. 134. The framework introduces a single-step process for the transition from the energetic minimum at region A to region B and the transition state at region R′ which represents a dividing surface at ξ'. The energetic minimum state A can be described by the partition function Z_0 valid for all $\xi < \xi'$ whereas Z_0' corresponds to the transition region R' at $\xi = \xi'$. If it is further assumed that no Gaussian hills are deposited at the transition region due to the rare event properties of the process, the mean transition time in a metadynamics simulation, denoted by the subscript MTD, can be approximated by:

$$\tau_{\mathrm{MTD}} = \frac{1}{\nu \kappa_{\mathrm{MTD}}} \frac{Z_{\mathrm{MTD}}}{Z'_{\mathrm{MTD}}} \tag{6.37}$$

with the characteristic frequency ν, the transition coefficient κ_{MTD}[138] and the partition functions Z_{MTD} and Z'_{MTD} as described above. With regard to the absence of deposited Gaussian hills in the transition region, it can be assumed that $\kappa_{\mathrm{MTD}} \approx \kappa$ and $Z'_{\mathrm{MTD}} \approx Z_0'$ in agreement with the behaviour found for an unbiased transition. The acceleration factor $\alpha = \tau/\tau_{\mathrm{MTD}}$ for the unbiased transition time τ and the transition time of the metadynamics simulation τ_{MTD}, which corresponds to the time needed for the first transition from A over R′ to B, can be obtained by:

$$\alpha \approx \frac{Z_0}{Z'_{\mathrm{MTD}}} = \langle \exp(\beta V_{\mathrm{B}}(\xi, t)) \rangle_{\mathrm{MTD}} \tag{6.38}$$

where the angular brackets denote the average over a metadynamics simulation. Finally, the rate constant for the transition can be estimated by

$$k_{\mathrm{AB}} = \frac{1}{\alpha \tau_{\mathrm{MTD}}}. \tag{6.39}$$

In fact, the transition time for the metadynamics simulation τ_{MTD} can be easily estimated by monitoring the time between the transitions whereas the corresponding actual value for $V_{\mathrm{B}}(\xi, t)$ is also needed for the calculation of α. A comparison between FFS and metadynamics indicates a good agreement for the rate constants obtained by both approaches.[89] The reasonable accuracy for the rate constants was also discussed in the original ref. 138

In summary, metadynamics can be applied to a broad range of biomolecular systems.[141] Several excellent reviews[17,142] discussed some successful applications of metadynamics. Recent approaches include docking or adsorption studies,[143–148] the study of ligand-translocation processes[149,150] and the determination of protein and DNA stable structures as well as the corresponding unfolding pathways.[11,151–154] The huge advantage of metadynamics can be clearly seen by its applicability for different simulation approaches ranging from *ab initio* MD[155,156] to coarse-grained dissipative particle dynamics simulations.[146] Furthermore, the broad applicability of metadynamics can be explained by the availability of efficient software tools to perform these simulations.[157]

6.3.8 Averaging Techniques in QM/MM Simulations

While full sampling is generally desirable to obtain free energy differences, computational considerations sometimes restrict the applicability. In particular, the QM/MM approach of coupling quantum mechanical (QM) treatment of the chemically active region to force field methods (MM) for the environment often pose severe restrictions on the achievable sampling time. As shown above, sampling can be restricted to the MM region. In many cases one has to approximate even further. In many QM/MM studies, MD sampling is avoided overall and only potential energies are considered. During the system preparation for enzymatic QM/MM calculations, the protein in water is often subjected to an MD simulation based solely on force fields in order to ensure proper solvation and relaxation of crystal artifacts. It became a standard procedure in the last decade to start the QM/MM geometry optimisation, *i.e.* the search for stationary points on the potential energy surface from a number of snapshots from such an MD run. While this obviously results in an improper sampling, at least it ensures that some flexibility of the protein structure is taken into account when reaction barriers are calculated. Such a procedure results in a set of reaction barriers $\Delta^{\ddagger}E_i$, one from each snapshot. The question arises as to how to average them. The arithmetic average fails to take into account transitions over lower barriers which are more likely than those over higher barriers. Thus, the average barrier (which corresponds to the desired average reaction rate) should be dominated by the lower values in the set of $\Delta^{\ddagger}E_i$. While there will be a spread between the values of the barriers, the energies of the reactant state minima of the different snapshots E_i^{RS} will vary as well. Typically, they vary much more than $\Delta^{\ddagger}E_i$ for large systems.[158] It was shown by comparison to full free energy calculations by umbrella sampling that a Boltzmann-weighted average (also called exponential average)[159]

$$\Delta^{\ddagger}E_{\mathrm{exp}} = -k_{\mathrm{B}}T\ln\left(\frac{1}{N}\sum_{i=1}^{N}\exp(-\Delta^{\ddagger}E_i/k_{\mathrm{B}}T)\right). \tag{6.40}$$

is the most promising approach to average between barriers obtained by different snapshots.[158] The exponential average is smaller than the arithmetic mean, but approaches the latter for large temperatures. In addition, it is also larger than the minimum of $\Delta^{\ddagger}E_i$ and approaches it for low temperatures. In many cases, however, the exponential average is close to the minimum of $\Delta^{\ddagger}E_i$ at ambient temperature. As a consequence, snapshots with small barriers dominate $\Delta^{\ddagger}E_{\mathrm{exp}}$. To improve sampling, snapshots are chosen for which a small $\Delta^{\ddagger}E_i$ is expected, *i.e.* those with a favourable environment for the reaction to occur. This can be estimated from the value of the reaction coordinate in the MD structure. In principle, one can separate eqn (6.40) as

$$\Delta^{\ddagger}E_{\mathrm{exp}} = k_{\mathrm{B}}T\ln N - k_{\mathrm{B}}T\ln\left(\sum_{i=1}^{N}\exp(-\Delta^{\ddagger}E_i/k_{\mathrm{B}}T)\right) \tag{6.41}$$

which allows one to manually choose the snapshots and to neglect $k_B T \ln N$. Thus, as long as this term is not too large compared to the second term, a proper sampling procedure outweighs the errors caused by neglecting $k_B T \ln N$. The exponential average has been used in a number of QM/MM simulation studies of enzymatic processes.[159–165] While it is obtained as a weighted average of potential energies, some relation to a full free energy barrier can be drawn. If $\Delta^{\ddagger}E_i$ are considered approximations to the work to drive the system over the barrier, and one assumes them to be drawn from a canonical ensemble, Jarzynski's equation[61] would relate the resulting exponential average to a free energy. Obviously, the two assumptions are only partially fulfilled. Nevertheless, using a series of snapshots and the exponential average of the resulting barriers should be the method of choice to obtain reaction barriers from QM/MM simulations of enzymatic processes.

6.4 Conclusions

Over recent decades, the number of free energy methods and rare event sampling techniques has grown rapidly. This can be related to the significant improvement of computational power and software development. The study of problems that were infeasible a few years ago, *e.g.* proteins with a large number of residues or protein docking, now become accessible thanks to improved hardware and software performance. The study of these systems by brute force MD simulations is still a challenging task[8] but some insight can be gained into the behaviour at long timescales. Nevertheless, for a detailed study of thermodynamic properties or the analysis of intermediate states and their transitions, rare event sampling techniques as well as free energy methods are still the methods of choice.

In this review, we described a few of the methods that have been developed. We have mostly focused on methods that were proven to be successful in the context of protein or enzyme simulations. For a description of additional methods, we refer the reader to excellent textbooks[25,26] and further review articles,[27,28] among others. Due to the vast amount of methods, it becomes clear that each problem demands its own approach to be solved. Hence, it cannot be assumed that a single method provides the best results for any arbitrarily chosen system. The study of thermodynamic properties usually demands free energy methods whereas transition pathways or further dynamical properties are best tackled by rare event techniques. Furthermore, it is evident that biasing techniques like metadynamics crucially influence the dynamics of the system, such that unbiased methods like TPS of FFS might be a better choice. The question of the chosen reaction coordinate is also important. As a possible option, high-temperature simulations as a prerequisite to estimate the important reaction coordinates or the accessible phase space as well as transition pathways might be useful.[11,36] Often the chosen method also depends on the length scale and the intrinsic timescale of the investigated system. A general recipe cannot be advised such that the machinery of published methods becomes reasonable. It can be assumed that additional and even more sophisticated methods for specific problems will be developed in the coming years.

In principle, most methods share some common points and can be interpreted as slight extensions of existing approaches. Further methods are also often combinations of formerly derived approaches. Typical examples are bias exchange metadynamics,[55] histogram reweighted metadynamics[35] or umbrella integration.[84] The powerful combination of existing approaches often offers the advantage of higher statistical accuracy or efficiency. It is therefore desirable to combine successful approaches in a smart way to extract the main advantages. In summary, the development of free energy methods or rare event sampling techniques will continue. The study of new systems will also drive the development of novel biased and unbiased methods. Fortunately, it can therefore be expected that the 'golden age' of free energy methods will at least last for the next decades.

Acknowledgements

This work was financially supported by the German Research Foundation (DFG) within the Cluster of Excellence in Simulation Technology (EXC 310/2) and within the collaborative research center SFB 716, both at the University of Stuttgart.

References

1. J. Šponer, X. Cang and T. E. Cheatham, *Methods*, 2012, **57**, 25.
2. A. Grossfield, P. Ren and J. W. Ponder, *J. Am. Chem. Soc.*, 2003, **125**, 15671.
3. R. Lonsdale, J. N. Harvey and A. J. Mulholland, *Chem. Soc. Rev.*, 2012, **41**, 3025.
4. P. E. Lopes, J. Huang, J. Shim, Y. Luo, H. Li, B. Roux and A. D. MacKerell Jr., *J. Chem. Theory Comput.*, 2013, **9**, 5430.
5. E. Brunk and U. Rothlisberger, *Chem. Rev.*, 2015, **115**, 6217.
6. S. Mishra and M. Meuwly *Kinetics and Dynamics*, Springer, 2010, p. 137.
7. V. S. Pande, I. Baker, J. Chapman, S. P. Elmer, S. Khaliq, S. M. Larson, Y. M. Rhee, M. R. Shirts, C. D. Snow, E. J. Sorin *et al.*, *Biopolymers*, 2003, **68**, 91.
8. D. E. Shaw, J. Grossman, J. A. Bank, B. Batson, J. A. Butts, J. C. Chao, M. M. Deneroff, R. O. Dror, A. Even and C. H. Fenton *et al. Proceedings of the International Conference for High Performance Computing, Networking, Storage and Analysis*, IEEE Press, p. 41.
9. J. D. Owens, M. Houston, D. Luebke, S. Green, J. E. Stone and J. C. Phillips, *Proc. IEEE*, 2008, **96**, 879.
10. J.-H. Prinz, H. Wu, M. Sarich, B. Keller, M. Senne, M. Held, J. D. Chodera, C. Schütte and F. Noé, *J. Chem. Phys.*, 2011, **134**, 174105.
11. J. Smiatek, C. Chen, D. Liu and A. Heuer, *J. Phys. Chem. B*, 2011, **115**, 13788.

12. S. Liu, S. Cao, K. Hoang, K. L. Young, A. S. Paluch and D. L. Mobley, *J. Chem. Theory Comput.*, 2016, DOI: 10.1021/acs.jctc.5b00934.
13. J. D. Bryngelson, J. N. Onuchic, N. D. Socci and P. G. Wolynes, *Proteins*, 1995, **21**, 167.
14. J. N. Onuchic, Z. Luthey-Schulten and P. G. Wolynes, *Annu. Rev. Phys. Chem*, 1997, **48**, 545.
15. N. D. Socci, J. N. Onuchic, P. G. Wolynes *et al.*, *Proteins*, 1998, **32**, 136.
16. J. Smiatek, D. Janssen-Müller, R. Friedrich and A. Heuer, *Phys. A*, 2014, **394**, 136.
17. A. Laio and F. L. Gervasio, *Rep. Prog. Phys.*, 2008, **71**, 126601.
18. C. M. Dobson, A. Šali and M. Karplus, *Angew. Chem., Int. Ed.*, 1998, **37**, 868.
19. A. Laio and M. Parrinello, *Proc. Natl. Acad. Sci. U. S. A.*, 2002, **99**, 12562.
20. E. Darve, D. Rodríguez-Gómez and A. Pohorille, *J. Chem. Phys.*, 2008, **128**, 144120.
21. Y. Sugita and Y. Okamoto, *Chem. Phys. Lett.*, 1999, **314**, 141.
22. C. Dellago, P. G. Bolhuis, F. S. Csajka and D. Chandler, *J. Chem. Phys.*, 1998, **108**, 1964.
23. R. J. Allen, C. Valeriani and P. R. ten Wolde, *J. Phys.: Condens. Mater.*, 2009, **21**, 463102.
24. H. Eyring, *J. Chem. Phys.*, 1935, **3**, 107.
25. C. Chipot and A. Pohorille, *Free Energy Calculations*, Springer, 2007.
26. D. Wales, *Energy Landscapes: Applications to Clusters, Biomolecules and Glasses*, Cambridge University Press, 2003.
27. C. D. Christ, A. E. Mark and W. F. van Gunsteren, *J. Comput. Chem.*, 2010, **31**, 1569.
28. S. Singh, M. Chopra and J. J. de Pablo, *Annu. Rev. Chem. Biomol. Eng.*, 2012, **3**, 369.
29. M. A. Rohrdanz, W. Zheng and C. Clementi, *Annu. Rev. Phys. Chem.*, 2013, **64**, 295.
30. N. S. Bieler and P. H. Hünenberger, *J. Chem. Phys.*, 2015, **142**, 165102.
31. A. Jain, R. Hegger and G. Stock, *J. Phys. Chem. Lett.*, 2010, **1**, 2769.
32. R. Hegger and G. Stock, *J. Chem. Phys.*, 2009, **130**, 034106.
33. S. V. Krivov, *J. Chem. Theory Comput.*, 2012, **9**, 135.
34. A. Amadei, A. B. Linssen and H. J. C. Berendsen, *Proteins*, 1993, **17**, 412.
35. J. Smiatek and A. Heuer, *J. Comput. Chem.*, 2011, **32**, 2084.
36. J. Smiatek, D. Liu and A. Heuer, *Curr. Phys. Chem*, 2012, **2**, 115.
37. A. Blondel, *J. Comput. Chem.*, 2004, **25**, 985.
38. R. Bitetti-Putzer, W. Yang and M. Karplus, *Chem. Phys. Lett.*, 2003, **377**, 633.
39. T. T. Pham and M. R. Shirts, *J. Chem. Phys.*, 2012, **136**, 124120.
40. C. Chipot *New Algorithms for Macromolecular Simulation*, Springer, 2006, p. 185.
41. P. Kollman, *Chem. Rev.*, 1993, **93**, 2395.
42. S. A. Martins, S. F. Sousa, M. J. Ramos and P. A. Fernandes, *J. Chem. Theory Comput.*, 2014, **10**, 3570.

43. B. L. Tembre and J. A. McCammon, *Comput. Chem.*, 1984, **8**, 281.
44. N. T. Southall, K. A. Dill and A. Haymet, *J. Phys. Chem. B*, 2002, **106**, 521.
45. P. Ball, *Chem. Rev.*, 2008, **108**, 74.
46. R. M. Levy and E. Gallicchio, *Annu. Rev. Phys. Chem*, 1998, **49**, 531.
47. S. Wan, R. H. Stote and M. Karplus, *J. Chem. Phys.*, 2004, **121**, 9539.
48. J. Smiatek, *J. Phys. Chem. B*, 2014, **118**, 771.
49. S. H. Fleischman and C. L. Brooks, *J. Chem. Phys.*, 1987, **87**, 3029.
50. R. H. Swendsen and J.-S. Wang, *Phys. Rev. Lett.*, 1986, **57**, 2607.
51. C. J. Geyer, *Am. Stat. Assoc.*, 1991, 156.
52. U. H. Hansmann, *Chem. Phys. Lett.*, 1997, **281**, 140.
53. A. Patriksson and D. van der Spoel, *Phys. Chem. Chem. Phys.*, 2008, **10**, 2073.
54. D. J. Earl and M. W. Deem, *Phys. Chem. Chem. Phys.*, 2005, **7**, 3910.
55. S. Piana and A. Laio, *J. Phys. Chem. B*, 2007, **111**, 4553.
56. H. Fukunishi, O. Watanabe and S. Takada, *J. Chem. Phys.*, 2002, **116**, 9058.
57. C. J. Woods, M. A. King and J. W. Essex in *New Algorithms for Macromolecular Simulation*, ed. B. Leimkuhler, C. Chipot, R. Elber, A. Laaksonen, A. E. Mark, T. Schlick, C. Schütte and R. Skeel, Springer, Berlin, 2006, p. 251.
58. I. V. Khavrutskii and A. Wallqvist, *J. Chem. Theory Comput.*, 2010, **6**, 3427.
59. R. Zhou, *Prot. Fold. Protoc.*, 2006, **350**, 205.
60. K. Ostermeir and M. Zacharias, *Biochim. Biophys. Acta*, 2013, **1834**, 847.
61. C. Jarzynski, *Phys. Rev. Lett.*, 1997, **78**, 2690.
62. G. Hummer, *J. Chem. Phys.*, 2001, **114**, 7330.
63. D. M. Zuckerman and T. B. Woolf, *Phys. Rev. Lett.*, 2002, **89**, 180602.
64. C. H. Bennett, *J. Comput. Phys.*, 1976, **22**, 245.
65. N. Lu, J. K. Singh and D. A. Kofke, *J. Chem. Phys.*, 2003, **118**, 2977.
66. S. Sakuraba and N. Matubayasi, *J. Chem. Phys.*, 2011, **135**, 114108.
67. J. Bentzien, R. P. Muller, J. Florián and A. Warshel, *J. Phys. Chem. B.*, 1998, **102**, 2293.
68. R. V. Stanton, M. Perakyla, D. Bakowies and P. A. Kollman, *J. Am. Chem. Soc.*, 1998, **120**, 3448.
69. Y. Zhang, H. Liu and W. Yang, *J. Chem. Phys.*, 2000, **112**, 3483.
70. J. Kästner, H. M. Senn, S. Thiel, N. Otte and W. Thiel, *J. Chem. Theory Comput.*, 2006, **2**, 452.
71. H. M. Senn, J. Kästner, J. Breidung and W. Thiel, *Can. J. Chem.*, 2009, **87**, 1322.
72. M. R. Shirts, E. Bair, G. Hooker and V. S. Pande, *Phys. Rev. Lett.*, 2003, **91**, 140601.
73. M. R. Shirts and J. D. Chodera, *J. Chem. Phys.*, 2008, **129**, 124105.
74. M. R. Shirts and V. S. Pande, *J. Chem. Phys.*, 2005, **122**, 144107.
75. A. de Ruiter, S. Boresch and C. Oostenbrink, *J. Comput. Chem.*, 2013, **34**, 1024.
76. I. R. McDonald and K. Singer, *J. Chem. Phys.*, 1967, **47**, 4766.

77. I. R. McDonald and K. Singer, *J. Chem. Phys.*, 1969, **50**, 2308.
78. G. M. Torrie and J. P. Valleau, *Chem. Phys. Lett.*, 1974, **28**, 578.
79. G. M. Torrie and J. P. Valleau, *J. Comput. Phys.*, 1977, **23**, 187.
80. S. Kumar, J. M. Rosenberg, D. Bouzida, R. H. Swendsen and P. A. Kollman, *J. Comput. Chem.*, 1992, **13**, 1011.
81. S. Kumar, J. M. Rosenberg, D. Bouzida, R. H. Swendsen and P. A. Kollman, *J. Comput. Chem.*, 1995, **16**, 1339.
82. B. Roux, *Comput. Phys. Commun.*, 1995, **91**, 275.
83. M. Souaille and B. Roux, *Comput. Phys. Commun.*, 2001, **135**, 40.
84. J. Kästner and W. Thiel, *J. Chem. Phys.*, 2005, **123**, 144104.
85. J. Kästner and W. Thiel, *J. Chem. Phys.*, 2006, **124**, 234106.
86. J. Kästner, *J. Chem. Phys.*, 2009, **131**, 034109.
87. J. Kästner, *J. Chem. Phys.*, 2012, **136**, 234102.
88. M. U. Bohner and J. Kästner, *J. Chem. Phys.*, 2012, **137**, 034105.
89. M. U. Bohner, J. Zeman, J. Smiatek, A. Arnold and J. Kästner, *J. Chem. Phys.*, 2014, **140**, 074109.
90. K. K. Han, *Phys. Lett. A*, 1992, **165**, 28.
91. N. F. Hansen and H. C. Andersen, *J. Phys. Chem.*, 1996, **199**, 1137.
92. Y. G. Chen and G. Hummer, *J. Am. Chem. Soc.*, 2007, **129**, 2414.
93. C. D. Christ and W. F. van Gunsteren, *J. Chem. Phys.*, 2007, **126**, 184110.
94. C. D. Christ and W. F. van Gunsteren, *J. Chem. Phys.*, 2008, **128**, 174112.
95. C. D. Christ and W. F. van Gunsteren, *J. Chem. Theory Comput.*, 2009, **5**, 276.
96. S. Riniker, C. D. Christ, N. Hansen, A. E. Mark, P. C. Nair and W. F. van Gunsteren, *J. Chem. Phys.*, 2011, **135**, 024105.
97. N. Hansen, J. Dolenc, M. Knecht, S. Riniker and W. F. van Gunsteren, *J. Comput. Chem.*, 2012, **33**, 640.
98. C. D. Christ and W. F. van Gunsteren, *J. Comput. Chem.*, 2009, **30**, 1664.
99. Z. Lin, H. Liu, S. Riniker and W. F. van Gunsteren, *J. Chem. Theory Comput.*, 2011, **7**, 3884.
100. Z. Lin and W. F. van Gunsteren, *J. Chem. Phys.*, 2013, **139**, 144105.
101. Z. Lin and W. F. van Gunsteren, *Protein Sci.*, 2016, **25**, 184.
102. N. Hansen, P. H. Hünenberger and W. F. van Gunsteren, *J. Chem. Theory Comput.*, 2013, **9**, 1334.
103. J. Gebhardt and N. Hansen, *Fluid Phase Equilib.*, 2016, **422**, 1–17.
104. N. Hansen, J. R. Allison, F. H. Hodel and W. F. van Gunsteren, *Biochemistry*, 2013, **52**, 4962.
105. T. Mori, R. J. Hamers, J. A. Pedersen and Q. Cui, *J. Phys. Chem. B*, 2014, **118**, 8210.
106. J. Lee, B. T. Miller, A. Damjanović and B. R. Brooks, *J. Chem. Theory Comput.*, 2014, **10**, 2738.
107. P. G. Bolhuis, D. Chandler, C. Dellago and P. Geissler, *Annu. Rev. Phys. Chem.*, 2002, **53**, 291.
108. C. Dellago, P. Bolhuis and P. L. Geissler, *Adv. Chem. Phys.*, 2002, **123**, 1.
109. P. Bolhuis and C. Dellago, *Eur. Phys. J.: Spec. Top.*, 2015, **224**, 2409.

110. C. Dellago, P. G. Bolhuis and P. L. Geissler, in *Computer Simulations in Condensed Matter Systems: From Materials to Chemical Biology Volume 1*, Springer, 2006, p. 349.
111. R. G. Mullen, J.-E. Shea and B. Peters, *J. Chem. Theory Comput.*, 2015, **11**, 2421.
112. R. Du, V. S. Pande, A. Y. Grosberg, T. Tanaka and E. S. Shakhnovich, *J. Chem. Phys.*, 1998, **108**, 334.
113. D. Chandler, *J. Chem. Phys.*, 1978, **68**, 2959.
114. C. H. Bennett, *ACS Symp. Ser.*, 1977, **46**, 63.
115. C. Dellago, P. G. Bolhuis and D. Chandler, *J. Chem. Phys.*, 1999, **110**, 6617.
116. P. G. Bolhuis, *Proc. Natl. Acad. Sci. U. S. A.*, 2003, **100**, 12129.
117. J. Juraszek and P. G. Bolhuis, *Biophys. J.*, 2008, **95**, 4246.
118. R. J. Dimelow, R. A. Bryce, A. J. Masters, I. H. Hillier and N. A. Burton, *J. Chem. Phys.*, 2006, **124**, 114113.
119. J. E. Basner and S. D. Schwartz, *J. Am. Chem. Soc.*, 2005, **127**, 13822.
120. B. C. Knott, M. H. Momeni, M. F. Crowley, L. F. Mackenzie, A. W. Götz, M. Sandgren, S. G. Withers, J. Ståhlberg and G. T. Beckham, *J. Am. Chem. Soc.*, 2014, **136**, 321.
121. D. Doron, A. Kohen, K. Nam and D. T. Major, *J. Chem. Theory Comput.*, 2014, **10**, 1863.
122. R. J. Allen, D. Frenkel and P. R. ten Wolde, *J. Chem. Phys.*, 2006, **124**, 194111.
123. C. Valeriani, R. J. Allen, M. J. Morelli, D. Frenkel and P. R. ten Wolde, *J. Chem. Phys.*, 2007, **127**, 114109.
124. K. Kratzer, A. Arnold and R. J. Allen, *J. Chem. Phys.*, 2013, **138**, 164112.
125. T. S. Van Erp and P. G. Bolhuis, *J. Comput. Phys.*, 2005, **205**, 157.
126. J. T. Berryman and T. Schilling, *J. Chem. Phys.*, 2010, **133**, 244101.
127. J. P. Hernandez-Ortiz, M. Chopra, S. Geier and J. J. de Pablo, *J. Chem. Phys.*, 2009, **131**, 044904.
128. E. Ribeiro Tzaras, F. Weik, C. Holm and J. Smiatek, submitted 2016.
129. W. Möbius, R. A. Neher and U. Gerland, *Phys. Rev. Lett.*, 2006, **97**, 208102.
130. E. E. Borrero and F. A. Escobedo, *J. Chem. Phys.*, 2006, **125**, 164904.
131. A. Taudt, A. Arnold and J. Pleiss, *Phys. Rev. E*, 2015, **91**, 033311.
132. F. A. Escobedo, E. E. Borrero and J. C. Araque, *J. Phys.: Condens. Mater.*, 2009, **21**, 333101.
133. T. Huber, A. E. Torda and W. F. Gunsteren, *J. Comput. -Aided. Mol. Des*, 1994, **8**, 695.
134. H. Grubmüller, *Phys. Rev. E*, 1995, **52**, 2893.
135. G. Bussi, A. Laio and M. Parrinello, *Phys. Rev. Lett.*, 2006, **96**, 090601.
136. A. Laio, A. Rodriguez-Fortea, F. L. Gervasio, M. Ceccarelli and M. Parrinello, *J. Phys. Chem. B*, 2005, **109**, 6714.
137. A. Barducci, G. Bussi and M. Parrinello, *Phys. Rev. Lett.*, 2008, **100**, 020603.

138. P. Tiwary and M. Parrinello, *Phys. Rev. Lett.*, 2013, **111**, 230602.
139. P. Tiwary and M. Parrinello, *J. Phys. Chem. B*, 2014, **119**, 736.
140. M. Salvalaglio, P. Tiwary and M. Parrinello, *J. Chem. Theory Comput.*, 2014, **10**, 1420.
141. V. Leone, F. Marinelli, P. Carloni and M. Parrinello, *Curr. Opin. Struct. Biol.*, 2010, **20**, 148.
142. A. Barducci, M. Bonomi and M. Parrinello, *WIRes Comput. Mol. Sci.*, 2011, **1**, 826.
143. F. L. Gervasio, A. Laio and M. Parrinello, *J. Am. Chem. Soc.*, 2005, **127**, 2600.
144. A. Kranjc, S. Bongarzone, G. Rossetti, X. Biarnés, A. Cavalli, M. L. Bolognesi, M. Roberti, G. Legname and P. Carloni, *J. Chem. Theory Comput.*, 2009, **5**, 2565.
145. M. Masetti, A. Cavalli, M. Recanatini and F. L. Gervasio, *J. Phys. Chem. B*, 2009, **113**, 4807.
146. J. Smiatek, A. Heuer, H. Wagner, A. Studer, C. Hentschel and L. Chi, *J. Chem. Phys.*, 2013, **138**, 044904.
147. C. Hentschel, H. Wagner, J. Smiatek, A. Heuer, H. Fuchs, X. Zhang, A. Studer and L. Chi, *Langmuir*, 2013, **29**, 1850.
148. F. S. Di Leva, E. Novellino, A. Cavalli, M. Parrinello and V. Limongelli, *Nucleic Acids Res.*, 2014, 247.
149. M. Ceccarelli, C. Danelon, A. Laio and M. Parrinello, *Biophys. J.*, 2004, **87**, 58.
150. M. Ceccarelli and P. Ruggerone, *Curr. Drug Targets*, 2008, **9**, 779.
151. M. Bonomi, D. Branduardi, F. L. Gervasio and M. Parrinello, *J. Am. Chem. Soc.*, 2008, **130**, 13938.
152. N. Todorova, F. Marinelli, S. Piana and I. Yarovsky, *J. Phys. Chem. B*, 2009, **113**, 3556.
153. F. Pietrucci and A. Laio, *J. Chem. Theory Comput.*, 2009, **5**, 2197.
154. V. Limongelli, S. De Tito, L. Cerofolini, M. Fragai, B. Pagano, R. Trotta, S. Cosconati, L. Marinelli, E. Novellino, I. Bertini *et al.*, *Angew. Chem., Int. Ed.*, 2013, **52**, 2269.
155. J. Blumberger, B. Ensing and M. L. Klein, *Angew. Chem., Int. Ed.*, 2006, **45**, 2893.
156. V. Stamenkovic, B. S. Mun, K. J. Mayrhofer, P. N. Ross, N. M. Markovic, J. Rossmeisl, J. Greeley and J. K. Nørskov, *Angew. Chem., Int. Ed.*, 2006, **118**, 2963.
157. G. A. Tribello, M. Bonomi, D. Branduardi, C. Camilloni and G. Bussi, *Comput. Phys. Commun.*, 2014, **185**, 604.
158. A. M. Cooper and J. Kästner, *ChemPhysChem*, 2014, **15**, 3264.
159. I. Logunov and K. Schulten, *J. Am. Chem. Soc.*, 1996, **118**, 9727.
160. R. Lonsdale, J. N. Harvey and A. J. Mulholland, *J. Phys. Chem. B*, 2010, **114**, 1156.
161. J. Kästner and P. Sherwood, *Mol. Phys.*, 2010, **108**, 293.
162. M. W. van der Kamp, J. Żurek, F. R. Manby, J. N. Harvey and A. J. Mulholland, *J. Phys. Chem. B*, 2010, **114**, 11303.

163. R. Lonsdale, K. T. Houghton, J. Żurek, C. M. Bathelt, N. Foloppe, M. J. de Groot, J. N. Harvey and A. J. Mulholland, *J. Am. Chem. Soc.*, 2013, **135**, 8001.
164. R. Lonsdale, S. Hoyle, D. T. Grey, L. Ridder and A. J. Mulholland, *Biochemistry*, 2012, **51**, 1774.
165. C. Z. Christov, A. Lodola, T. G. Karabencheva-Christova, S. Wan, P. V. Coveney and A. J. Mulholland, *Biophys. J.*, 2013, **104**, L5.

Section II: Methods

CHAPTER 7

Methods to Trace Conformational Transitions

PEDRO SFRISO*[a,b] AND MODESTO OROZCO*[a,b,c]

[a] Institute for Research in Biomedicine (IRB Barcelona), The Barcelona Institute of Science and Technology, Barcelona, Spain; [b] Joint IRB-BSC Program in Computational Biology, Barcelona, Spain; [c] Department of Biochemistry and Molecular Biology, University of Barcelona, Spain
*Email: pedro.sfriso@irbbarcelona.org; modesto.orozco@irbbarcelona.org

7.1 Proteins are Molecular Machines

Most protein functions rely on their ability to deform or change shape in a precise and regulated way.[1–3] Protein flexibility is encoded in the structure that in turn is imprinted in the sequence.[2,4] Therefore, in addition to the average structure, evolution used protein dynamics to optimise function, from mechanical work, to transport or enzymatic activity.[2,4–8] The range and types of enzymatic motions is very broad: from local side-changes fluctuations to dramatic domain translocations including local folding/unfolding events.[9] Such motions are critical to enzyme catalysis particularly to ligand binding/release and formation of the reactive cavity.[3,10] Thus, a complete understanding of the structure and function of proteins requires a deep knowledge on their dynamics.[11]

Structural biology experiments are filling databases with static snapshots of protein conformations.[12] In some cases, when the protein is resolved in different conditions by X-ray diffraction, alternative conformations can be obtained, providing an indirect picture of protein flexibility (see Figure 7.1). For small proteins NMR experiments can provide also indirect information

RSC Theoretical and Computational Chemistry Series No. 9
Simulating Enzyme Reactivity: Computational Methods in Enzyme Catalysis
Edited by Iñaki Tuñón and Vicent Moliner

Published by the Royal Society of Chemistry, www.rsc.org

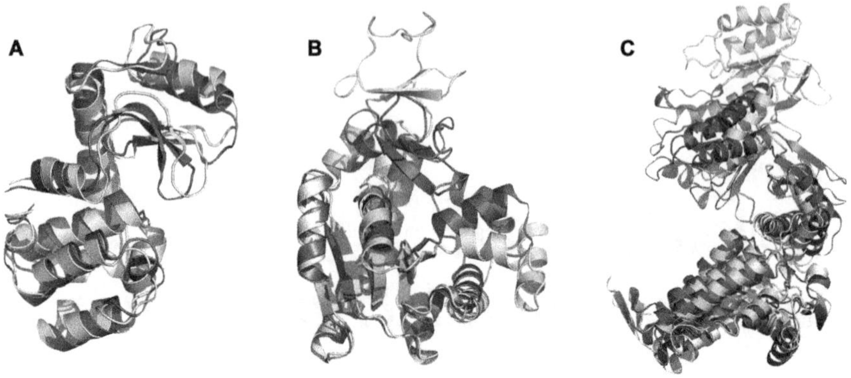

Figure 7.1 Examples of conformational transitions represented by a pair of crystal structures. (A) Lyzosyme displays subtle conformational change between (PDB 149L (gray) and 2LZM (white). (B) Adenylate kinase open (PDB AKE, white) and closed (PDB 1AKE, gray) is probably the most studied conformational transitions, with more than 2000 devoted publications. (C) GroEL monomer (PDB 1SX4, white) and (PDB 1OEL, gray). This large conformational transition occurs for all subunits of the GroEL complex.

on protein flexibility, but the cases where full dynamic information can be derived are scarce, reflecting the intrinsic problems of experimental techniques to provide a detailed picture of protein flexibility.[13,14] Consequently, the use of theoretical models is required to dissect the causes and consequences of protein's motion.

Computationally, the simulation of protein motions should follow the conformational changes from one stable state (A) to a second one (B). The sequence of events occurring along these transitions (in the adopted notation, A→B) is known as the conformational transition: the lowest free energy path connecting both states. A large variety of computational methods have been designed to find these conformational transitions; the aim of this chapter is to provide a review of the most representative ones. These methods are helping to explain longstanding questions in enzyme dynamics, such as: how the chemical energy and mechanical motion are interconverted?[15–18] how are allosteric movements regulated?;[19–22] what is the prevalence of conformational selection *vs.* inducted fit recognitions,[23–26] or what are the protein's moonlighting mechanisms.[27,28] Ultimately, computational methods will help us to understand, predict and engineer molecular machines, including new enzymes.[17,29–31]

7.2 Computational Methods to Trace Transition Paths

As introduced, computational models are the only practical option to follow transition pathways. The myriad of methods available is presented here based on the formalism in which generate intermediate conformations. A full description of all methods available is unfeasible. Yet, we sought to outline the assumptions and ideas behind various approaches, show their utility, providing with a non-exhaustive, hopefully insightful, introduction to the relevant literature.

7.2.1 Interpolation Schemes

The conformational motion between two known states can be visualised by interpolation methods that connect A→B with intermediate structures. These methods determine the type, site and magnitude of the conformational change, for instance, for classification purposes. Generally, interpolation methods assume that the protein moves in rigid blocks, following the transition path by pivoting over few points or screw over an axis.[32,33] Using these geometrical approaches, Gerstein and coworkers built a database of molecular motions[34] that was the first attempt to classify protein flexibility. Nonetheless, paths derived with interpolation methods should be treated with caution, since steric clashes and alterations of bond distances and angles can occur, leading to structures which are not chemically meaningful. To alleviate these problems, Kim *et al.*[35,36] presented a model that follows the transition between two conformations using distances between the nodes, in Elastic Network Models (see below), rather than the position vectors. In an alternative approach, Seeliger *et al.*[37] presented the tCONCOORD method for geometric exploration of conformational transition paths. tCONCOORD analyses internal interactions in the structure and identifies, based on geometrical considerations, the essential degrees of freedom. Special attention is dedicated to hydrogen bonds, whose solvation probability is used to score their stability. Less stable hydrogen bonds are considered mobile while stable ones and other restrains such as bond distances, angles and planarity are kept along the transition path. Ensembles generated with tCONCOORD display a good overlap with experimental structures and MD simulations.[37] Again careful inspection of generated paths is required, since the method does not feel any energetic bias along the motion. Weiss and Levitt[38] improved conformational pathways by avoiding (through the use of geometrical restraints) the population of high-energy areas of the conformational landscape. Despite its simplicity, the resulting method, CLIMBER, has been able to derive conformational pathways visiting known intermediates.

7.2.2 Methods Based on Normal Modes

Normal Mode Analysis (NMA) extracts equilibrium motions directly from the energy surface.[39–43] NMA assumes that the initial structure corresponds to the (free) energy minimum and its conformational energy can be approximated as a multidimensional parabola. Consequently, the potential energy surface (V) for N particles can be approximated as:

$$V(r) = \frac{1}{2}\sum_{i,j}^{3N} \frac{\partial^2 V}{\partial r_i \partial r_j} \Delta r_i \Delta r_j \tag{7.1}$$

where r_i are the coordinates for the ith atom and Δr_i its relative displacement (identical for j index). The second derivative term in eqn (7.1) is the element i–j of the Hessian matrix of the system. Eigenvectors and eigenvalues obtained by diagonalisation of the mass-weighted Hessian matrix, lead to $3N-6$ concerted displacements referred as normal modes (NM). Under the NMA formalism, it is assumed that large-amplitude NM are functional movements, coded in the structure to aid protein function.[43–46] NMA can work with any description of the protein (including the atomistic one), and with any representation of the protein Hamiltonian. However, for the purpose of following transitions paths NMA has been mainly used in combination with simple protein models like Elastic Network Models (ENM).[47–50] ENM describe protein dynamics based on the three-dimensional position of particles (typically Cα), which are connected to neighbouring atoms with springs. The approach may seem a priori simplistic. However, it has been shown that a combination 2 NM (selected within the 15 first modes of structure A) often contains key information on the conformational transition to the target structure (B).[35,51–53] This fact, which supports the idea that equilibrium structure is somehow pre-coded to explore relevant transitions, explains the success of many of ENM-NMA methods cited below.

Tama and Sanejouand[51] used NMA to explore the motion of 20 proteins and demonstrated the ability of NM to overlap with known conformational transitions. The authors showed that NMA methods perform better if started with the open forms of the proteins, linking the quality of NMA results with the shape of the protein. They also quantified that collective motions are, as expected, better captured by NMA than local ones. Since this seminal work, a large amount of research was dedicated to study conformational changes by combination of NM.[54,55] Here we selected some illustrative examples.

Xu *et al.*[56] followed NM of haemoglobin to generate structures from which atomistic details of the T→R2 transition were recovered, including the increase in stiffness at hinge points or the burial of some residues. The same group applied their adaptive anisotropic network model to the study the functional dynamics of GroEL.[57] The protocol required two end structures that are both starting points for an iterative advance over the conformational space. A double-minima energy function[58] was used to compute the NM connecting both end conformations generating an energy landscape from which intermediate states are retrieved. Using another variant of the ENM-NMA Maragakis and Karplus studied the transition path of adenylate kinase.[59] In their contribution, they calibrated the force constants of the elastic network to reproduce experimental B-factors of end structures.

This strategy led to a very good agreement with known X-ray structures along the transition path and to accurate identification of hinge points.

Recent advances in ENM-NMA methods moved in different directions. For example, some models replaced Cartesian coordinates with internal coordinates yielding more natural protein motions.[60] Besides, the dimensionality of the Hessian matrix is drastically reduced allowing the fast computation of NM for huge molecular machines, with more than 100k residues.[61] Another direction explored was to perturb the elastic network, mimicking disease-related mutations. The impact of on dynamics is strikingly captured in the altered NM.[62,63] A different approach consists on following transition paths by updating and adapting NM of the reference structure to the current state in the path,[64–66] which guarantees smoother and more reliable transition paths.

NMA analysis can accurately capture protein dynamics near equilibrium,[67] providing impressive amount of information on conformational transitions. However, fine details of mechanisms of motion are out of the capabilities of such simple methods. Also, when the conformational change is not collective or requires large changes in the topology more advanced sampling engines are needed.

7.2.3 Minimum Energy Paths

Minimum energy path (MEP) approaches connect two known end states with conformers being at the minimum potential energy (not free energy) transition path.[68–76] We will use the popular Conjugate Peak Refinement (CPR) method by the Karplus' group[77] as a representative example to illustrate the idea of MEP. CPR assumes that both initial and target states are low energy conformations and also that the conformational transition occurs through the saddle points of the valley connecting them. Schematically, the algorithm works as following:

1) CPR starts with a rough estimation of the transition path (A → B), for instance a linear interpolation.
2) The highest energy conformation in the path is identified.
3) From the conjugate direction of (2) the local minimum energy state (C) is identified as well as its conjugate gradient.
4) The transition path is now split in two: A → C and C → B. The procedure continues in (2) for the two halves.
5) The path is refined until all maxima in the transition path are saddle points, yielding a MEP.

Additional steps are required for further refinement. During the iterative process, intermediate structures are added, removed or enhanced in order to avoid high-energy conformers. Limiting atomic displacements to a few Armstrongs impedes uncontrolled movements far from the initial path. Typically, energy functions evaluating the intermediate structures are obtained from atomistic classical force fields. Although some exceptions,[78,79] MEP approaches yield zero temperature pathways, precluding the study of temperature effects. Nowadays, the main application of MEP techniques is to search for good estimations of transition paths to start higher-level simulations.

Including temperature effects in the search for transition paths leads to minimum free energy paths (MFEP).[80–82] Similarly, free energy paths derivation starts with an initial guess of the reactive trajectory. Improvements to account for temperature effects come at the relaxation step of the intermediate structures. For instance, in the elegant string method such relaxation is done with local molecular dynamics (MD) trajectories that guarantee the correct Boltzmann weighting of states.[73,83–85] Although computationally very intensive, MFEP are smoother than single trajectories; indeed they are reaction tubes that average out unimportant motion for the conformational transition.[86] MFEPs expressed as a function of selected collective variables – whose correct identification is critical – show great promise for modeling dynamics at enzymatic processes.[68,87–90]

7.3 Transition Paths from Atomistic Simulations

The principal technique to study conformational transitions with atomistic detail is molecular dynamics (MD). The technique implements a sampling engine based on the integration of Newton's equation of motion defined by an atom-detailed classical Hamiltonian (the force field). Over the last decades classical force fields have been extensively refined, and MD codes have been largely improved, transforming MD in a very powerful technique to reproduce fine details of protein flexibility.[7,91] In principle, given one initial conformation, the motions dictated by physical forces will reveal the entire conformational space of a protein, tracing spontaneously any conformational transition. In practice, however, inaccuracies in the classical force field, and timescale limitations in the trajectories make it difficult to find always the transition pathway. There are however, numerous successful examples, some of them discussed below.

7.3.1 MD: Unbiased

The accessibility to specific-purpose chips[92] allowed Shaw's group to study the transitions of the EGFR kinase domain. They found spontaneous transitions from the active to the Src-like inactive conformation. The authors observed that the full conformational change took 7.0 and 8.7 μs in trajectories of 23 and 12 μs respectively. During the transition, two long-lived intermediate conformations were identified; their signature match hydrogen/deuterium exchange experiments better than intermediate X-ray structures. Interestingly, intermediate conformations differ to both end states more than end structures between themselves, stressing the need for unbiased sampling to reproduce non-linear transitions. An alternative to study slow conformational motions without introducing an external bias to the simulations is to use Markov State models[93] (MSM) to process unbiased trajectories. MD-MSM has the advantage that avoids the need for a long single trajectory, substituting it by an ensemble of shorter trajectories, easier to obtain in commonly available supercomputers. A recent example is the study by Noé's group of conformational motions upon binding of trypsin.[94] In this example, 543 MD trajectories were run accumulating roughly 150 μs of protein dynamics. The study showed that the conformational transition has intermediate features between the conformational-selection and induced-fit paradigms. Also, it identified six metastable states with conformational rearrangements necessary for the binding pathway. Similar approaches were used by De Fabritiis' group to describe the effects of phosphorylation, as modulator of an intrinsically disordered domain.[95] Coupling MD to Markov State Models permits a robust assessment of conformational kinetics, arguably the most challenging magnitude to estimate in conformational changes.[96] MD-MSM opens the possibility to use a variety of computational resources, not exclusively specific hardware, or supercomputers to study protein flexibility.[97,98] In this field, Pande and coworkers pioneered the use of cloud computing,[99–101] obtaining interesting results on slow conformational transitions, for example folding of proteins, or the activation pathway of Src kinase.[93,101]

The increase in the power of computers, and the development of interfaces adapted to couple MD with MSM are increasing the range of applicability of unbiased MD simulations. There are however, several cases where biasing the trajectory towards the transition path is required, as unbiased trajectories do not achieve spontaneous sampling of the transition. There are two basal ways of biasing simulations: using a predefined coordinate or simply by its energy.

7.3.2 MD: Biased by a Predefined Coordinate

Basically, these procedures lead to the transition path more or less accurately by forcing the trajectories to explore desired conformations along a privileged coordinate. Given two conformations of the same protein (A and B) the simplest measure to quantify their deviation is RMSD. Targeted molecular dynamics[102] (TMD) and variants[103–105] use the RMSD towards the target structure as a magnitude to minimise along the trajectory. The bias potential introduced is (eqn (7.2)):

$$U_{\text{TMD}} = \frac{1}{2}K(\text{RMSD}(t) - \rho(t))^2 \tag{7.2}$$

where RMSD(t) is the instantaneous RMSD, after superposing current state to target state. K is the spring constant. $\rho(t)$ is a function, often linear, that sets the desired RMSD to target at a given time t. TMD is computationally appealing, allowing to perturb of only a part of the system. However, since

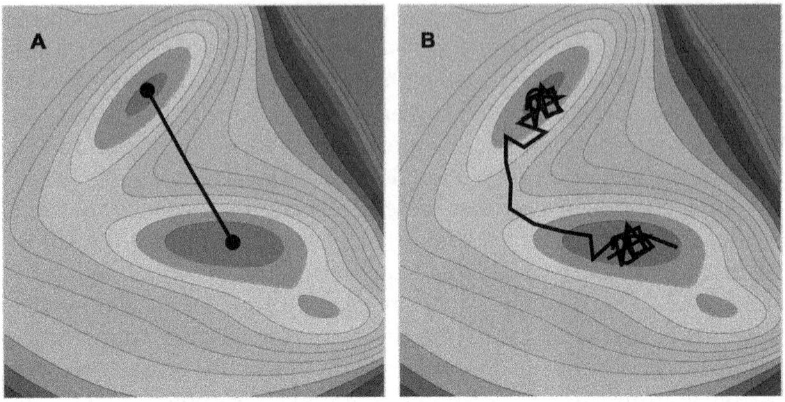

Figure 7.2 Potential problems of biased simulations. (A) A linear trajectory connecting two stable conformers of the protein. The trajectory, too biased, advances directly towards target state incurring in high-energy areas. This can happen in TMD or SMD if the biasing force is too strong. (B) An example of a transition path that correctly respects the energy landscape of the protein.

the perturbation is based on RMSD and not on the energy, there is no guarantee that high-energy barriers are avoided (Figure 7.2), and might be inaccurate in cases of non-linear transitions. Nevertheless, despite its simplicity, TMD can provide accurate results, especially when combined with experimental measurements. In this line, Gardino *et al.*[106] studied TMD pathways of the conformational change of nitrogen regulatory protein C combined with NMR observations. The authors detected two transient hydrogen bonds that, by distorting temporally a α-helix, decreased the energy barrier from 54 to 6 kcal mol^{-1}, in agreement with experimental data. In another study, Weng *et al.*[107] used short TMD simulations to investigate the conformational motion that allows the trans-membrane domain of the cassette transporter BtuCD to translocate vitamin B12 across the membrane.

A technique conceptually related to TMD, which is becoming increasingly popular is steered molecular dynamics (SMD).[108–110] Inspired by the original work of Grubmüller *et al.*,[111] SMD introduces a harmonic force between two points (either a fixed point and a particle of the system or two particles of the system) to encourage the simulation to sample the desired transition. In an application example, Böckmann and Grubmüller simulated the torque leading to F0 rotation, induced in the cell by proton motion, of the FoF1-ATP synthase.[112] The mechanism of early motion was revealed, exemplifying the power of force-based methods in the study of conformational transitions.[113]

SMD is gaining increasing interest since potential mean force profile can be computed coupling SMD to Jarzynski equation.[114] Accordingly, an ensemble of SMD trajectories can reveal not only transition pathways but also free energy variations associated to them. This was exploited to investigate the energetics of intermediate states of an extremely challenging transition:[115] the transcription anti-terminator RfaH. RfaH C-terminal domain detaches from the N-terminal domain, and in a second stage re-arranges from a α-helical motive to a β-barrel. The authors found that the critical energetic barrier is the disruption of inter-domain interactions. Other examples of conformational changes recently studied with SMD can be found elsewhere.[116–118]

Maxwell-Demon molecular dynamics,[119] also called dynamic importance sampling (DIMS),[120,121] biases trajectories by introducing information that help them to sample the transition path. It does not distort the energy landscape but selects slices of trajectories that move towards the target state by using a ratchet-like Metropolis acceptance filter. If $\Delta\varphi$ is an observable that captures the motion of slice of trajectory generated, the probability of accepting it (p_{acc}) is defined as:

$$p_{\text{acc}}(\Delta\varphi) = \begin{cases} 1 & \text{if } \Delta\varphi \leq 0 \\ \mathrm{e}^{-|\gamma\Delta\varphi|^2} & \text{if } \Delta\varphi > 0 \end{cases} \tag{7.3}$$

where γ is a parameter tuned to control the acceptance rate, typically defined by the initial distance between two known structures. Here, $\Delta\varphi < 0$ means that the proposed slice of trajectory is moving towards the target structure and is consequently accepted. Backwards steps, $\Delta\varphi > 0$, can be also accepted, allowing the algorithm to trace non-linear trajectories. If the slice is not accepted, a new one is generated. Repeatedly, the algorithm guarantees net motion towards the target structure.

Finally, an independent strategy to bias the simulation is to follow the essential space of the initial structure. Although there is no guarantee that such motions will lead to a concerted conformational transition, simple paths often overlap with the intrinsic deformations.[35,52] This strategy extends the sampling of standard MD,[122,123] being particularly suitable to deformations upon ligand binding.[124]

All the techniques noted above are powerful and have been successfully applied in many cases, but are far from being of universal use, and finding artifactual pathways is not rare. Two common strategies to improve accuracy that have been proposed are: (i) to integrate experimental information and (ii) to use the predicted pathway as an starting point for more exhaustive sampling, for instance, by means of umbrella sampling simulations[125,126] (US). US divides the initial path in windows, and enforces the sampling inside each window by independent MD trajectories. The effect of the biasing potential can be rigorously neutralised in the final free energy estimate by using statistical techniques,[126,127] allowing the user to obtain a realistic picture of the free energy change associated to the explored transition.

7.3.3 MD: Biased by Energy

Processes, like conformational changes, can be often seen as fast, but rare events.[8,82] This observation indicates that the limiting step for such

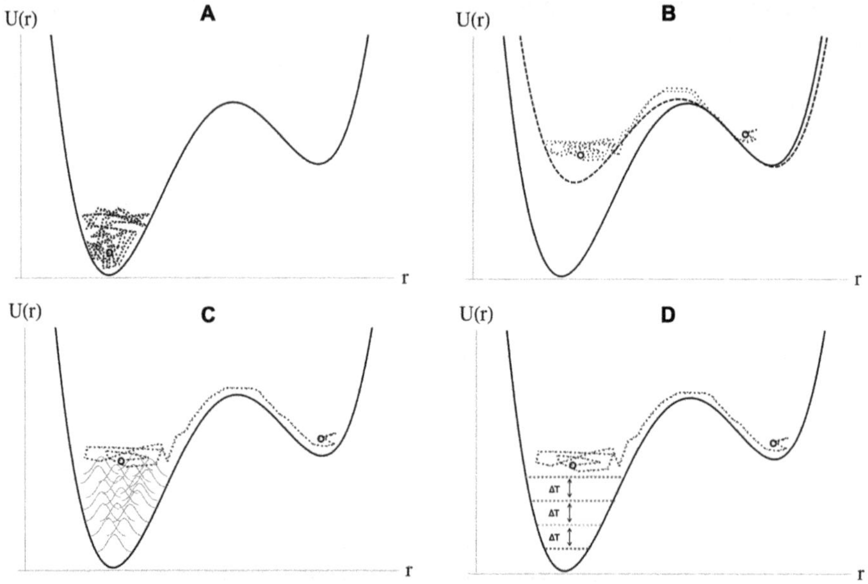

Figure 7.3 Strategies to reduce the energetic barriers. Energy profiles (*y*-axes) following a reaction coordinate (*r*, *x*-axes). (A) A two-minima energy profile. A trajectory starts on the left-hand side basin, being unable to escape from it due to the high energy barrier. (B) A scaling of the Hamiltonian will increase the potential energy of the left state to artificially reduce the energetic barrier. (C) Flooding algorithms schematic example. The trajectory again started in the left-hand side basin, but penalty functions (grey Gaussians) are added at visited states. At one point the barrier is apparently lower and the exploration of the conformational space continues in the right-hand side basin. (D) Effect of raising the temperature of the system. In REMD, one replica will be running at each temperature (represented as different horizontal dashed lines). The kinetic energy at a given temperature is plotted as horizontal lines, schematically showing how deep the energy minimum feels at a given temperature.

processes is abandoning the initial energy minima (Figure 7.3A). Several computational tricks were proposed to favor the system to abandon the starting basin. In most of the available methods this is done by raising the temperature of the system and/or by modifying the potential energy functional. Although not specific to conformational transitions, it is worth mentioning replica exchange family methods (or parallel tempering), since their increased sampling could push the system out of minima and trigger large-scale conformational changes.[128,129] When coupled to MD simulations (replica-exchange molecular dynamics; REMD) several independent copies (replicas) of the system are simulated in parallel. Replicas differ in either the temperature or in the Hamiltonian. In the standard application of the technique, a selected pair of replicas, typically contiguous in the replica variable, is interchanged with a transition probability (eqn (7.4)) based on

Metropolis criterion, which at the end guarantees the statistical significance of the obtained ensemble. For example, in a temperature REMD simulation:

$$p = \min\left\{1,\ \exp\left[\Delta E_{ij}\left(\frac{1}{k_{\mathrm{b}}T_i} - \frac{1}{k_{\mathrm{b}}T_j}\right)\right]\right\} \tag{7.4}$$

where i and j stands for two protein configurations obtained a temperatures T_i and T_j. ΔE_{ij} is the difference of potential energy between i and j configurations. Conformational variability arises from the walk between replicas, overcoming kinetic traps faster than traditional MD (Figure 7.3D). Recently, REMD was used to study the conformational dynamics of the small heat shock protein family,[130] a group of molecular chaperones that prevent unfolded proteins from aggregating. The authors found that the intrinsic disorder of the N-terminal domain is crucial for substrate recognition, and that there is a close relation between chaperone activity and the conformational state of the small heat shock protein dimer. Similarly, Collepardo-Guevara *et al.*[131] used REMD to describe motions of unstructured histone tails. They found that gains in structure related to epigenetic modifications alter nucleosome compaction, modulating gene activity. In order to encourage conformational changes in REMD simulations, Zacharias proposed a sampling strategy designed specifically to that purpose.[132] Concisely, an ENM model was used to identify the main deformations of the protein, often indicative of functional transition paths.[52] Combination of NM – weighted by their eigenvalue – defined a coordinate (Δd_{ij}), which through a bias potential (eqn (7.5)) drives the trajectory away from the current state:

$$V(d_{ij}) = \begin{cases} k\left[(d_{ij} - d_{ij0})^2 - \Delta d_{ij}^2\right]^2, & |d_{ij} - d_{ij0}| \leq 0 \\ 0, & |d_{ij} - d_{ij0}| > 0 \end{cases} \tag{7.5}$$

where d_{ij}, d_{ij0} are internal distances in the present and reference state (structure used for ENM analysis). Again, k is the force constant. $V(d_{ij})$ has its maximum value at the reference structure, and from there it decreases both at smaller and larger distances, approaching zero as the change in distance is exactly Δd_{ij}. Here, a Hamiltonian-REMD is used to modulate the strength of the bias potential, where the initial replica is an unbiased simulation, followed by a linear increase of the bias term. The method showed promising results in modeling domain motions, outperforming unbiased MD sampling.[133]

An alternative to REMD based protocols that pursue the same objectives but running only one replica is accelerated molecular dynamics (AMD).[134] AMD uses a potential energy threshold below which the Hamiltonian is scaled to raise potential energy (Figure 7.3B). Dihedrals terms are particularly targeted to enhance the likelihood of visiting new conformations, often across high-energy barriers. Since the bias in the potential is known,

Boltzmann re-weighting techniques can be applied to recover the original free energy profile.

Other methods introduce penalty potentials to visited structures to facilitate escaping energy barriers (Figure 7.3C). Generally referred as flooding algorithms, several variants have been proposed: Local Elevation,[135] Hyperdynamics,[136] Conformational Flooding[137,138] and Metadynamics.[139] These techniques are widely used to escape free energy minima and boost sampling capabilities of simulations; we will discuss Metadynamics as an illustrative example.

Metadynamics eases trajectories to new regions of the conformational space by filling the visited potential energy surface with Gaussian functions, which penalise already sampled conformations. The algorithm assumes that the system can be described by few collective variables *l* that fully locate the trajectory in the conformational space. Each time that similar conformations are sampled (in *l* space) a positive Gaussian function is added to the original Hamiltonian (eqn (7.11) and (7.12)), discouraging the system to come back to this point. More Gaussians sum up with the evolution of the simulation until the energy landscape is full, point at which, the real free energy landscape is the opposite of the sum of all Gaussians.

$$V(\lambda,t)=\int_0^t \mathrm{d}t'\mathrm{w}\exp\left(-\sum_i^d\frac{(\lambda_i-\lambda_i(t))^2}{2\sigma^2}\right) \tag{7.6}$$

$$H_{\text{total}}=H_{\text{original}}+V(\lambda,\,t) \tag{7.7}$$

where *l* are the collective variables, *t* is the time, *d* the number of collective variables, *w* and *σ* are the height and width of the Gaussian added at time *t*. Metadynamics ensure that trajectories escape the initial basin, forcing motions beyond equilibrium.[140] If the collective variables are chosen to connect two conformers or to encourage the motion of a certain motif of the protein, then transitions pathways can be easily recovered. The main problem of Metadynamics in its original formulation is to identify the stopping point of a simulation (*i.e.* when to stop adding Gaussians), which leads to convergence problems as useless Gaussian are randomly added, distorting the energetic description. To overcome this problem the well-tempered Metadynamics was introduced where the height of the Gaussian is history-dependent, being smaller as conformations are revisited.[141,142] Sutto and Gervasio[143] showed a nice example of the power of Metadynamics to dissect the free energy landscape of the EGFR kinase. The authors quantified the effect of oncogenic mutations of the kinase domain combining well-tempered Metadynamics simulations with temperature replica exchange method. They confirmed the link between active ← → inactive transition and the oncogenic mutations: mutants showed a remarked propensity towards active forms. The study showed that with massive computing and advance simulations schemes, molecular simulations are in the point of revealing the mechanistic effect of mutations directly from free energy computations.[144]

All the methods commented in this section help to find transition pathways, but the way in which they work masks kinetic aspects of such transitions, which in some cases can be recovered by advanced methods as those described below.

7.3.4 Advanced Methods

7.3.4.1 Milestoning

Milestoning[145] was designed by vanden-Eijnden *et al.*[146] to follow processes where the long-timescale diffusive motion face continuous barriers, like conformational transitions of biomolecules. The approach is based on milestones over a predetermined reaction coordinate, defining the transition pathway. A milestone can be seen as a checkpoint in the pathway; for instance, a set of structures spread along the transition path or desired values of one (or more) dihedral angles. Once defined, the transition path is decomposed into a sequence of sub-transitions in between milestones. The advantage is that the trajectory can be split in independent segments. Computations are therefore done in parallel very efficiently, enlarging the spanned timescale. Milestoning reveals not only equilibrium thermodynamic properties like free energy but, but also kinetic rates of conformational changes. Applications examples span from large molecular machines motions, such as myosin,[147] to membrane permeation events.[148] Finally, with recent improvements, energy estimations can be as accurate as desired in the exact formulation.[149,150] Despite the parallel approach, the method is still computationally intensive and requires careful inspection of milestones; sampling can be very good if the intermediates are well chosen, but incorrect if the intermediates are poorly selected.

7.3.4.2 Transition Path Sampling

Transition Path Sampling (TPS) was designed by Chandler and coworkers specifically to find conformational transitions in complex energy surfaces.[151,152] TPS assumes that both end states (A and B) are stable enough to be free energy minima and that the transition is dominated by a single dominant energy barrier. TPS requires to work:

An unequivocally characterisation of the energy basins of both end structures by an order parameter (*i.e.* dihedral angles, hydrogen bond distances or interaction energies)

An initial transition path that connect states A and B. It can be obtained with TMD, SMD or MEPs methods.

The initial path is path is inspected to identify the transition state (TS), or the top of the energy barrier. Since end structures (A and B) act as attractors – they are energy minima, is likely that any perturbation of the initial path near the energy barrier will generate a trajectory that still connects A and B. At this point is worth noting the separation between time-scales in activated conformational changes. On one hand, long times (t_{long}) are required to get a trajectory from a stable structure to the top of the energy barrier. On the other hand, short times (t_{short}) are required to observe events near the barrier region or going back to an energy minimum. Consequently, short MD trajectories (but larger than t_{short}) are started near the TS and integrated forward and backwards, generating an ensemble of transition paths. Despite the unphysical nature of the initial sampling trajectories, the protocol converges to physically meaningful transitions passing through the TS. TPS benefit from tailored methods to perturb the initial path efficiently,[153,154] and does not require a reaction coordinate to study the conformational change. Furthermore, since TPS is based on a solid statistic formulation, all relevant information can be extracted from the ensemble paths, such as the reaction mechanism, the TSs, and the rate constants.[155–158] Again, the method is computationally expensive and needs robust convergence tests.

7.4 Methods Based on Coarse-grained Simulations

In the previous section, we presented biasing methods, which, while universal in nature, were designed to work coupled to atomistic simulations with explicit water molecules interacting through refined energy functions. Unfortunately, even using the most efficient algorithms, atomistic methods are often too slow to trace, within reasonable computation time, conformational transitions, especially in those cases where no clear idea on the potential pathway exists. In this section, we extend our discussion to methods aiming to reduce the complexity of energy landscape by reducing the number of particles and/or simplifying the energetic interactions. Coarse-grained (CG) models reduce details of the system gaining not only efficiency but also clarity to interpret the results with a rapid feedback loop with experiments.[7,159–162] Leaving the detail of CG formalisms derivation to previous publications,[39,163,164] we will discuss the applications on conformational transitions.

The first method to discuss conserves atomistic (or quasi-atomistic) detail for proteins but reduces solvent representation to an implicit model,[165] and uses simple step potentials for treating interactions. The method, named MDdMD, uses as input two known structures (A and B) to estimate the transition.[166] MDdDM is based on highly efficient discrete molecular dynamics (DMD) simulations coupled to an Maxwell-Devil biasing scheme[119,121] that bias the trajectory towards a target state. In DMD simulations,[167–169] the system moves in ballistic regime; particles advance at constant velocity until an event (collision) occurs. At this point, velocities of colliding particles are immediately updated following the energy and momentum conservation rules. The method efficiency is suitable for online calculations, being freely available at http://mmb.irbbarcelona.org/MDdMD/.

The next family of methods leaves out the atomic detail, representing proteins residues by a few particles (often a single one centered at Cα position). With such a limited resolution, the most accurate way of modeling residues interactions is using a reference structure to parameterise the model. This strategy, pioneered by Gō and coworkers[42,170,171] is referred as structure-based models (SBM) or Gō-like models. This formulation allows studying protein flexibility solely based on the experimental 3D position of the atoms. For equilibrium fluctuations a single structure is used as reference. Instead, for modeling conformational transitions combinations of at least two structures (A and B) are usually adopted, for example, in Onuchic's functional.[172,173] Basically, an attracting potential is used for native contacts interactions either in structure A or B. The energy contributions are separated in bonded (H_b) and non-bonded (H_{nb}) interactions for clarity.

$$H_{SBM} = H_b + H_{nb} \tag{7.8}$$

The detailed description for the bonded terms is presented below. The superscript 0 refers to values in the initial structures. Usually, these values coincide between A and B states, if not the average or a range of values are considered.

$$\begin{aligned} H_b = & \sum_{i=1}^{N-1} K_s\left(r_i - r^0_{i,i+1}\right)^2 + \sum_{i=1}^{N-2} K_\theta\left(\theta_i - \theta^0_{i,i+2}\right)^2 \\ & + \sum_{i=1}^{N-3} \sum_{n=1,3} K^n_\phi\left(1 - \cos\left[n(\phi_i - \phi^0_i)\right]\right) \end{aligned} \tag{7.8a}$$

where i runs for all N particles in the system. K_s, K_θ and K_ϕ are the stretching, bending and torsional force constants. Here, the first term in (eqn (7.8a)) ensures that bond distances for particles i and $i+1$ oscillate around the reference distance $r^0_{i,i+1}$. The second term has identic aim but for bending of angles, and the third one for dihedrals rotations. For the non-bonded interaction terms:

$$\begin{aligned} H_{nb} = \sum_{\substack{1 \le i < N-4 \\ i+4 < j < N}} \Bigg[& \varepsilon^A\left(\left(\frac{r^{0,A}_{i,j}}{r_{ij}}\right)^{12} - 2\left(\frac{r^{0,A}_{i,j}}{r_{ij}}\right)^6\right)\delta^A_{ij} \\ & + \varepsilon^B\left(\left(\frac{r^{0,B}_{i,j}}{r_{ij}}\right)^{12} - 2\left(\frac{r^{0,B}_{i,j}}{r_{ij}}\right)^6\right)\delta^B_{ij}\left(1-\delta^A_{ij}\right) + \varepsilon^r\left(\frac{\sigma}{r_{ij}}\right)^{12}\left(1-\delta^A_{ij}\right)\left(1-\delta^B_{ij}\right)\Bigg] \end{aligned} \tag{7.8b}$$

where the summation runs for all non-local pair of particles. The first term in the summation represents an attractive Lennard–Jones potential, regulated by the Kronecker delta δ^A_{ij} that is valued 1 if i–j is a native contact in A

and 0 otherwise. The second term is analogous for B structure, but only for exclusive contacts of B. Finally, for non-native interactions a repulsion term is added, with σ as exclusion radii. ε^A, ε^B and ε^r are the respective energy terms for the Lennard–Jones potential.

SBM are gaining popularity as fast and accurate tools to model protein dynamics.[159,174] SBM are rooted in the principle of minimal frustration, that states that evolution tuned protein sequences to avoid stabilising non-native interactions that would disturb the folding pathways.[1,175,176] In other words, native interactions are the optimal ones for the protein and that changes in such interactions are penalised. Hyeon and Onuchic[177,178] used recently their functional to study the striking coordination and regulation of kinesin motor heads that results in fast motion along the protofilament of microtubule. Kinesin motor heads are aligned one leading and the other one at rear. Both domains are allosterically regulated, despite interacting through a distant stalk domain. The study revealed that changes in the catalytic core of leading kinesin motor upon binding make nucleotide binding in the leading head not favorable while the rear domain is still bound to microtubule.[179] Moreover, premature ATP binding is inhibited before the chemical reaction (ATP hydrolysis) in the rear domain is completed, during which the leading head is still tightly bound.[180,181] Following a similar SBM approach Whitford *et al.*[172] demonstrated that a partial cracking of a α-helix reduces the energy barrier in adenylate kinase conformational transition and probably also in other systems.[182–184] The latter being another lesson learned from CG simulations: some transition pathways elicit local unfolding to avoid higher energy strained structures.[185]

An alternative and significantly faster method to trace conformational transitions is to survey double-minima SBM with DMD simulations.[186] As discussed, DMD framework demands step potentials with flat energy minima, which can easily accommodate multiple minima[167,169] (Figure 7.4). The transition between the two minima is favored by using a smooth enhanced sampling engine. The method finds transition paths in an average

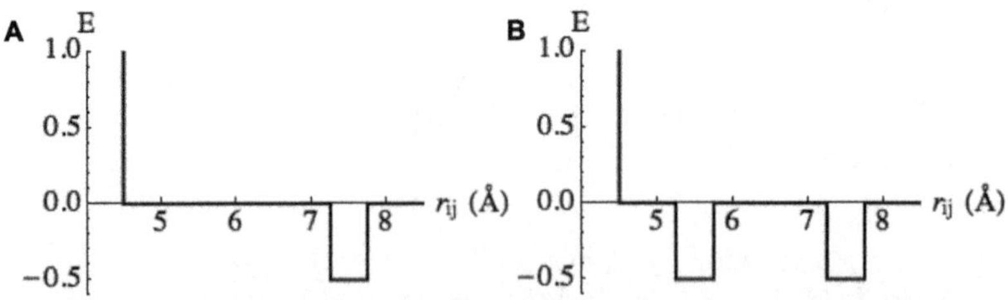

Figure 7.4 Modelling conformational changes with simple SBM step-potentials. (A) For a pair of particles that do not change their relative distance (r_{ij}) in the conformational transition, a single energy minimum is placed at the reference distance of both known structures. (B) When the distance between particles changes along with the conformational transition two minima are used. Minima positions correspond to each distance in the end conformers, typically two PDB structures.

time of two minutes in a laptop computer, allowing us to move the protein flexibility analysis to the proteome scale. Conformational transition pathways can be obtained from the online server (http://mmb.irbbarcelona.org/GOdMD/).

The idea to use multiple minima SBM potential has been explored also in other formalisms. For example, Thirumalai group used a potential similar to that in eqn (7.8) (for dual minima at T and R states) coupled to Brownian dynamics[187] to study the mechanistic details of the GroEL/GroES conformational transition. In an initial stage, an ensemble of T structures is generated in a standard Brownian dynamic simulation. Then, a smooth adaptation of the Hamiltonian occurs that weights in the interactions of the R structure, assumed to be initiated by ATP binding. Finally, once the Hamiltonian is fully shifted towards the R state the structure is allowed to oscillate until relaxation.

7.5 Predicting Conformational Transition Pathways

Most of the pathfinder methods described above demand the previous knowledge of at least two experimental structures. When only one conformer of the protein is known, modeling conformational changes becomes a greater challenge, especially if such transition does not follow the direction of principal NM. A few strategies to obtain estimates of alternative structures in these cases have been developed. One of these strategies consists on searching for homologous proteins that have a similar target conformation available. This search, that could be rather arduous, is facilitated by available servers, like ConTemplate (http://bental.tau.ac.il/contemplate) that suggests alternative conformations for a query protein of known structure.[188] When the desired conformer is not available from homology modelling, molecular simulations are the only alternative. Since timescale limitations makes pure-force atomistic simulations not applicable when the target structure is far away from the known conformation, physical models are combined with experimental data in hybrid strategies.

7.5.1 Experimentally Biased Simulation Methods

Molecular simulations can incorporate any kind of experimental information, even very low-resolution data to bias trajectory towards unknown conformations. The protocols used to model conformational changes are not conceptually very different from the ones used to resolve protein structures. Indeed, innovative X-ray designs[189–191] yield protein structures together with insights of dynamics – mainly local, although combination of such data with NMR spectroscopy data allows the exploration of longer-range motions.[192] Particularly for conformational motions, NMR chemical shifts were introduced in a penalty function for Metadynamics simulations.[193,194] The latter allowed the exploration of alternative conformations in proteins: the Proline isomerase transition and free energy landscape were studied with considerable structural and energetic detail.[195]

Combining atomistic simulations with low-resolution methods like small angle X-ray scattering (SAXS) yields promising results. For instance, Chen and Hub[196,197] used the SAXS profile of the Leucine Binding Protein to bias MD trajectories from the open to the closed states, showing the synergy between both techniques.[198,199] SAXS has also been recently used to define collective variables for Metadynamics simulations,[200] and to refine NMA ensembles of transition pathways.[201] More detailed examples on the richness of SAXS data can be found elsewhere.[202] Other low-resolution data, such as the radius of gyration derived from chromatography or sedimentation experiments has been used by de Groot and coworkers to bias MD trajectories connecting closed and open states of proteins.[203]

7.5.2 Coevolution Biased Simulation Methods

Sequence-based contacts were reported to capture indirectly protein dynamics.[204] Coevolution analysis performed over the history of the protein sequence reveals residues showing correlated mutations, which could be indicative of a proximity in the 3D space.[205–210] Sfriso *et al.*[211] showed in a systematic study that coevolution contacts capture several conformational states of the protein. By assuming that such contacts should be brought together in space, a multi-minima SBM is constructed; DMD simulations are used to explore the multiple minima surface revealing complex traits of protein dynamics. Functional alternative conformers can be predicted using clustering techniques in a high-throughput manner. Very recently, Sutto *et al.*[212] presented similar ideas implemented in the context of atomistic molecular dynamics finding again the excellent biasing capabilities of coevolution data.

7.6 Discussion

Modelling conformational transitions is a very challenging problem, which has encouraged many researchers to develop complex methodology. Unbiased atomistic simulations are undoubtedly the best choice, but current limitations in force fields and computer power restrict their application to small proteins undergoing simple conformational changes.[213] Next step down in the ladder are enhanced sampling algorithms coupled to atomistic simulations, which can offer an accurate energetic picture. However, in one way or another, they need some estimation of the transition path, showing performance sensitive to the nature of the conformational transition.[182,184,214–217] When the transition necessarily requires ligand binding, or chemical alterations in the protein (for example phosphorylation), an extra layer of complexity is added[218] (Figure 7.5). Some protocols exist to model ligand binding effects,[218–221] but explicitly including ligands to in pathfinder methods is still an open field of research.

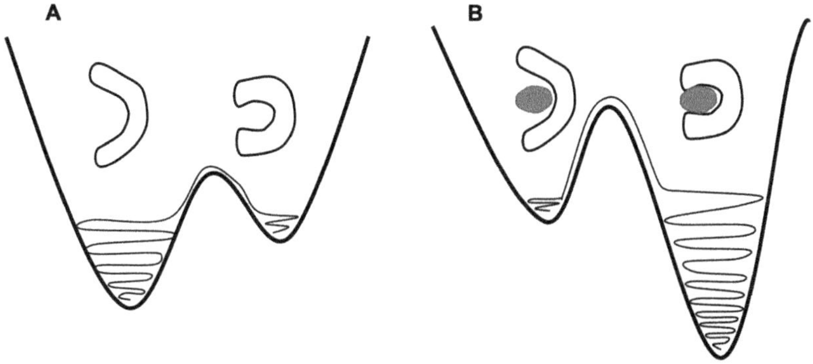

Figure 7.5 Energetic contribution of ligand binding. When simulating a conformational change that is coupled to ligand binding, the energy landscape obtained is distorted compared with the complete (protein–ligand) biological scenario. (A) shows the energy landscape without considering ligand effects while (B) includes those effects. In the presence of the ligand the closed conformation (right-hand side) is the stable one, opposite to the ligand-free situation. This occurs with many proteins including adenylate kinase.[227] An advantage of structure-based models is that they accept the reference (or several reference) structure as energy minima, regardless of the environment, being ligand effects automatically accounted.

Coarse-grained protocols are a natural alternative to atomistic models. They are faster, easier to tune, and can naturally incorporate SBMs.[222,223] The caveat of these methods is the lack of resolution and the use of unphysical potentials, which in the absence of experimental data are often unable to distinguish realistic from artifactual structures. Incorporation of experimental restraints in either atomistic or CG methods seems a natural choice to improve predictions of conformational transitions, but such information is emerging slowly.

Finally, when the reader decides to choose between one of the methods explained here, or the myriad of other alternative approaches available in the literature, he/she must be aware of the need to validate the predicted transition, as results strongly depend on the method used. For example, Seyler and Beckstein[224] showed, using the very well-studied adenylate kinase as test system, that paths obtained with different methods can be quite diverse. Therefore, some validation of predicted path is usually required. In dubious situations, statistical sampling over ensembles of paths is a robust strategy to evaluate the results.[225,226]

Overall, it is clear that we have entered in the 'dynamics age', where proteins are not longer being considered as static entities. Simulation techniques alone, or coupled to experimental measures are going to provide a new, detailed view on the mechanism of complex conformational motions. Eventually, advancing towards the final goal of structural approaches to biology: understand the principles of life from the basic rules of physics.

References

1. J. Bryngelson, J. Onuchic, N. Socci and P. Wolynes, *Proteins: Struct., Funct., Bioinf.*, 1995, **21**, 167–195.
2. K. Henzler-Wildman and D. Kern, *Nature*, 2007, **450**, 964–972.
3. E. Z. Eisenmesser, D. A. Bosco, M. Akke and D. Kern, *Science*, 2002, **295**, 1520–1523.
4. A. Leo-Macias, P. Lopez-Romero, D. Lupyan, D. Zerbino and A. R. Ortíz, *Biophys. J.*, 2005, **88**, 1291–1299.
5. C. Micheletti, *Phys. Life Rev.*, 2013, **10**, 1–26.
6. M. Orozco, *Phys. Life Rev.*, 2012, **10**, 29–30.
7. M. Orozco, *Chem. Soc. Rev.*, 2014, **43**, 5051–5066.
8. J. Villali and D. Kern, *Curr. Opin. Chem. Biol.*, 2010, **14**, 636–643.
9. K. Moritsugu, T. Terada and A. Kidera, *J. Am. Chem. Soc.*, 2012, **134**, 7094–7101.
10. D. R. Glowacki, J. N. Harvey and A. J. Mulholland, *Nat. Chem.*, 2012, **4**, 169–176.
11. A. van der Vaart, *Theor. Chem. Acc.*, 2006, **116**, 183–193.
12. P. W. Rose, A. Prli, C. Bi, W. F. Bluhm, C. H. Christie, S. Dutta, R. K. Green, D. S. Goodsell, J. D. Westbrook, J. Woo, J. Young, C. Zardecki, H. M. Berman, P. E. Bourne and S. K. Burley, *Nucleic Acids Res.*, 2015, **43**, D345–D356.
13. R. Elber, *Curr. Opin. Struct. Biol.*, 2011, **21**, 167–172.
14. P. I. Zhuravlev and G. A. Papoian, *Q. Rev. Biophys.*, 2010, **43**, 295–332.
15. R. Elber and S. Kirmizialtin, *Curr. Opin. Struct. Biol.*, 2013, **23**, 206–211.
16. F. Tama and C. L. Brooks, *Annu. Rev. Biophys. Biomol. Struct.*, 2006, **35**, 115–133.
17. J. Czub and H. Grubmüller, *J. Am. Chem. Soc.*, 2014, **136**, 6960–6968.
18. K. Nam, J. Pu and M. Karplus, *Proc. Natl. Acad. Sci. U. S. A.*, 2014, **111**, 17851–17856.
19. T. C. B. McLeish, T. L. Rodgers and M. R. Wilson, *Phys. Biol.*, 2013, **10**, 056004.
20. V. J. Hilser, J. O. Wrabl and H. N. Motlagh, *Annu. Rev. Biophys.*, 2012, **41**, 585–609.
21. J. F. Swain and L. M. Gierasch, *Curr. Opin. Struct. Biol.*, 2006, **16**, 102–108.
22. R. Nussinov and C.-J. Tsai, *Curr. Opin. Struct. Biol.*, 2015, **30**, 17–24.
23. T. Wlodarski and B. Zagrovic, *Proc. Natl. Acad. Sci. U. S. A.*, 2009, **106**, 19346–19351.
24. D.-A. Silva, G. R. Bowman, A. Sosa-Peinado and X. Huang, *PLoS Comput. Biol.*, 2011, **7**, e1002054.
25. H.-X. Zhou, *Biophys. J.*, 2010, **98**, L15–L17.
26. P. Csermely, R. Palotai and R. Nussinov, *Trends Biochem. Sci.*, 2010, **35**, 539–546.
27. D. H. E. W. Huberts and I. J. van der Klei, *Biochim. Biophys. Acta*, 2010, **1803**, 520–525.

28. C. Gancedo and C. L. Flores, *Microbiol. Mol. Biol. Rev.*, 2008, **72**, 197–210.
29. I. R. Vetter and A. Wittinghofer, *Science*, 2001, **294**, 1299–1304.
30. L. V. Bock, C. Blau, G. F. Schröder, I. I. Davydov, N. Fischer, H. Stark, M. V. Rodnina, A. C. Vaiana and H. Grubmüller, *Nat. Struct. Mol. Biol.*, 2013, **20**, 1390–1396.
31. R. Molina, S. Stella, P. Redondo, H. Gomez, M. J. Marcaida, M. Orozco, J. Prieto and G. Montoya, *Nat. Struct. Mol. Biol.*, 2015, **22**, 65–72.
32. J. Franklin, P. Koehl, S. Doniach and M. Delarue, *Nucleic Acids Res.*, 2007, **35**, W477–W482.
33. Y. Ye and A. Godzik, *Nucleic Acids Res.*, 2004, **32**, W582–W585.
34. W. G. Krebs and M. B. Gerstein, *Nucleic Acids Res.*, 2000, **28**, 1665–1675.
35. M. K. Kim, R. L. Jernigan and G. S. Chirikjian, *Biophys. J.*, 2002, **83**, 1620–1630.
36. M. K. Kim, G. S. Chirikjian and R. L. Jernigan, *J. Mol. Graphics Modell.*, 2002, **21**, 151–160.
37. D. Seeliger, J. Haas and B. L. de Groot, *Structure*, 2007, **15**, 1482–1492.
38. D. R. Weiss and M. Levitt, *J. Mol. Biol.*, 2009, **385**, 665–674.
39. M. Orozco, L. Orellana, A. Hospital, A. N. Naganathan, A. Emperador, O. Carrillo and J. L. Gelpí, *Adv. Protein Chem. Struct. Biol.*, 2011, **85**, 183–215.
40. A. Emperador, O. Carrillo, M. Rueda and M. Orozco, *Biophys. J.*, 2008, **95**, 2127–2138.
41. M. Levitt, C. Sander and P. S. Stern, *J. Mol. Biol.*, 1985, **181**, 423–447.
42. N. Go, T. Noguti and T. Nishikawa, *Proc. Natl. Acad. Sci. U. S. A.*, 1983, **80**, 3696–3700.
43. B. Brooks and M. Karplus, *Proc. Natl. Acad. Sci. U. S. A.*, 1983, **80**, 6571–6575.
44. J. A. Velazquez-Muriel, M. Rueda, I. Cuesta, A. Pascual-Montano, M. Orozco and J.-M. Carazo, *BMC Struct. Biol. [Online]*, 2009, **9**, 6.
45. D. Tobi and I. Bahar, *Proc. Natl. Acad. Sci. U. S. A.*, 2005, **102**, 18908–18913.
46. D. A. Liberles, S. A. Teichmann, I. Bahar, U. Bastolla, J. Bloom, E. Bornberg-Bauer, L. J. Colwell, A. P. J. de Koning, N. V. Dokholyan, J. Echave, A. Elofsson, D. L. Gerloff, R. A. Goldstein, J. A. Grahnen, M. T. Holder, C. Lakner, N. Lartillot, S. C. Lovell, G. Naylor, T. Perica, D. D. Pollock, T. Pupko, L. Regan, A. Roger, N. Rubinstein, E. Shakhnovich, K. Sjölander, S. Sunyaev, A. I. Teufel, J. L. Thorne, J. W. Thornton, D. M. Weinreich and S. Whelan, *Protein Sci.*, 2012, **21**, 769–785.
47. M. Tirion, *Phys. Rev. Lett.*, 1996, **77**, 1905–1908.
48. A. R. Atilgan, S. R. Durell, R. L. Jernigan, M. C. Demirel, O. Keskin and I. Bahar, *Biophys. J.*, 2001, **80**, 505–515.
49. J. A. Kovacs, P. Chacón and R. Abagyan, *Proteins: Struct., Funct., Bioinf.*, 2004, **56**, 661–668.

50. L. Orellana, M. Rueda, C. Ferrer-Costa, J. R. Lopez-Blanco, P. Chacón and M. Orozco, *J. Chem. Theory Comput.*, 2010, **6**, 2910–2923.
51. F. Tama and Y. H. Sanejouand, *Protein Eng.*, 2001, **14**, 1–6.
52. H. G. Dos Santos, J. Klett, R. Mendez and U. Bastolla, *Biochim. Biophys. Acta*, 2013, **1834**, 836–846.
53. E. Lindahl, C. Azuara, P. Koehl and M. Delarue, *Nucleic Acids Res.*, 2006, **34**, W52–W56.
54. S. E. Dobbins, V. I. Lesk and M. J. E. Sternberg, *Proc. Natl. Acad. Sci. U. S. A.*, 2008, **105**, 10390–10395.
55. M. Delarue and Y. H. Sanejouand, *J. Mol. Biol.*, 2002, **320**, 1011–1024.
56. C. Xu, D. Tobi and I. Bahar, *J. Mol. Biol.*, 2003, **333**, 153–168.
57. Z. Yang, P. Májek and I. Bahar, *PLoS Comput. Biol.*, 2009, **5**, e1000360.
58. J.-W. Chu and G. A. Voth, *Biophys. J.*, 2007, **93**, 3860–3871.
59. P. Maragakis and M. Karplus, *J. Mol. Biol.*, 2005, **352**, 807–822.
60. R. Mendez and U. Bastolla, *Phys. Rev. Lett.*, 2010, **104**, 228103.
61. J. R. López-Blanco, R. Reyes, J. I. Aliaga, R. M. Badia, P. Chacón and E. S. Quintana-Ortí, *J. Comput. Phys.*, 2013, **246**, 275–288.
62. U. Bastolla, *Wiley Interdiscip. Rev.: Comput. Mol. Sci.*, 2014, **4**, 488–503.
63. L. Orellana, A. Hospital and M. Orozco, Oncogenic mutations of the EGF-Receptor ectodomain reveal an unexpected mechanism for ligand-independent activation, bioRxiv, 2014.
64. O. Miyashita, *Proc. Natl. Acad. Sci. U. S. A.*, 2003, **100**, 12570–12575.
65. T. R. Lezon, A. Sali and I. Bahar, *PLoS Comput. Biol.*, 2009, **5**, e1000496.
66. T. Haliloglu and I. Bahar, *Curr. Opin. Struct. Biol.*, 2015, **35**, 17–23.
67. M. Rueda, P. Chacón and M. Orozco, *Structure*, 2007, **15**, 565–575.
68. V. Ovchinnikov, M. Karplus and E. Vanden-Eijnden, *J. Chem. Phys.*, 2011, **134**, 085103.
69. M. Berkowitz, J. D. Morgan, J. A. McCammon and S. H. Northrup, *J. Chem. Phys.*, 1983, **79**, 5563–5565.
70. R. Elber and M. Karplus, *Chem. Phys. Lett.*, 1987, **139**, 375–380.
71. R. Czerminski and R. Elber, *J. Chem. Phys.*, 1990, **92**, 5580–5601.
72. L. R. Pratt, *J. Chem. Phys.*, 1986, **85**, 5045–5048.
73. E. Weinan, W. Ren and E. Vanden-Eijnden, *Phys. Rev. B: Condens. Matter Mater. Phys.*, 2002, **66**, 052301.
74. E. Weinan and E. Vanden-Eijnden, *Annu. Rev. Phys. Chem.*, 2010, **61**, 391–420.
75. G. Henkelman, B. P. Uberuaga and H. Jónsson, *J. Chem. Phys.*, 2000, **113**, 9901–9904.
76. J.-W. Chu, B. L. Trout and B. R. Brooks, *J. Chem. Phys.*, 2003, **119**, 12708–12717.
77. S. Fischer and M. Karplus, *Chem. Phys. Lett.*, 1992, **194**, 252–261.
78. W. Quapp, *J. Theor. Comput. Chem.*, 2003, **2**, 385–417.
79. S. Huo and J. E. Straub, *J. Chem. Phys.*, 1997, **107**, 5000–5006.
80. Y. Matsunaga, H. Fujisaki, T. Terada, T. Furuta, K. Moritsugu and A. Kidera, *PLoS Comput. Biol.*, 2012, **8**, e1002555.
81. A. C. Pan, D. Sezer and B. Roux, *J. Phys. Chem. B*, 2008, **112**, 3432–3440.

82. R. Elber, *Curr. Opin. Struct. Biol.*, 2005, **15**, 151–156.
83. L. Maragliano and E. Vanden-Eijnden, *Chem. Phys. Lett.*, 2007, **446**, 182–190.
84. L. Maragliano, A. Fischer, E. Vanden-Eijnden and G. Ciccotti, *J. Chem. Phys.*, 2006, **125**, 024106.
85. E. Weinan, W. Ren and E. Vanden-Eijnden, *J. Phys. Chem. B*, 2005, **109**, 6688–6693.
86. E. Vanden-Eijnden and M. Venturoli, *J. Chem. Phys.*, 2009, **130**, 194103.
87. M. Sanchez-Martinez, M. Field and R. Crehuet, *J. Phys. Chem. B*, 2015, **119**, 1103–1113.
88. H. Hu, Z. Lu and W. Yang, *J. Chem. Theory Comput.*, 2007, **3**, 390–406.
89. D. Branduardi and J. D. Faraldo-Gómez, *J. Chem. Theory Comput.*, 2013, **9**, 4140–4154.
90. G. Andrés Cisneros and W. Yang, in *Multi-scale Quantum Models for Biocatalysis: Modern Techniques and Applications*, ed. D. M. York and T.-S. Lee, Springer, Netherlands, Dordrecht, 2009, pp. 57–78.
91. K. Lindorff-Larsen, P. Maragakis, S. Piana, M. P. Eastwood, R. O. Dror and D. E. Shaw, *PLoS One*, 2012, **7**, e32131.
92. D. E. Shaw, J. C. Chao, M. P. Eastwood, J. Gagliardo, J. P. Grossman, C. R. Ho, D. J. Lerardi, I. Kolossváry, J. L. Klepeis, T. Layman, C. McLeavey, M. M. Deneroff, M. A. Moraes, R. Mueller, E. C. Priest, Y. Shan, J. Spengler, M. Theobald, B. Towles, S. C. Wang, R. O. Dror, J. S. Kuskin, R. H. Larson, J. K. Salmon, C. Young, B. Batson and K. J. Bowers, *Commun. ACM*, 2008, **51**, 91.
93. V. S. Pande, K. Beauchamp and G. R. Bowman, *Methods*, 2010, **52**, 99–105.
94. N. Plattner and F. Noé, *Nat. Commun.*, 2015, **6**, 7653.
95. N. Stanley, S. Esteban-Martín and G. De Fabritiis, *Nat. Commun.*, 2014, **5**, 1–8.
96. J. D. Chodera and F. Noé, *Curr. Opin. Struct. Biol.*, 2014, **25**, 135–144.
97. V. S. Pande, I. Baker, J. Chapman, S. P. Elmer, S. Khaliq, S. M. Larson, Y. M. Rhee, M. R. Shirts, C. D. Snow and E. J. Sorin, *Biopolymers*, 2003, **68**, 91–109.
98. A. L. Beberg, D. L. Ensign, G. Jayachandran, S. Khaliq and V. S. Pande, *Parallel & Distributed Processing, IEEE International Symposium*, 2009, 1–8.
99. D. L. Ensign, P. M. Kasson and V. S. Pande, *J. Mol. Biol.*, 2007, **374**, 806–816.
100. G. R. Bowman and V. S. Pande, *Proc. Natl. Acad. Sci. U. S. A.*, 2010, **107**, 10890–10895.
101. D. Shukla, Y. Meng, B. I. T. Roux and V. S. Pande, *Nat. Commun.*, 2014, **5**, 1–11.
102. J. Schlitter, M. Engels and P. Krüger, *J. Mol. Graphics*, 1994, **12**, 84–89.
103. P. Krüger, S. Verheyden, P. J. Declerck and Y. Engelborghs, *Protein Sci.*, 2001, **10**, 798–808.
104. A. van der Vaart and M. Karplus, *J. Chem. Phys.*, 2005, **122**, 114903.

105. M. Madhusoodanan and T. Lazaridis, *Biophys. J.*, 2003, **84**, 1926–1939.
106. A. K. Gardino, J. Villali, A. Kivenson, M. Lei, C. F. Liu, P. Steindel, E. Z. Eisenmesser, W. Labeikovsky, M. Wolf-Watz, M. W. Clarkson and D. Kern, *Cell*, 2009, **139**, 1109–1118.
107. J. Weng, K. Fan and W. Wang, *PLoS One*, 2012, **7**, e30465.
108. S. Izrailev, S. Stepaniants, M. Balsera, Y. Oono and K. Schulten, *Biophys. J.*, 1997, **72**, 1568–1581.
109. B. Isralewitz, M. Gao and K. Schulten, *Curr. Opin. Struct. Biol.*, 2001, 224–230.
110. H. Grubmüller and P. Tavan, *J. Chem. Phys.*, 1994, **101**, 5047.
111. H. Grubmüller, B. Heymann and P. Tavan, *Science*, 1996, **271**, 997–999.
112. R. A. Böckmann and H. Grubmüller, *Nat. Struct. Biol.*, 2002, **9**, 198–202.
113. J. Gullingsrud and K. Schulten, *Biophys. J.*, 2003, **85**, 2087–2099.
114. C. Jarzynski, *Phys. Rev. Lett.*, 1997, **78**, 2690–2693.
115. J. B. GC, B. S. Gerstman and P. P. Chapagain, *J. Phys. Chem. B*, 2015, **119**, 12750–12759.
116. M. Shen, J. Guan, L. Xu, Y. Yu, J. He, G. W. Jones and Y. Song, *J. Biomol. Struct. Dyn.*, 2012, **30**, 652–661.
117. M. A. Cuendet and O. Michielin, *Biophys. J.*, 2008, **95**, 3575–3590.
118. T. Giorgino and G. De Fabritiis, *J. Chem. Theory Comput.*, 2011, **7**, 1943–1950.
119. M. Rueda, E. Cubero, C. A. Laughton and M. Orozco, *Biophys. J.*, 2004, **87**, 800–811.
120. O. Beckstein, E. J. Denning, J. R. Perilla and T. B. Woolf, *J. Mol. Biol.*, 2009, **394**, 160–176.
121. J. R. Perilla, O. Beckstein, E. J. Denning and T. B. Woolf, *J. Comput. Chem.*, 2010, **32**, 196–209.
122. B. L. de Groot, A. Amadei, R. M. Scheek, N. A. van Nuland and H. J. Berendsen, *Proteins: Struct., Funct., Bioinf.*, 1996, **26**, 314–322.
123. J. Sonne, C. Kandt, G. H. Peters, F. Y. Hansen, M. Ø. Jensen and D. P. Tieleman, *Biophys. J.*, 2007, **92**, 2727–2734.
124. R. Chaudhuri, O. Carrillo, C. A. Laughton and M. Orozco, *J. Chem. Theory Comput.*, 2012, **8**, 2204–2214.
125. G. M. Torrie and J. P. Valleau, *J. Comput. Phys.*, 1977, **23**, 187–199.
126. S. Krumar, D. Bouzida, R. H. Swendsen, P. Kollman and J. Rosenberg, *J. Comput. Chem.*, 1992, **13**, 1011–1021.
127. F. Zhu and G. Hummer, *J. Comput. Chem.*, 2011, **33**, 453–465.
128. Y. Sugita and Y. Okamoto, *Chem. Phys. Lett.*, 1999, **314**, 141–151.
129. D. J. Earl and M. W. Deem, *Phys. Chem. Chem. Phys.*, 2005, **7**, 3910.
130. S. Patel, E. Vierling and F. Tama, *Biophys. J.*, 2014, **106**, 2644–2655.
131. R. Collepardo-Guevara, G. Portella, M. Vendruscolo, D. Frenkel, T. Schlick and M. Orozco, *J. Am. Chem. Soc.*, 2015, **137**, 10205–10215.
132. M. Zacharias, *J. Chem. Theory Comput.*, 2008, **4**, 477–487.
133. K. Ostermeir and M. Zacharias, *Proteins: Struct., Funct., Bioinf.*, 2014, **82**, 3410–3419.

134. D. Hamelberg, J. Mongan and J. A. McCammon, *J. Chem. Phys.*, 2004, **120**, 11919.
135. T. Huber, A. E. Torda and W. F. van Gunsteren, *J. Comput. -Aided Mol. Des.*, 1994, **8**, 695–708.
136. A. F. Voter, *Phys. Rev. Lett.*, 1997, **78**, 3908.
137. Grubmuller, *Phys. Rev. E: Stat. Phys., Plasmas, Fluids, Relat. Interdiscip. Top.*, 1995, **52**, 2893–2906.
138. E. M. Müller, A. de Meijere and H. Grubmüller, *J. Chem. Phys.*, 2002, **116**, 897–905.
139. A. Laio and M. Parrinello, *Proc. Natl. Acad. Sci. U. S. A.*, 2002, **99**, 12562–12566.
140. V. Leone, F. Marinelli, P. Carloni and M. Parrinello, *Curr. Opin. Struct. Biol.*, 2010, **20**, 148–154.
141. A. Barducci, G. Bussi and M. Parrinello, *Phys. Rev. Lett.*, 2008, **100**, 020603.
142. A. Barducci, M. Bonomi and M. Parrinello, *Wiley Interdiscip. Rev.: Comput. Mol. Sci.*, 2011, **1**, 826–843.
143. L. Sutto and F. L. Gervasio, *Proc. Natl. Acad. Sci. U. S. A.*, 2013, **110**, 10616–10621.
144. A. Berteotti, A. Cavalli, D. Branduardi, F. L. Gervasio, M. Recanatini and M. Parrinello, *J. Am. Chem. Soc.*, 2009, **131**, 244–250.
145. P. Májek and R. Elber, *J. Chem. Theory Comput.*, 2010, **6**, 1805–1817.
146. E. Vanden-Eijnden, M. Venturoli, G. Ciccotti and R. Elber, *J. Chem. Phys.*, 2008, **129**, 174102.
147. R. Elber and A. West, *Proc. Natl. Acad. Sci. U. S. A.*, 2010, **107**, 5001–5005.
148. A. E. Cardenas, G. S. Jas, K. Y. DeLeon, W. A. Hegefeld, K. Kuczera and R. Elber, *J. Phys. Chem. B*, 2012, **116**, 2739–2750.
149. J. M. Bello-Rivas and R. Elber, *J. Chem. Phys.*, 2015, **142**, 094102.
150. J. M. Bello-Rivas and R. Elber, *J. Comput. Chem.*, 2015, 1–12.
151. P. G. Bolhuis, D. Chandler, C. Dellago and P. L. Geissler, *Annu. Rev. Phys. Chem.*, 2002, **53**, 291–318.
152. C. Dellago, P. G. Bolhuis and D. Chandler, *J. Chem. Phys.*, 1999, **110**, 6617.
153. P. G. Bolhuis, C. Dellago and D. Chandler, *Faraday Discuss.*, 1998, **110**, 421–436.
154. C. Dellago, P. G. Bolhuis and D. Chandler, *J. Chem. Phys.*, 1998, **108**, 9236–9245.
155. P. G. Bolhuis, *Proc. Natl. Acad. Sci. U. S. A.*, 2003, **100**, 12129–12134.
156. J. Juraszek, J. Vreede and P. G. Bolhuis, *Chem. Phys.*, 2012, **396**, 30–44.
157. R. Radhakrishnan and T. Schlick, *Proc. Natl. Acad. Sci. U. S. A.*, 2004, **101**, 5970–5975.
158. D. Moroni, P. G. Bolhuis and T. S. van Erp, *J. Chem. Phys.*, 2004, **120**, 4055.
159. W. G. Noid, *J. Chem. Phys.*, 2013, **139**, 090901.

160. W. G. Noid, J.-W. Chu, G. S. Ayton, V. Krishna, S. Izvekov, G. A. Voth, A. Das and H. C. Andersen, *J. Chem. Phys.*, 2008, **128**, 244114.
161. T. Head-Gordon, *Curr. Opin. Struct. Biol.*, 2003, **13**, 160–167.
162. S. Takada, *Curr. Opin. Struct. Biol.*, 2012, **22**, 130–137.
163. H. I. Ingólfsson, C. A. Lopez, J. J. Uusitalo, D. H. de Jong, S. M. Gopal, X. Periole and S. J. Marrink, *Wiley Interdiscip. Rev.: Comput. Mol. Sci.*, 2013, **4**, 225–248.
164. V. Tozzini, *Curr. Opin. Struct. Biol.*, 2005, **15**, 144–150.
165. T. Lazaridis and M. Karplus, *Proteins: Struct., Funct., Bioinf.*, 1999, **35**, 133–152.
166. P. Sfriso, A. Emperador, L. Orellana, A. Hospital, J. L. Gelpí and M. Orozco, *J. Chem. Theory Comput.*, 2012, **8**, 4707–4718.
167. E. A. Proctor, F. Ding and N. V. Dokholyan, *Wiley Interdiscip. Rev.: Comput. Mol. Sci.*, 2011, **1**, 80–92.
168. P. Sfriso, A. Emperador, J. Gelpí and M. Orozco, *Series in Computational Biophysics*, CRC Press, 2014, pp. 339–362.
169. E. A. Proctor and N. V. Dokholyan, *Curr. Opin. Struct. Biol.*, 2016, **37**, 9–13.
170. Y. Ueda, H. Taketomi and N. Go, *Biopolymers*, 1978, **17**, 1531–1548.
171. H. Taketomi, Y. Ueda and N. Gō, *Int. J. Pept. Protein Res.*, 1975, **7**, 445–459.
172. P. C. Whitford, O. Miyashita, Y. Levy and J. N. Onuchic, *J. Mol. Biol.*, 2007, **366**, 1661–1671.
173. K.-I. Okazaki, N. Koga, S. Takada, J. N. Onuchic and P. G. Wolynes, *Proc. Natl. Acad. Sci. U. S. A.*, 2006, **103**, 11844–11849.
174. S. Yang, S. S. Cho, Y. Levy, M. S. Cheung, H. Levine, P. G. Wolynes and J. N. Onuchic, *Proc. Natl. Acad. Sci. U. S. A.*, 2004, **101**, 13786–13791.
175. R. B. Best, G. Hummer and W. A. Eaton, *Proc. Natl. Acad. Sci. U. S. A.*, 2013, **110**, 17874–17879.
176. V. Muñoz and W. A. Eaton, *Proc. Natl. Acad. Sci. U. S. A.*, 1999, **96**, 11311–11316.
177. C. Hyeon and J. N. Onuchic, *Proc. Natl. Acad. Sci. U. S. A.*, 2007, **104**, 2175–2180.
178. C. Hyeon and J. N. Onuchic, *Proc. Natl. Acad. Sci. U. S. A.*, 2007, **104**, 17382–17387.
179. C. Hyeon and J. N. Onuchic, *Biophys. J.*, 2011, **101**, 2749–2759.
180. B. Jana, C. Hyeon and J. N. Onuchic, *PLoS Comput. Biol.*, 2012, **8**, e1002783.
181. C. Hyeon, S. Klumpp and J. N. Onuchic, *Phys. Chem. Chem. Phys.*, 2009, **11**, 4899.
182. P. C. Whitford, K. Y. Sanbonmatsu and J. N. Onuchic, *Rep. Prog. Phys.*, 2012, **75**, 076601.
183. Y. Shan, A. Arkhipov, E. T. Kim, A. C. Pan and D. E. Shaw, *Proc. Natl. Acad. Sci. U. S. A.*, 2013, **110**, 7270–7275.
184. P. C. Whitford, J. N. Onuchic and P. G. Wolynes, *HFSP J.*, 2008, **2**, 61–64.

185. P. C. Whitford and J. N. Onuchic, *Curr. Opin. Struct. Biol.*, 2015, **30**, 57–62.
186. P. Sfriso, A. Hospital, A. Emperador and M. Orozco, *Bioinformatics*, 2013, **29**, 1980–1986.
187. C. Hyeon, G. H. Lorimer and D. Thirumalai, *Proc. Natl. Acad. Sci. U. S. A.*, 2006, **103**, 18939–18944.
188. A. Narunsky, S. Nepomnyachiy, H. Ashkenazy, R. Kolodny and N. Ben-Tal, *Structure*, 2015, 1–10.
189. H. van den Bedem, G. Bhabha, K. Yang, P. E. Wright and J. S. Fraser, *Nat. Methods*, 2013, **10**, 896–902.
190. M. E. Wall, P. D. Adams, J. S. Fraser and N. K. Sauter, *Structure*, 2014, **22**, 182–184.
191. H. van den Bedem and J. S. Fraser, *Nat. Methods*, 2015, **2**, 307–318.
192. R. B. Fenwick, H. van den Bedem, J. S. Fraser and P. E. Wright, *Proc. Natl. Acad. Sci. U. S. A.*, 2014, **111**, E445–E454.
193. G. Bussi, F. L. Gervasio, A. Laio and M. Parrinello, *J. Am. Chem. Soc.*, 2006, **128**, 13435–13441.
194. D. Granata, C. Camilloni, M. Vendruscolo and A. Laio, *Proc. Natl. Acad. Sci. U. S. A.*, 2013, **110**, 6817–6822.
195. E. Papaleo, L. Sutto, F. L. Gervasio and K. Lindorff-Larsen, *J. Chem. Theory Comput.*, 2014, **10**, 4169–4174.
196. P.-C. Chen and J. S. Hub, *Biophys. J.*, 2015, **108**, 2573–2584.
197. P.-C. Chen and J. S. Hub, *Biophys. J.*, 2014, **107**, 435–447.
198. R. P. Rambo and J. A. Tainer, *Biophys. J.*, 2015, **108**, 2421–2423.
199. R. P. Rambo and J. A. Tainer, *Nature*, 2014, **496**, 477–481.
200. D. Kimanius, I. Pettersson, G. Schluckebier, E. Lindahl and M. Andersson, *J. Chem. Theory Comput.*, 2015, **11**, 3491–3498.
201. A. Panjkovich and D. I. Svergun, *Phys. Chem. Chem. Phys.*, 2015, 1–13.
202. R. P. Rambo and J. A. Tainer, *Annu. Rev. Biophys.*, 2013, **42**, 415–441.
203. D. Seeliger and B. L. de Groot, *PLoS Comput. Biol.*, 2010, **6**, e1000634.
204. D. de Juan, F. Pazos and A. Valencia, *Nat. Rev. Genet.*, 2013, **14**, 249–261.
205. F. Morcos, B. Jana, T. Hwa and J. N. Onuchic, *Proc. Natl. Acad. Sci. U. S. A.*, 2013, **110**, 20533–20538.
206. F. Morcos, A. Pagnani, B. Lunt, A. Bertolino, D. S. Marks, C. Sander, R. Zecchina, J. N. Onuchic, T. Hwa and M. Weigt, *Proc. Natl. Acad. Sci. U. S. A.*, 2011, **108**, E1293–E1301.
207. A. Bakan, A. Dutta, W. Mao, Y. Liu, C. Chennubhotla, T. R. Lezon and I. Bahar, *Bioinformatics*, 2014, **30**, 2681–2683.
208. D. S. Marks, L. J. Colwell, R. Sheridan, T. A. Hopf, A. Pagnani, R. Zecchina and C. Sander, *PLoS One*, 2011, **6**, e28766.
209. D. S. Marks, T. A. Hopf and C. Sander, *Nat. Biotechnol.*, 2012, **30**, 1072–1080.
210. T. A. Hopf, L. J. Colwell, R. Sheridan, B. Rost, C. Sander and D. S. Marks, *Cell*, 2012, **149**, 1607–1621.
211. P. Sfriso, M. Duran-Frigola, R. Mosca, A. Emperador, P. Aloy and M. Orozco, *Structure*, 2016, **24**, 116–126.

212. L. Sutto, S. Marsili, A. Valencia and F. L. Gervasio, *Proc. Natl. Acad. Sci. U. S. A.*, 2015, **112**, 13567–13572.
213. Z. Fan, R. O. Dror, T. J. Mildorf, S. Piana and D. E. Shaw, *Proc. Natl. Acad. Sci. U. S. A.*, 2015, **112**, 7454–7459.
214. B. Ma, C.-J. Tsai, T. Haliloglu and R. Nussinov, *Structure*, 2011, **19**, 907–917.
215. D. M. Zuckerman, *Annu. Rev. Biophys.*, 2011, **40**, 41–62.
216. R. O. Dror, R. M. Dirks, J. P. Grossman, H. Xu and D. E. Shaw, *Annu. Rev. Biophys.*, 2012, **41**, 429–452.
217. R. C. Bernardi, M. C. R. Melo and K. Schulten, *Biochim. Biophys. Acta*, 2015, **1850**, 872–877.
218. D. L. Mobley and K. A. Dill, *Structure*, 2009, **17**, 489–498.
219. L.-W. Yang, A. Kitao, B.-C. Huang and N. Gō, *Biophys. J.*, 2014, **107**, 1415–1425.
220. M. Ikeguchi, J. Ueno, M. Sato and A. Kidera, *Phys. Rev. Lett.*, 2005, **94**, 078102.
221. K.-I. Okazaki and S. Takada, *Proc. Natl. Acad. Sci. U. S. A.*, 2008, **105**, 11182–11187.
222. C. Hyeon, P. A. Jennings, J. A. Adams and J. N. Onuchic, *Proc. Natl. Acad. Sci. U. S. A.*, 2009, **106**, 3023–3028.
223. W. Li, W. Wang and S. Takada, *Proc. Natl. Acad. Sci. U. S. A.*, 2014, **111**, 10550–10555.
224. S. L. Seyler and O. Beckstein, *Mol. Simul.*, 2014, 1–23.
225. S. Yang, L. Salmon and H. M. Al-Hashimi, *Nat. Methods*, 2014, **11**, 552–554.
226. D. D. Boehr, R. Nussinov and P. E. Wright, *Nat. Methods*, 2009, **5**, 789–796.
227. H. D. Song and F. Zhu, *PLoS One*, 2013, **8**, e68023.

CHAPTER 8

Key Concepts and Applications of ONIOM Methods

HAJIME HIRAO,*[a] KAI XU,[a] PRATANPHORN CHUANPRASIT,[a] ADHITYA MANGALA PUTRA MOELJADI[a] AND KEIJI MOROKUMA*[b]

[a] Division of Chemistry and Biological Chemistry, School of Physical and Mathematical Sciences, Nanyang Technological University, 21 Nanyang Link, 637371 Singapore; [b] Fukui Institute for Fundamental Chemistry, Kyoto University, 34-4 Takano Nishihiraki-cho, Sakyo, Kyoto 606-8103, Japan
*Email: hirao@ntu.edu.sg; morokuma@fukui.kyoto-u.ac.jp

8.1 Introduction

The ever-growing intellectual curiosity of chemists, both experimental and computational, to understand the chemistry of complex molecular systems such as enzymes at the quantum mechanical level has led to the formulation of the hybrid quantum mechanics and molecular mechanics (QM/MM) method.[1–4] Although the Schrödinger equation cannot be solved exactly for molecules containing many electrons, if an accurate approximation of quantum mechanics (QM) is applied, their electronic properties can be described in a reliable manner that accounts for electron correlation effects. Many QM methods have been put forward over the years, and through the application of these methods, quantum mechanical aspects of molecules have been explored and elucidated. However, in general, the more accurate a QM approximation is, the more demanding and thus more time-consuming

RSC Theoretical and Computational Chemistry Series No. 9
Simulating Enzyme Reactivity: Computational Methods in Enzyme Catalysis
Edited by Iñaki Tuñón and Vicent Moliner

Published by the Royal Society of Chemistry, www.rsc.org

the computation tends to be. In fact, the number of atoms that accurate QM methods can handle within a short time has not increased so rapidly as to reach that of enzymes (*i.e.* ~10^3) at the time of writing. Therefore, even with the fast growth of computer power achieved so far, quantum mechanical treatment of large molecules is still a difficult challenge. The QM/MM approach exploits the advantages of QM and MM methods: the former is more accurate while the latter is computationally much less demanding. In essence, it is with this feature that the QM/MM method can describe quantum mechanical events of large molecular systems such as enzymatic reactions at practically manageable computational costs. Given that several QM methods with different levels of accuracy are now available, there is no reason to restrict the coupling to the QM-MM pair. For example, combining two different levels of QM methods (QM/QM) should also be possible. However, no matter which two methods are chosen for the coupling, the resultant model is more or less heterogeneous. An important technical issue is how to combine the two parts as naturally as possible.

There are two major ways of evaluating the total energy of an entire molecular system in the QM/MM method, namely, the 'additive' and 'subtractive' schemes. The additive version of QM/MM is detailed in another chapter. In this chapter, we intend to discuss mainly the subtractive scheme called ONIOM (Our own N-layer Integrated molecular Orbital and molecular Mechanics).[5–13] ONIOM encompasses not only the two-layer QM/MM treatment (ONIOM2(QM:MM)), but also hybrid QM/QM (ONIOM2(QM:QM′)) and other types of multilayer, multilevel calculations. Here, QM′ stands for a lower-level QM method. Thus, ONIOM is a powerful and flexible method that offers many different ways of dealing with large molecular systems such as enzymes. Nevertheless, we herein focus mainly on the ONIOM2(QM:MM) method, which is particularly useful in exploring enzymatic reactions computationally. In-depth description of the theoretical background of ONIOM has been given in a recent review paper.[12] Therefore, in what follows, at first only the key concepts of ONIOM are discussed, *i.e.* those that are deemed pedagogically and pragmatically useful for students as well as researchers who are interested in using the method. Subsequently, several recent applications of ONIOM methods, made mainly by the authors, are showcased to explain how ONIOM can be applied to specific problems of enzyme reactivity.

8.2 Methodological Aspects of ONIOM

8.2.1 Energy

As in any other QM or MM method, the potential energy or (total) energy of a molecule is the most practically useful quantity in ONIOM. When applying the ONIOM method to calculate the energy, one must first define 'layers' so that the entire system can be described by multilevel approximations.[5–12] In the case of ONIOM2(QM:MM), two layers are defined that are usually called

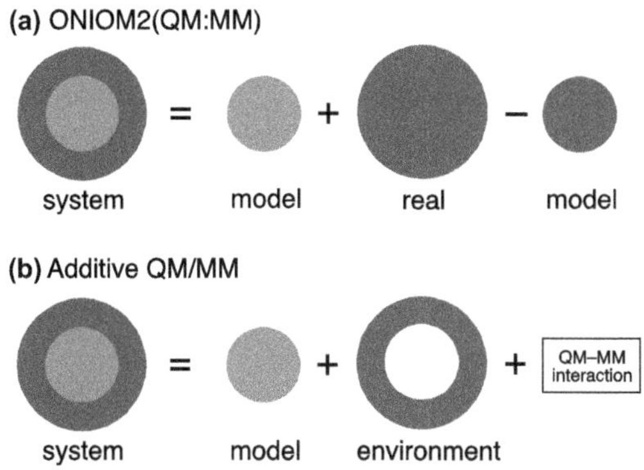

Figure 8.1 Comparison of the hierarchical regions used in (a) the ONIOM2(QM:MM) method and (b) the additive QM/MM method. QM and MM calculations are denoted by red and green.

'model' and 'real' systems (Figure 8.1a). A more accurate but more expensive QM method and a less accurate but much faster MM method are used in combination. A QM method is used to describe the model system quantum mechanically, whereas an MM method is used for classical descriptions of both the real and model systems. The energy of the system is defined in ONIOM2(QM:MM) as

$$E_{\text{ONIOM}} = E_{\text{QM,model}} + E_{\text{MM,real}} - E_{\text{MM,model}} \tag{8.1}$$

This formula may be compared with the energy expression for the additive QM/MM:

$$E_{\text{QM/MM}} = E_{\text{QM}} + E_{\text{MM}} + E_{\text{QM-MM}} \tag{8.2}$$

One notices a formal similarity between the two energy expressions, in that they both consist of three terms. However, there is a clear conceptual distinction between the subtractive and additive schemes of QM/MM; this may be understood intuitively by comparing how hierarchical regions or systems are defined in the two schemes (see Figure 8.1). As shown in Figure 8.1a, the real system in ONIOM2(QM:MM) is not the complement of the model system in describing the entire system. As such, just applying QM and MM methods to the model and real systems, respectively, to get the first two terms in eqn (8.1), necessarily leads to double counting of the energy of the model system, which is evaluated in both the QM and MM calculations. However, just by adding the third term in eqn (8.1), the computational demand of which is generally much lower than that for the first two terms, the double-counting problem is eliminated. In other words, out of the two types of energies calculated for the model system by the QM and MM methods, the unnecessary and less important MM energy is 'subtracted' here. This way of defining two layers has a great technical advantage because it does not require the preparation of a hollowed molecule (shown in green on the right-hand side in Figure 8.1b) that is necessary in the additive QM/MM (*vide infra*). As such, ONIOM allows for straightforward exploitation of a variety of QM and MM methods that have already been implemented in a software package. In addition, this feature allows the hybridisation of two different QM methods, *i.e.* ONIOM2(QM:QM′), to be achieved in a straightforward manner, and more than two methods can also be combined more easily than in the additive scheme of QM/MM.

On the other hand, when applying the additive QM/MM, again the atoms in the system are prioritised first to define two regions, but here, the entire system is split into, say, ‘model’ and ‘environment’ systems (see Figure 8.1b). It is clear that the environment is essentially the complement to the model system within the entire system, and thus using the first two terms in eqn (8.2) does not lead to overcounting of the energy contribution of the model system. However, these two terms do not account for the interaction energy between the two systems, and to compensate for this ‘shortage’, we must add the third term in eqn (8.2). At this point, readers may wonder how this sort of interaction between the two regions is accounted for in the ONIOM2(QM:MM) method. As far as the mechanical embedding (ME) scheme of ONIOM is concerned, this effect is included in the second term of eqn (8.1), which evaluates the energy of the real system using MM, whereas strictly speaking, when applying the electronic embedding (EE) scheme of ONIOM2(QM:MM), it is not accurate to say that only the second term accounts for the interaction between the model and environment systems. More details about the EE energy will be discussed later. Given that the model and environment systems complement each other in making up the entire system in the additive QM/MM scheme, and that different methods (QM and MM) are applied to these systems, it is no wonder that these systems are often called the ‘QM region’ and ‘MM region’, respectively. It is indeed a common practice to use these terms, especially in additive QM/MM methods. However, some care needs to be exercised in the case of ONIOM2(QM:MM). The model system may safely be called the ‘QM region’ in the same way as in the additive QM/MM. However, addressing the outer environment part as the ‘MM region’ could be misleading because MM is applied to both the model and real systems (the second and third terms in eqn (8.1)); neither of these systems is the same as the environment system or the MM region of the additive QM/MM (Figure 8.1). Therefore, the term ‘MM region’ should be used with some care to mean the environment system in ONIOM with this caveat in mind.

8.2.2 Treatment of the Boundary

The essential meanings of the layers used in ONIOM2(QM:MM) have already been explained well in Figure 8.1a, but in practice, one must care about another technical issue that emerges when defining a model system. Because the model system is a limited portion within the entire system, in many applications of ONIOM to enzymatic systems, a few covalent bonds

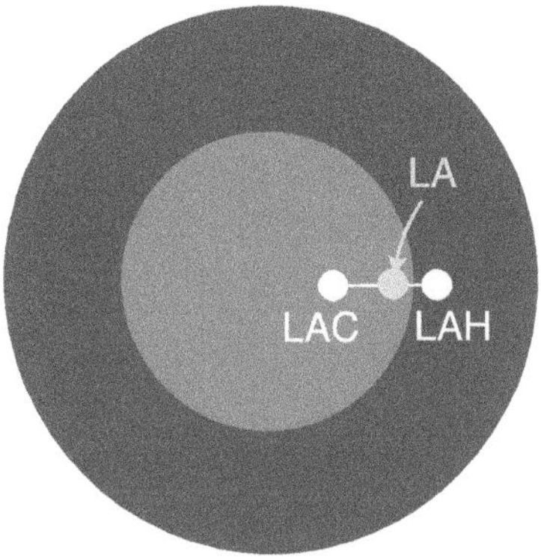

Figure 8.2 Schematic illustration of several key atoms at the core–environment boundary in the ONIOM2(QM:MM) method.

must be cut to create a distinct model system, which in turn leads to the generation of unsaturated bonds. Therefore, the boundary between the model and environment systems must be treated with some care. Several different methods have been proposed to make such boundaries chemically natural. In the standard approach adopted in ONIOM, a link atom (LA)[14] is placed according to

$$\mathbf{R}_{\mathrm{LA}} = \mathbf{R}_{\mathrm{LAC}} + g(\mathbf{R}_{\mathrm{LAH}} - \mathbf{R}_{\mathrm{LAC}}) \tag{8.3}$$

where LAC and LAH stand for 'link atom connection' and 'link atom host', respectively (see Figure 8.2), and g is a scale factor. Here, the LAC–LAH bond must be cut to define a model system; consequently, the LAC atom in the model system (or the QM region) will become a dangling bond. In other words, the LAC atom will not satisfy the valency requirement unless some special amendment is made, and thus a QM calculation on the resultant model system (say, the 'core' system that has not undergone the amendment of LA addition) may give a spurious result, especially in terms of the electronic distribution within the model system. To rectify this problem, a LA, which is typically a hydrogen (H-link) atom, is placed between the LAC and LAH according to eqn (8.3). The LA is now included in the model system, although not in the real system, which leads to a minor inconsistency between the number of atoms in the original, real system and the number of atoms in the ONIOM model. Nevertheless, the position of the LA is restricted by eqn (8.3), and thus this inconsistency does not lead to a redundancy of the internal degrees of freedom in the coordinates of the real system **R**. Therefore, even frequency calculations can be performed meaningfully within **R**.

8.2.3 Energy Gradients

Derivatives of potential energy with respect to certain variables play indispensable roles in computational studies of molecules. In particular, geometry optimisation and frequency analysis are two major types of calculation in the study of chemical reactions, and these calculations can be run efficiently if derivatives of energy with respect to the atomic coordinates of the real system **R** can be calculated analytically. In ONIOM2(QM:MM), the first derivative of the energy of the entire system with respect to **R** is simply

$$\frac{\partial E_{\text{ONIOM}}}{\partial \mathbf{R}} = \frac{\partial}{\partial \mathbf{R}}(E_{\text{QM,model}} + E_{\text{MM,real}} - E_{\text{MM,model}}) \tag{8.4}$$

This equation suggests that the $\partial E/\partial \mathbf{R}$ values can be calculated separately for the three different steps, *i.e.* the QM calculation on the model system 'QM, model' and the MM calculations on the real system 'MM, real' and the model system 'MM, model'. However, for this purpose, it is practically more convenient if the derivatives of $E_{\text{QM,model}}$ and $E_{\text{MM,model}}$ can be calculated with respect to the coordinates of the model system ($\mathbf{R}_{\text{model}}$) instead of **R**, because many software packages can calculate such quantities. This is indeed possible by introducing the Jacobian (**J**) for a coordinate transformation:

$$\frac{\partial E_{\text{ONIOM}}}{\partial \mathbf{R}} = \frac{\partial E_{\text{QM,model}}}{\partial \mathbf{R}_{\text{model}}} \cdot \mathbf{J} + \frac{\partial E_{\text{MM,real}}}{\partial \mathbf{R}} - \frac{\partial E_{\text{MM,model}}}{\partial \mathbf{R}_{\text{model}}} \cdot \mathbf{J} \tag{8.5}$$

The Hessian can also be calculated similarly as

$$\frac{\partial^2 E_{\text{ONIOM}}}{\partial \mathbf{R}^2} = \mathbf{J}^{\text{T}} \cdot \frac{\partial^2 E_{\text{QM,model}}}{\partial \mathbf{R}_{\text{model}}{}^2} \cdot \mathbf{J} + \frac{\partial^2 E_{\text{MM,real}}}{\partial \mathbf{R}^2} - \mathbf{J}^{\text{T}} \cdot \frac{\partial^2 E_{\text{MM,model}}}{\partial \mathbf{R}_{\text{model}}{}^2} \cdot \mathbf{J} \tag{8.6}$$

With these derivatives in hand, efficient geometry optimisation and frequency calculations are possible with the ONIOM2(QM:MM) method.

8.2.4 Geometry Optimisation

Geometry optimisation on potential energy surfaces of molecules is one of the most frequently used techniques in ONIOM[15] as well as pure QM and MM calculations. Technically, geometry optimisation boils down to a non-linear function minimisation problem, in which the minimised target function is the potential energy and the variables are nuclear coordinates. Thus, it is made more efficient if analytical derivatives of energy with respect to the nuclear arrangements are available. In general, the optimisation converges more rapidly if a vector containing the exact first derivatives (**g**, *i.e.* negative of the forces) and a matrix containing the exact second derivatives (**H**, *i.e.* Hessian) of energy with respect to some appropriate coordinates **q** for the molecule are available. In particular, second-order methods, which make use of the Hessian, permit a faster convergence of geometry optimisation, because the energy of the system after a new step ($\Delta\mathbf{q}$) is taken can be predicted more accurately by

$$E(\mathbf{q}+\Delta\mathbf{q}) \approx E(\mathbf{q}) + \mathbf{g}\Delta\mathbf{q} + \frac{1}{2}\Delta\mathbf{q}^{\mathrm{T}}\mathbf{H}\Delta\mathbf{q} \tag{8.7}$$

However, exact evaluation of the Hessian matrix elements is much more time-consuming than that of energy and gradient calculations; this is therefore often sidestepped. When optimising a minimum-energy geometry, the Hessian is usually not calculated exactly even once throughout the entire process; instead, an approximate Hessian matrix is prepared and updated by using the gradient information. Geometry optimisation of a transition state is more challenging, and necessitates location of the geometry at a saddle point of a potential energy surface. In this case, it is necessary to use a more reliable Hessian matrix than in the case of energy minimisation. In a commonly used procedure for the geometry optimisation of transition states in Gaussian 09 software,[13] the Hessian is calculated exactly at the first step of geometry optimisation, and it is updated without executing exact evaluation of the Hessian in subsequent steps. It should also be noted that in general, a good initial geometry must be used to ensure that the geometry optimisation of a transition state is successful. To this end, a series of relaxed energy scanning calculations is usually performed before starting transition state optimisation, with a certain bond distance or bond angle used as a reaction coordinate.

As we have seen above, energies and energy gradients for the three different calculations can be computed separately in the ONIOM2(QM:MM) method. When only the energy E_{ONIOM} is required, the three calculations usually proceed from 'MM, real', to 'QM, model', and then to 'MM, model'. A primitive approach to geometry optimisation is therefore to do the same for gradient calculations in a sequential manner and obtain the ONIOM energy gradients for the coordinate system **R**. However, the computational demands for energies and gradients are typically much lower for the MM calculations on the real system than for the QM calculations on the model system. Given these differences, sequentially evaluating the gradients for the three steps to move the geometry of the entire system one step forward is not a sensible procedure. Instead, in ONIOM2(QM:MM) calculations, geometry optimisation is usually performed by employing the so-called microiteration scheme,[15] in which the geometry of the environment system is fully optimised at a given geometry of the model system. After achieving zero forces for the environment system, the atoms in the model system can be moved one step forward (macroiteration). These are iterated until the system reaches a stationary point on the potential energy surface. For example, in Gaussian 09 software, the progress of geometry optimisation can be monitored by constantly checking the values shown in Table 8.1. At every geometry optimisation step, Gaussian makes a judgment about whether the calculation has converged or not. As shown in the table, the atomic forces

Table 8.1 Example of the progress of forces, displacements and energies during geometry optimisation (the final three steps) with Gaussian 09.

Cycle	Item	Value	Threshold	Converged?
N-2	Maximum force	0.000084	0.000450	YES
	RMS Force	0.000018	0.000300	YES
	Maximum displacement	0.003625	0.001800	NO
	RMS Displacement	0.000600	0.001200	YES
	ONIOM: extrapolated energy	−2411.92895189751		
N-1	Maximum force	0.000704	0.000450	NO
	RMS Force	0.000056	0.000300	YES
	Maximum displacement	0.003091	0.001800	NO
	RMS Displacement	0.000591	0.001200	YES
	ONIOM: extrapolated energy	−2411.92895152356		
N	Maximum force	0.000168	0.000450	YES
	RMS Force	0.000019	0.000300	YES
	Maximum displacement	0.001292	0.001800	YES
	RMS Displacement	0.000267	0.001200	YES
	ONIOM: extrapolated energy	−2411.92895225989		

and displacements are of particular importance for the judgment. Once a geometry is obtained that has maximum force, root-mean-square (RMS) force, maximum displacement, and RMS displacement values smaller than the threshold values, the geometry optimisation is regarded as converged.

Despite the proven efficiency of the microiteration/macroiteration method, one sometimes encounters cases in which geometry optimisation does not converge, with unwanted oscillation behavior in the energy, even after running many optimisation cycles. One simple, practically useful solution to this problem may be to extract the XYZ coordinates that have the smallest 'maximum force' value (see Table 8.1) from the output file and then restart a new geometry optimisation calculation, because having small forces is a good indication that the geometry is close to the targeted stationary point. In addition, the current updated Hessian may not be very reliable anymore and thus may have to be reset. This is a convenient recipe for resolving the issue of slow convergence, especially in optimising a transition state geometry.

The so-called quadratically coupled optimisation method improves the reliability of optimisation steps, thereby expediting the convergence of geometry optimisation.[16] The above-mentioned alternating microiteration/macroiteration procedure works fine in many cases. However, this method does not necessarily guarantee a well-behaved change in the Hessian during updates. In addition, the cross terms of the Hessian, which represent the coupling between the coordinates of the model and environment systems, are neglected. This omission sometimes gives rise to unpleasant oscillation behavior in the energy during geometry optimisation. The quadratic coupling method amends the deficiencies of the updated Hessian, thereby allowing a more reliable geometry optimisation step to be taken. In many cases, the oscillation behavior during geometry optimisation can be resolved significantly by this method, and a very fast convergence can be attained.

8.2.5 Embedding Schemes

There are several different ways to deal with the electrostatic interactions between the model and environment systems in QM/MM methods. Two major approaches are the mechanical embedding (ME) and electronic embedding (EE) schemes. Furthermore, a polarisable embedding (PE) version of ONIOM has also been examined recently.[17] ONIOM2(QM:MM)-ME is the original form of ONIOM2(QM:MM), in which the core–environment electrostatic interaction is evaluated fully classically. Fully classical evaluation of the interaction means that the electrostatic interactions are evaluated using the Coulomb formula that is included within the second term of eqn (8.1). Consequently, the total energy, $E_{\text{ONIOM-ME}}$, is affected significantly by the atomic point charges used for the model system. When conducting reaction pathway analyses, point charges of the model system may be determined for a reactant state, and this charge set is usually used throughout for all subsequent species on the reaction pathway. However, the electronic state of the model system could essentially be altered depending on its geometry, and accordingly, the point charges should differ at different geometries. Therefore, a more sophisticated procedure for ONIOM-ME is to determine charges for different states. However, even for one particular intermediate state, determining a fully consistent charge set cannot be achieved straightaway. For example, certain initial charges may be assigned to the model system, and a new charge set may be obtained for the optimised geometry. However, another run of geometry optimisation with the new charge set may lead to a slightly different optimised geometry, which in turn could have somewhat different optimal point charges. In this context, the best procedure is to repeat the charge update and geometry optimisation alternately until the energy does not change anymore,[9] although this procedure requires additional computational costs compared with the primitive ONIOM-ME calculation.

ONIOM-ME is a computationally very efficient approach for studying large molecular systems. However, it does not account for the polarisation of the model system, which occurs under the perturbation exerted by the environment system, or more specifically, by the point charges placed on atoms in the environment system. By contrast, the EE version of ONIOM (ONIOM-EE) can describe this sort of polarisation effect.[12] In the ONIOM2(QM:MM)-EE method, the energy of the system is approximated as

$$E_{\text{ONIOM-EE}} = E^{\text{V}}_{\text{QM,model}} + E_{\text{MM,real}} - E^{\text{V}}_{\text{MM,model}} \tag{8.8}$$

which looks similar to eqn (8.1) for the ONIOM-ME method; however, the first and third terms are modified so that the effect of the polarisation of the model system is included effectively. The first term is calculated using the Hamiltonian:

$$H^{V}_{\text{QM,model}} = H_{\text{QM,model}} - \sum_{i}\sum_{A_{\text{env}}} \frac{s_{A_{\text{env}}} q_{A_{\text{env}}}}{r_{iA_{\text{env}}}} + \sum_{A_{\text{QM}}}\sum_{A_{\text{env}}} \frac{Z_{A_{\text{QM}}} s_{A_{\text{env}}} q_{A_{\text{env}}}}{r_{A_{\text{QM}}A_{\text{env}}}} \tag{8.9}$$

Here, q is a point charge in the environment system, r is the distance between an electron in the model system and a point charge in the environment system or between a nucleus in the model system and a point charge in the environment system, and the subscripts QM and env refer to the QM (model) and environment systems, respectively. The first term on the right-hand side in eqn (8.9) is also used in the ONIOM-ME method, but here, two additional terms are used. In fact, it is due to the second term of eqn (8.9) that the electronic state of the model system undergoes polarisation. For example, when a density functional theory (DFT) method is used for the 'QM, model' calculation, this term is appended to the one-electron operator of the Kohn–Sham equations. Iterative self-consistent field (SCF) optimisation of Kohn–Sham orbitals results in polarised electronic distributions within the model system, as compared with the case in which this term is not included (*i.e.* ONIOM-ME or gas-phase calculation of the model system). One notices that a scale factor is used in the second term. The scale factor is essentially 1 (no scaling) for most of the point charges, but it could also be 0. The use of the scale factor originates from the fact that the resultant electronic state of the model system is likely to suffer from overpolarisation if all of the point charges from the environment system are included in the QM calculation. Thus, if no scaling is applied, the interaction between the electrons in the model system and a nearby atom, for example, LAH in Figure 8.2, could be too large. In particular, the LAH and LA are very close to each other, and thus inclusion of the point charge of the LAH could affect the QM region too strongly, but not meaningfully. Therefore, some of the point charges in the environment system, typically those for atoms that are within three bonds away from the LAC, are left out by using a scale factor of 0.[10] The third term in eqn (8.9) accounts for the classical electrostatic interactions between the nuclei in the model system and the point charges in the environment system. For consistency, the same scale factor as in the second term is used here. We can now realise that the second and third terms in eqn (8.9) describe the electrostatic interaction between the model system and the environment system.

van der Waals (vdW) interactions constitute another important form of nonbonded interactions in ONIOM2(QM:MM). Where are the vdW interactions between the two systems calculated in the EE scheme? This is done in the second term of eqn (8.8), which evaluates all of the MM energy contributions within the entire system. Thus, the vdW term is calculated in the same way as in ONIOM-ME. However, this also means that the (coulombic) electrostatic interactions between the two systems are again calculated here, despite the fact that the electrostatic interactions have already been calculated semiclassically in the first term of eqn (8.8). Thus, the electrostatic interactions have been calculated twice up to this point. Out of these two electrostatic contributions, we want to eliminate the somewhat less accurate coulombic electrostatic interactions. This is done during the MM calculation on the model system by the third term of eqn (8.8). More specifically, this term has the form

$$E^{V}_{\mathrm{MM,model}} = E_{\mathrm{MM,model}} + \sum_{A_{\mathrm{QM}}} \sum_{A_{\mathrm{env}}} \frac{q_{A_{\mathrm{QM}}} s_{A_{\mathrm{env}}} q_{A_{\mathrm{env}}}}{r_{A_{\mathrm{QM}} A_{\mathrm{env}}}} \tag{8.10}$$

where the first term is the same as that used in eqn (8.1), while the second term has the role of rectifying the double counting of electrostatic interactions.

We have seen above that the electrostatic interactions in ONIOM-EE are calculated in the first term of eqn (8.8), and in so doing, the point charges for the nearby atoms in the environment system are zeroed out. It is important to note, however, that this does not mean that the contributions of these nearby atoms to the electrostatic interactions are completely neglected, because when calculating the second term in eqn (8.8), such charge elimination is not applied to these nearby atoms. In addition, these contributions are not eliminated by eqn (8.10). Therefore, the nearby atoms are still allowed to have nonzero electrostatic interactions with the assigned point charges of the model system in a classical sense through the Coulomb formula.

We have also seen above that geometry optimisation can be performed efficiently by using the technique of microiteration. In an ONIOM2(QM:MM)-EE calculation, during the microiteration step, the positions of the atoms in the environment system change, and this in turn should alter the electronic state of the model system through the second term of eqn (8.9). However, determining the electronic state of the model system quantum mechanically for every different configuration of the environment system will cause a significant increase in the number of QM calculations, and thus micro-iteration will not accelerate the geometry optimisation as markedly as it does in ONIOM2(QM:MM)-ME. Therefore, instead of running too many QM calculations, point charges are determined and assigned to the model system at a given geometry of the model system, to approximate the core–environment interaction classically during the microiteration for the geometry optimisation with the ONIOM2(QM:MM)-EE method.

8.2.6 Set-up for ONIOM Calculations

ONIOM2(QM:MM) is frequently used when studying enzymatic reactions, and its fundamental theories have already been explained above. Once an input file for an enzyme–substrate system is prepared, the ONIOM calculations can be carried out more or less routinely to investigate reaction mechanisms, just by following the theories described above. However, in practice, one of the most time-consuming steps in ONIOM calculations is often the prior set-up of a structure model that is necessary before starting ONIOM calculations. Even though the set-up process may not involve any demanding calculations, it often requires careful checking. As such, the set-up is technically a very important step in ONIOM calculations and has been explained in detail in a recently published book chapter.[18] As mentioned there, different research groups have developed their own workflows for the set-up, and there is no single perfect method. Thus, what we explain below is just one of many choices.

Typically, an initial structure of the enzyme of interest is taken from the Protein Data Bank (PDB) webpage, where a large number of protein structures are deposited that were determined mainly by X-ray diffraction techniques.[19] The XYZ coordinates in the downloaded PDB files sometimes have missing atoms or amino acid residues. Missing atoms may be appended by use of software such as Modeller.[20] Hydrogen atoms of proteins are not easily determined by X-ray diffraction and thus are usually not included in PDB files. Although the positions of hydrogen atoms on carbon atoms can be straightforwardly determined on the basis of hybridisation states and chemical intuitions, the positions of hydrogen atoms on titratable residues must be determined with care. For this purpose, web tools for predicting local pK_a values such as PROPKA are useful.[21] Nevertheless, it is advised that visual inspection be carried out after any automated protonation of an enzyme.

The organic substrate molecule that undergoes a chemical reaction in the active site of an enzyme can be built with molecular visualisation software such as GaussView.[22] In contrast to amino acid residues for which atomic charges have already been determined and implemented in force-field libraries, substrates usually do not have predetermined charges. To fill this gap, the geometry of a substrate in its isolated state may be optimised first using a QM method, and the geometry-optimised substrate may then be subjected to an additional calculation to determine atomic charges with, for example, the restrained electrostatic potential (RESP) method.[23] The Antechamber module of AMBER software can be used to determine RESP atomic charges based on the electrostatic potential as calculated by Gaussian 09.[13,24] As we have seen above, the atomic charges determined here will have a significant influence on the energies in the ONIOM2(QM:MM)-ME method. vdW parameters should also be given to the substrate so that the nonbonded vdW interactions between the core and environment systems can be calculated by the second term of eqn (8.1) or eqn (8.8). The vdW parameters may be taken, for example, from the existing parameter libraries for a certain version of AMBER. Parameters for bonded interactions should also be given to the substrate. However, if the entire substrate is included in the model system in ONIOM calculations, the bonded parameters used for the substrate will not affect the ONIOM energy, because the energies for bonded interactions will be calculated twice in the two MM calculations in ONIOM2(QM:MM), and the opposite signs of these energy terms result in their mutual cancellation. Therefore, the result will remain unaffected even if some spurious bonded parameters are given to the atoms in the model system for which parameters are not available.

The geometry of the substrate in a certain pose may already be included in a PDB file. In such cases, this geometry may be used directly in the ONIOM calculation because the PDB geometry is likely to reflect the plausible configuration of the substrate during reactions. Still, this substrate geometry is not complete because hydrogen atoms are not included in the PDB file. Therefore, hydrogen atoms must be added to the substrate, *e.g.* by using GaussView, before starting any geometry optimisation calculations for the substrate. When a crystal structure is not available for an enzyme in complex with the substrate, the geometry of the substrate must be built with some software like GaussView, as mentioned above. Importantly, however, in this case, there will be uncertainties regarding its configuration within the active site of the enzyme. Therefore, a few plausible configurations of the substrate may be produced first by performing docking simulations using, *e.g.* AutoDock Vina.[25]

The resultant protein–substrate structure may be subjected either to a molecular dynamics (MD) simulation or directly to an ONIOM calculation. By determining the reaction potential energy profile in the latter approach, one can already gain valuable insight into the enzyme reactivity. However, enzymes generally have many more degrees of freedom than organic molecules, and thus using potential energies as calculated by eqn (8.1) or eqn (8.8) for assessing the relative stability of different states does not necessarily provide sufficient information about the reactivity. A more reliable quantity in such situations is free energy, which can be evaluated by taking into account a large number of possible conformations of the enzyme and substrate. To generate many different conformational samples for free energy calculations, MD is the method of choice. However, combining MD sampling and ONIOM2(QM:MM) energy calculation could lead to prohibitively high computational demands, although calculations become feasible if appropriate approximations are introduced to reduce the cost.[26] Therefore, in many cases, MD simulation is used to produce a handful of 'snapshots' of different enzyme–substrate geometries. Potential energy calculations may be performed for each of the selected snapshots, to see how the shapes of the reaction potential energy surfaces can differ for different snapshots. Such calculations, which involve additional MD simulations and ONIOM calculations on different snapshots, are expected to provide more reliable enzyme–substrate geometries than the direct ONIOM calculation using a crystal structure, and a more reliable picture of the reaction pathway may be obtained. However, it should be kept in mind that here, MD simulation is performed with MM force fields that have rather limited accuracy, which means that MD simulation might not always improve the enzyme–substrate geometry compared with the original crystal structure. Thus, there is no guarantee that the snapshots reflect likely enzymatic configurations. In addition, a number of snapshots in the order of 10^0 or 10^1 is not at all sufficient for obtaining statistically meaningful interpretations. Therefore, even if very different barrier heights are obtained for two different snapshots, one cannot immediately generalise from this result to argue, for instance, that the system will encounter these distinct situations frequently.

Thus, these calculations should be carried out with the inherent limitations in mind. A possible set-up procedure is summarised in Figure 8.3, which one may follow when running ONIOM calculations of an enzyme–substrate complex.

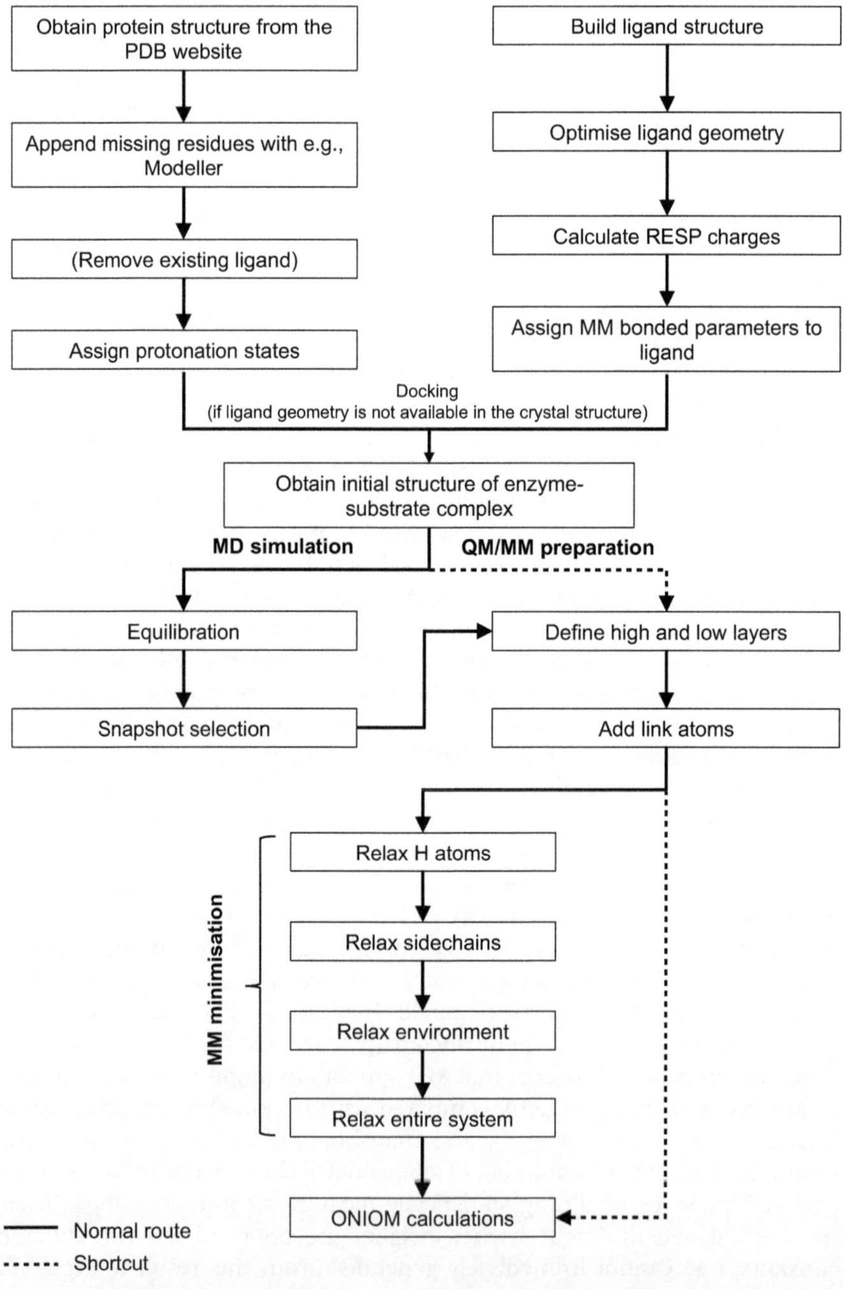

Figure 8.3 A possible set-up procedure for an ONIOM calculation of an enzyme–substrate complex.

Thus, ONIOM2(QM:MM) calculations are performed, in many cases, to calculate potential energy surfaces of enzymes. Although this procedure is regarded as a practically reasonable compromise, one must be aware of one problematic consequence that could be observed during determination of reaction pathways. When using potential energy for the analysis of chemical reactions, it is important to make sure that this potential energy is smoothly connected along the reaction coordinate. One typical case in which this necessary requirement is violated is encountered when the conformation of an amino acid residue in the environment system changes significantly at some point on the reaction pathway. This kind of conformational change could result in a decrease in the total energy (*i.e.* E_{ONIOM}) by, say, 10 kcal mol^{-1}, but this energy decrease should not be regarded as being associated with the exothermicity of the reaction. For example, this sort of conformational change could make a transition state even lower in energy than the prior intermediate state, but this energy diagram does not allow one to make a meaningful interpretation of the enzyme reactivity. When such an energy drop has happened, the energy of the prior intermediate should be newly determined. If the combination of the second and third terms of eqn (8.1) ($E_{\mathrm{MM,real}} - E_{\mathrm{MM,model}}$), which is also called the S_{low} value,[12] changes significantly on a reaction pathway, it is possibly a telltale sign of this kind of unwanted conformational change in the environment system.

When starting ONIOM calculations, one must make sure that the system's 'connectivity' is defined correctly for all atoms. In pure QM calculations, the connectivity does not have to be defined, because such information is not used when solving the Kohn–Sham equations or any other QM equations. However, in ONIOM2(QM:MM) calculations, one must declare clearly how atoms in the system are covalently linked. The reason is that such connectivity information is necessary when applying an MM force-field method for the second and third terms of eqn (8.1) or eqn (8.8). An example is given in Figure 8.4, which describes the connectivity for the first few atoms in an

```
1 3 1.0 4 1.0 2 1.0 5 1.0
2
3
4
5 6 1.0 7 1.0 12 1.0
6
7 10 1.0 8 1.0 9 1.0
8
9
10 11 1.0
11
12 13 2.0 14 1.5
13
14 15 1.0 16 1.0
15
16 17 1.0 23 1.0 18 1.0
17
18 19 1.0 20 1.0 21 1.0
19
20
```

Figure 8.4 Example of connectivity specification in Gaussian 09 software for the first 20 atoms in an enzyme–substrate complex.

enzyme–substrate complex in Gaussian 09 format. The first line states that the first atom is connected to the third, fourth, second, and fifth atoms. The number '1.0' denotes the bond order, but as long as an AMBER-type force field is used, the bond order does not affect the resultant energy because the energy expression does not contain any contributions from bond order. Again, GaussView helps determine the connectivity of large enzymes. The second to fourth lines just mention the atom ID for a single atom, but this does not necessarily mean that these atoms are not connected to any atom. In this particular case, these atoms are connected to the first atom, as described in the first line. In other words, the connectivity specification can be done only once.

8.2.7 Preparation of a Decent Initial Orbital Guess for the Model System

Especially when a transition metal such as iron is included in the model system, the QM calculation on the model system, namely, the first term in eqn (8.1) or eqn (8.8), does not always converge to the desired electronic state. This is particularly true when a DFT method is used in a broken-symmetry fashion to describe the electronic state of the model system. When the calculation has converged to an undesired electronic state, a conventional approach for rectifying the problem is to alter the electronic occupancy of orbitals. When a Kohn–Sham DFT method is used, a single determinant of occupied orbitals (Kohn–Sham orbitals) is used to describe the noninteracting electrons. To change the electronic occupancy of Kohn–Sham orbitals, one must first identify those orbitals that are supposed to be occupied and unoccupied, by using graphics software such as GaussView. Subsequently, the ordering of these orbitals is altered so that some of the initially unoccupied and occupied orbitals are made occupied and unoccupied, respectively. To prepare a decent initial orbital guess for ONIOM calculations, one reasonable procedure may be to extract only the model system from the entire system and perform pure DFT calculations to get a desired orbital occupancy, because for the orbital manipulation, thousands of MM atoms could be distracting. Once a desired orbital occupancy is achieved, the orbital data may be printed as a text file by using the 'punch = mo' keyword of Gaussian 09. The text-format orbital guess may be read by the 'guess = cards' keyword when performing an ONIOM calculation, so that the specified initial guess is read in the QM calculation step on the model system. A more convenient approach that has become available recently is the fragment guess option of Gaussian 09, which allows intuitive constructions of initial orbitals and is thus highly useful.

8.3 Application of ONIOM2(QM:MM) to the Reactions of Iron Enzymes

ONIOM calculations help resolve mechanistic conundrums of nonheme-type and heme-type iron enzymes, which are interesting research targets of computational chemistry in general.[27] Therefore, let us first discuss some of our applications of ONIOM calculations to iron enzymes.

8.3.1 *myo*-Inositol Oxygenase

myo-Inositol oxygenase (MIOX) is a metalloenzyme that has a coupled dinuclear iron cluster containing nonheme ligands at its active site.[26–31] The major catalytic role of MIOX is to assist the first committed step of the catabolism of *myo*-inositol (MI), *i.e.* the conversion of MI into D-glucuronate (DG) (Figure 8.5). There are nine isomers of cyclohexanehexol that differ in how the hydroxyl groups are attached to cyclohexane, and MI is the most abundant form. Careful inspection of crystal structures shows that one of the iron ions in the diiron cluster is used to anchor the MI substrate, while the other is used to form a ferric superoxide species that effects C–H bond activation, which is the first step of the MI-to-DG conversion.[29,30]

Hirao and Morokuma undertook extensive ONIOM(B3LYP:AMBER) studies to gain a detailed understanding of the ferric superoxide reactive species and to elucidate how MI is converted to DG in the active site of MIOX. The SDD effective core potential basis set was used for Fe and the 6-31G* basis set was employed for the other atoms. For single-point calculations, the TZVP basis set was used. As for the AMBER MM calculations, the default parameters in Gaussian 09 were used. Their ONIOM calculations suggested that the ferric superoxide species has a side-on geometry and an $S = 1/2$ spin state. Both of the iron ions have high-spin states, and these are antiferromagnetically coupled. The potential energy diagram for the entire reaction pathway was also determined (Figure 8.6), which suggested that the barrier for the O–O bond cleavage step is higher than that for the initial C–H activation step (Figure 8.7). They also computed theoretical ^{2}H-KIE values for these steps at 5 °C and obtained 6.6 (transition state theory) and 8.8 (with the Wigner correction) for the C–H cleavage step and 1.1 (with both approximations) for the O–O cleavage step. Indeed, a preceding experimental study by Xing *et al.* showed that the KIE for the C–H cleavage step is 8–16. Interestingly, they further demonstrated that the C–H cleavage step is not the slowest step in the reaction, because the ^{2}H-KIE on steady-state turnover was not significantly larger than 1 (≈1.35). The ONIOM computational study

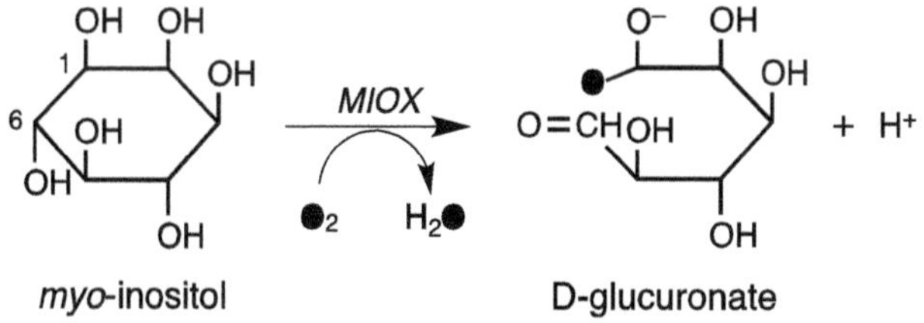

Figure 8.5 Conversion of MI to DG catalysed by MIOX.
Reprinted with permission from H. Hirao and K. Morokuma, *J. Am. Chem. Soc.*, 2009, **131**, 17206.[31] Copyright (2009) American Chemical Society.

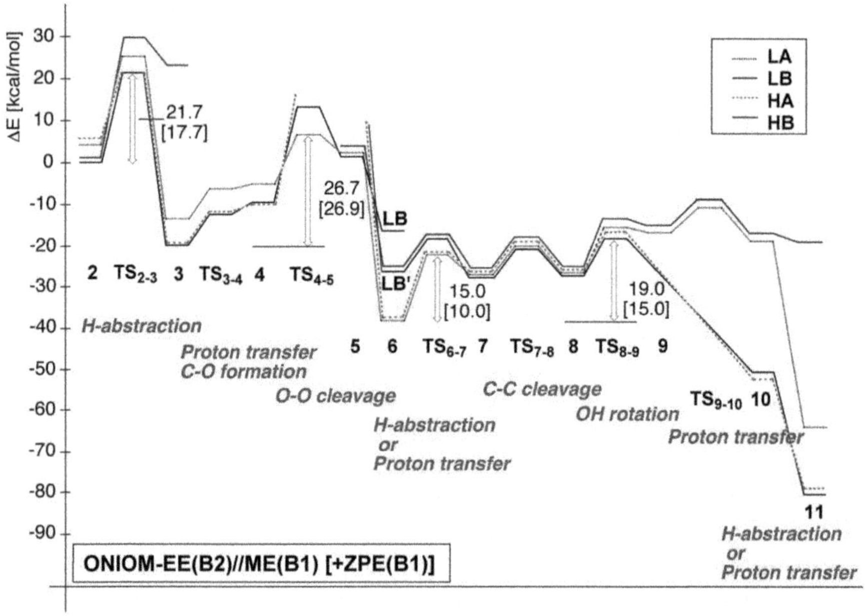

Figure 8.6 ONIOM-calculated energy diagram for the MIOX-catalysed reaction. Reprinted with permission from H. Hirao and K. Morokuma, *J. Am. Chem. Soc.*, 2009, **131**, 17206.[31] Copyright (2009) American Chemical Society.

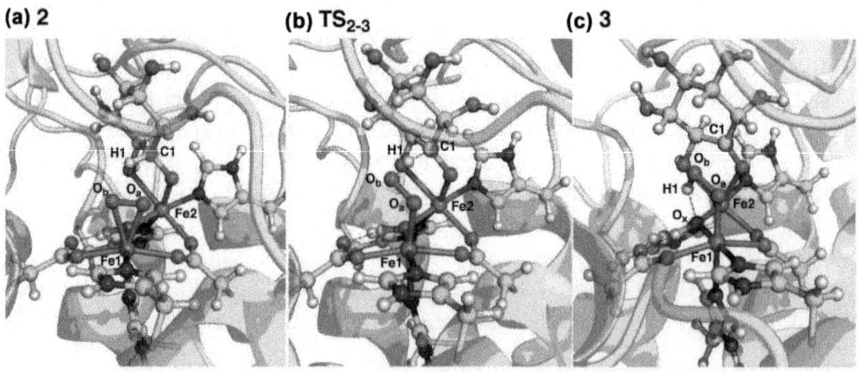

Figure 8.7 Geometric change during the C–H cleavage step of the MIOX-catalysed reaction.
Reprinted with permission from H. Hirao and K. Morokuma, *J. Am. Chem. Soc.*, 2009, **131**, 17206.[31] Copyright (2009) American Chemical Society.

supported this conclusion, offering the additional microscopic view that the insignificant KIE is due to the higher barrier for the O–O cleavage process during the decay.[28,31]

Another interesting finding with regard to C–H bond cleavage by the ferric superoxide was that one of the two electrons of the C–H bond, which is left behind on the MI substrate, goes down to the ferric ion, instead of remaining on the substrate (Figure 8.8). As a result, the original Fe(III) site is reduced to Fe(II), and the substrate does not keep an unpaired electron. This electron reorganisation is important in driving the C–H cleavage process. In fact, in the highest-spin state (HB in Figure 8.8), the unpaired electron remained on the substrate, and the barrier for C–H cleavage was high, because the unpaired electron on the substrate and the electrons on Fe(III) repelled each other.

After the C–H cleavage, an Fe(III)–OOH intermediate (**3** in Figures 8.6 and 8.7) is generated, and its distal oxygen subsequently forms a C–O bond with the substrate to form an Fe(III)–OOR species (**4** in Figure 8.6). During this step, the distal proton of Fe(III)–OOH was transferred to the hydroxide ligand between two iron ions, and to mitigate the accumulation of positive charges, the hydroxide donates its proton to the nearby aspartate ligand. Thus, the ONIOM calculations pointed out the non-innocent role of the bridging ligand in the MI-to-DG conversion. A comparison of the stability of the Fe(III)–OOH and Fe(III)–OOR species, based on ONIOM-EE calculations, showed that the former should be more stable. Therefore, the ONIOM calculations suggested that the experimentally trapped but uncharacterised intermediate formed after the C–H cleavage should be the Fe(III)–OOH species.

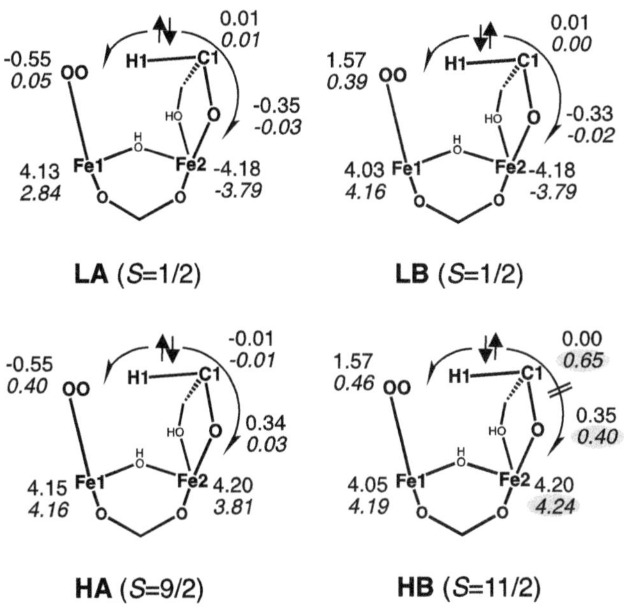

Figure 8.8 Calculated spin population values before and after the C–H cleavage step of the MIOX-catalysed reaction in various spin states.
Reprinted with permission from H. Hirao and K. Morokuma, *J. Am. Chem. Soc.*, 2009, **131**, 17206.[31] Copyright (2009) American Chemical Society.

The ONIOM calculations further identified the essential role played by the protein environment around the active site. When DFT calculations were employed for a truncated model without including the protein environment, the active site cluster collapsed after the O–O bond cleavage (*via* $\mathbf{TS}_{4\text{–}5}$ in Figure 8.6). To prevent this unwanted structural collapse, the MI substrate must be anchored in an appropriate position; this can be done through H-bonding interactions between the hydroxyl groups of MI and the surrounding residues as well as the steric restraint imposed by the protein.

8.3.2 2-Hydroxyethylphosphonate Dioxygenase

2-Hydroxyethylphosphonate dioxygenase (HEPD) is a mononuclear non-heme iron enzyme responsible for the biosynthesis of a particular type of compound containing a C–P bond (C–P compounds).[32–39] Given that C–P compounds widely exist in natural products such as lipids and glycoproteins, and artificial chemicals such as herbicides and drugs, HEPD is an enzyme of practical importance. HEPD is also a mechanistically interesting enzyme in bioinorganic chemistry. For the catalysis of the reaction, HEPD utilises Fe(II) and O_2 to cleave the C–C bond of 2-hydroxyethylphosphonate (2-HEP), eventually yielding hydroxymethylphosphonate (HMP) and formic acid (Figure 8.9a). HEPD can also catalyse the conversion of 1-HEP to acetylphosphate (AcP) and H_2O; in this case, the P–C bond of the substrate is cleaved. With Fe(II) and O_2, HEPD can form a ferric superoxide reactive species in the active site, as in the case of MIOX above, to effect the initial C–H activation step. Because HEPD has only one iron ion in the reaction center, the ferric superoxide formation and the substrate binding occur at the same iron. Thus, 2-HEP is bound to the iron in a bidentate fashion (Figure 8.10), and H-abstraction by the ferric superoxide occurs from C2 of 2-HEP or C1 of 1-HEP.

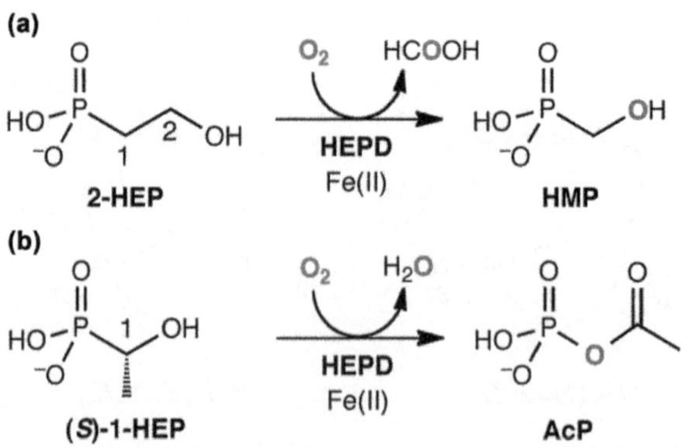

Figure 8.9 HEPD-catalysed reactions.
Reprinted with permission from H. Hirao and K. Morokuma, *J. Am. Chem. Soc.*, 2011, **133**, 14550.[38] Copyright (2011) American Chemical Society.

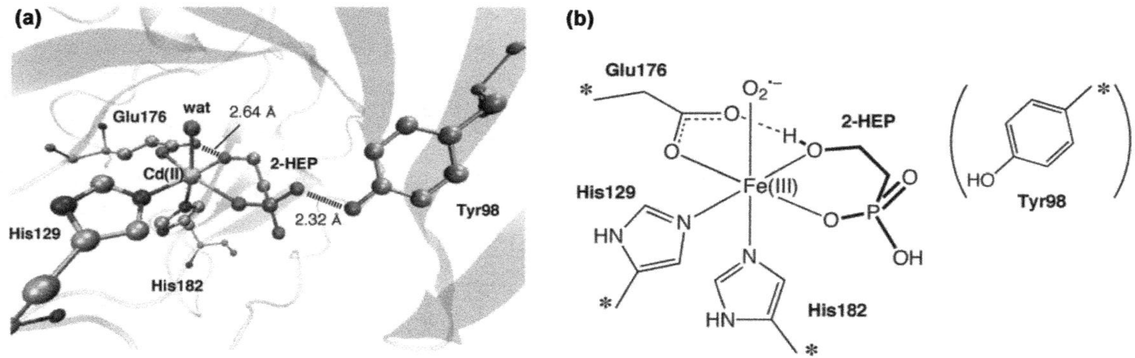

Figure 8.10 (a) Active site of HEPD (PDB code 3GBF). (b) Active site model used for DFT calculations. Tyr98 was included for a few test calculations. The atoms marked with asterisks were frozen in geometry optimisations with DFT. The same atoms (except Tyr98) were also selected to define the model system in ONIOM calculations.
Reprinted with permission from H. Hirao and K. Morokuma, *J. Am. Chem. Soc.*, 2010, **132**, 17901.[39] Copyright (2010) American Chemical Society.

Thus, the HEPD-catalysed reactions of seemingly analogous substrates yield different types of products (Figure 8.9), which may imply that different mechanisms could operate within the same enzyme. To better understand the mechanisms of HEPD reactions, Hirao and Morokuma performed DFT and ONIOM(DFT:MM)-EE calculations.[38,39] In their DFT study, the potential energy surface of the reaction of 2-HEP was broadly explored. To maintain the active site geometry at the geometry of the enzyme, several atoms were kept fixed during geometry optimisation (Figure 8.10b). Geometry optimisation was performed at the B3LYP/[SDD(Fe),6-31G*(others)] level, and the 6-311+G(d,p) basis set was used for single-point energy calculations. The DFT calculations suggested that the reaction begins with H-abstraction from the C2 atom of 2-HEP by the ferric superoxide reactive species. This H-abstraction is accompanied by a proton-coupled electron transfer (PCET) in which the proton of the hydroxyl group of 2-HEP transfers to the adjacent Glu176, while an electron on the hydroxyl oxygen transfers to the ferrous center. Subsequently, there is a homolytic O–O bond cleavage that leads to a diol radical species that undergoes facile C–C bond cleavage, which was one of the most interesting findings. Another interesting finding was that as a result of O–O bond cleavage, a ferric hydroxide species is formed, and a substrate radical species, which is generated upon C–C cleavage, attacks the hydroxide of the ferric hydroxide, to yield the HMP product (Figure 8.11).

Hirao and Morokuma went on to perform ONIOM calculations to make the theoretical model more realistic and to elucidate the origin of different reaction patterns observed for 2-HEP and 1-HEP.[38,39] Geometry optimisation was carried out at the ONIOM(B3LYP/[SDD(Fe),6-31G*(others)]:AMBER)-EE level, and single-point energy calculations were performed with the 6-311+G(d,p) basis set. AMBER is the one implemented in Gaussian 09. The overall mechanistic picture for the reaction of 2-HEP, as delineated based on the ONIOM calculations (Figure 8.12a), was essentially the same as that obtained by DFT calculations; thus, the reaction is initiated by the H-abstraction from the alpha carbon of the hydroxyl group to produce an Fe(II)OOH intermediate, and the final HMP product is formed *via* the formation of a ferric hydroxide species (Figure 8.11). Nevertheless, it turns out that this mechanism, involving a substrate radical and ferric hydroxide, can explain the preceding experimental results reported by Whitteck *et al.*[36] In their experiment, they introduced a chiral center to the C1 position of 2-HEP by replacing one of the two hydrogen atoms with deuterium, but this stereochemical information was not retained after its enzymatic conversion to HMP. The computationally derived ferric hydroxide mechanism illustrated in Figure 8.11 is consistent with this experimental outcome, because the barrier for the C–O formation from **Int-rad** *via* **TSco** is >13 kcal mol^{-1} (Figure 8.12a), which is much higher than the barrier to rotation about the single P–C bond of the substrate radical. Thus, the substrate radical can attack the hydroxide from either face, leading to the loss of stereochemistry at C1 (Figure 8.13).

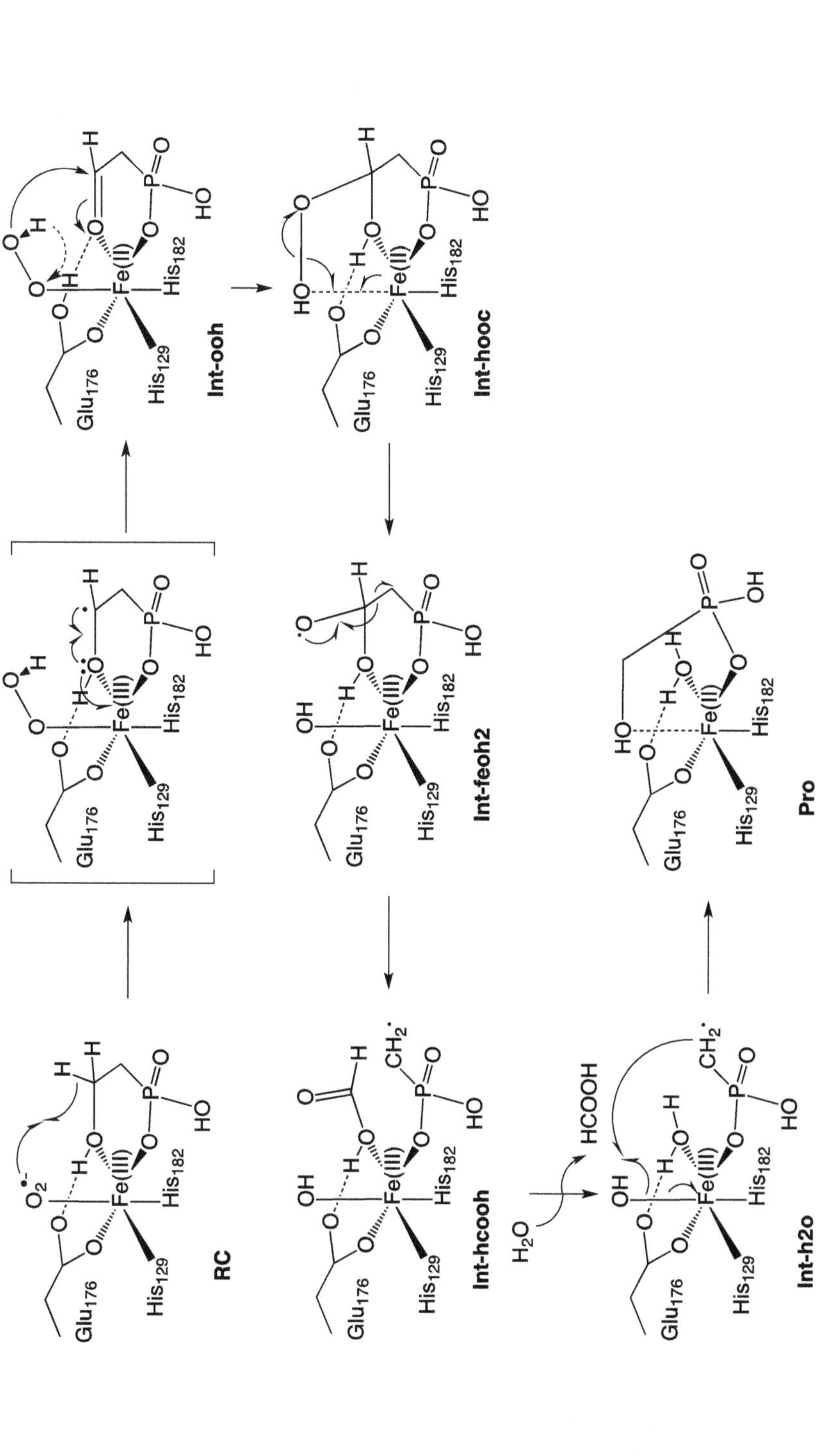

Figure 8.11 Proposed mechanism of the HEPD-catalysed reaction. Reprinted with permission from H. Hirao and K. Morokuma, *J. Am. Chem. Soc.*, 2010, **132**, 17901.[39] Copyright (2010) American Chemical Society.

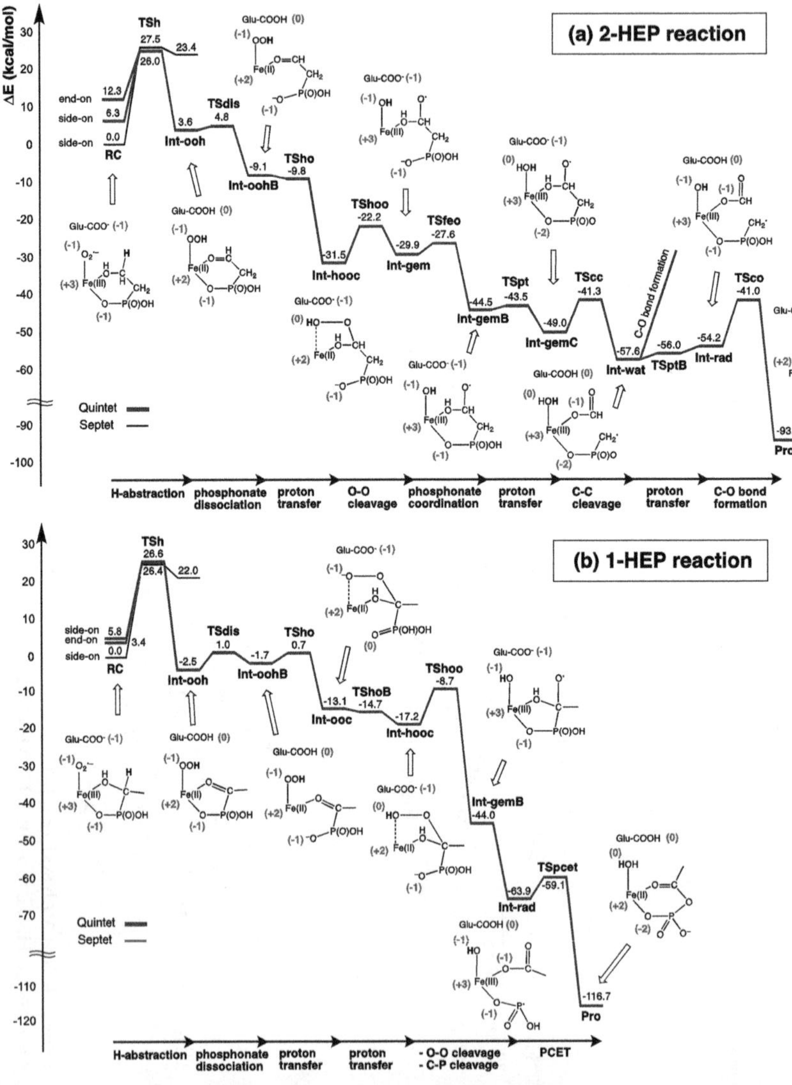

Figure 8.12 ONIOM-calculated energy profiles for the HEPD-catalysed reactions of (a) 2-HEP and (b) 1-HEP. To help trace how electron reorganisations occur, formal group charges are shown in red.
Reprinted with permission from H. Hirao and K. Morokuma, *J. Am. Chem. Soc.*, 2011, **133**, 14550.[38] Copyright (2011) American Chemical Society.

The ONIOM calculations also provided valuable insight into the question of why 2-HEP and 1-HEP exhibit different reactivity patterns. As mentioned above, the final step of the reaction of 2-HEP involves the attack of a substrate radical on the OH moiety of the ferric hydroxide. In the substrate radical, there is an unpaired electron on C1 in this case. By contrast, the

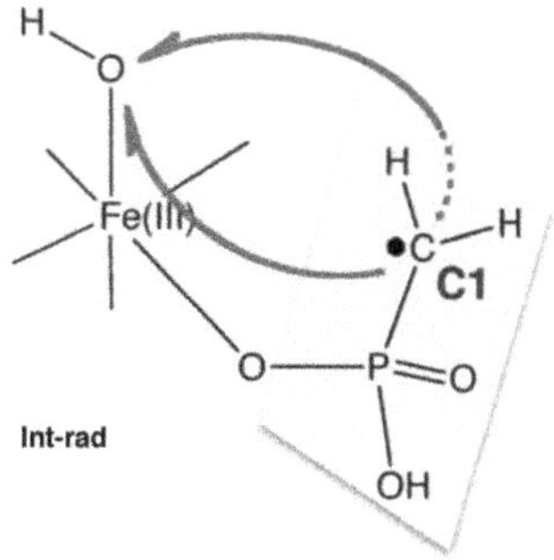

Figure 8.13 Our proposed mechanism that rationalises the experimental finding on the loss of stereochemistry at C1 of 2-HEP in the HEPD-catalysed reaction.
Reprinted with permission from H. Hirao and K. Morokuma, *J. Am. Chem. Soc.*, 2011, **133**, 14550.[38] Copyright (2011) American Chemical Society.

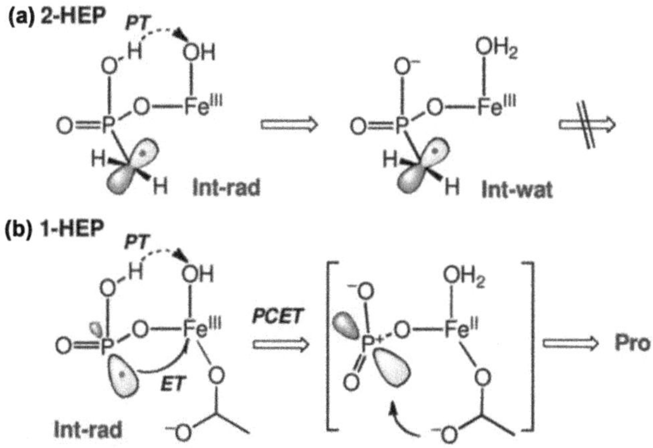

Figure 8.14 Possibility of PCET in HEPD-catalysed reactions: (a) 2-HEP and (b) 1-HEP. Reprinted with permission from H. Hirao and K. Morokuma, *J. Am. Chem. Soc.*, 2011, **133**, 14550.[38] Copyright (2011) American Chemical Society.

corresponding radical species in the reaction of 1-HEP has an unpaired electron on the phosphorus atom of the phosphite radical. In this case, the computation showed that the unpaired electron is easily transferred to the ferric center *via* PCET (*via* **TSpcet** in Figure 8.12b), in which the electron shift from the phosphorus to the ferric center is promoted by the proton transfer from the phosphonate OH group to the OH group of the ferric hydroxide (Figure 8.14b). This type of PCET is not permitted in the reaction of 2-HEP (Figure 8.14a).

8.3.3 Aromatase

Aromatase, or CYP19A1, is an isoform of the cytochrome P450 enzyme that plays a critical role in a hormone biosynthesis pathway in which androgens are converted into their corresponding estrogens (Figure 8.15a).[40–42] Among postmenopausal women, decreasing the level of estrogens is expected to prevent effectively the occurrence of breast cancer and to suppress already established tumors. Therefore, inhibition of aromatase has been considered one of the most promising therapeutic strategies for the treatment of such breast cancer.[41,43] Apart from the pharmaceutical interest, aromatase is an enzyme that offers mechanistically interesting conundrums. It is known that aromatase involves three distinct reaction steps, each of which requires O_2 for catalysis. In contrast to the first and second steps, which have already been clearly understood, the mechanism of the third step has been highly controversial over the past several decades. For this controversial reaction step, the ferric peroxide and Compound I (Cpd I) intermediates[44–47] are the most likely candidates that may act as the reactive species. In the ferric peroxide mechanism, ferric peroxide first reacts with the aldehyde form of an androgen to form a C–O bond, and subsequent steps involving O–O bond cleavage yields an estrogen and formic acid (Figure 8.15b). Importantly, in this mechanism, one of the two oxygen atoms from O_2 is inevitably incorporated into formic acid. On the other hand, in the Cpd I mechanism, it has been thought that the first step should be H-abstraction from the 1β-carbon of the substrate. Until recently, there had been a general consensus that the ferric peroxide mechanism was more plausible. However, a few exciting papers were published in 2014 that implicated Cpd I instead of ferric peroxide.[48–50] For example, Yoshimoto and Guengerich performed $^{18}O_2$-labeling experiments and showed that the oxygen from O_2 is not incorporated into the formic acid product.[50] They also showed that *gem*-diol is another possible form of the substrate. In addition, they observed a carboxylic acid product, but it was composed of a mixture of two types with and without $^{18}O_2$.

Very recently, Xu *et al.*[51] have performed ONIOM2(B3LYP:AMBER) calculations to address the question of how Cpd I can effect the aromatisation, especially in the third step in Figure 8.15a. To build an ONIOM model, an X-ray crystal structure of aromatase (PDB code 3EQM, 2.90 Å resolution) was used as a starting point.[52] After removing two phosphate ions included in the crystal structure, the protonation states of histidines were checked and adjusted, and the androgen substrate was modified to its *gem*-diol and aldehyde forms. After some preparatory MM calculations, ONIOM-ME calculations were performed to explore reaction pathways. For the QM calculations within ONIOM, the B3LYP/[SDD(Fe),6-31G*] method was used, while for MM calculations, the AMBER03 force field was used. For the latter, the 'amber = softfirst' keyword of Gaussian 09 was used. Single-point energy refinement was performed using a larger basis set (6-311+G(d,p)) and the ONIOM-EE scheme. To avoid drastic geometric changes of proteins during reaction pathway analyses and reduce computational costs, all amino acid residues located more than 10 Å away from the Fe ion were frozen during the geometry optimisation. The side chains of Asp309 and Thr310 were included in the model system. Asp309 is probably protonated, and its carboxyl group

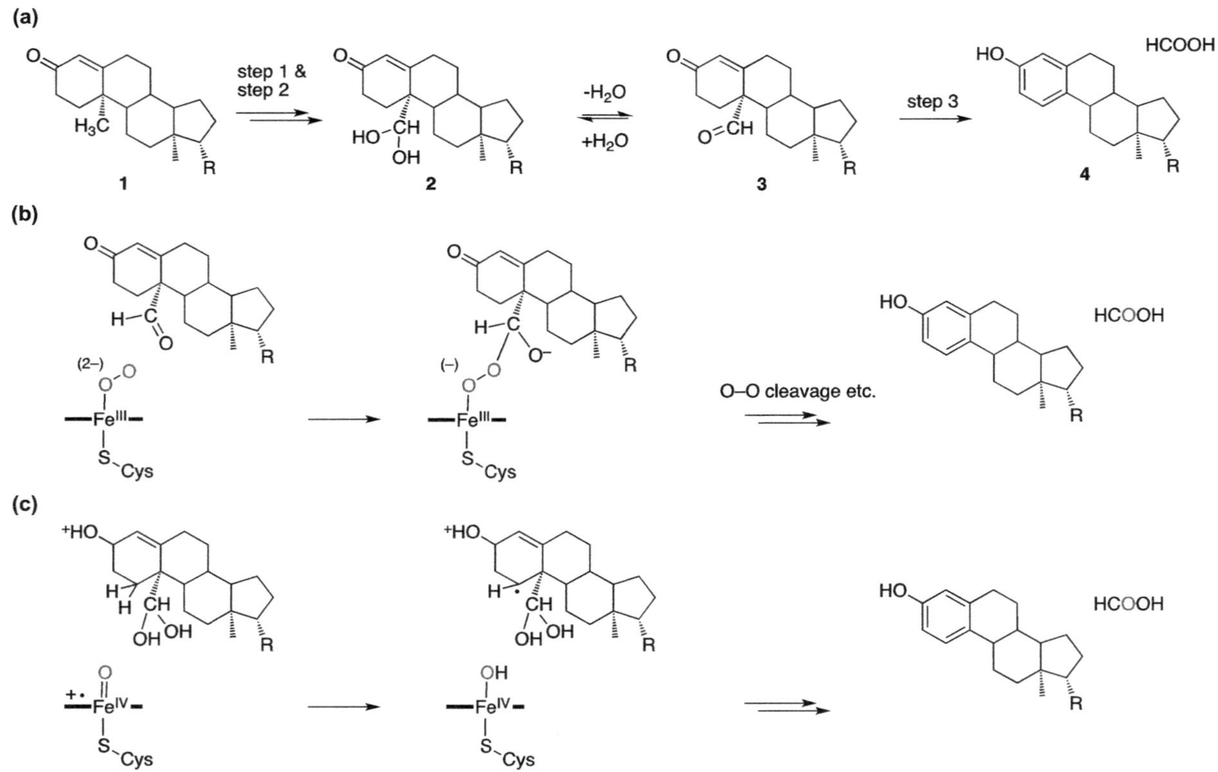

Figure 8.15 (a) A possible reaction sequence during the three-step conversion of androgens into oestrogens catalysed by aromatase. (b) A possible mechanism driven by ferric peroxide. (c) A possible mechanism driven by Cpd I. The oxygen atoms coming from the O_2 molecule are coloured red.

forms a hydrogen bond with the carbonyl group of the substrate, thus playing an important role in anchoring the substrate in the active site pocket in a specific orientation. In general, cytochrome P450 enzymes possess a threonine residue in the vicinity of the iron center, and this threonine residue is located within a hydrogen-bonding network that enables proton transfer from a distant Asp or Glu residue to the distal oxygen of Fe(III)OOH in Cpd 0. This proton transfer results in heterolytic O–O bond cleavage and subsequent formation of Cpd I, which is an oxoiron(IV) porphyrin π-cation radical species.[53,54] As this essential threonine, Thr310 is likely to play an essential role in generating Cpd I from the precursor species Cpd 0.

Surprisingly, the ONIOM calculations showed that the *gem*-diol form of the substrate can undergo H-abstraction from its O–H bond to form eventually an aromatised product (Figure 8.16). Although O–H bonds are generally stronger than aliphatic C–H bonds, this trend was rationalised in terms of the PCET effect, which provides some stabilisation to the reaction.[55] Thus, according to the spin natural orbital (SNO) analysis, in the transition state for O–H cleavage, an unpaired electron accumulates in the lone-pair orbital of the oxygen rather than on the O–H bond, whereas in the case of C–H cleavage, an unpaired electron is localised on the C–H bond (Figure 8.17). The different distributions of unpaired electrons are indicative of different modes of electron reorganisation. Importantly, as a consequence of this O–H abstraction mechanism, the O atom in Cpd I is eventually incorporated into the produced water molecule, rather than into the formic acid product (Figure 8.18); this is consistent with the experimental observation regarding the fate of oxygen in the ^{18}O-labeling experiments performed by Yoshimoto and Guengerich.[50] The calculations also showed that another experimentally observed product, carboxylic acid, can be formed from either the *gem*-diol substrate or the aldehyde substrate *via* H-abstraction from their C–H bonds by Cpd I. However, the fate of the oxo oxygen was predicted to be different in these two reactions (Figure 8.18). Thus, the reaction of the *gem*-diol substrate results in the eventual incorporation of the oxo oxygen in the produced water molecule, whereas the reaction of the aldehyde form leads to a carboxylic acid group including the oxo oxygen. These results were also consistent with the experimental observation of these two different types of carboxylic acid product.

8.3.4 Fe-MOF-74, a Metal–Organic Framework that has Similarities to Iron Enzymes

Metal–organic frameworks (MOFs) have emerged as an important class of porous materials that can be applied to heterogeneous catalysis.[56–61] We can see from the above examples of iron enzymes that redox-active transition metals can play key roles in the catalytic machineries of metalloenzymes. Interestingly, in contrast to metalloenzymes, which typically have one or a few more metals in the active site, there are numerous metal nodes in a single MOF, which are sustained by coordination bonding with organic

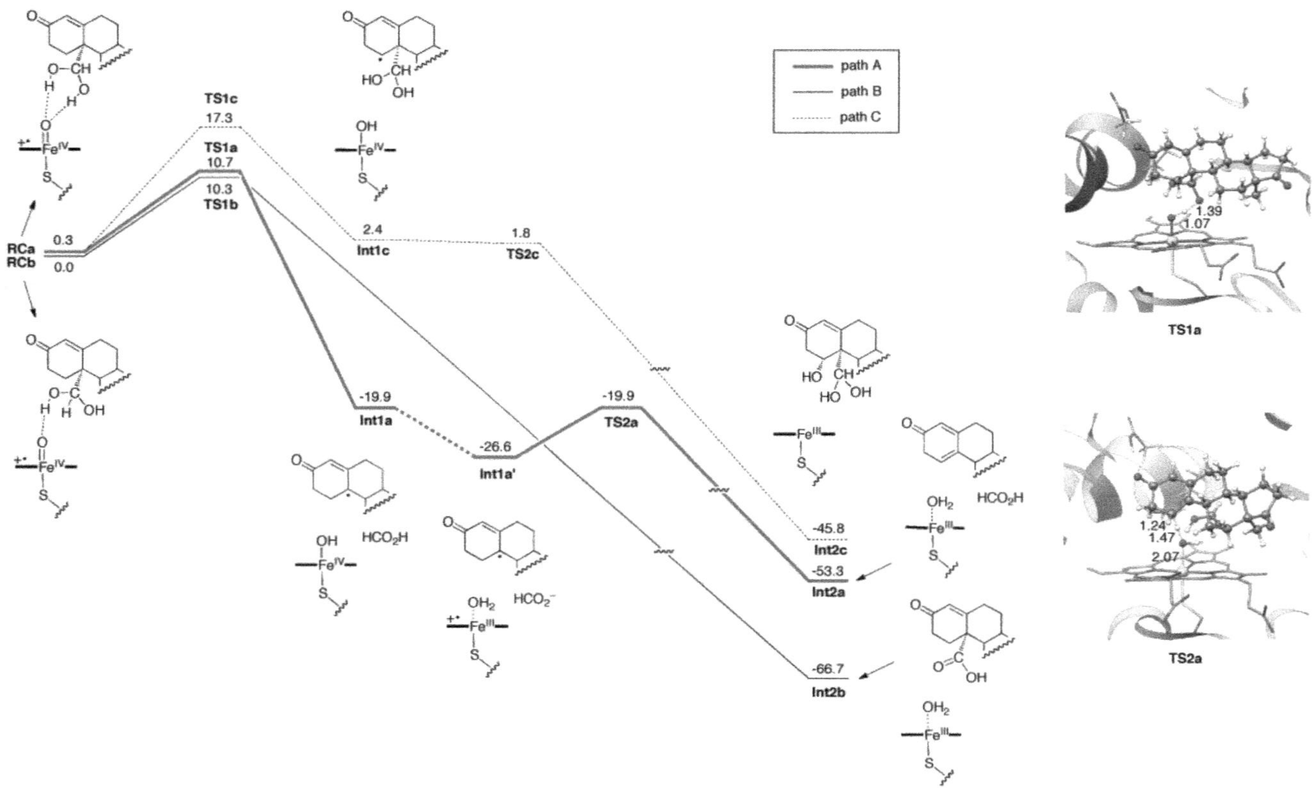

Figure 8.16 Energy diagram (kcal mol^{-1}) for the reactions of the *gem*-diol type of substrate with aromatase Cpd I. Adapted with permission from K. Xu, Y. Wang and H. Hirao, *ACS Catal.*, 2015, **5**, 4175.[51] Copyright (2015) American Chemical Society.

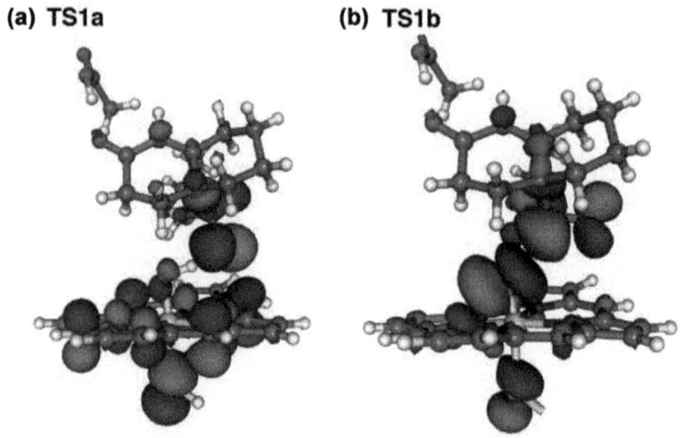

Figure 8.17 SNOs for the transition states for (a) O–H cleavage and (b) C–H cleavage. Reprinted with permission from K. Xu, Y. Wang and H. Hirao, *ACS Catal.*, 2015, 5, 4175.[51] Copyright (2015) American Chemical Society.

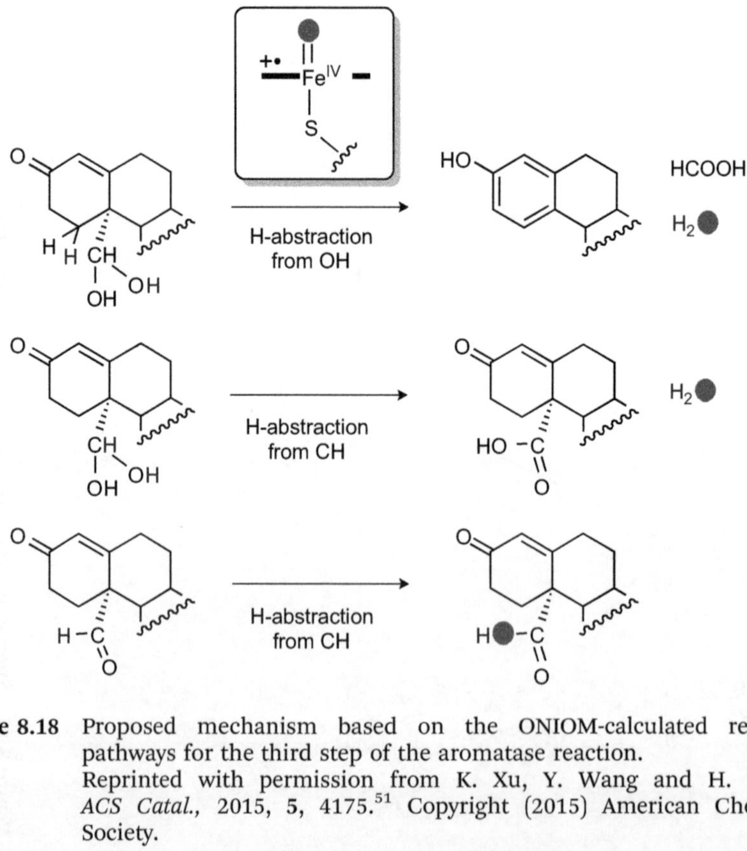

Figure 8.18 Proposed mechanism based on the ONIOM-calculated reaction pathways for the third step of the aromatase reaction.
Reprinted with permission from K. Xu, Y. Wang and H. Hirao, *ACS Catal.*, 2015, 5, 4175.[51] Copyright (2015) American Chemical Society.

linkers. Thus, many metals could be used simultaneously for catalysis. While the shape of the active site pocket of a metalloenzyme may be altered by the mutation of amino acids, the size and shape of the inner pores of MOFs, which affect the substrate selectivity, can be adjusted by varying the metal nodes and organic linkers used to construct the framework.[62] By exploiting the great variety of existing or new organic linkers and redox-active metals, it may be possible to design MOFs that exhibit powerful metalloenzyme-like catalytic activity. In addition, given that MOFs generally have much larger molecular size than metalloenzymes, one sensible first step toward fully understanding the catalytic mechanisms of MOFs is to apply QM/MM methods.[63–67]

Recently, Hirao *et al.* have studied computationally the mechanism for the hydroxylation of ethane with Fe_2(dobdc) (dobdc^{4-} = 2,5-dioxido-1,4-benzenedicarboxylate), which is also known as Fe-MOF-74 or CPO-27-Fe. For the computational investigation, they built an ONIOM-based multiscale model.[68] In its native state, Fe-MOF-74 has open Fe(II) centers and was previously shown, with appropriate treatments, to allow the hydroxylation of phenol,[69] conversion of methanol into formaldehyde,[66] oxidation of 1,4-cyclohexadiene,[70] and hydroxylation of ethane[71] to proceed. In particular, the hydroxylation of ethane was proposed to occur *via* the generation of a short-lived oxoiron(IV) species,[71] which is reminiscent of the Cpd I species of P450s and oxoiron(IV) species of nonheme enzymes. Interestingly, in contrast to many of the known synthetic oxoiron(IV) complexes, which tend to have low-spin ground states, the oxoiron(IV) species of Fe-MOF-74 was predicted to have a high-spin ($S = 2$) ground state because of the weak ligand field in Fe-MOF-74.[71]

A realistic ONIOM model was built from a published crystal structure of Fe-MOF-74, and the Fe(III)OH centers were modified appropriately (Figure 8.19). Before the modification, the model system included three Fe(III)OH ions, and the central Fe(III)OH was replaced with Fe(IV)O, which reacts with ethane. To simplify the electronic structure of the model system, the other two Fe(III)OH units in the QM region were replaced with Mg(II). In fact, a magnesium-diluted analog of Fe-MOF-74, *i.e.* $Fe_{0.1}Mg_{1.9}$(dobdc), also showed catalytic activity,[71] and thus it could be postulated that the neighboring metals do not significantly affect the reactivity of Fe(IV)O. In the geometry optimisation calculations, the QM calculations within ONIOM were performed at the B3LYP/[SDD(Fe),6-31G*(others)] level, while the universal force field (UFF) was used for the MM calculations. The 6-311+G(df,p) basis set was used for the single-point energy refinement with ONIOM.

The ONIOM calculations showed that the initial C–H bond cleavage leads to the formation of an intermediate composed of an ethyl radical and an Fe(III)OH species, which subsequently undergoes a radical rebound to form an alcohol product. It should be noted that the position and orientation of the substrate are not tightly restricted in the pore space; this situation is distinctly different from that of well-caged active sites of enzymes.[72] Thus, it might also be possible for the ethyl radical species to react with another

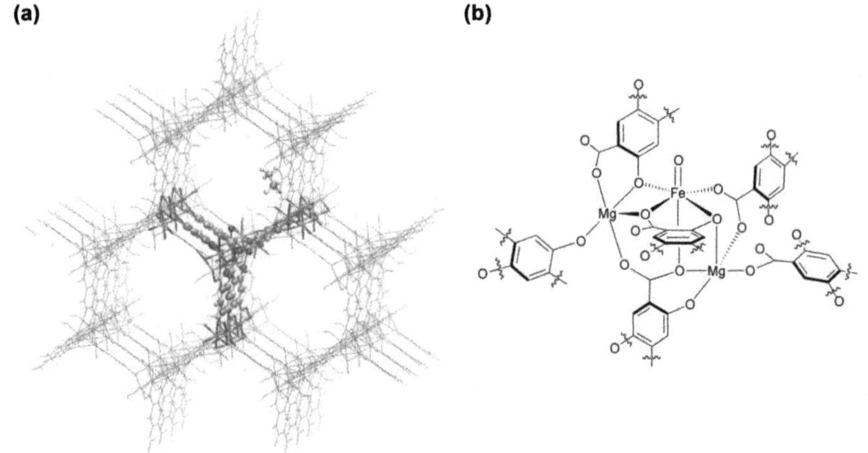

Figure 8.19 (a) ONIOM model, with the core system, the optimised atoms in the environment system, and the frozen atoms in the environment system shown in ball-and-stick, stick and wire representations, respectively. (b) Atoms in the core system. Wavy lines highlight covalent bonds at which H-link atoms are attached.
Reprinted with permission from H. Hirao, W. K. H. Ng, A. M. P. Moeljadi and S. Bureekaew, *ACS Catal.*, 2015, **5**, 3287.[68] Copyright (2015) American Chemical Society.

metal site having an Fe(III)OH or Fe(IV)O unit of the MOF, instead of undergoing the rebound. Consistent with the previous computational results,[71] the high-spin state was found to be more stable than the low-spin state in the initial reactant complex (**RC**) state. Furthermore, the ONIOM calculations showed that the lower energy of the quintet state than the triplet state is preserved throughout the reaction pathway (Figure 8.20). This trend is significantly different from that for the reactions of P450 Cpd I and most of the synthetic nonheme oxoiron(IV) complexes, which have a low-spin ground state in the reactant state but later may undergo spin-crossover to a higher-spin state.[73–75] The reaction of the Fe(III)OH species, *i.e.* another possible reactive species, with ethane was also examined, but the calculated barrier was much higher (34.9 kcal mol^{-1}) than in the case of oxoiron(IV), thus ruling out this possibility (Figure 8.20).

In addition, acetaldehyde was experimentally observed in the product mixture, which is presumably due to a secondary oxidation of the above ethanol product by another oxoiron(IV) in the MOF.[71] Having shown that the quintet state was more stable in the reaction of ethane, Hirao *et al.* examined two possible pathways for the reaction of ethanol in the quintet state (Figure 8.21). The two pathways differed in whether the initial H-abstraction occurs from the O–H bond or the C–H bond. The pathway involving the O–H bond cleavage was found to have a lower energy barrier, despite the fact that the G4-calculated bond dissociation enthalpies for the C–H and O–H bonds

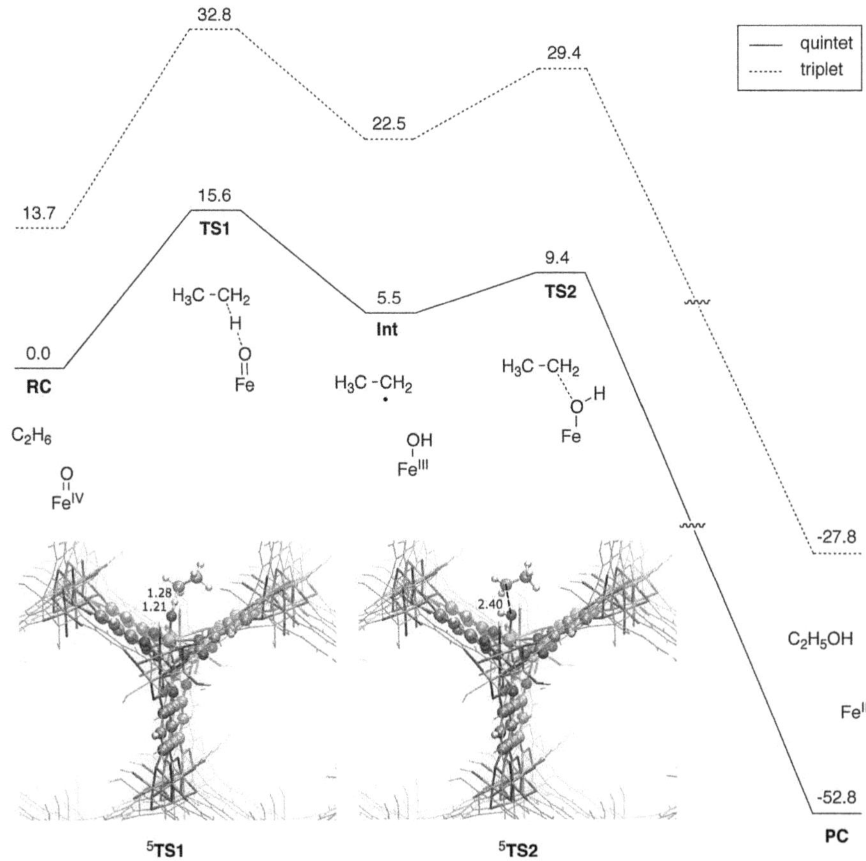

Figure 8.20 Energy diagram (kcal mol^{-1}) for the hydroxylation of ethane effected by an oxoiron(IV) species of Fe-MOF-74.
Reprinted with permission from H. Hirao, W. K. H. Ng, A. M. P. Moeljadi and S. Bureekaew, *ACS Catal.*, 2015, 5, 3287.[68] Copyright (2015) American Chemical Society.

of ethanol were 94.5 and 104.1 kcal mol^{-1}, respectively. The authors rationalised this finding in terms of a PCET effect operating only in the O–H cleavage pathway. In fact, the SNO for the transition state of the O–H cleavage pathway was not localised onto the O–H bond; rather, it had an amplitude in a perpendicular direction (Figure 8.21). This SNO has a significant character of the oxygen lone-pair orbital, and in such cases, PCET can stabilise the transition state.[55] By contrast, the SNO for the transition state of the C–H bond cleavage pathway was well localised on the breaking C–H bond, indicative of the character of hydrogen-atom transfer. A similar trend was previously observed experimentally by Wang *et al.*,[76] who discovered that a synthetic nonheme diiron(IV) complex cleaved the O–H bonds of methanol and *tert*-butyl alcohol rather than their weaker C–H bonds.

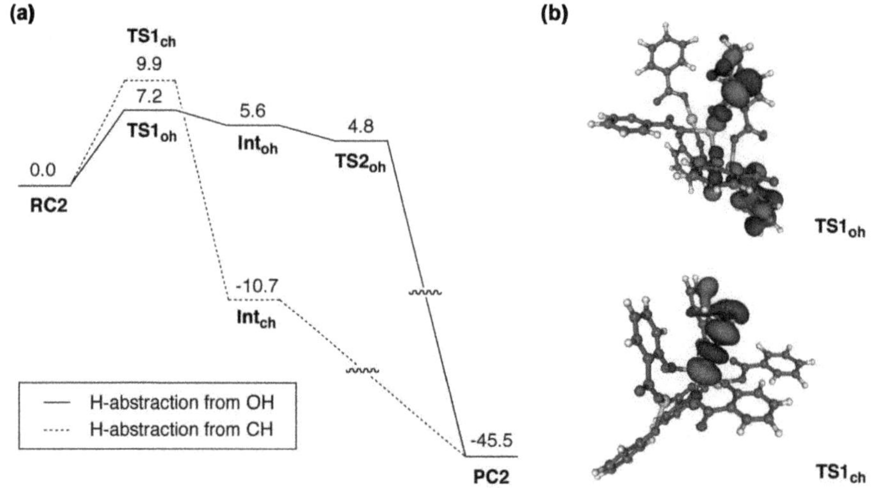

Figure 8.21 (a) Energy diagram (kcal mol^{-1}) for the reaction of ethanol with an oxoiron(IV) species of Fe-MOF-74. (b) SNOs for transition states. Reprinted with permission from H. Hirao, W. K. H. Ng, A. M. P. Moeljadi and S. Bureekaew, *ACS Catal.*, 2015, **5**, 3287.[68] Copyright (2015) American Chemical Society.

8.4 Energy Decomposition Analysis of the Core–Environment Interactions Within Enzymes

The remarkable catalytic activity of iron enzymes is primarily due to the formation of reactive species such as oxoiron(IV) and ferric superoxide at the iron center. As described above, ONIOM calculations help us derive detailed insights into the reactivity of these species. Another important aspect of iron enzymes that influences the catalytic activity is the protein environmental effect. In the active site of an enzyme, the configuration and orientation of a substrate are judiciously regulated sterically and electrostatically by the surrounding amino acid residues, and the protein environment can also lower activation barriers through electrostatic interactions. These protein environmental effects can be quantified by ONIOM calculations. As we have seen earlier, there is a fundamental difference in how ONIOM-ME and ONIOM-EE deal with the electrostatic interaction between the core and environment systems. The latter method takes into account the polarisation of the model system caused by the point charges of the surrounding amino acid residues. By contrast, the electronic state of the model system, as described by ONIOM-ME, could be viewed as a special, artificial situation that is created by suppressing the polarisation. This difference can be exploited fruitfully to derive theoretical insights into protein environmental effects.

Hirao took advantage of the difference between ONIOM-ME and ONIOM-EE to perform an energy decomposition analysis (EDA) of the interaction

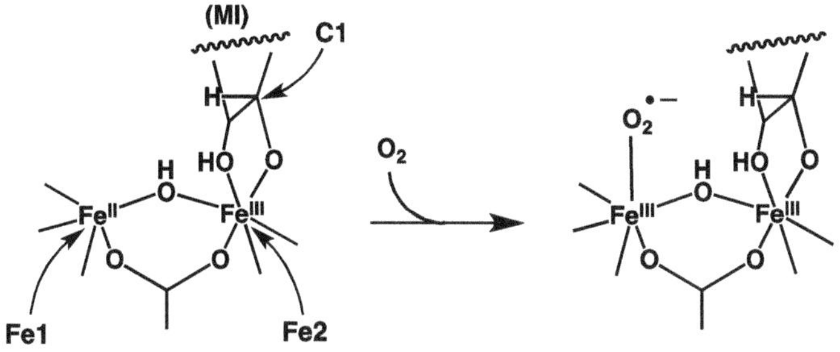

1 2

Figure 8.22 O_2-binding to the ferrous centre to form the ferric superoxide reactive intermediate in MIOX.
Reprinted with permission from H. Hirao, *J. Phys. Chem. B*, 2011, **115**, 11278.[77] Copyright (2011) American Chemical Society.

between core and environment systems in the dioxygen-binding process of MIOX (Figure 8.22).[77] He defined the electrostatic (E_{es}), vdW (E_{vdW}), and polarisation (E_{pol}) energies as follows:

$$E_{es} = \sum_i \sum_j \frac{s_{ij}^{es}(q_i q_j)}{r_{ij}} \tag{8.11}$$

$$E_{vdW} = \sum_i \sum_j s_{ij}^{vdW} \left(\frac{A_{ij}}{r_{ij}^{12}} - \frac{B_{ij}}{r_{ij}^{6}} \right) \tag{8.12}$$

$$E_{pol} = E_{ONIOM\text{-}EE} - E_{ONIOM\text{-}ME} \tag{8.13}$$

where i and j are indices for the atoms in the core and environment systems, respectively. The AMBER energy expression was used to evaluate the nonbonded electrostatic and vdW interaction energies. Therefore, scale factors used are according to the AMBER convention, and nonbonded energies are scaled down when i and j are in bonded 1–2, 1–3, or 1–4 arrangements. The polarisation energy was defined as the difference between the energies from ONIOM-EE and ONIOM-ME calculations. The point charges used in eqn (8.11) can be determined by the Merz–Kollman (MK) method at the ONIOM-ME geometry without including the polarisation effect.[78,79] However, geometry optimisation can cause a change in charge distribution with the model system (see Section 9.2.5). Therefore, geometry optimisation and charge determination were iterated alternately until the change in total energy became less than 0.025 kcal mol^{-1}. The effect of dispersion within the model system was also considered. According to the EDA, the O_2-binding process is enhanced by the protein effects, *i.e.* the electrostatic, vdW, and polarisation effects. Moreover, the intracluster dispersion effect was also shown to facilitate the O_2-binding and is larger in magnitude than the other effects. A similar EDA study was also performed for the Cpd I species of cytochrome P450cam to quantify the protein environmental effects.[80]

It is instructive to consider the analogy between this ONIOM-based EDA method and traditional EDA methods in quantum chemistry.[81–83] Quantum chemical EDA calculations usually proceed in a stepwise manner and employ intermediate wave functions. For example, when evaluating the electrostatic effect, two fragments are assumed to have their own wave functions in their isolated states. Such intermediate wave functions include Hartree and exchange effects within the fragments, but the exchange of electrons between fragments and polarisation are not permitted. Thus, electrons within each fragment are not allowed to relax in the presence of the other fragment. Similarly, when E_{es} is calculated by eqn (8.11), ONIOM-ME is used in which polarisation of the QM wave function is not allowed. The use of such constrained, hypothetical conditions allows one to evaluate the electrostatic interaction energy between the core and environment systems. When the polarisation energy is calculated by conventional EDA approaches, electrons are allowed to polarise within the fragments under the influence of the other fragment. From the difference in energies after and before polarisation, the polarisation energy can be evaluated. A conceptually similar procedure can be adopted by performing ONIOM-EE and ONIOM-ME calculations in eqn (8.13), to calculate the polarisation energy caused by the protein environment.

8.5 Application of ONIOM2(QM:MM) to the Reactions of other Types of Enzymes

Needless to say, the application of ONIOM is not limited to iron enzymes; it can be applied to a broad range of enzymatic processes to derive valuable insights. We now turn to computational studies of other types of enzymes with ONIOM.

8.5.1 *myo*-Inositol Monophosphatase

As we have seen in the MIOX subsection, MI is a molecule of physiological importance. In contrast to MIOX, which catabolises MI, the role of *myo*-Inositol monophosphatase (IMPase) is to assist the production of MI.[84–90] More specifically, IMPase catalyses the conversion of *myo*-inositol-1-phosphate (Ins(1)P) to MI and inorganic phosphate in the central nervous system (CNS) (Figure 8.23). In fact, MI and its derivatives play vital roles in a variety of essential physiological processes in the brain. However, the poor permeability of MI across the blood–brain barrier makes it difficult to supply MI to the brain. Therefore, the IMPase-catalysed biosynthesis of MI is regarded as an important supply route with which to maintain the level of MI. IMPase is also thought to have a mechanistic link to the so-called 'inositol depletion hypothesis', in which the efficacy of lithium therapy in

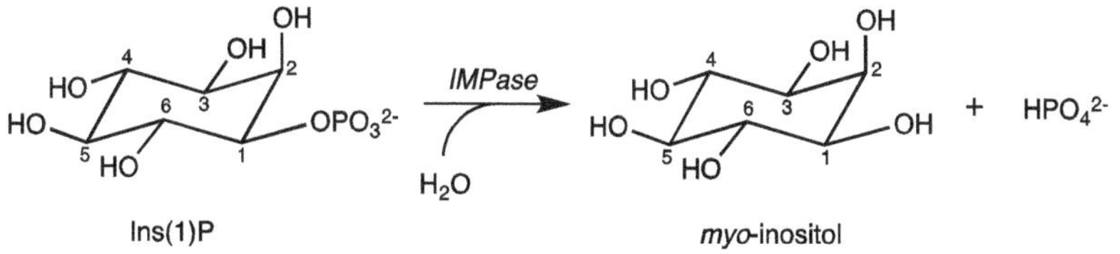

Figure 8.23 IMPase-catalysed conversion of Ins(1)P to MI and inorganic phosphate. Reprinted with permission from X. Wang and H. Hirao, *J. Phys. Chem. B*, 2013, **117**, 833.[91] Copyright (2013) American Chemical Society.

the treatment of bipolar disorder is assumed to be a result of a decrease in the MI level; thus, the decreased level may be caused by inhibition of IMPase by lithium ions.

Despite the apparent importance of IMPase in the context of CNS pathology, its detailed catalytic mechanism has remained largely unclear and controversial. In particular, one important question that remains to be answered is how many magnesium ions (Mg^{2+}) are used for the catalytic conversion of Ins(1)P in the active site of the enzyme. Unlike MIOX, IMPase does not use iron ions for the catalysis, and the electronic structure of the active site is not as complicated as in the case of MIOX. Nevertheless, the enzyme offers another type of intriguing conundrum regarding the number of Mg^{2+} ions. As shown in Figure 8.24, there are three Mg^{2+} binding sites in the active site of IMPase. It is not clear whether these sites are fully occupied or not during the catalysis. To address this question, Wang and Hirao applied ONIOM2(QM:MM)-EE calculations to the IMPase-catalysed reaction.[91] They compared the case in which all three metal sites (Mg1–Mg3) are fully occupied (Model-I) and the case in which only the first and second sites (Mg1 and Mg2) are occupied (Model-II). Geometry optimisation was performed at the ONIOM2(B3LYP/6-31G*:AMBER)-EE level, and single-point energy calculations were performed at the ONIOM2(B3LYP/6-311+G(d,p):AMBER)-EE level. The MPWB1K functional was also tested. The AMBER force field used for MM calculations is based on the parm96 parameter set implemented in Gaussian 09. Figure 8.25 shows the initial **RC** state for the two mechanisms. In the reaction, the water molecule (w86) between the second and third metal-binding sites is considered to attack the phosphate group from the back, to trigger an S_N2 reaction. The ONIOM calculations suggested that, when the metal-binding sites are fully occupied, the S_N2 reaction is highly endothermic and does not provide a low-energy reaction pathway. By contrast, in the two-metal mechanism, a reasonably low-energy reaction pathway was found to be available, suggesting that the number of Mg^{2+} ions used during the catalysis is two. However, the barrier for the second step on this pathway, in which a proton from w86 is donated to the substrate to form a complete MI molecule, was too high (28.5 kcal mol^{-1} with MPWB1K and 30.6 kcal mol^{-1} with B3LYP). Interestingly, this barrier was lowered to 26.0 kcal mol^{-1} with MPWB1K and 28.6 kcal mol^{-1} with B3LYP when a water molecule near the second metal-binding site (w184) was allowed to

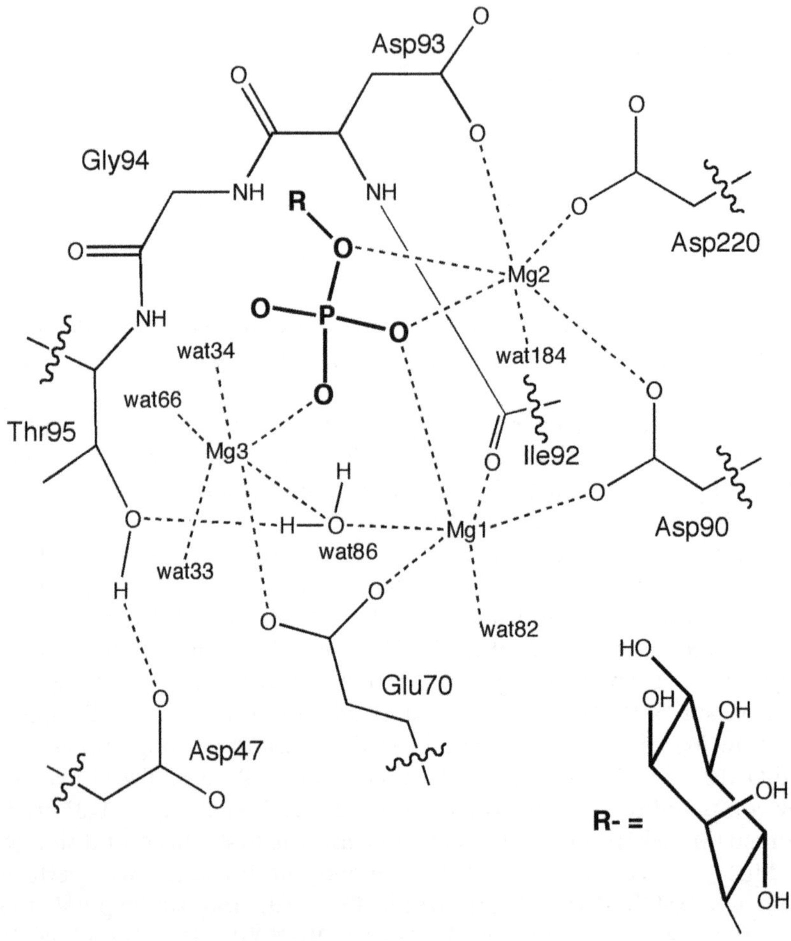

Figure 8.24 The active site of IMPase defined as the core system. Reprinted with permission from X. Wang and H. Hirao, *J. Phys. Chem. B*, 2013, **117**, 833.[91] Copyright (2013) American Chemical Society.

participate in the proton transfer to the substrate. In this pathway, w86 first donates its proton to the nearby Glu70, which relays the proton to w184, before w184 donates the proton to the substrate. This result underlined the important role of w184, and the additional MD simulations showed that this water molecule has a pronounced propensity to stay near the second metal-binding site (Figure 8.26). Thus, the results of MD simulation lent additional support to the essential role of w184 in catalysis.

8.5.2 QueF Nitrile Reductase

QueF nitrile reductase catalyses the conversion of 7-cyano-7-deazaguanine ($preQ_0$) into 7-aminomethyl-7-deazaguanine ($preQ_1$) in a NADPH-dependent

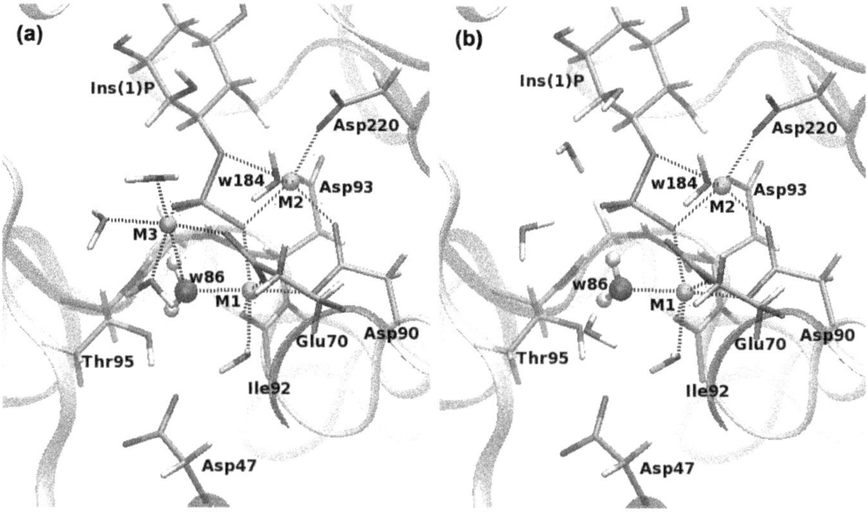

Figure 8.25 ONIOM-optimised geometries of the **RC** species of IMPase in (a) Model-I and (b) Model-II.
Reprinted with permission from X. Wang and H. Hirao, *J. Phys. Chem. B*, 2013, **117**, 833.[91] Copyright (2013) American Chemical Society.

fashion (Figure 8.27).[92] During this conversion, the nitrile group of $preQ_0$ is transformed into a primary amine group. Nitrile reductase has two practically important implications. First, the enzyme can be a potential target for the development of antibacterial drugs. The enzymatic $preQ_0$-to-$preQ_1$ conversion is a key step in the biosynthetic pathway of quenosine. Suppressing the bacterial pathogenicity by inhibiting the QueF-catalysed reaction could be a promising strategy for inhibitor design. Second, it may be possible to exploit engineered nitrile reductase enzymes for industrial nitrile-to-amine conversion processes. Indeed, microbial nitrile hydratase, nitrilase, and amidase have already been utilised as useful biocatalysts for the conversion of nitrile-containing compounds into amides and carboxylates.[92] The nitrile-to-amine conversion is also an important transformation especially in the synthesis of drug candidates; for this, nitrile reductase may be used. However, many of the current synthetic approaches to the nitrile-to-amine conversion rely on complex metal hydrides. Because such catalytic processes generate toxic by-products and solvent waste, the development of environmentally more sustainable biocatalysts is highly warranted. For effective applications of the QueF enzyme to be made in the context of these practical implications, one must have the molecular details of the enzyme catalysis.

Ribeiro *et al.* have recently studied the catalytic mechanism of *Vibrio cholerae* QueF nitrile reductase, by applying the ONIOM2(QM:MM)-EE approach.[93] The enzymatic reaction is considered to consist of four distinct reaction stages, as illustrated in Figure 8.28, but only some of them have been experimentally verified. The QM calculation in ONIOM was performed at the B3LYP/6-31G(d) level, while the AMBER force field

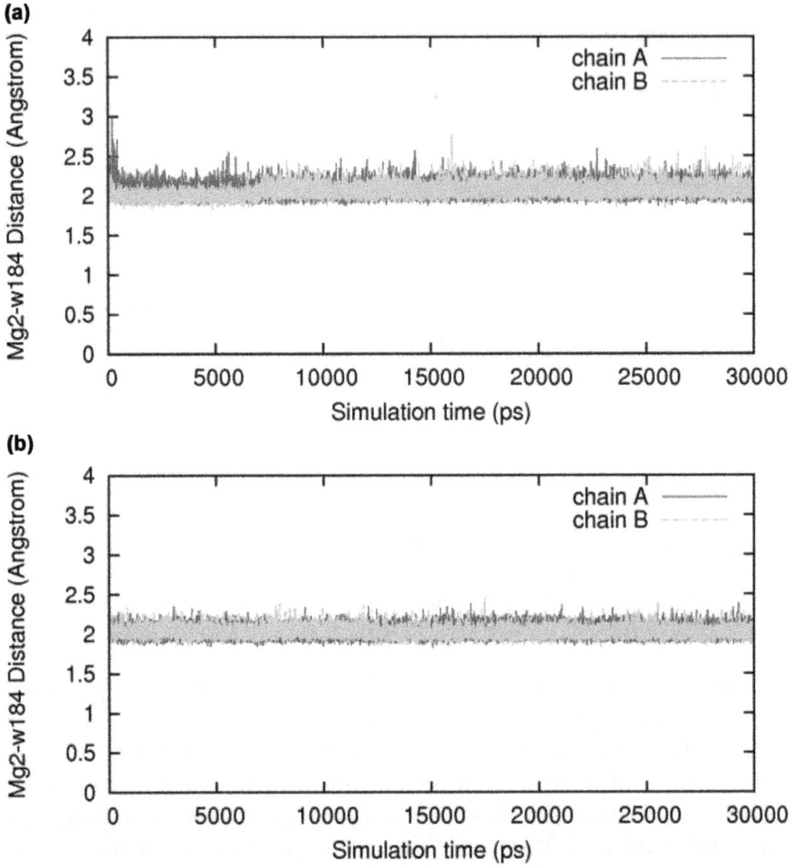

Figure 8.26 Change in the distance between the magnesium ion at the Mg2 site and the oxygen atom of w184 during MD simulation. (a) Model-I and (b) Model-II.
Reprinted with permission from X. Wang and H. Hirao, *J. Phys. Chem. B*, 2013, **117**, 833.[91] Copyright (2013) American Chemical Society.

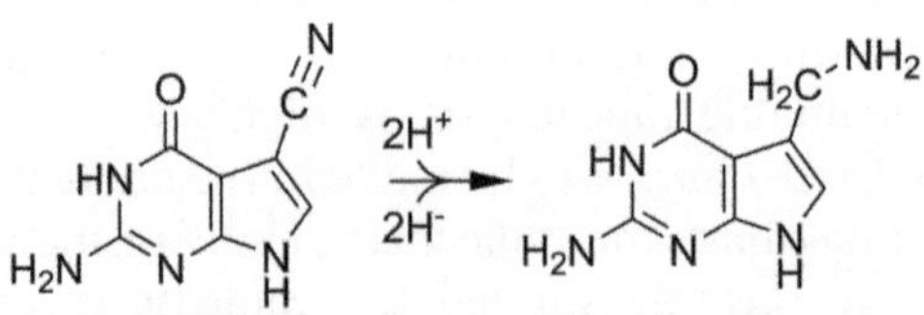

$PreQ_0$ $PreQ_1$

Figure 8.27 Conversion of $PreQ_0$ to $PreQ_1$ catalysed by nitrile reductase.
Reprinted with permission from A. J. M. Ribeiro, L. Yang, M. J. Ramos, P. A. Fernandes, Z.-X. Liang and H. Hirao, *ACS Catal.*, 2015, **5**, 3740.[93] Copyright (2015) American Chemical Society.

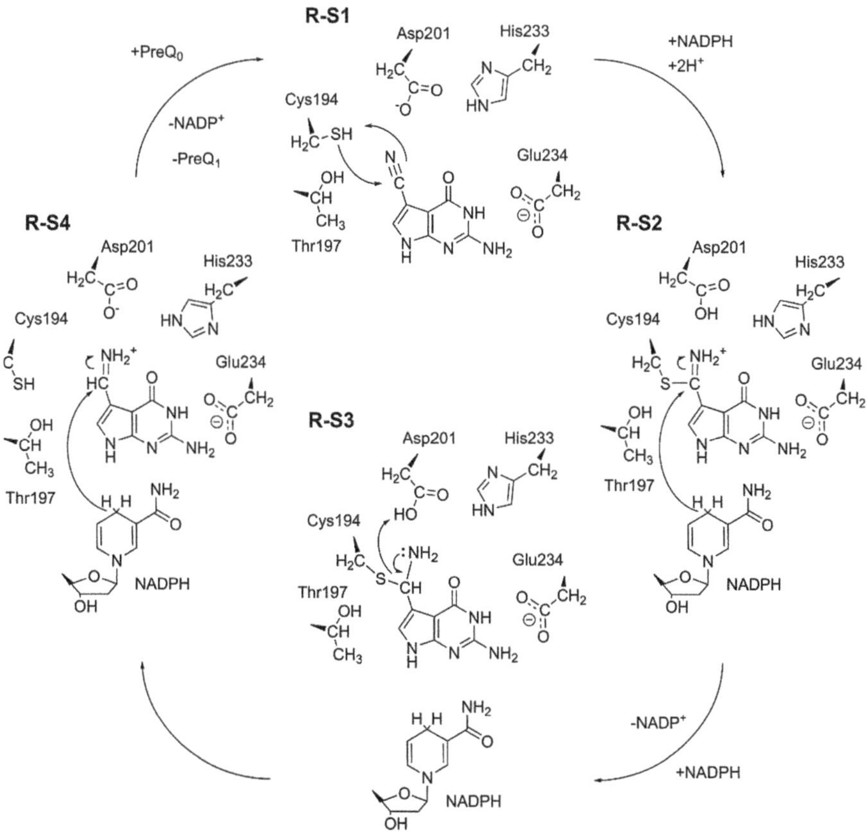

Figure 8.28 Proposed four stages within the catalytic reaction of nitrile reductase. Reprinted with permission from A. J. M. Ribeiro, L. Yang, M. J. Ramos, P. A. Fernandes, Z.-X. Liang and H. Hirao, *ACS Catal.*, 2015, **5**, 3740.[93] Copyright (2015) American Chemical Society.

(parm96) was used for the MM calculations. This ONIOM level was used for geometry optimisation calculations, and single-point energy calculations were performed using the B3LYP/6-311+G(2d,2p) method during the QM calculation step. Because some of the atoms in NADPH were missing in the X-ray crystal structure, the missing atoms were appended using GaussView software.

The extensive ONIOM calculations provided detailed insights into the four stages of the QueF-catalysed reaction (Figure 8.28). In the first stage (Figure 8.29), Cys194 donates its proton to the nearby Asp201 residue to form a cysteinate. As indicated by the red color in Figure 8.29, this deprotonation results in the accumulation of a negative charge on the sulfur atom of Cys194. Thus, Cys194 becomes more reactive and can attack the nitrile group of the substrate to form a C–S bond. This C–S bond formation is highly feasible with an energy barrier of 19.0 kcal mol^{-1}.

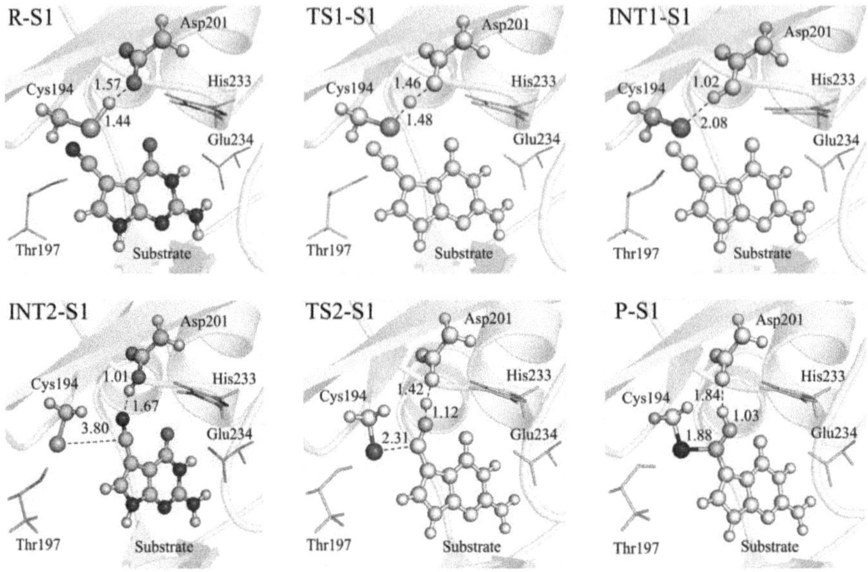

Figure 8.29 First stage of the nitrile reductase reaction. Reprinted with permission from A. J. M. Ribeiro, L. Yang, M. J. Ramos, P. A. Fernandes, Z.-X. Liang and H. Hirao, *ACS Catal.*, 2015, **5**, 3740.[93] Copyright (2015) American Chemical Society.

The thioimidate intermediate, formed in the first stage (Figure 8.29), undergoes reduction by accepting a hydride from NADPH in the second stage. This hydride transfer step has an energy barrier of 19.1 kcal mol^{-1}, and thus the barrier is as high as that for the first stage. Subsequently, Cys194 undergoes dissociation from the substrate in the third stage, because otherwise another reduction of the substrate is difficult. Calculations show that the third stage is quite facile, with the highest energy barrier here being only 4.9 kcal mol^{-1}. In the final stage (stage 4), there is a second hydride transfer from NADPH to the substrate. This stage has an energy barrier of 20.8 kcal mol^{-1}, which is higher than the other barriers. From all of these results, it was suggested that the fourth stage, which is responsible for the second hydride transfer, is the rate-determining step of the enzymatic reaction (Figure 8.30). The highest barrier, 20.8 kcal mol^{-1}, was consistent with the available experimental kinetic data for several nitrile reductase enzymes (Figure 8.31).

8.6 Application of ONIOM2(QM:QM′) to Enzymatic Reactions

Although our focus herein has been mainly on the ONIOM2(QM:MM) method, as described in Section 9.2, ONIOM also allows two (or more)

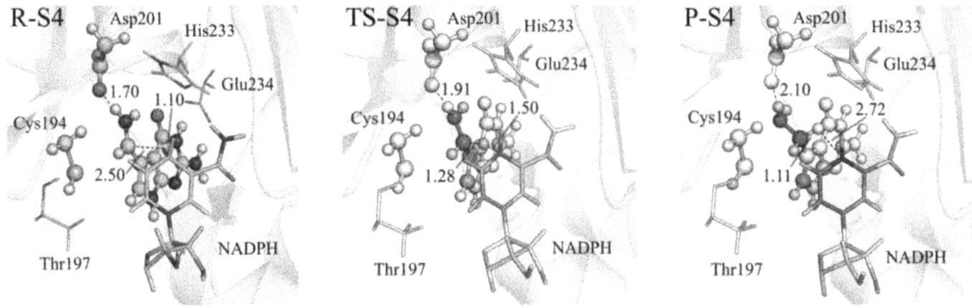

Figure 8.30 Fourth stage of the nitrile reductase reaction.
Reprinted with permission from A. J. M. Ribeiro, L. Yang, M. J. Ramos, P. A. Fernandes, Z.-X. Liang and H. Hirao, *ACS Catal.*, 2015, **5**, 3740.[93] Copyright (2015) American Chemical Society.

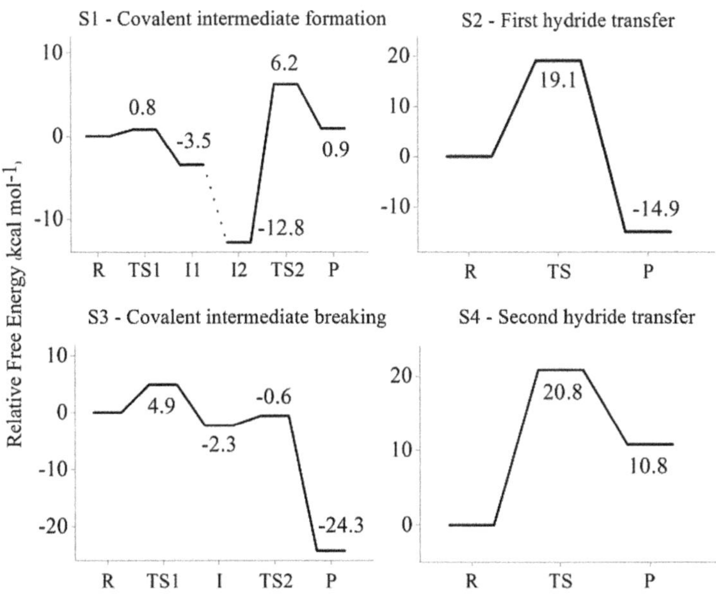

Figure 8.31 Summary of the computationally determined energy profiles for the four reaction stages.
Reprinted with permission from A. J. M. Ribeiro, L. Yang, M. J. Ramos, P. A. Fernandes, Z.-X. Liang and H. Hirao, *ACS Catal.*, 2015, **5**, 3740.[93] Copyright (2015) American Chemical Society.

different levels of QM methods to be combined. Here, let us look at a successful application of the ONIOM2(QM:QM′) method to an enzymatic reaction.

8.6.1 Asparaginase *Erwinia chrysanthemi* (L-asparaginase II)

Asparaginase *Erwinia chrysanthemi* (L-asparaginase II) is an FDA-approved chemotherapeutic enzyme for the treatment of acute lymphoblastic leukemia.[94] This enzyme catalyses the hydrolysis reaction that converts L-asparagine into L-asparate and ammonia. The therapeutic mechanism of this treatment stems from the fact that certain types of tumor cells cannot produce a sufficient amount of asparagine; thus, their growth depends on external asparagine sources. By contrast, normal cells can generate asparagine on their own. Administration of L-asparaginase II decreases the amount of asparagine in the bloodstream, thereby leading to the selective suppression of the growth of tumor cells.[95]

Despite the therapeutic value of L-asparaginase II, the mechanism of enzyme catalysis is still unclear. Therefore, Gesto *et al.*[96] undertook ONIOM2(QM:QM′) calculations to elucidate the reaction mechanism. They used ONIOM(B3LYP/6-31G(d):AM1) and ONIOM(M06-2X/6-311++G(2d,2p):M06-2X/6-31G(d)) for geometry optimisation and single-point energy calculations, respectively. On the basis of ONIOM2(QM:QM′) calculations, they proposed a new mechanism, shown in Figure 8.32. In the first step, a water molecule attacks the C_γ atom

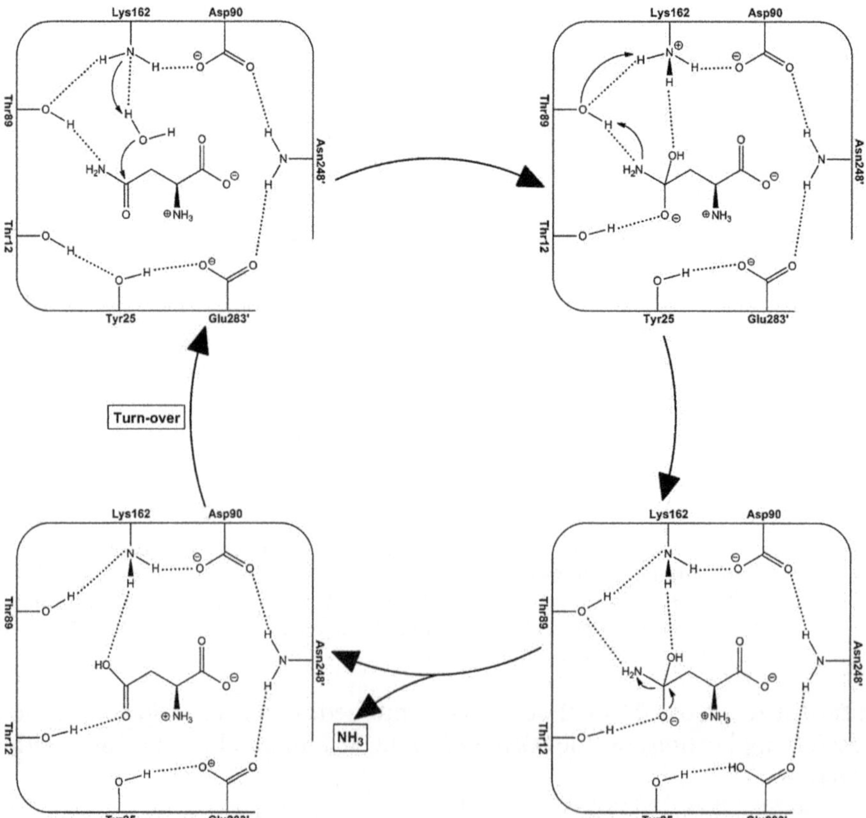

Figure 8.32 Proposed mechanism of the reaction catalysed by L-asparaginase II. Reprinted with permission from D. S. Gesto, N. M. F. S. A. Cerqueira, P. A. Fernandes and M. J. Ramos, *J. Am. Chem. Soc.*, 2013, **135**, 7146.[96] Copyright (2013) American Chemical Society.

of L-asparagine to form a tetrahedral intermediate, which subsequently undergoes a concerted process involving proton transfers from Lys162 to Thr89 and from Thr89 to the substrate, before generating an ammonia molecule and L-asparate. The calculations showed that the first step is the rate-limiting step, having the highest barrier (20.2 kcal mol^{-1}).

8.7 Conclusion

In this chapter, we have discussed key concepts and a few recent applications of ONIOM, with particular emphasis placed on the ONIOM2(QM:MM) method. Enzymatic reactions can be studied by pure QM calculations, but in that approach, the active site must be extracted to simplify the calculation. The QM approach allows one to study the intrinsic aspects of enzymatic reactions, but the protein environment sometimes exerts considerable influence on the reaction. In comparison with the pure QM treatment, ONIOM models contain a much larger number of atoms in the system, thereby providing more realistic pictures of enzymes. Significant progress has been made in various technical aspects of ONIOM, and hence, enzymatic reactions can be explored efficiently and effectively using realistic models. Offering atomic-level insights into experimentally inaccessible fast processes and short-lived species within complex enzymes, such ONIOM-based computational studies will continue to stimulate experimentalists and enhance fruitful collaborations between computational and experimental chemists.

Acknowledgements

H. H. gratefully acknowledges financial support in the forms of a Nanyang Assistant Professorship, an AcRF Tier 1 grant (RG116/15), and a JST-PRESTO grant.

References

1. B. Honig and M. Karplus, *Nature*, 1971, **229**, 558.
2. A. Warshel and M. Levitt, *J. Mol. Biol.*, 1976, **103**, 227.
3. H. M. Senn and W. Thiel, *Angew. Chem., Int. Ed.*, 2009, **48**, 1198.
4. M. W. van der Kamp and A. J. Mulholland, *Biochemistry*, 2013, **52**, 2708.
5. E. Maseras and K. Morokuma, *J. Comput. Chem.*, 1995, **16**, 1170.
6. S. Humbel, S. Sieber and K. Morokuma, *J. Chem. Phys.*, 1996, **105**, 1959.
7. M. Svensson, S. Humbel, R. D. J. Froese, T. Matsubara, S. Sieber and K. Morokuma, *J. Phys. Chem.*, 1996, **100**, 19357.
8. S. Dapprich, I. Komaromi, K. S. Byun, K. Morokuma and M. J. Frisch, *J. Mol. Struct.: THEOCHEM*, 1999, **461**, 1.
9. P. Tao, J. F. Fisher, Q. Shi, T. Vreven, S. Mobashery and H. B. Schlegel, *Biochemistry*, 2009, **48**, 9839.
10. T. Vreven, K. S. Byun, I. Komaromi, S. Dapprich, J. A. Montgomery Jr., K. Morokuma and M. J. Frisch, *J. Chem. Theory Comput.*, 2006, **2**, 815.

11. L. W. Chung, H. Hirao, X. Li and K. Morokuma, *WIREs Comput. Mol. Sci.*, 2012, **2**, 327.
12. L. W. Chung, W. M. C. Sameera, R. Ramozzi, A. J. Page, M. Hatanaka, G. P. Petrova, T. V. Harris, X. Li, Z. Ke, F. Liu, H.-B. Li, L. Ding and K. Morokuma, *Chem. Rev.*, 2015, **115**, 5678.
13. M. J. Frisch, G. W. Trucks, H. B. Schlegel, G. E. Scuseria, M. A. Robb, J. R. Cheeseman, G. Scalmani, V. Barone, B. Mennucci, G. A. Petersson, H. Nakatsuji, M. Caricato, X. Li, H. P. Hratchian, A. F. Izmaylov, J. Bloino, G. Zheng, J. L. Sonnenberg, M. Hada, M. Ehara, K. Toyota, R. Fukuda, J. Hasegawa, M. Ishida, T. Nakajima, Y. Honda, O. Kitao, H. Nakai, T. Vreven, J. A. Montgomery Jr., J. E. Peralta, F. Ogliaro, M. Bearpark, J. J. Heyd, E. Brothers, K. N. Kudin, V. N. Staroverov, R. Kobayashi, J. Normand, K. Raghavachari, A. Rendell, J. C. Burant, S. S. Iyengar, J. Tomasi, M. Cossi, N. Rega, J. M. Millam, M. Klene, J. E. Knox, J. B. Cross, V. Bakken, C. Adamo, J. Jaramillo, R. Gomperts, R. E. Stratmann, O. Yazyev, A. J. Austin, R. Cammi, C. Pomelli, J. W. Ochterski, R. L. Martin, K. Morokuma, V. G. Zakrzewski, G. A. Voth, P. Salvador, J. J. Dannenberg, S. Dapprich, A. D. Daniels, Ö. Farkas, J. B. Foresman, J. V. Ortiz, J. Cioslowski and D. J. Fox, *Gaussian 09 Revision D.01*, Gaussian Inc., Wallingford, CT, 2009.
14. U. C. Singh and P. A. Kollman, *J. Comput. Chem.*, 1986, **7**, 718.
15. T. Vreven, K. Morokuma, Ö. Farkas, H. B. Schlegel and M. J. Frisch, *J. Comp. Chem.*, 2003, **24**, 760.
16. T. Vreven, M. J. Frisch, K. N. Kudin, H. B. Schlegel and K. Morokuma, *Mol. Phys.*, 2006, **104**, 701.
17. S. Caprasecca, S. Jurinovich, L. Viani, C. Curutchet and B. Mennucci, *J. Chem. Theory Comput.*, 2014, **10**, 1588.
18. J. B. Foresman and Æ. Frisch, *Exploring Chemistry with Electronic Structure Methods*, Gaussian Inc., Pittsburgh, PA, 3rd edn, 2015.
19. H. M. Berman, J. Westbrook, Z. Feng, G. Gilliland, T. N. Bhat, H. Weissig, I. N. Shindyalov and P. E. Bourne, *Nucleic Acids Res.*, 2000, **28**, 235.
20. A. Sali and T. L. Blundell, *J. Mol. Biol.*, 1993, **234**, 779.
21. H. Li, A. D. Robertson and J. H. Jensen, *Proteins*, 2005, **61**, 704.
22. R. Dennington, T. Keith and J. Millam, *GaussView, Version 5, Semichem Inc.*, Shawnee Mission, KS, 2009.
23. C. I. Bayly, P. Cieplak, W. D. Cornell and P. A. Kollman, *J. Phys. Chem.*, 1993, **97**, 10269.
24. D. A. Case, J. T. Berryman, R. M. Betz, D. S. Cerutti, T. E. Cheatham III, T. A. Darden, R. E. Duke, T. J. Giese, H. Gohlke, A. W. Goetz, N. Homeyer, S. Izadi, P. Janowski, J. Kaus, A. Kovalenko, T. S. Lee, S. LeGrand, P. Li, T. Luchko, R. Luo, B. Madej, K. M. Merz, G. Monard, P. Needham, H. Nguyen, H. T. Nguyen, I. Omelyan, A. Onufriev, D. R. Roe, A. Roitberg, R. Salomon-Ferrer, C. L. Simmerling, W. Smith, J. Swails, R. C. Walker, J. Wang, R. M. Wolf, X. Wu, D. M. York and P. A. Kollman, *AMBER 2015*, University of California, San Francisco, CA, 2015.

25. O. Trott and A. J. Olson, *J. Comput. Chem.*, 2010, **31**, 455.
26. T. Kawatsu, M. Lundberg and K. Morokuma, *J. Chem. Theory Comput.*, 2011, **7**, 390.
27. H. Hirao, N. Thellamurege and X. Zhang, *Front. Chem.*, 2014, **2**, 1.
28. G. Xing, Y. Diao, L. M. Hoffart, E. W. Barr, K. S. Prabhu, R. J. Arner, C. C. Reddy, C. Krebs and J. M. Bollinger Jr., *Proc. Natl. Acad. Sci. U. S. A.*, 2006, **103**, 6130.
29. P. M. Brown, T. T. Caradoc-Davies, J. M. J. Dickson, G. J. S. Cooper, K. M. Loomes and E. N. Baker, *Proc. Natl. Acad. Sci. U. S. A.*, 2006, **103**, 15032.
30. A. G. Thorsell, C. Persson, N. Voevodskaya, R. D. Busam, M. Hammarström, S. Gräslund, A. Gräslund and B. M. Hallberg, *J. Biol. Chem.*, 2008, **283**, 15209.
31. H. Hirao and K. Morokuma, *J. Am. Chem. Soc.*, 2009, **131**, 17206.
32. R. M. Cicchillo, H. Zhang, J. A. V. Blodgett, J. T. Whitteck, G. Li, S. K. Nair, W. A. van der Donk and W. W. Metcalf, *Nature*, 2009, **459**, 871.
33. W. W. Metcalf and W. A. van der Donk, *Annu. Rev. Biochem.*, 2009, **78**, 65.
34. W. A. van der Donk, C. Krebs and J. M. Bollinger Jr., *Curr. Opin. Struct. Biol.*, 2010, **20**, 673.
35. J. T. Whitteck, R. M. Cicchillo and W. A. van der Donk, *J. Am. Chem. Soc.*, 2009, **131**, 16225.
36. J. T. Whitteck, P. Malova, S. C. Peck, R. M. Cicchillo, F. Hammerschmidt and W. A. van der Donk, *J. Am. Chem. Soc.*, 2011, **133**, 4236.
37. S. C. Peck, H. A. Cooke, R. M. Cicchillo, P. Malova, F. Hammerschmidt, S. K. Nair and W. A. van der Donk, *Biochemistry*, 2011, **50**, 6598.
38. H. Hirao and K. Morokuma, *J. Am. Chem. Soc.*, 2011, **133**, 14550.
39. H. Hirao and K. Morokuma, *J. Am. Chem. Soc.*, 2010, **132**, 17901.
40. E. A. Thompson and P. K. Siiteri, *J. Biol. Chem.*, 1974, **249**, 5364.
41. P. A. Cole and C. H. Robinson, *J. Med. Chem.*, 1990, **33**, 2933.
42. E. R. Simpson, M. S. Mahendroo, G. D. Means, M. W. Kilgore, M. M. Hinshelwood, S. Graham-Lorence, B. Amarneh, Y. Ito, C. R. Fisher and M. D. Michael, *Endocr. Rev.*, 1994, **15**, 342.
43. W. R. Miller, *Endocr. -Relat. Cancer*, 1996, **3**, 65.
44. J. T. Groves, *J. Inorg. Biochem.*, 2006, **100**, 434.
45. J. Rittle and M. T. Green, *Science*, 2010, **330**, 933.
46. S. Shaik, H. Hirao and D. Kumar, *Nat. Prod. Rep.*, 2007, **24**, 533.
47. S. Shaik, S. Cohen, Y. Wang, H. Chen, D. Kumar and W. Thiel, *Chem. Rev.*, 2010, **110**, 949.
48. P. J. Mak, A. Luthra, S. G. Sligar and J. R. Kincaid, *J. Am. Chem. Soc.*, 2014, **136**, 4825.
49. Y. Khatri, A. Luthra, R. Duggal and S. G. Sligar, *FEBS Lett.*, 2014, **588**, 3117.
50. F. K. Yoshimoto and F. P. Guengerich, *J. Am. Chem. Soc.*, 2014, **136**, 15016.
51. K. Xu, Y. Wang and H. Hirao, *ACS Catal.*, 2015, **5**, 4175.

52. D. Ghosh, J. Griswold, M. Erman and W. Pangborn, *Nature*, 2009, **457**, 219.
53. B. Meunier, S. P. de Visser and S. Shaik, *Chem. Rev.*, 2004, **104**, 3947.
54. I. G. Denisov, T. M. Makris, S. G. Sligar and I. Schlichting, *Science*, 2005, **105**, 2253.
55. D. Usharani, D. C. Lacy, A. S. Borovik and S. Shaik, *J. Am. Chem. Soc.*, 2013, **135**, 17090.
56. S. Kitagawa, R. Kitaura and S.-I. Noro, *Angew. Chem., Int. Ed.*, 2004, **43**, 2334.
57. H.-C. Zhou, J. R. Long and O. M. Yaghi, *Chem. Rev.*, 2012, **112**, 673.
58. Z.-Y. Gu, J. Park, A. Raiff, Z. Wei and H.-C. Zhou, *ChemCatChem*, 2013, **6**, 67.
59. G. Férey, *Chem. Soc. Rev.*, 2008, **37**, 191.
60. J. Lee, O. K. Farha, J. Roberts, K. A. Scheidt, S. T. Nguyen and J. T. Hupp, *Chem. Soc. Rev.*, 2009, **38**, 1450.
61. M. Yoon, R. Srirambalaji and K. Kim, *Chem. Rev.*, 2012, **112**, 1196.
62. M. Fujita, Y. J. Kwon, S. Washizu and K. Ogura, *J. Am. Chem. Soc.*, 1994, **116**, 1151.
63. S. Choomwattana, T. Maihom, P. Khongpracha, M. Probst and J. Limtrakul, *J. Phys. Chem. C*, 2008, **112**, 10855.
64. G. A. E. Oxford, R. Q. Snurr and L. J. Broadbelt, *Ind. Eng. Chem. Res.*, 2010, **49**, 10965.
65. M. Zheng, Y. Liu, C. Wang, S. Liu and W. Lin, *Chem. Sci.*, 2012, **3**, 2623.
66. S. Yadnum, S. Choomwattana, P. Khongpracha, J. Sirijaraensre and J. Limtrakul, *ChemPhysChem*, 2013, **14**, 923.
67. C. Raksakoon, T. Maihom, M. Probst and J. Limtrakul, *J. Phys. Chem. C*, 2015, **119**, 3564.
68. H. Hirao, W. K. H. Ng, A. M. P. Moeljadi and S. Bureekaew, *ACS Catal.*, 2015, **5**, 3287.
69. S. Bhattacharjee, J.-S. Choi, S.-T. Yang, S. B. Choi, J. Kim and W.-S. Ahn, *J. Nanosci. Nanotechnol.*, 2010, **10**, 135.
70. M. Matthias, R. E. Johnsen, P. D. C. Dietzel and H. Fjellvåg, *Microporous Mesoporous Mater*, 2012, **157**, 62.
71. D. J. Xiao, E. D. Bloch, J. A. Mason, W. L. Queen, M. R. Hudson, N. Planas, J. Borycz, A. L. Dzubak, P. Verma, K. Lee, F. Bonino, V. Crocellà, J. Yano, S. Bordiga, D. G. Truhlar, L. Gagliardi, C. M. Brown and J. R. Long, *Nat. Chem.*, 2014, **6**, 590.
72. K.-B. Cho, H. Hirao, S. Shaik and W. Nam, *Chem. Soc. Rev.*, 2016, **45**, 1197.
73. H. Hirao, D. Kumar, L. Que and S. Shaik, *J. Am. Chem. Soc.*, 2006, **128**, 8590.
74. H. Hirao, D. Kumar, W. Thiel and S. Shaik, *J. Am. Chem. Soc.*, 2005, **127**, 13007.
75. S. Shaik, H. Hirao and D. Kumar, *Acc. Chem. Res.*, 2007, **40**, 532.
76. D. Wang, E. R. Farquhar, A. Stubna, E. Münck and L. Que, *Nat. Chem.*, 2009, **1**, 145.

77. H. Hirao, *J. Phys. Chem. B*, 2011, **115**, 11278.
78. U. C. Singh and P. A. Kollman, *J. Comput. Chem.*, 1984, **5**, 129.
79. B. H. Besler, K. M. Merz and P. A. Kollman, *J. Comput. Chem.*, 1990, **11**, 431.
80. H. Hirao, *Chem. Lett.*, 2011, **40**, 1179.
81. K. Kitaura and K. Morokuma, *Int. J. Quantum Chem.*, 1976, **10**, 325.
82. Y. Mo, J. Gao and S. D. Peyerimhoff, *J. Chem. Phys.*, 2000, **112**, 5530.
83. H. Hirao, *Chem. Phys. Lett.*, 2007, **443**, 141.
84. I. W. Chen and F. C. Charalampous, *J. Biol. Chem.*, 1966, **241**, 2194.
85. F. Eisenberg, *J. Biol. Chem.*, 1967, **242**, 1375.
86. S. J. Pollack, M. R. Knowles, J. R. Atack, H. B. Broughton, C. I. Ragan, S. A. Osborne and G. McAllister, *Eur. J. Biochem.*, 1993, **217**, 281.
87. S. J. Pollack, J. R. Atack, M. R. Knowles, G. McAllister, C. I. Ragan, R. Baker, S. R. Fletcher, L. L. Iversen and H. B. Broughton, *Proc. Nat. Acad. Sci. U. S. A.*, 1994, **91**, 5766.
88. J. R. Atack, H. B. Broughton and S. J. Pollack, *Trends Neurosci.*, 1995, **18**, 343.
89. J. R. Atack, H. B. Broughton and S. J. Pollack, *FEBS Lett.*, 1995, **361**, 1.
90. J. R. Atack, *Brain Res. Rev.*, 1996, **22**, 183.
91. X. Wang and H. Hirao, *J. Phys. Chem. B*, 2013, **117**, 833.
92. L. Yang, S. L. Koh, P. W. Sutton and Z.-X. Liang, *Catal. Sci. Technol.*, 2014, **4**, 2871.
93. A. J. M. Ribeiro, L. Yang, M. J. Ramos, P. A. Fernandes, Z.-X. Liang and H. Hirao, *ACS Catal.*, 2015, **5**, 3740.
94. P. A. Dinndorf, J. Gootenberg, M. H. Cohen, P. Keegan and R. Pazdur, *Oncologist*, 2007, **12**, 991.
95. K. B. McCredie, D. H. W. Ho and E. J. Freireich, *CA Cancer J. Clin.*, 2008, **23**, 220.
96. D. S. Gesto, N. M. F. S. A. Cerqueira, P. A. Fernandes and M. J. Ramos, *J. Am. Chem. Soc.*, 2013, **135**, 7146.

CHAPTER 9

First Principles Methods in Biology: From Continuum Models to Hybrid Ab initio *Quantum Mechanics/Molecular Mechanics*

JENS DREYER,[a] GIUSEPPE BRANCATO,[b] EMILIANO IPPOLITI,[a] VITO GENNA,[c] MARCO DE VIVO,[a,c] PAOLO CARLONI*[a,d] AND URSULA ROTHLISBERGER[e]

[a] Computational Biomedicine Institute (IAS-5/INM-9), Forschungszentrum Jülich, Wilhelm-Johnen-Straße, 52428 Jülich, Germany; [b] Scuola Normale Superiore (SNS), Piazza dei Cavalieri 7, 56126 Pisa, Italy; [c] Laboratory of Molecular Modeling and Drug Discovery, Istituto Italiano di Tecnologia, Via Morego 30, 16163 Genoa, Italy; [d] Computational Biophysics, German Research School for Simulation Sciences, 52425 Jülich, Germany; [e] Ecole polytechnique Fédérale de Lausanne, Laboratoire de chimie et biochimie computationnelles, ISIC, FSB-BSP, CH-1015 Lausanne, Switzerland
*Email: p.carloni@grs-sim.de

9.1 Introduction

First principles (*i.e.* non-empirical) simulations constitute nowadays a key tool in the investigation of biomolecular reactivity, photochemistry and spectroscopy.[1–7] They can be broadly divided in two categories, hybrid

RSC Theoretical and Computational Chemistry Series No. 9
Simulating Enzyme Reactivity: Computational Methods in Enzyme Catalysis
Edited by Iñaki Tuñón and Vicent Moliner

Published by the Royal Society of Chemistry, www.rsc.org

quantum mechanics/molecular mechanics (QM/MM) and continuum-based methods.

First principles QM/MM approaches successfully combine the advantages of both the accuracy and the general applicability of QM methods with the efficiency of classical, force field-based MM methods.

First principles QM/MM approaches reconcile the possibility of describing electronic processes such as chemical reactions, charge transfer or photo-induced electronic excitations with a proper description of extended environmental effects, *e.g.* arising from ambient solvents and/or biomolecular environments. The system of interest is separated into two parts: one (the QM part) comprises the chemically/photophysically active region treated by computationally demanding electronic structure methods. The remainder (the MM part) is described efficiently at a lower level of theory by classical force fields. Special attention, obviously, has to be paid to the coupling of both regions.

The combination of first principles QM/MM methods with statistical mechanics *via* molecular dynamics (MD) simulations at finite temperature provides a powerful approach to investigate dynamics, reactivity and thermodynamics of biological molecules, such as enzymes, photoreceptors, receptor/drug complexes. The size of the QM part is routinely restricted to a few hundred atoms. The MM part, in contrast, often extends to some hundred thousands of atoms, allowing for increasingly realistic simulations of biological systems.

First principles QM/MM MD simulations require considerable computational resources, such that affordable simulation times are often restricted to sub-nanosecond time scales. This time scale might not be sufficient to capture rare events or ensure adequate conformational sampling. For circumventing this drawback, QM/MM methods may be coupled with enhanced sampling methods such as free energy perturbation,[8] thermodynamic integration,[8] umbrella sampling[9] and metadynamics.[10,11] These methods have been successfully applied to adequately describe chemical reactions and biomolecular processes.[12–18]

The basic idea of the QM/MM approaches was first introduced by Warshel and Levitt in 1976.[19] It was then further developed by Singh and Kollmann[20] and Field *et al.*[21] and the diverse and widespread applications have been documented in a vast number of papers (for recent reviews, see for example ref. 22–42).

An important point concerns the treatment of boundary conditions due to the typical finite and small size of atomistic simulations of bio-systems, which are in the order of 10^5–10^6 atoms as compared to experimental samples usually larger by more than 10 orders of magnitude. The usual choice is to adopt periodic boundary conditions (PBC), that is to simulate an explicit molecular system virtually replicated in all three directions of the Cartesian space. The main advantage of PBC consists of avoiding the presence of any physical boundary that may unrealistically perturb the system.

An alternative approach, particularly suitable for solute–solvent systems is represented by the use of continuum models, according to which an explicit molecular system is embedded into a cavity of a polarisable dielectric medium mimicking the natural bio-environment.[43–48] Such an approach has been less exploited in the context of first principles QM/MM MD simulations,[49–54] especially because it is not obvious how to set up an optimal interaction potential between the discrete and continuum regions of the system. Indeed, implicit solvent models, such as the polarisable continuum model (PCM)[55–57] and the conductor-like screening model (COSMO)[58] successfully capture bulk solution properties, yet they miss specific interactions. Here, we have included in the final part of the present chapter a brief review of recently proposed hybrid discrete/continuum methods.

9.2 First Principles QM/MM Methods

9.2.1 Introduction

In the most frequently used QM/MM schemes, the molecular system is described by a single hybrid Hamiltonian

$$H = H_{QM} + H_{MM} + H_{QM/MM}, \tag{9.1}$$

where H_{QM} is the quantum Hamiltonian, H_{MM} is the molecular mechanics Hamiltonian, and $H_{QM/MM}$ is the interaction Hamiltonian between QM and MM regions. The lowest eigenvalue of the Hamiltonian in eqn (9.1) determines the total ground state energy E of the mixed quantum/classical system:

$$E = E_{QM} + E_{MM} + E_{QM/MM}. \tag{9.2}$$

The advantage of such an additive scheme is that the QM calculation can be directly executed in the presence of the classical environment in such a way that the electron density of the QM system is optimised in (and polarised by) the external electrostatic field of the surroundings. The prize for this is that the real system is replaced by a somewhat artificial, heterogeneous construct, in which different parts of the system are described at largely disparate levels, *i.e.* one part of the system is represented in electronic detail, whereas all the surroundings is reduced to a purely classical (mechanical and electrostatic) description.

In this way, an abrupt QM/MM border is created. One of the drastic consequences of this approach is the fact that when passing from the QM to the MM zone of the system the electrons' description suddenly changes from an explicit to an implicit one (*via* force field). Such a simplified description can necessarily only constitutes a somewhat crude representation of the true uniform system.

9.2.2 The QM Part

The QM part is often described by density functional theory (DFT) methods due to the favourable ratio between computational cost and accuracy.†

Hartree–Fock (HF) methods are rarely applied as they require similar computational resources as DFT, while providing a less accurate description due to the lack of electron correlation.[24]

Wavefunction-based electronic structure post-HF methods include a certain degree of electron correlation with respect to HF methods and thus provide a more accurate description of molecular systems. However, post-HF methods imply high computational costs caused by the need to calculate and transform four-index two-electron interaction integrals. To cope with this problem, methods have been devised to reduce the number of relevant integrals by sophisticated integral screening techniques[62] or by simplifying the four-index integrals to less computationally demanding quantities with three or two indices.[63–66] The latter techniques, referred to as density fitting (DF) or resolution of the identity (RI), fit orbital overlap densities with auxiliary basis functions, thereby reducing the number of indices and consequently the number of integrals.

Canonical post-HF methods are applied for energy calculations at fixed geometries, but their unfavourable scaling behaviour with system size (measured by $\mathcal{N}$) prevents their direct application in MD schemes.[25] In fact, the unfavourable scaling ranges from $O(\mathcal{N}^5)$ for second-order Møller–Plesset perturbations theory (MP2), to $O(\mathcal{N}^6)$ for coupled-cluster methods with singles and doubles (CCSD) or configuration interaction methods with single and double excitations (CISD), and to $O(\mathcal{N}^7)$ for CCSD with perturbative treatment of triple excitations (CCSD(T)), the latter is often referred to as the "gold standard of quantum chemistry". The unfavorable scaling is known, however, to be an artifact of canonically delocalised molecular orbitals (MOs). Dynamical electron correlation effects, which are primarily addressed by post-HF methods, are in fact associated with short-range, dispersive-like interactions, which fall off with a factor of $1/r^6$, with r being the interelectronic distance. Local correlation methods,[67–70] denoted by 'L' in front of the method acronym, exploit the locality of localised orbitals by introducing a hierarchical scheme for selecting important electron pair interactions. Nearby pairs are treated on the highest level of theory, the ones further apart on a lower level of theory, and interactions of distant pairs are neglected entirely. Introducing localised virtual orbitals likewise lowers the number of excitations considerably due to truncation. Local correlation methods capture a large fraction of the dynamical correlation energy and reach asymptotically linear scaling with system size $O(\mathcal{N})$. A drawback is the necessity to properly define the parameters of the hierarchical scheme.

---

†Semiempirical methods, such as MNDO, AM1, PM3, OM2/3 are computationally highly efficient and they are frequently used.[24] Approximate (semiempirical) DFT methods such as the self-consistent-charge density functional tight binding (SCC-DFTB) method,[59–61] which comes at a considerably lower cost than a regular DFT calculations, are also popular.

To further increase efficiency, local correlation methods have been combined with DF methods.[71,72] In order to improve the convergence of correlation energy with the size of the basis set, so-called explicit electron correlation methods introduce an additional term into the wavefunction with an explicit interelectronic distance dependence,[73,74] which improves the description of the cusp region of the wavefunction for $r_{12}\rightarrow 0$. Explicitly correlated methods have also been combined with local correlation approaches.[75]

Frequently, post-HF methods are used for energy calculations or optimisations on snapshots obtained from QM/MM MD simulations on lower level of theories, *e.g.* semi-empirical methods, to include proper phase space sampling.[24,25,33,76] Linear scaling quantum chemical methods[77,78] making use of integral screening,[62] DF[71] or post-HF local correlation methods (*e.g.* LMP2,[75] LCCSD[79,80]) have expanded the scope of full QM simulations to larger systems containing many hundred or even a few thousand atoms, but as yet haven't been widely explored for QM/MM MD simulations.[78,81] A later paragraph of this section will provide more details on post-HF methodologies in combination with QM/MM approaches.

9.2.3 The MM Part

The MM part is commonly described by a mechanical force field.[82] Popular choices for mechanical force fields are AMBER,[83–85] GROMOS[86–89] or CHARMM.[82,90–93] The basic functional form of, for example, the AMBER force field is given by

$$E_{\mathrm{MM}}(R) = \sum_{\mathrm{bonds}} k_b(R-b_0)^2 + \sum_{\mathrm{angles}} k_\theta(\theta-\theta_0)^2 + \sum_{\mathrm{torsions}} \frac{1}{2}V_\phi[1-\cos(n\phi-\phi_0)] + \sum_{j=1}^{N-1}\sum_{i=j+1}^{N}\left\{\epsilon_{ij}\left[\left(\frac{\sigma_{ij}}{R_{ij}}\right)^{12} - 2\left(\frac{\sigma_{ij}}{R_{ij}}\right)^{6}\right] + \frac{q_i q_j}{4\pi\epsilon_0 R_{ij}}\right\} \tag{9.3}$$

including bonded terms as sums over bonds and angles represented by harmonic potentials, and a sum over torsional angles given by a periodic function, as well as non-bonded terms modeled by a Lennard–Jones potential for van der Waals forces and by electrostatic Coulomb point charge interactions. The CHARMM force field in addition introduces two other harmonic terms, one for improper dihedral angles and a limited set of so-called Urey–Bradley $1-3$-distance terms important to properly describe vibrations.[93] Later, so-called CMAP dihedral angle cross terms, a grid-based energy correction map designed to improve the description of protein backbone conformational properties, were introduced.[94]

In contrast to the all-atom AMBER and CHARMM force fields, the GROMOS force fields are in fact united atom force fields without explicit aliphatic (nonpolar) hydrogen atoms.

Water and other solvents are very commonly described through rigid models in order to reduce the degrees of freedom and speed up the calculations without significantly affecting the accuracy. Frequently applied water force fields include TIP3P,[95] TIP4P,[96] SPC[97] or SPC/E[98] models. There are some indications that for QM/MM simulations, particularly with higher levels of QM theory, the TIP4P model is preferable.[99]

We close this section by mentioning that there have been some recent attempts to introduce polarisable force fields in combination with *ab initio* methods.[100,101] Traditionally, the most popular approaches used to include polarisation effects into standard force fields are based on the induced point dipole method,[102] the classical Drude oscillator model,[103] and the fluctuating charges model.[104] By allowing molecular polarisation, the mutual electrostatic interactions between the QM and the MM partitions are in principle more properly described, especially in cases when a highly inhomogeneous electric field is involved due to strongly polar or charged chemical groups or in the case of interfaces.

As an example, in various computational studies[105,106] polarisation has been shown to play a crucial role in describing the liquid/vapor interfaces at molecular level.

However, from a computational viewpoint, such a mutual polarisation of both classical and quantum parts requires iterations until self-consistency is reached[100,101] with a noticeable increase of computing time.

Also, it should be considered that polarisable force fields usually necessitate an *ad hoc* parametrisation that has not yet been extensively carried out in conjunction with typical quantum mechanical models, such as HF and DFT methods. While it is desirable to use polarisable force fields more systematically within the context of QM/MM approaches, this represents at the moment an active area of research that requires further developments and validation studies.

9.2.4 The $E_{\mathrm{QM/MM}}$ Coupling Term

In general, the coupling term includes bonded, electrostatic, and van der Waals interactions between the QM and MM atoms:

$$E_{\mathrm{QM/MM}} = E^{\mathrm{bond}}_{\mathrm{QM/MM}} + E^{\mathrm{el}}_{\mathrm{QM/MM}} + E^{\mathrm{vdW}}_{\mathrm{QM/MM}} \tag{9.4}$$

The coupling of the electrostatic interactions $E^{\mathrm{el}}_{\mathrm{QM/MM}}$ can be grouped into three different schemes:[24,107] mechanical, electrostatic, and polarisation coupling. In the simple mechanical scheme, the interaction is described by MM terms only, for example by using a point charge distribution for the QM atoms. The QM calculation itself is thus performed in the absence of an external field. Electrostatic embedding schemes introduce the MM part as an external charge distribution into the QM Hamiltonian, for example through point charges from the force field. This scheme, in which the QM electronic structure is polarised by the environment, is used most often in current QM/MM applications. If polarisable force fields are used the interaction could be further extended to include polarisation of the environment as well, which is referred to as polarised embedding.

Both the bonded $E^{\text{bond}}_{\text{QM/MM}}$ and van der Waals $E^{\text{vdW}}_{\text{QM/MM}}$ interaction energies are commonly described through the force field terms as in eqn (9.3) (see also next section):

$$E^{\text{bond}}_{\text{QM/MM}} = \sum_{I'}\sum_{I} 4\varepsilon_{I'I}\left(\left(\frac{\sigma_{I'I}}{R_{I'I}}\right)^{12} - \left(\frac{\sigma_{I'I}}{R_{I'I}}\right)^{6}\right) \tag{9.5}$$

$$\begin{aligned} E^{\text{vdW}}_{\text{QM/MM}} = & \sum_{b}\frac{1}{2}k_b(R_{I'I} - b_0)^2 + \sum_{\theta}\frac{1}{2}k_\theta(\theta_{I''J''K''} - \theta_0)^2 \\ & + \sum_{\phi}\sum_{n} k_n[1 + \cos(n\phi_{I''J''K''L''}) - \phi_0] \end{aligned} \tag{9.6}$$

where now I runs over QM and I' over MM atoms and at least one atom of the triple ($I''J''K''$) and quadruples ($I''J''K''L''$) of bonded atoms is a MM atom.

9.3 *Ab initio* QM/MM MD Simulation Techniques

9.3.1 DFT Car–Parrinello MD Approach

DFT is a widely applied quantum chemical method for the investigation of biological systems. It scales favorably with the number of electrons and the accuracy of the employed exchange-correlation functionals, which contains all the intricacies of the many-body problem, is constantly improving.[108–113] Its scope was further enlarged in 1985, when Car and Parrinello (CP) proposed a unified scheme for DFT and MD.[114] By treating the electronic degrees of freedom as dynamical variables, they managed to describe the time evolution of molecular systems (presently up to almost 2000 atoms)[7] without resorting to a force field.[115–117] The method enabled new types of realistic simulations for many different kinds of systems. Born–Oppenheimer approaches to first principles MD[115,117] are also widely and efficiently used.[118]

Most systems of biological relevance are large: for instance, a system containing a protein in aqueous solution may consist of 10^4 to 10^5 atoms. To deal with these systems, hybrid Car–Parrinello molecular dynamics/molecular mechanics (CPMD/MM) schemes have been introduced. Most current CPMD/MM applications in biophysics employ the approach developed by Rothlisberger and coworkers,[119,120] in which the CPMD program (CPMD) is used for the QM part and the classical part, calculated with routines from Gromos96 code,[121] is described either by the GROMOS or AMBER force field.[‡]

[‡]A similar hybrid approach[122,123] has also been implemented in the CP2K code.[124]

9.3.2 Comparison between Full QM and QM/MM Calculations

To identify where the main approximations in a QM/MM approach enter and see how severe they are, let us consider the case when the entire system (QM + MM) is described uniformly at the DFT level. The total (electronic plus core–core interaction)[§] energy of such a system is given by the density functional[126]

$$E = T[\rho] + \int_{\Omega} V^{\mathrm{ex}}(\mathbf{r})\rho(\mathbf{r}) + \frac{1}{2}\iint \frac{\rho(\mathbf{r}_1)\rho(\mathbf{r}_2)}{r_{12}}\mathrm{d}\mathbf{r}_1\mathrm{d}\mathbf{r}_2 + E_{\mathrm{xc}}[\rho] + \frac{1}{2}\sum_I\sum_J \frac{Z_I Z_J}{R_{IJ}} \tag{9.7}$$

where T and E_{xc} are the kinetic and the exchange-correlation energy functionals, respectively; V^{ex} is the external electrostatic potential created by the positively charged nuclei (or cores);[¶] $\mathbf{r}$, $\mathbf{r}_1$, $\mathbf{r}_2$ refer to the electronic degrees of freedom, while r_{12} represents interelectronic and R_{IJ} internuclear distances; Z_I and Z_J represent the nuclear (or core) charge of atom I and J, respectively.

Now, we partition the system into two parts, A and B, with respective densities ρ_{A} and ρ_{B}. The total density ρ can be expressed (see also ref. 127) as

$$\rho(\mathbf{r}) = \rho_{\mathrm{A}}(\mathbf{r}) + \rho_{\mathrm{B}}(\mathbf{r}). \tag{9.8}$$

Analogous to eqn (9.2) the total energy is given by

$$E = E_{\mathrm{A}} + E_{\mathrm{B}} + E_{\mathrm{A/B}} \tag{9.9}$$

with

$$\begin{aligned} E = {} & T[\rho_{\mathrm{A}}] + T[\rho_{\mathrm{B}}] + T^{\mathrm{NL}} + \int_{\Omega} V^{\mathrm{ex}}(\mathbf{r})\rho_{\mathrm{A}}(\mathbf{r})\mathrm{d}\mathbf{r} + \int_{\Omega} V^{\mathrm{ex}}(\mathbf{r})\rho_{\mathrm{B}}(\mathbf{r})\mathrm{d}\mathbf{r} \\ & + \frac{1}{2}\iint \frac{\rho_{\mathrm{A}}(\mathbf{r}_1)\rho_{\mathrm{A}}(\mathbf{r}_2)}{r_{12}}\mathrm{d}\mathbf{r}_1\mathrm{d}\mathbf{r}_2 + \frac{1}{2}\iint \frac{\rho_{\mathrm{B}}(\mathbf{r}_1)\rho_{\mathrm{B}}(\mathbf{r}_2)}{r_{12}}\mathrm{d}\mathbf{r}_1\mathrm{d}\mathbf{r}_2 \\ & + \frac{1}{2}\iint \frac{\rho_{\mathrm{A}}(\mathbf{r}_1)\rho_{\mathrm{B}}(\mathbf{r}_2)}{r_{12}}\mathrm{d}\mathbf{r}_1\mathrm{d}\mathbf{r}_2 + E_{\mathrm{xc}}[\rho_{\mathrm{A}}] + E_{\mathrm{xc}}[\rho_{\mathrm{B}}] + E_{\mathrm{xc}}^{\mathrm{NL}} \\ & + \frac{1}{2}\sum_I\sum_J \frac{Z_I Z_J}{R_{IJ}}. \end{aligned} \tag{9.10}$$

[§]Here, we are implicitly assuming that the Born–Oppenheimer approximation[125] is valid and that only the electrons are dealt with at quantum level while the nuclei are still described as point-like charges moving according to the classical Newtonian laws. This level of approximation turns out to be adequate for most biophysical applications.

[¶]In some schemes only the outermost electrons of each atom (usually the valence electrons) are described by a wave function. All the other electrons are described implicitly by introducing pseudopotentials (see later) and the nuclear charge is replaced by the core charge, *i.e.* the difference between the atomic number and the number of explicit valence electrons. Often this modified nucleus is referred to as "core".

The terms T^{NL} and $E_{\mathrm{xc}}^{\mathrm{NL}}$ are:

$$\begin{aligned} T^{\mathrm{NL}} &= T[\rho_{\mathrm{A}} + \rho_{\mathrm{B}}] - T[\rho_{\mathrm{A}}] - T[\rho_{\mathrm{B}}] \\ E_{\mathrm{xc}}^{\mathrm{NL}} &= E_{\mathrm{xc}}[\rho_{\mathrm{A}} + \rho_{\mathrm{B}}] - E_{\mathrm{xc}}[\rho_{\mathrm{A}}] - E_{\mathrm{xc}}[\rho_{\mathrm{B}}]. \end{aligned} \tag{9.11}$$

These terms account for the nonlinearity of the kinetic energy and the exchange-correlation density functionals, respectively.[‖] They are only zero if ρ_{A} and ρ_{B} are spatially well separated.

For the particular case where we describe part A of the system with another approach than part B, it is useful to separate also the external potential V^{ex} into contributions from the nuclear cores of A and those of B:

$$V^{\mathrm{ex}}(r) = V_{\mathrm{A}}^{\mathrm{ex}}(r) + V_{\mathrm{B}}^{\mathrm{ex}}(r). \tag{9.12}$$

E_{A} and E_{B} in eqn (9.9) are given by the terms

$$\begin{aligned} E_Y = T[\rho_Y] &+ \int_\Omega V^{\mathrm{ex}}(\mathbf{r})\rho_Y(\mathbf{r})\mathrm{d}\mathbf{r} + \frac{1}{2}\iint \frac{\rho_Y(\mathbf{r}_1)\rho_Y(\mathbf{r}_2)}{r_{12}}\mathrm{d}\mathbf{r}_1\mathrm{d}\mathbf{r}_2 \\ &+ E_{\mathrm{xc}}[\rho_Y] + \frac{1}{2}\sum_{I\in Y}\sum_{J\in Y}\frac{Z_I Z_J}{R_{IJ}} \end{aligned} \tag{9.13}$$

where $Y = \mathrm{A}$ or B, respectively. Often the core charges Z_I and Z_J are expanded into Gaussian shaped charge distributions with width R_{c} of the form

$$Z_I = \int_\Omega \rho_I^{\mathrm{nucl}}(\mathbf{r} - \mathbf{R}_I)\mathrm{d}\mathbf{r} = \int_\Omega \frac{Z_I}{R_{\mathrm{c}}^3}\pi^{-2/3}\exp\left[-\frac{|\mathbf{r} - \mathbf{R}_I|^2}{R_{\mathrm{c}}^2}\right]\mathrm{d}\mathbf{r} \tag{9.14}$$

and the three Coulomb terms can be summarised into one expression, which depends on the combined nuclear and electronic charge density $\rho^{\mathrm{el+nucl}} = \rho^{\mathrm{el}} + \rho^{\mathrm{nucl}}$

$$\begin{aligned} &\int_\Omega V_Y^{\mathrm{ex}}(\mathbf{r})\rho_Y(\mathbf{r})\mathrm{d}\mathbf{r} + \frac{1}{2}\iint \frac{\rho_Y(\mathbf{r}_1)\rho_Y(\mathbf{r}_2)}{r_{12}}\mathrm{d}\mathbf{r}_1\mathrm{d}\mathbf{r}_2 + \frac{1}{2}\sum_{I\in Y}\sum_{J\in Y}\frac{Z_I Z_J}{R_{IJ}} \\ &= \frac{1}{2}\iint \frac{\rho_Y^{\mathrm{el+nucl}}(\mathbf{r}_1)\rho_Y^{\mathrm{el+nucl}}(\mathbf{r}_2)}{r_{12}}\mathrm{d}\mathbf{r}_1\mathrm{d}\mathbf{r}_2. \end{aligned} \tag{9.15}$$

‖ The term $E_{\mathrm{xc}}^{\mathrm{NL}}$ in eqn (9.11) arises in the construction of *ab initio* atomic pseudopotentials when the system has to be partitioned into valence and core densities. In this case $E_{\mathrm{xc}}^{\mathrm{NL}}$ corresponds to the *nonlinear core correction.*[128]

The interface term $E_{\text{A-B}}$ describes the interaction between A and B and therefore contains all the remaining terms

$$
\begin{aligned}
E_{\text{A-B}} = T^{\text{NL}} + \int_{\Omega} V_{\text{B}}^{\text{ex}}(\mathbf{r})\rho_{\text{A}}(\mathbf{r})d\mathbf{r} + \int_{\Omega} V_{\text{A}}^{\text{ex}}(\mathbf{r})\rho_{\text{B}}(\mathbf{r})d\mathbf{r} \\
+\frac{1}{2}\iint \frac{\rho_{\text{A}}(\mathbf{r}_1)\rho_{\text{B}}(\mathbf{r}_2)}{r_{12}} d\mathbf{r}_1 d\mathbf{r}_2 + E_{\text{xc}}^{\text{NL}} + \frac{1}{2}\sum_{I\in\text{A}}\sum_{J\in\text{B}} \frac{Z_I Z_J}{R_{IJ}}.
\end{aligned}
\tag{9.16}
$$

For the special case where part A is treated with a QM and part B with an MM method, the first two energy terms in eqn (9.2) correspond to

$$
E_{\text{QM}} = T[\rho_{\text{QM}}] + \frac{1}{2}\iint \frac{\rho_{\text{QM}}^{\text{el+nucl}}(\mathbf{r}_1)\rho_{\text{QM}}^{\text{el+nucl}}(\mathbf{r}_2)}{r_{12}} d\mathbf{r}_1 d\mathbf{r}_2 + E_{\text{xc}}[\rho_{\text{QM}}] \tag{9.17}
$$

$$
E_{\text{MM}} = T[\rho_{\text{MM}}] + \frac{1}{2}\iint \frac{\rho_{\text{MM}}^{\text{el+nucl}}(\mathbf{r}_1)\rho_{\text{MM}}^{\text{el+nucl}}(\mathbf{r}_2)}{r_{12}} d\mathbf{r}_1 d\mathbf{r}_2 + E_{\text{xc}}[\rho_{\text{MM}}]. \tag{9.18}
$$

E_{MM} is delegated to the classical force field. Clearly, none of the current force fields for biomolecular simulations can provide an exact match of the terms in eqn (9.18). As electrons are not considered explicitly, force fields are parameterised to single (or to the average of several) configurations with fixed electron density distributions. Therefore, the kinetic energy term in eqn (9.18) can be considered as an additive constant that is not taken explicitly into account.

The effect of the exchange-correlation energy functional is often replaced by a pair-additive van der Waals term:

$$
E_{\text{xc}} \approx E^{\text{vdW}} = \sum_{I'}\sum_{J'} 4\varepsilon_{I'J'}\left(\left(\frac{\sigma_{I'J'}}{R_{I'J'}}\right)^{12} - \left(\frac{\sigma_{I'J'}}{R_{I'J'}}\right)^{6}\right). \tag{9.19}
$$

The electrostatic potential due to the combined electronic and ionic charge distribution is approximated *via* effective point charges, usually located at atomic positions:

$$
\frac{1}{2}\iint \frac{\rho_{\text{MM}}^{\text{el+nucl}}(\mathbf{r}_1)\rho_{\text{MM}}^{\text{el+nucl}}(\mathbf{r}_2)}{r_{12}} d\mathbf{r}_1 d\mathbf{r}_2 \approx \frac{1}{2}\sum_{I'}\sum_{J'} \frac{q_{I'}q_{J'}}{R_{I'J'}}. \tag{9.20}
$$

The set of effective (often empirical) point charges commonly used in biomolecular force fields cannot be expected to faithfully reproduce the left hand side of eqn (9.20), *i.e.* to be fully consistent with the electronic structure method used for the QM part. However, due to the extremely cumbersome work involved in the development of a general and transferable force field for complex biological systems, people usually prefer to employ existing parameterisations instead of constructing a fully *ab initio* derived force field. In addition, it turns out that although the magnitude of effective point charges used in different force fields can vary largely, the average electrostatic potentials seem to be in surprisingly good agreement with each other as well as with DFT descriptions.[129]

In spite of this somewhat reassuring caveat, the fact remains that real electronic charge distributions are far from mere assemblies of point charges. The point charge approximation breaks completely down in the description of covalent chemical bonds that are characterised by highly inhomogeneous and highly directional distributions of the electron density. Clearly, simple van der Waals/electrostatic descriptions such as those in eqn (9.19) and (9.20) cannot reproduce the intricacies of chemical bonding. In most force fields, the interaction between nearest, second nearest and third nearest neighbor atoms linked by chemical bonding are therefore mimicked by mechanical bond, angle and torsional angle terms given in the first line of eqn (9.3). For the atoms connected *via* bonded terms, the non-bonded (electrostatic and van der Waals) interactions are either omitted or scaled down (so-called exclusion rules).

Replacing eqn (9.5) and (9.6) in eqn (9.4), and using eqn (9.15), the interaction energy in eqn (9.16) becomes:

$$
\begin{aligned}
E_{\mathrm{QM/MM}} = & \sum_{I'} \int_{\Omega} \frac{q_{I'}}{\mathbf{R}_{I'} - \mathbf{r}} \rho_{\mathrm{QM}}^{\mathrm{el+nucl}}(\mathbf{r}) \mathrm{d}\mathbf{r} \\
& + \sum_{I'} \sum_{I} 4\varepsilon_{I'I} \left(\left(\frac{\sigma_{I'I}}{R_{I'I}} \right)^{12} - \left(\frac{\sigma_{I'I}}{R_{I'I}} \right)^{6} \right) \\
& + \sum_{b} \frac{1}{2} k_b (R_{I'I} - b_0)^2 + \sum_{\theta} \frac{1}{2} k_\theta (\theta_{I''J''K''} - \theta_0)^2 \\
& + \sum_{\phi} \sum_{n} k_n [1 + \cos(n\phi_{I''J''K''L''}) - \phi_0]
\end{aligned}
\tag{9.21}
$$

where I runs over QM and I' over MM atoms and at least one atom of the triple $(I''J''K'')$ and quadruples $(I''J''K''L'')$ of bonded atoms is a QM atom. In this formulation, the effective classical point charges act as an external field to the QM calculation, *i.e.* the electron density of the QM part is polarised by the classical environment (in contrast to the mechanical coupling scheme described earlier).

Both the van der Waals term and the bonded terms are acting on atomic positions only, *i.e.* are not part of the total electronic potential and thus are not directly felt by the electrons. If we want to achieve a closer model of a full QM description, the deviations caused by the actual MM representation have to be compensated by a correction term ΔV in the total potential that the electrons of the QM part experience

$$V_{\text{tot}} = V_{\text{QM}} + V_{\text{QM/MM}} + \Delta V \tag{9.22}$$

$$\Delta V = \Delta V^{\text{NL}} + \Delta V^{\text{el}} \tag{9.23}$$

where ΔV^{el} accounts for the error in the electrostatic terms (deviation of the classical electrostatic potential from the QM reference and reduction of the electronic density distribution to a point charge representation) whereas the nonlinear correction term ΔV^{NL} results from the non-linearity corrections in eqn (9.11). Thus this term is a mere artifact of the density partitioning and is not present in a system treated at the uniform level. To keep this term minimal, the somewhat trivial but important condition has to be fulfilled that the QM part has to be chosen in such a way that the electronic wave functions are localised in this region. If this condition cannot be fulfilled, the correction term ΔV^{NL} gains in importance.

How can we assess the importance of the correction term ΔV in practice? Ideally, one would like that the electron density in the QM region, ρ_{QM}, matches as closely as possible the electron density in the same region produced by a full QM representation of the system (ρ_{true}). According to the Hohenberg–Kohn theorem,[126] if the two densities are identical, all the properties we calculate for the QM region are identical to those of the real system. In other words, if we determine the correction potential ΔV in such a way that the total electronic potential in a QM/MM simulation V_{tot} minimises the density difference

$$\int_{\Omega'} (\rho_{\text{true}}(\mathbf{r}) - \rho_{\text{QM}}(\mathbf{r}))^2 \mathrm{d}\mathbf{r} \tag{9.24}$$

where Ω' is a suitably chosen volume of the QM region, our QM/MM simulation approaches the full QM reference results in an optimal way.

9.3.3 CPMD/MM Method: Basics

The Car–Parrinello method[114] can be extended into a QM/MM scheme using a mixed Lagrangian of the form:[119]

$$\begin{aligned}\mathcal{L} = {} & \frac{1}{2}\mu \sum_i \int \mathrm{d}\mathbf{r} \dot{\psi}_i^*(\mathbf{r}) \dot{\psi}_i(\mathbf{r}) + \frac{1}{2} \sum_I M_I \dot{\mathbf{R}}_I^2 - E_{\text{MM}} - E_{\text{QM/MM}} - E_{\text{QM}} \\ & + \sum_{i,j} \Lambda_{i,j} \left(\int \mathrm{d}\mathbf{r} \psi_i^*(\mathbf{r}) \psi_j(\mathbf{r}) - \delta_{i,j} \right)\end{aligned} \tag{9.25}$$

where μ is the fictitious mass associated with the electronic degrees of freedom, ψ_i are the Kohn–Sham one-particle orbitals, M_I is the mass of atom I and $\Lambda_{i,j}$ are Lagrange multipliers that enforce orthonormality of the

Kohn–Sham orbitals. The energy of the QM system E_{QM} is given by the Kohn–Sham energy density functional:[130]

$$E_{QM} = E_{KS}[\psi_i, \mathbf{R}_I] = -\frac{1}{2}\int d\mathbf{r}\psi_i^*(\mathbf{r})\nabla^2\psi_i(\mathbf{r}) + \int d\mathbf{r}V^{ex}(\mathbf{r})\rho_{QM}(\mathbf{r})$$
$$+\frac{1}{2}\int d\mathbf{r}d\mathbf{r}'\rho_{QM}(\mathbf{r})\frac{1}{|\mathbf{r}-\mathbf{r}'|}\rho_{QM}(\mathbf{r}') + E_{xc}[\rho_{QM}(\mathbf{r})] \tag{9.26}$$

where for the spin unpolarised case, the electron density $\rho_{QM}(r)$ is given by the sum of the densities of the doubly occupied one-particle states:

$$\rho_{QM} = 2\sum_i \psi_i^*(\mathbf{r})\psi_i(\mathbf{r}). \tag{9.27}$$

The purely classical part E_{MM} is described by a standard biomolecular force field (eqn (9.3)). The interaction between the QM and MM parts, $E_{QM/MM}$, is included in the form of eqn (9.21) with the only exception of the harmonic bond interactions between QM and MM atoms which are omitted from the classical description and treated at the QM level with the help of specially designed empirical pseudopotentials.[131]

Standard implementations of Car–Parrinello MD simulations use plane wave basis sets. In this case, due to the high intrinsic flexibility of a plane wave basis set (in contrast to, for example, the minimal basis sets used in semi-empirical QM/MM calculations), special care has to be taken in order that the CPMD/MM interface is described in an accurate and consistent way. In the Rothlisberger implementation of the CPMD/MM,[119] the quantum/classical correction term ΔV in eqn (9.22) consists of specifically designed monovalent pseudopotentials to represent bonds between QM and MM parts of the system[131] and of modified screened Coulomb potentials for the interaction of the quantum electron density with close by classical point charges.[119]

In the context of a plane wave based Car–Parrinello scheme, a direct evaluation of the first term of eqn (9.21) is prohibitive as it involves order of $N_r \times N_{MM}$ operations, where N_r is the number of real space grid points (typically $\sim 100^3$) and N_{MM} is the number of classical atoms (usually of the order of 10 000 or more in systems of biochemical relevance). Therefore, the interaction between the QM system and the more distant MM atoms is included *via* a Hamiltonian term explicitly coupling the multipole moments of the quantum charge distribution with the classical point charges. This two level electrostatic coupling scheme can also be refined to an intermediate third layer that makes efficient use of variational D-RESP charges.[132,133] Highly efficient schemes based on a dual grid approach[134] or a multigrid approach with Gaussian expansion[122] have also been proposed in this context.

It should be mentioned that the QM system must be treated as a finite cluster by decoupling it from the artificial periodic images, which is achieved by the Hockney[135] or the Martyna–Tuckerman[136] decoupling schemes.

The Rothlisberger implementation of the hybrid CPMD/MM[119] establishes an interface between the Car–Parrinello code CPMD and the classical force fields GROMOS96 and AMBER in combination with a particle–particle–particle mesh (P3M) treatment of the long-range electrostatic interactions.[137] With this implementation, efficient and consistent simulations of complex systems (of the order of 10^5 atoms) can be performed. In these calculations, the steric and electrostatic effects of the surroundings are taken into account explicitly.

The most stringent current limitation is the short timescale accessible *via* CPMD/MM simulations of the order of tens to hundreds of picoseconds, which severely restricts the accuracy of time-averaged properties, such as binding free energies. Possible remedies for this problem are: (i) resorting to semi-empirical methods that allow sampling for hundreds of picoseconds,[138] (ii) employing multiple time step sampling for the QM and MM parts,[139] (iii) using enhanced sampling approaches such as metadynamics,[140] introducing either a classical[14,141] or electronic[142,143] bias potential, (iv) exploiting a linear response approximation with respect to a reference potential.[19]

The issue of the accuracy of DFT is also very important. A particular problem is the adequate description of London dispersion forces. Several methods have been developed to cure this problem, for example the addition of an effective atom-centred non-local term to the exchange-correlation potential may cure this significant drawback without additional computational cost.[144] The dispersion correction most commonly used with the CPMD/MM scheme are dispersion corrected atom-centred potentials (DCACPs) that are directly included in the electronic Hamiltonian, so that they are highly transferable and able to capture dispersion effects with an accuracy close to CCSD(T) quality.[145–147]

Another well-known issue of most DFT calculations is the underestimation of energy barriers associated with proton transfer events and other chemical reactions. Very recent highly efficient exact exchange implementations with plane waves[148] might make the use of hybrid exchange-correlation functionals, such as B3LYP[149–152] or meta-hybrid functionals such as, for example, the affordable MXX suite,[111,153] which in turn might help improve the accuracy of the results.

The main intrinsic approximations of a QM/MM approach lie in the reduction of the real electron density distribution of the MM part to a mere point charge representation and the neglect of the kinetic energy and exchange-correlation corrections (9.11) on the electronic level. All of these three terms are particularly severe in the neighborhood of a covalent chemical bond, where the electron density distribution is far from isotropic and the densities of the QM and MM part are strongly overlapping. In force field approximation, these deficiencies in the description of chemical bonding are remedied by including the special bonding terms given in the first line of eqn (9.3). However, these terms are a function of atomic coordinates only and do not in influence the electronic potential in a direct way. One of the most frequently encountered problems in QM/MM simulations thus occurs when the border between QM and MM parts has to run across a chemical bond: this is usually referred to as link atom problem. For QM/MM simulations of biological systems this is essentially always the case. In fact, a typical QM/MM partitioning for such systems includes only a portion of a biological macromolecule. The latter must then be cut into a QM and a MM region. As electronic degrees of freedom are explicitly described only in the QM region, when passing from the QM to the MM region the QM system contains unsaturated valences and has to be made chemically inert. This can be done by introducing an explicit correction term in the total electronic potential felt by the QM electrons. For the case of a QM/MM bond cut, the simplest way is to use a monovalent pseudopotential situated at the position of the first MM atom. This pseudopotential is usually constructed in such a way that the electrons of the QM region are scattered correctly by the classical environment. It is a common choice in CPMD/MM simulations to employ analytic, non-local pseudopotentials of *e.g.* the von Barth–Car[154] or the Goedecker[155] type to represent the MM atoms involved in QM/MM bond cuts. In case of the latter

$$V^{\text{eff}}(\mathbf{r},\mathbf{r}') = V^{\text{loc}}(\mathbf{r})\delta(\mathbf{r}-\mathbf{r}') + \sum_l V_l^{\text{nloc}}(\mathbf{r},\mathbf{r}') \tag{9.28}$$

where

$$V^{\text{loc}}(\mathbf{r}) = -\frac{Z_I}{r}\,\text{erf}\left[\frac{\mathbf{r}}{r_{\text{loc}}\sqrt{2}}\right] + \exp\left[-\frac{r^2}{2r_{\text{loc}}^2}\right]\Theta \tag{9.29}$$

$$\Theta = C_1 + C_2\left(\frac{r}{r_{\text{loc}}}\right)^2 + C_3\left(\frac{r}{r_{\text{loc}}}\right)^4 + C_4\left(\frac{r}{r_{\text{loc}}}\right)^6 \tag{9.30}$$

$$V_l^{\text{nloc}}(\mathbf{r},\mathbf{r}') = \sum_{m=-l}^{+l} Y_{l,m}(\mathbf{r}) \sum_{i,j=1}^{3} p_i^l(\mathbf{r}) h_{i,j}^l p_j^l(\mathbf{r}') Y_{l,m}^*(\mathbf{r}') \tag{9.31}$$

$$p_{\text{lh}} \propto r^{\,l+2(h-1)} \exp[-r^2/2r_l^2] \tag{9.32}$$

to represent the MM atoms involved in QM/MM bond cuts. The adjustable parameters r_{loc}, r_l, $h_{i,j}^l$ and C_1 to C_4 can be determined in analogy to eqn (9.24) by minimising the density penalty

$$F[\rho_{\text{QM}}(\mathbf{r}),\{\sigma_i\}] = \int_{\Omega'} \left|\rho_{\text{ref}}(\mathbf{r}) - \rho_{\text{QM}}(\mathbf{r},\{\sigma_i\})\right|^2 d\mathbf{r} \tag{9.33}$$

where the σ_i's are the set of adjustable parameters and ρ_{ref} is a reference density that approximates ρ_{true} in eqn (9.24). ρ_{ref} is usually determined from a QM/MM calculation of the system with extended QM part.[131]

There are many other *ad hoc* procedures in use to cure the link atom problem. Commonly used strategies are to add capping atoms (hydrogen or fluorine) or to represent the last QM atom with frozen frontier orbitals.[156] A naïve implementation of hydrogen capping introduces new atoms into the QM system that are not present in the real system. As a consequence, the QM portion is chemically not identical with the real system (*e.g.* the true system may contain C–C bonds at the boundary that are now described with C–H bonds that clearly have different electronic and chemical properties). Furthermore, additional degrees of freedom have been introduced and interactions of these nonexistent ghost atoms with the classical environment have to be carefully removed. To avoid this problem, capping atoms are placed on the line of the bond in a constant distance or ratio to the MM or QM atom and, thus, do not introduce additional degrees of freedom. An alternative approach to the link atom problem is the frozen frontier orbitals for the boundary atoms.[19,156] In this way, no additional physical interactions and degrees of freedom are introduced and the QM part retains its original composition. However, frozen orbitals have to be determined *via* calculations on small model systems and, as the name says, they remain frozen when transferred into the real environment. Specially parameterised pseudopotentials such as the ones described above, on the other hand, have the additional flexibility to adjust to changes in the environment.

Another possible artifact in QM/MM simulations, in particular in combination with extended and highly flexible basis sets (such as, for example, plane waves) is the electron spill-out problem. As shown in eqn (9.21), the exchange interactions between QM and MM part are taken into account on the level of atomic pair interactions only. Once again, these terms do not directly affect the electrons of the QM part. For a proper description of the electronic structure of the QM region an electronic correction term $\Delta V_{\mathrm{xc}}^{\mathrm{NL}}$ has to be included. This term is especially important for regions with overlapping or nearly overlapping densities between QM and MM parts, which is particularly the case for the nearby atoms surroundings the QM region. Due to the fact that the MM part contains no explicit electrons, the electrons of the QM part are no longer repelled by the closed-shell cores of the MM region. As a result of this missing Pauli repulsion, the electrons of the QM part can artificially localise on nearby positively charged classical point charges. This phenomenon is called electron spill-out. This effect can be avoided by using Gaussian smeared (screened) classical charges** or by replacing the classical point charge potential by suitably constructed ionic pseudopotentials with screened electrostatic interactions.[119] The latter solution is the one implemented in CPMD. In particular

$$E_{\mathrm{QM/MM}}^{\mathrm{el}} = \sum_{I \in \mathrm{MM}} q_I \int \mathrm{d}\mathbf{r}\, \rho(\mathbf{r}) v_I(|\mathbf{r} - \mathbf{R}_I|) \tag{9.34}$$

where q_I is the classical point charge located at $\mathbf{R}_I$ and

$$v_I(|\mathbf{r} - \mathbf{R}_I|) = \frac{r_{\mathrm{c}}^4 - r^4}{r_{\mathrm{c}}^5 - r^5} \tag{9.35}$$

**Drastic artifacts are possible by choosing too large widths for the Gaussian broadening.

(with r_c chosen as the covalent radius of atom I) is a Coulomb interaction potential modified at short range in such a way as to avoid spill-out of the electron density to nearby positively charged classical point charges.

Other potential sources of problems are possible incompatibilities between the QM and MM descriptions, such as imbalances in the electrostatic interactions that can lead to artificial preferences of, for example, substrate–QM, respectively substrate–MM interactions. Another problem is the consistent application of the classical exclusion rules for non-bonded interactions. In most force field definitions, non-bonded interactions (such as van der Waals and electrostatics) are not taken into account for nearby bonded neighbours. Such a selective neglect of a particular pair of interactions is not easily transferable to a many-body QM description. A consistent approach is however possible *via* mapping of the many-body electronic Hamiltonian to a pair-additive point charge representation.[133]

9.3.4 Applications to Biological Systems

CPMD/MM approach can effectively simulate complex reactions from first principles. This approach, which includes temperature effects, can benefit from the use of statistical mechanics methods[10,12–14] to investigate rare events. Recent reviews report for example enzymatic reactions[157–164] or selectivity properties of ion channels and porins[165–168] investigated with this approach.

CPMD/MM applications are also of importance to study drug action. Indeed, the interaction between a ligand and its target might at times depend on the electronic structure in such a subtle way that is difficult to capture with force field-based MD. A recent review reports CPMD/MM applications that address this issue.[169] The method has also recently provided valuable insights on DNA damage[16,170–175] and on the binding of anticancer drugs to DNA[176–180] or to the copper transport protein,[181–183] which is supposed to function as a transporter of cisplatin.

Ab initio methods find nowadays a wide application in biochemistry since they can help understanding the interaction between the ligand and its receptor, often shedding new lights into the mechanism for biological function, at the atomic detail. Importantly, the detailed understanding of the ligand–receptor interaction can also be used in the rational design of inhibitors, aiding the structure-based design of new drugs.[41,184] In this regard, QM-based studies can be used, for example, to evaluate potent enzyme inhibitors through the application of QM-corrected scoring functions for docking calculations and QM-derived binding affinity calculations.[185–188] Other examples are QM-based electrostatic potential maps used to characterise the binding site of the receptor;[189–191] QM calculations to determine the protonation states of key residues of the binding pocket;[192,193] and QM-based studies of the enzymatic reaction mechanism of targets for drug discovery.[41,42,164,194–201]

In this scenario, here we report some representative CPMD/MM studies on enzymatic chemical reactions carried out by mono-, bi- or tri-metallic core enzymes (*i.e.* metalloproteins) of pharmaceutical interest. These reported examples are just some of the several informative CPMD/MM studies demonstrating how this hybrid QM/MM approach can treat the metal–protein interaction in metalloenzymes, shedding lights into the mechanistic details for catalysis. In particular, we will report computational studies where *ab initio* CPMD/MM simulations have been used to determine enzymatic reaction mechanisms, revealing possible enzymatic transition state (TS) structures, which can be used as a template for the design of TS analogues drugs.[202,203] TS analogues are indeed considered as one of the most potent inhibitors due to their similarity in the geometry and physico-chemical properties with the enzymatic TS.[204]

The first example of CPMD/MM simulations, which contributed to show the effectiveness of this computational approach for understanding metalloenzyme function, is the study of De Vivo *et al.*[205] in 2008 that described the dynamics and energetics of the reaction catalysed by Ribonuclease H (RNase H, Figure 9.1).

RNase H, constitutively present in prokaryotic and eukaryotic cells, belongs to nucleotidyl-transferase (NT) superfamily and hydrolyses the phosphodiester bond (P–O3′) of the RNA strand. To do so, RNase H uses a two-metal-ion mechanism,[206] with two magnesium ions that aid RNase H catalysis. A series of so-called 'Blue-Moon' ensemble simulations[207] were carried out along a selected degree of freedom (*i.e.* reaction coordinate). These simulations, performed on a system formed by a total of ~37 000 atoms of which a hundred were treated at the QM level, were coupled with the thermodynamics integration (TI) approach to return the free energy profile for the reaction under study. By collecting ~200 ps of CPMD/MM dynamics, key aspects of the enzymatic activity of RNase H such as the mechanism and energetics of nucleophile formation and the role of each of the two metals during catalysis were clarified. Overall, these simulations described the continuous pathway from reagents to products along catalysis, revealing the existence of metastable intermediates and high-energy TSs that can hardly be captured by static X-ray structures, alone. In particular, the concerted motion of the two catalytic metal ions has been shown to aid both TS formation and leaving group departure, highlighting a dynamic rearrangement of the metal ions in an extended neighbourhood of the catalytic pocket. Interestingly, classical MD simulations coupled to adaptive biasing force (ABF) for enhanced sampling have later suggested that the concerted motions of the catalytic ions could be affected by the presence of a third ion, which might therefore contribute in the modulation of RNase H catalysis.[199] At different ionic concentrations, in MD simulations a third Mg(II) persistently locates nearby the native two-Mg(II) catalytic site, at times affecting the location of the nucleophilic water molecule and, consequently, the enzymatic catalysis. This computational result offers a rational explanation for the experimental evidence that Mg(II) concentration modulate RNase H function.[208,209]

In 2005 and 2007, De Vivo *et al.* have described a two-step reaction mechanism for the phosphatase activity catalysed by soluble epoxide

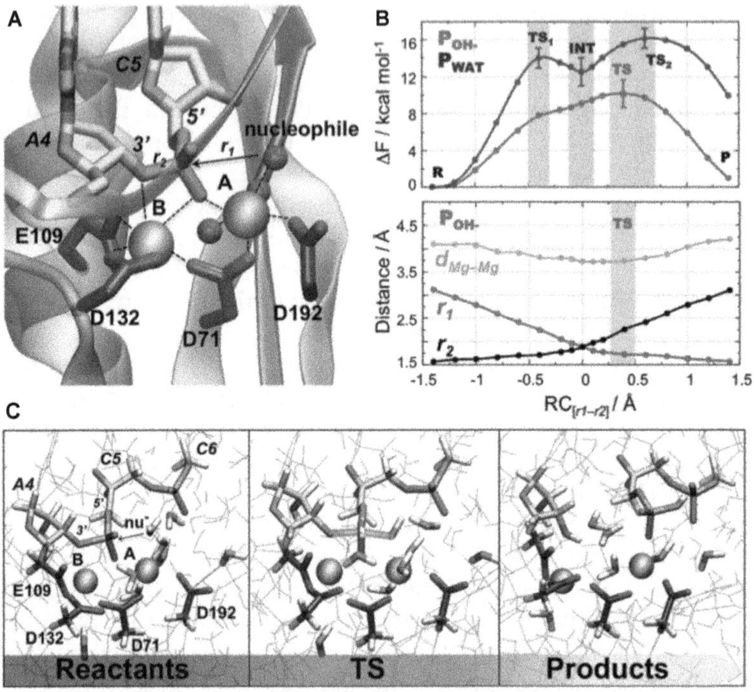

Figure 9.1 (A) Catalytic site of the RNase H in complex with a RNA/DNA hybrid, showing a typical 2M architecture. (B) Energetic (upper graph) and structural (lower graph) properties for the RNA phosphodiester cleavage catalysed by RNase H *via* the 2M mechanism, as from CPMD/MM. The free energy profile, obtained through thermodynamic integration, for two investigated pathways (PWAT or one hydroxide ion, POH) are shown. The reaction coordinate (RC) used represents the difference between the forming (r_1) and breaking (r_2) bonds. Transition state (TS) and the metastable intermediate (INT) are evidenced. Average bond distances [r_1, r_2, and the metal–metal distance (dMg–Mg)], along the lower free energy path (POH) are reported in the lower graph. (C) Representative snapshots from the QM/MM MD, showing the reactants, TS and product states of RNase H catalysis. QM atoms are shown explicitly; MM atoms are in thin lines.

hydrolase (sEH),[198,210] an Mg(II)-dependent enzyme (monometal core) of pharmaceutical interest involved in the fatty acid metabolism.[211–214] CPMD/MM simulations (~100 ps in total) were used to explore two main steps along the reaction mechanism: (i) a nucleophilic attack on the phosphate group of the phosphoester substrate by Asp9 and (ii) hydrolysis of the phosphoenzyme intermediate *via* a nucleophilic attack at the scissile phosphorus atom by a water molecule. Simulations show an in-line nucleophilic substitution during both steps, highlighting a rather dissociative character in the TS regions, with planar metaphosphate-like TS that does not structurally resemble a phosphorane species, as detected in the RNase H catalysis.[205]

Notably, the comparison of the QM/MM results on sEH (single Mg(II) catalytic ion) and those on RNase H (two Mg(II) ions) for the metal dependence in phosphoryl transfer reactions is quite instructive. Based on these QM/MM studies of metalloenzymes[197,198,205,210] different mechanisms for phosphoryl transfers (associative *vs.* dissociative) seem to be induced according to the metal(s) geometry and stoichiometry during catalysis.

During sEH catalysis, the metaphosphate group that is transferred is stabilised by its apical coordination to the only Mg(II) ion present in the catalytic site. In the RNase H enzymatic reaction, instead, the two Mg(II) stabilise the attacking and leaving groups, while the metaphosphate group is in between the two ions, showing a phosphorane-like TS. Hence, this comparison supports the hypothesis, reported by De Vivo *et al.*[205] in 2008, that two ions can more easily facilitate the formation of a metastable intermediate during phosphoryl transfers, as in the case of RNase H.

A recent computational study[215] of the metalloenzyme topoisomerase type II (Topo II) further remarks the functional role of two metal ions for phosphodiester bond cleavage for DNA processing. Recent X-ray structures of Topo II have shown that two Mg(II) ions are likely placed in the catalytic site.[216] During Topo II function, the DNA strand is cleaved and rejoined to allow DNA topology control. Hybrid Born–Oppenheimer QM/MM MD simulations have been used to simulate a catalytically competent Topo II state, where the two catalytic ions spontaneously relax into a two-metal-ion architecture, as that in RNase H.[199,210] The position of the two Mg(II) ions in the reactants seems therefore similar to several other metalloenzymes,[192,205,217,218] suggesting that Topo II likely cleaves the substrate DNA with a mechanism that might be analogous to RNase H.[199,205,219]

A second example of CPMD/MM calculations of metalloproteins is the investigation of metallo β-lactamases (MβLs). MβLs is a bacterial enzymatic family that use a mono- or binuclear Zn(II) core to hydrolyse β-lactam antibiotics, preventing therefore their beneficial effects.[163,192,220] The metal aided catalytic site of MβLs is designed to activate a nucleophilic hydroxide for the hydrolysis of the β-lactam ring. Due to their central role in antibacterial digestion and because of a potential risk of massively disseminated infections, MβLs structures are nowadays widely studied to conceive new anti-bacterial drugs. In this regard, Dal Peraro *et al.*[192,221] used CPMD/MM simulations to clarify the reaction mechanism of binuclear CcrA (B1 MβLs family) of Bacteroides fragilis, employing a model system formed by ~38 000 atoms, with a QM-shell of ~50 atoms. Seemingly to the RNase H study discussed before, a series of constrained CPMD/MM simulations coupled to TI returned the free energy profile associated to the reaction mechanism for β-lactam ring opening. These CPMD/MM simulations reveal the evolutionary insertion of the second equivalent zinc ion in the catalytic site, passing from BcII to CcrA species in B1 MβL subclass. The binuclear CcrA stabilises better the negative charge developed at the β-lactam N5 atom upon nucleophilic attack and its consequent detachment from the second zinc ion, favoring catalysis compared to the mononuclear species (BcII from *Bacillus cereus*).[163,192]

The presence of a second Zn(II) in the catalytic mechanism further lowers the energetic barrier for β-lactam hydrolysis initially found in mononuclear BcII B1 enzymes,[221] generating a more efficient single-step mechanism. Taken together, these computational studies of MβL catalysis, integrated with the experimental data, suggest that Zn-bound water is a shared and essential chemical feature across B1 MβLs.

Finally, we report CPMD/MM simulations to investigate fatty acid metabolism, as regulated by the membrane-bound fatty acid amide hydrolase (FAAH) enzyme, which is a serine hydrolase that modulates the endocannabinoid system, being primarily responsible for regulating the level of its main cannabinoid substrate anandamide (Figure 9.2).

Increased levels of endogenous anandamide might help to cure a variety of diseases including pain, inflammation, and cancer. For this reason, pharmacological inhibition of FAAH is a promising strategy for drug discovery. The computational investigation of FAAH catalysis *via* CPMD/MM

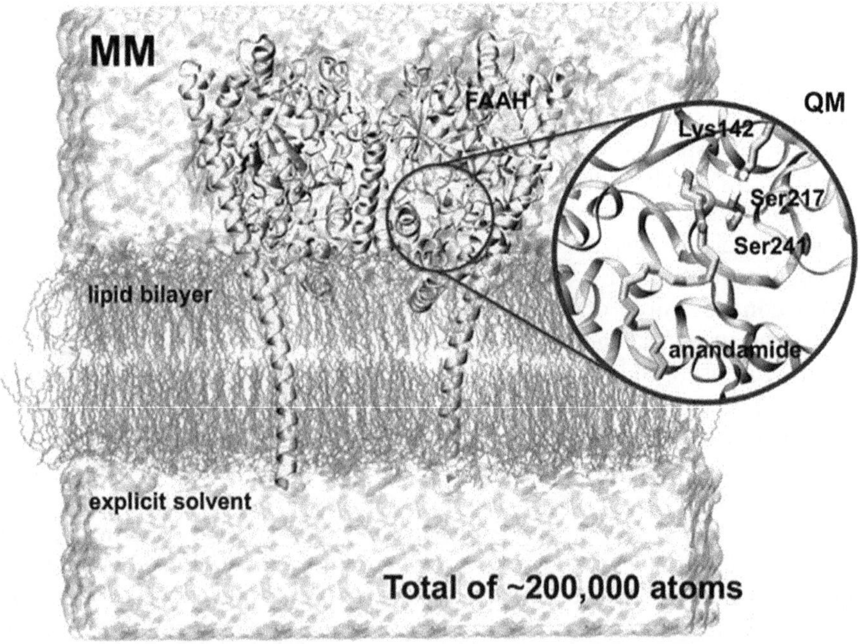

Figure 9.2 Representative QM/MM partitioning of a biological system (the enzyme FAAH in complex with anandamide embedded in a realistic membrane/water environment). While the enzyme (gold), the solvent (grey surface), and the lipid bilayer (violet) are treated at the MM level of theory, only the atoms directly involved in the reactive process (shown as cyan sticks) are treated at the QM level. The circle on the right is a close view of the reactive region of the complex, where anandamide and part of the catalytic triad residues Ser241, Ser217 and Lys142 (81 atoms in total) are treated at the QM level. A total of ~200 000 atoms are explicitly considered.

started with microsecond MD simulations of the enzyme in a realistic membrane/water environment, which provided a detailed understanding, as well as a robust model, for the reactant state of FAAH complex.[222] Full quantum mechanics and hybrid CPMD/MM approaches were then used for the characterisation of FAAH catalysis and inhibition.[200,201,219,223] In this case, the investigated a model system (FAAH enzyme bound to the membrane in explicit solvent) formed by ∼200 000 atoms with the catalytic core composed by ∼80 atoms treated at the DFT/BLYP level. Once again, TI coupled with Blue-moon ensemble simulations was used to investigate and compute free energy profiles for the catalytic hydrolysis of the substrate anandamide. These simulations (for a total of ∼120 ps of CPMD/MM simulations) have depicted a highly concerted two-step catalytic mechanism characterised by (i) acyl-enzyme formation after hydrolysis of the substrate amide bond and (ii) deacylation reaction with the reestablishment of the catalytic machinery. In particular, these simulations clarified a crucial event for anandamide hydrolysis, which is the inversion of the reactive nitrogen of the scissile amide bond occurring during the acylation rate-limiting step. To do so, FAAH uses an elegant catalytic strategy to induce amide bond distortion, reactive nitrogen inversion, and amide bond hydrolysis, promoting catalysis to completion.[200,223] Taken together, these few representative examples demonstrate how *ab initio* QM/MM simulations can provide informative insights into the enzymatic reaction mechanisms of enzymes of pharmaceutical relevance. These findings, for instance, could inspire more practical structure-based drug design (SBDD) efforts, with the ultimate goal to design potent enzyme inhibitors such as TS analogues. Nowadays, QM-based studies have seen an increasing applicability to practical drug design programs. Given the constant growth of computational power associated with a continuously improvement of algorithms for computations, QM-based methods, including hybrid QM/MM simulations, are destined to become a routinely used and powerful tool for effective SBDD, in the next future.

9.3.5 Post-HF Approaches

As mentioned at the beginning of this chapter, many methods have been devised to reduce high computational costs of the wavefunction-based electronic structure post-HF approaches. In particular, these methods reduce the number of relevant integrals to calculate by sophisticated integral screening techniques[62] or by simplifying the four-index integrals to less computationally demanding quantities with three or two indices.[63–66] These are the density fitting (DF), the resolution of the identity (RI) techniques that fit orbital overlap densities with auxiliary basis functions. These approaches have been successfully implemented and applied in QM/MM approaches.[224,225]

High-level post-HF methods have been applied in QM/MM schemes for investigating enzymatic catalysis in order to get accurate data for activation free energy barriers and reaction rates, *e.g.* through transition state theory.[33,76,226–230] With this high-level approach convergence of the QM contribution to the computed barriers in the order of ~1 kcal mol^{-1} ('chemical accuracy') is achieved, as assessed by comparison to experimental data. For instance, *para*-hydroxybenzoate hydroxylase and chorismate mutase[231] were investigated in a multilevel scheme. Statistical sampling has been carried out at the semi-empirical level, supplemented by reaction path DFT B3LYP/MM optimisations and final QM/MM energy calculations with advanced post-HF methods such as MP2, LMP2 and LCCSD(T). Free energies were calculated by umbrella sampling or thermodynamic integration. This multilevel scheme is based on the assumption that geometries, as sampled in the MD simulation and obtained by geometry optimisation, are less sensitive to the level of theory than energies, which are finally used to evaluate activation free energies and reaction rates. The hydroxylase system has been reinvestigated with explicit correlation methods and making use of DF.[232] Similarly, proton abstraction from acetyl-coenzyme A in citrate synthase[233] or cytochrome P45076,[234,235] and soluble epoxide hydrolase,[236] both of which are involved in drug metabolism, have been investigated by QM/MM calculations with high levels of correlated *ab initio* methods. Energy barriers depend on the size of the QM region, the choice of the basis set and the QM method, whereby only correlated QM methods are found to be in nearly quantitative agreement with experiments. For DFT-based QM/MM considerable improvements for the description of activation energies were found by including dispersion corrections, even though effects on optimised geometries are relatively small.[33,234,237] In this case, dispersion effects are conveniently included by empirical R-6 corrections according to Grimme[144,238,239] (denoted, for example, D2, D3), by empirical fitting of density functionals, as pursued by Zhao and Truhlar[111] with the Minnesota-functionals series, by direct inclusion in the functional, *e.g.* DF07[240] or by optimisation of atom-centred potentials.[145]

9.3.6 Excited States

Spectroscopy and photochemical reactions in biological systems are decisively influenced by effects of the solvent and/or biomolecular environment. Spectroscopic signals usually depend critically on the environment due to specific interactions such as hydrogen bonding as well as conformational fluctuations at finite temperatures. Understanding how the environment of spectroscopic probes determines their signatures, *i.e.* frequency positions, intensities or band shapes, is therefore crucial to derive structural and mechanistic conclusions from spectroscopy.[55,241] Likewise photochemical reaction pathways are critically determined by protein environments due to strong correlations with structural and dynamical protein properties such as conformational flexibility.

As yet, most QM/MM investigations on optical chromophore properties have been focused on solvent effects on absorption spectra,[242–244] whereas studies on biomolecular environmental shifts on optical spectra are still relatively rare.[245–252] That is, the full conformational flexibility of large protein environments at ambient temperature based on MD and its influence on absorption or emission band positions and shapes, *e.g.* inhomogeneous broadening of spectra, is only rudimentary investigated.

QM/MM simulations are well suited to investigate the factors that control photochemical excitation and reactivity. In excited-state QM/MM schemes, the QM part is either described *via* multiconfigurational wave function-based quantum chemical methods,[253–256] or by many-body perturbation theory[243,257–259] or through excited-state extensions of DFT.[242,243,259–261] Whereas the former two types of approaches are still limited to fairly small systems and relatively small basis sets, the latter one is also extendible to rather large systems.

Modeling excited-state decay requires consideration of non-adiabatic dynamics,[262] *e.g.* through trajectory-based methods such as surface hopping techniques. As photochemical reactions are often fast, *i.e.* in the femto to picosecond time range, modeling the excited-state dynamics is within reach of current QM/MM MD approaches. Indeed, QM/MM methods have also been explored for excited-state dynamics governing photochemical reactions.[242,253–256,261,263–269] DFT for ground states or time-dependent DFT (TD-DFT)[270] for the description of electronically excited states are computationally very efficient and have been used successfully, in particular in conjunction with Car–Parrinello MD methods[243,249,250,271–275] including non-adiabatic dynamics.[251,265,266,276–279] Nevertheless, the quality of results obtained by TD-DFT calculations depends on the system under investigation and on the functional used to reproduce the exchange and correlation interactions.[280] TD-DFT is known to be particularly problematic for excited states involving charge transfer and thus, the method is not generally applicable.[244,281–283] However, range-separated hybrid functionals have been shown to mitigate such errors.[284–287]

Accurate and more generally applicable wavefunction-based quantum chemical techniques for excited states such as coupled-cluster (CC) or multiconfigurational complete active space – self-consistent field (CASSCF) methods have been mostly limited to small molecules in the gas phase.[260]

In recent years, however, CASSCF methods[288] were increasingly applied in the CASPT2//CASSCF/MM schemes,[254–256] where the reaction coordinate is calculated at the CASSCF level, which takes the static correlation of the wave function into account. Energies are re-evaluated at the more accurate CASPT2 level,[289–291] which accounts for dynamical correlation effects by a second-order multireference perturbation theory implementation of CASSCF.

A CASPT2//CASSCF/AMBER (or CHARMM) electrostatic embedding approach has been implemented in the groups of Olivucci[292–295] or Robb[256,292] with classical or semi-classical trajectory methods to map reaction pathways. The method has been applied to excited-state properties of several light-sensitive proteins, for instance rhodopsin,[293,295–298] photoactive yellow protein,[256,292] or the fluorescent protein asFP595.[299] A decisive role for photochemical reactions is played by conical intersections, which serve as photochemical funnels.[300] Conical intersections in proteins have been investigated by QM/MM methods.[299]

The calculation of single finite trajectories does not provide real statistical sampling. This would require the computation of statistical properties averaged over tens or hundreds of trajectories.[255] The development of *ab initio* multiple spawning (AIMS) and the multi-configuration time-dependent Hartree (MCTDH) method try to address this issue.[301]

9.4 Continuum Models

9.4.1 Introduction

In recent years, there has been a growing interest in hybrid explicit/implicit methods especially for simulating solute–solvent systems.[43–48,302–306] According to such models, part of the solvent, typically including a few solvation shells, is treated explicitly along with the solute, while bulk effects are taken into account effectively by means of a surrounding dielectric continuum. On one hand, this allows reducing significantly the number of degrees of freedom of the system with respect to standard methods based on PBC, and, on the other hand, to retain the specific interactions between the solute and the solvent (*e.g.* hydrogen bonds), which are often necessary to describe accurately structural and dynamical features of complex systems.[307–310] Besides, in some cases solvent molecules need to be included explicitly in quantum mechanical calculations of spectroscopic parameters in order to better match experimental data.[311–314]

Another key point of continuum models is that they allow quite naturally the use of localised basis sets for the quantum mechanical treatment of the core region along with well-trusted and newly developed DFT methods based on hybrid functionals, *e.g.* B3LYP[149–152] and PBE0.[315,316] Indeed, explicit/implicit models rooted into DFT are attractive for the computation of averaging effects brought about by dynamics when a spectroscopic transition is fast with respect to the time scale of a dynamical phenomenon. In such circumstances, MD simulations are able to take into account at the same time solvent librations and solute vibrations, both of which can lead to non-negligible averaging effects of spectroscopic parameters.

The use of non-PBC also might be more consistent, from a physical point of view, for studying solute–solvent systems, avoiding possible correlation effects[317–324] and other possible problems with charged systems[136,325] that may plague simulations using PBC. Generally, they require a large number of explicit solvent molecules with respect to theoretical approaches using continuum boundary conditions.

However, a major problem that typically affects explicit/implicit solvation methods is related to the appearance of spurious boundary effects at the interface between the molecular systems and the continuum, which perturb the explicit solvent dynamics with respect to bulk behavior. A number of different approaches have been proposed[50,304,326–329] specifically for the treatment of the physical boundaries. The most encouraging results were obtained by fitting a reduced number of parameters, while enforcing at the same time proper boundary conditions.[304,327–329]

9.4.2 QM/MM MD Simulations with GLOB Approach

The general liquid optimised boundary (GLOB)[53,54] model belongs to the family of discrete/continuum methods mainly designed for simulating non-periodic systems, such as biomolecules in solution and at room temperature. The potentiality of the model have been already shown in several applications ranging from simple organic molecules in aqueous or non-aqueous solutions[54,314] to radical species,[330–332] and from metal cations in solution[333–335] to transient excited-state chromophores,[336] including also an extended liquid system, such as water.[333] In most cases, QM/MM MD simulations have been elegantly combined with a posteriori calculations of several spectroscopic parameters, while retaining the same molecular non-periodic model, which have shown a remarkable agreement with the corresponding experimental counterparts.[337,338]

According to the GLOB model, an explicit molecular system, typically a solute surrounded by a limited number of solvation shells, is embedded into a cavity of a structureless solvent, keeping fixed both the number of molecules and the volume during the simulation. The continuum medium, which mimics the response of solvent bulk, is completely specified by a few parameters, *e.g.* the dielectric permittivity, and depends on the nature of the solvent as well as on the physical conditions, such as density and temperature. Within the framework of mono-electronic QM methods, such as HF and DFT approaches, the free energy of an explicit molecular system can be expressed as:

$$A(\mathbf{x}) = E(\mathbf{P},\mathbf{x}) + W(\mathbf{P},\mathbf{x}) \tag{9.36}$$

where $E(\mathbf{P},\mathbf{x})$ is the potential energy of the explicit system in the nuclear configuration $\mathbf{x}$, and $W(\mathbf{P},\mathbf{x})$ is the 'mean field' contribution that accounts for the interactions with the environment. The explicit presence of the one-electron density matrix $\mathbf{P}$ in both terms on the r.h.s. points out that explicit system and implicit bulk are mutually polarised, *i.e.* the mean field response is always considered at equilibrium. The same expression holds also when the system is partitioned according to a QM/MM scheme. In the latter case, the QM region, and consequently the density matrix, $\mathbf{P}$, representing only the core region of the system is affected by both the bulk, approximated by the dielectric continuum, and the remaining explicit system treated at MM level. When non-polarisable MM force fields are considered, the one-electron density matrix, $\mathbf{P}$, can be obtained *via* the self-consistent field (SCF) solution of an electronic Hamiltonian which includes both the mean field operator as an effective term,[43,339] and the electronic embedding of the MM atomic charges. According to a usual partitioning scheme of the solvation process, originally proposed by Ben-Naim,[340] the mean field contribution, W, can be decomposed into three terms:

$$W = W_{\text{elec}} + W_{\text{disp-rep}} + W_{\text{cav}} \tag{9.37}$$

where W_{elec} is a long-range electrostatic term, generally referred to as the 'reaction field' representing the linear response of a dielectric continuum to the electric field generated by the system; $W_{\text{disp-rep}}$ is a short-range term that accounts mainly for the dispersion–repulsion interactions, and finally W_{cav} is a cavitation free energy, which is the reversible work spent to form the cavity into the liquid, regardless of the chemical nature of the solute. At a given density and temperature, W_{cav} does depend only on the size and shape of the cavity, hence it is constant for a fixed cavity. For the case of fixed volume cavities, typically considered in QM/MM simulations, W_{cav} has no effects on the dynamics of the system and can be safely ignored.

W_{elec} is induced by the charge distribution of the whole explicit system, which includes both the quantum core and the MM region, *via* the set of the atomic charges, q_I. Such a distribution is generally represented by a charge density of the electronic and nuclear charges, $\rho(\mathrm{r})$, plus the density of the partial point charges describing the Coulomb interactions at the MM level, which can be formally expressed in terms of a Dirac delta function as $\sum_I \delta(\mathbf{r} - \mathbf{r}_I)q_I$. Hence, the mean field contributions can be easily integrated in QM/MM algorithms which describe the hybrid electrostatic interactions at QM level. On the other hand, in the same spirit of well-established methodologies,[341–348] $W_{\text{disp-rep}}$ can be considered as a classical mean force potential that does not affect the electronic distribution of the system, hence it does not depend on P.

In the GLOB model, the mean field contributions have been included as a modification of the ONIOM[349–351] scheme.[53] In particular, W_{elec} is provided by the conductor-like model (CPCM),[352–354] corresponding to the PCM limit for infinite dielectric constant, but used with satisfactory results also for solvents of rather low polarity. The dielectric boundary, which can be thought as an average solvent accessible area, is defined on the basis of the cavity surface embedding the explicit system. The CPCM is a boundary element method (BEM), *i.e.* the response is evaluated on finite surface elements of the boundary called tesserae, and the reaction field is represented by apparent surface charges located onto such tesserae. In particular, the volume accessible to the molecular system is kept constant in a NVT ensemble, and under such conditions the dielectric boundary is given by a fixed regular surface representing the solvent accessible area. In the practical implementation, the molecular system is confined into a regular cavity, typically a sphere with radius R_{cav} where the dielectric continuum surface is given by a spherical surface with radius $R_{\text{cav}} = R_{\text{cav}} + \Delta R$, where ΔR is an optimised parameter. The use of the CPCM, together with the choice of a constant topology of the apparent charges (fixed boundary), is particularly attractive when *ab initio* MD is considered, due to the computational efficiency of both energy and gradient calculations.

At variance with previous approaches, the modeling of the short–medium range interactions, $W_{\text{disp-rep}}$, is specifically designed to properly perform NVT MD simulations of molecular liquids and solutions.[304,305] Moreover, $W_{\text{disp-rep}}$ may also include terms of electrostatic origin (reaction field) not fully accounted for by intrinsically approximated continuum models,[54] due to the lack of specific solvent(explicit)–solvent(implicit) hydrogen-bond interactions that could induce anisotropic distributions of the molecular orientation in proximity of the cavity surface.[50] As a result, such an additional potential effectively removes unwanted physical anisotropies of the solvent, such as the non-uniform distribution of the density and the artificial solvent polarisation due to the partial alignment of the molecular dipoles at the boundary.[54]

Altogether, the free energy contributions described above define a QM/MM potential well-suited for *ab initio* MD simulations in condensed phase using non-PBC and atomic basis sets to describe the electronic density. Forces provided by the CPCM energy derivatives are analytically defined as functions of the explicit molecular system coordinates, as well as energy and forces from the mean field contributions not explicitly dependent upon the density matrix ($W_{\text{disp-rep}}$). Therefore, conservative dynamics can be safely performed, as extensively validated in previous studies.[53,54] It is worth recalling that the SCF process in the PCM variational formulation accounts also for the mutual polarisation between the explicit system and the bulk (dielectric continuum), therefore the optimised density P describes a minimum of the free energy functional, *i.e.* an electronic density simultaneously relaxed with respect to the nuclei and equilibrated with solvent bulk. As a matter of fact, time-dependent solvation effects rely on the explicit representation of the closest solvent layers, while it can be assumed that dynamical effects at the continuum interface are less important for an accurate description of the solute.

In this framework, the atom-centred matrix propagation (ADMP)[355–357] model is exploited to perform first-principle and QM/MM simulations. ADMP is based on an extended Lagrangian scheme, such as Car–Parrinello MD,[358] which allows mimicking a MD on an electronically adiabatic surface by evolving the electronic structure on the nuclear time scale. In particular ADMP adopts a localised picture, *i.e.* the density matrix P of the atomic basis set collects fictitious dynamic variables propagated together with nuclear coordinates. Hence, the computational burden of the SCF process, which should be, in principle, carried out at each time step of the simulation, is avoided. As a consequence, the electronic density is, rigorously speaking, not fully relaxed about each nuclear configuration: instead, the energy potential calculated on-the-fly oscillates about the true Born–Oppenheimer surface. When we consider ADMP simulations in the condensed phase, the calculation of energy, energy derivatives and in particular of the Fock matrix, F, include terms accounting for the reaction field[43] and adjusted on-the-fly according to the propagated density matrix, P. In this case the deviation with respect to the adiabatic surface along the trajectory also contains a mean field contribution due to the deviation with respect to equilibrium of the electrostatic term, W_{elec}, *i.e.* the value one would obtain at full convergence of the variational process for a given nuclear configuration. However, the adiabaticity of the system during a GLOB/ADMP MD simulation can be effectively monitored through the same parameters introduced for the gas-phase ADMP formalism, namely the magnitude of the commutator [F,P] and the rate of change of the fictitious kinetic energy,[355,356] and test simulations have shown that the resulting dynamics ensures a correct adiabatic behavior, similar to the corresponding ADMP simulations in the gas phase.[54]

9.4.3 Applications to Open-shell Systems in Solution

In open-shell systems, such as transition metals or radical species, intrinsic quantum mechanical properties, such as spin delocalisation and spin transfer, may play a decisive role. No general and well-trusted force fields suitable for treating open-shell systems in solution and one often has to rely on first-principle methodologies. Only in a few cases, classical models, usually derived from *ad hoc* fitting procedures *versus* QM data, have been tailored towards at least a class of chemical compounds, as in the case of nitroxide based radicals.[359–361] Especially within the molecules of biological interest, there are many examples of radical species of amino acids and DNA or RNA bases that are involved in relevant enzymatic reactions, electron transfer processes,[362] ultra-violet radiation-mediated DNA damages.[363,364] One wants to describe, at molecular level, the role of stereo-electronic, dynamical, and environmental effects in tuning the structure and magnetic properties of such radicals in aqueous solution. While experimental techniques, such as high-field EPR spectroscopy, may promptly unravel many features of complex radical species in the condensed phase, quite often a complementary theoretical approach is needed in order to fully extract the molecular information encoded into such spectroscopic signals. In this case, accurate molecular modeling comes into play to confirm or, better, interpret the quite involved spectroscopic response. Moreover, through the use of theoretical approaches, different and subtle contributions that may affect the observed spectroscopic properties can be disentangled and evaluated separately, such as the effect of the solvent, the thermal effects and the effect of specific intermolecular interactions (*e.g.* hydrogen bondings). As a further source of uncertainty, it is often the case that radical compounds can exist in different isomeric forms and/or protonation states, almost isoenergetic to one another, depending on pH values and physical conditions: when this occurs, theoretical investigations of both structural and spectroscopic properties of radicals generally allow one to unambiguously identify the chemical species experimentally detected.

For illustrative purposes, we consider here two examples of open-shell biomolecules recently studied by the GLOB QM/MM approach previously described: the glycine radical[330] and the uracil radical anion[332] in aqueous solution. In particular, the glycine radical is the simplest example of an aliphatic radical issuing from elimination reactions involving amino acids, which are usually α-carbon-centred radicals.

Interestingly, already glycine and its derivatives show a number of unexpected features (*e.g.* preference for neutral or zwitterionic structure in different condensed phases, anomalously low hydrogen hyperfine coupling in aqueous solution), which have stimulated a large number of experimental[365–368] and theoretical[369–372] studies. An important issue that has inspired a number of theoretical studies consists in the analysis of the relative stability of the different glycine radical forms, *e.g.* NH_2–CH–COOH *vs.* NH_3^+–CH–COO^-, along with the calculation of the corresponding EPR hyperfine coupling constants.[369,373,374] It is well known that the zwitterionic form of the glycine radical is unstable in the gas phase (molecular optimisation leads to the neutral form), but it can exist in a more favorable polar environment, *e.g.* it has been detected experimentally in the solid state.[375] However, in aqueous solution, a satisfactory interpretation of the EPR parameters in structural terms requires the inclusion of intramolecular vibrational averaging and solvent effects.[369] Indeed, by including, at the same time, all short-term dynamical and solvent effects through GLOB QM/MM simulations of the glycine radical in aqueous solution and considering all the possible chemical species, namely the zwitterionic, (neutral) non-zwitterionic and anionic forms, it was recently possible to assign unambiguously the preferred chemical species in two different pH ranges, respectively $pH = 1–10$ and $pH > 10$.[330] Briefly, it has been observed that the neutral and non-zwitterionic form of the glycine radical is the only one compatible with EPR experiments at low pH, whereas at basic pH the observed EPR hyperfine coupling constants agree with those issuing from calculations on the anionic form.

Analogously, in the case of the uracil radical anion, as generated by the attachment of an excess electron, the combination of EPR data with quantum mechanical calculations and GLOB QM/MM MD simulations[332] have unraveled the electronic and structural nature of the only chemical species detected in aqueous solution and at room temperature. In fact, by applying QM/MM simulations to the study of five different tautomeric forms of the uracil radical anion in aqueous solution, chosen among those previously identified by gas-phase experiments[376] or predicted theoretically,[377] it has been possible to identify and characterise unequivocally the canonical form of the uracil radical as the prevalent tautomer in aqueous solution, contrary to both gas-phase results[378] and a recently reported thermodynamic analysis.[377]

References

1. L. M. S. Shepherd and C. A. Morrison, *J. Phys. Chem. B*, 2010, **114**, 7047–7055.
2. B. F. Habenicht, S. J. Paddison and M. E. Tuckerman, *Phys. Chem. Chem. Phys.*, 2010, **12**, 8728–8732.

3. T. D. Kühne, T. A. Pascal, E. Kaxiras and Y. Jung, *J. Phys. Chem. Lett.*, 2011, **2**, 105–113.
4. C. Schiffmann and D. Sebastiani, *Phys. Status Solidi B*, 2012, **249**, 368–375.
5. R. Glaves, G. Mathias and D. Marx, *J. Am. Chem. Soc.*, 2012, **134**, 6995–7000.
6. L. Vilčiauskas, M. E. Tuckerman, G. Bester, S. J. Paddison and K. D. Kreuer, *Nat. Chem.*, 2012, **4**, 461–466.
7. C. Zhang, D. G. Knyazev, Y. A. Vereshaga, E. Ippoliti, T. H. Nguyen, P. Carloni and P. Pohl, *Proc. Natl. Acad. Sci. U. S. A.*, 2012, **109**, 9744–9749.
8. J. Kästner, H. M. Senn, S. Thiel, N. Otte and W. Thiel, *J. Chem. Theory Comput.*, 2006, **2**, 452–461.
9. J. Kästner, *Wiley Interdiscip. Rev.: Comput. Mol. Sci.*, 2011, **1**, 932–942.
10. A. Barducci, M. Bonomi and M. Parrinello, *Wiley Interdiscip. Rev.: Comput. Mol. Sci.*, 2011, **1**, 826–843.
11. L. Sutto, S. Marsili and F. L. Gervasio, *Wiley Interdiscip. Rev.: Comput. Mol. Sci.*, 2012, **2**, 771–779.
12. B. Ensing, M. De Vivo, Z. Liu, P. Moore and M. L. Klein, *Acc. Chem. Res.*, 2006, **39**, 73–81.
13. X. Biarnés, S. Bongarzone, A. Vargiu, P. Carloni and P. Ruggerone, *J. Comput. -Aided Mol. Des.*, 2011, **25**, 395–402.
14. V. Leone, F. Marinelli, P. Carloni and M. Parrinello, *Curr. Opin. Struct. Biol.*, 2010, **20**, 148–154.
15. Y. Zhang and G. A. Voth, *J. Chem. Theory Comput.*, 2011, **7**, 2277–2283.
16. J. Garrec, C. Patel, U. Rothlisberger and E. Dumont, *J. Am. Chem. Soc.*, 2012, **134**, 2111–2119.
17. P. Söderhjelm, G. A. Tribello and M. Parrinello, *Proc. Natl. Acad. Sci. U. S. A.*, 2012, **109**, 5170–5175.
18. V. Limongelli, L. Marinelli, S. Cosconati, C. La Motta, S. Sartini, L. Mugnaini, F. Da Settimo, E. Novellino and M. Parrinello, *Proc. Natl. Acad. Sci. U. S. A.*, 2012, **109**, 1467–1472.
19. A. Warshel and M. Levitt, *J. Mol. Biol.*, 1976, **103**, 227.
20. U. C. Singh and P. A. Kollman, *J. Comput. Chem.*, 1986, **7**, 718–730.
21. M. J. Field, P. A. Bash and M. Karplus, *J. Comput. Chem.*, 1990, **11**, 700–733.
22. H. Lin and D. Truhlar, *Theor. Chem. Acc.*, 2007, **117**, 185–199.
23. H. M. Senn and W. Thiel, in *Atomistic Approaches in Modern Biology*, ed. M. Reiher, Springer, Berlin-Heidelberg, 1st edn, 2007, vol. 268, ch. 5, pp. 173–290.
24. H. M. Senn and W. Thiel, *Angew. Chem., Int. Ed.*, 2009, **48**, 1198–1229.
25. R. A. Mata, *Phys. Chem. Chem. Phys.*, 2010, **12**, 5041–5052.
26. F. H. Wallrapp and V. Guallar, *Wiley Interdiscip. Rev.: Comput. Mol. Sci.*, 2011, **1**, 315–322.
27. D. Xu, M. Zheng and S. Wu, in *Quantum Simulations of Materials and Biological Systems*, ed. J. Zeng, R. Q. Zhang and H. R. Treutlein, Springer, Berlin-Heidelberg, 1st edn, 2012, ch. 9, pp. 155–168.

28. J. Sgrignani and A. Magistrato, *Curr. Comput. -Aided Drug. Des.*, 2012, **9**, 15–34.
29. T. Steinbrecher and M. Elstner, in *Biomolecular Simulations*, ed. L. Monticelli and E. Salonen, Humana Press, 1st edn, 2013, vol. 924, ch. 4, pp. 91–124.
30. G. Groenhof, in *Biomolecular Simulations*, ed. L. Monticelli and E. Salonen, Springer, Berlin-Heidelberg, 1st edn, 2013, vol. 924, ch. 2, pp. 43–66.
31. K. Meier, A. Choutko, J. Dolenc, A. P. Eichenberger, S. Riniker and W. F. van Gunsteren, *Angew. Chem., Int. Ed.*, 2013, **52**, 2820–2834.
32. H. M. Senn and W. Thiel, *Curr. Opin. Chem. Biol.*, 2007, **11**, 182–187.
33. R. Lonsdale, K. E. Ranaghan and A. J. Mulholland, *Chem. Commun.*, 2010, **46**, 2354–2372.
34. O. Acevedo and W. L. Jorgensen, *Acc. Chem. Res.*, 2010, **43**, 142–151.
35. S. F. Sousa, P. A. Fernandes and M. J. Ramos, *Phys. Chem. Chem. Phys.*, 2012, **14**, 12431–12441.
36. A. Lodola and A. Mulholland, in *Biomolecular Simulations*, ed. L. Monticelli and E. Salonen, Humana Press, 1st edn, 2013, vol. 924, ch. 3, pp. 67–89.
37. M. W. van der Kamp and A. J. Mulholland, *Biochemistry*, 2013, **52**, 2708–2728.
38. B. Kirchner, F. Wennmohs, S. Ye and F. Neese, *Curr. Opin. Chem. Biol.*, 2007, **11**, 134–141.
39. C. Bo and F. Maseras, *Dalton Trans.*, 2008, **22**, 2911–2919.
40. J. A. G. Lochana and C. Menikarachchi, *Curr. Top. Med. Chem.*, 2010, **10**, 46–54.
41. M. De Vivo, *Front. Biosci.*, 2011, **16**, 1619–1633.
42. A. Lodola and M. De Vivo, *Adv. Protein Chem. Struct. Biol.*, 2012, **87**, 337–362.
43. J. Tomasi and M. Persico, *Chem. Rev.*, 1994, **94**, 2027.
44. M. Orozco and F. J. Luque, *Chem. Rev.*, 2000, **100**, 4187.
45. B. Roux and T. Simonson, *Biophys. Chem.*, 1999, **78**, 1.
46. P. A. Kollman, I. Massova, C. Reyes, B. Kuhn, S. Huo, L. Chong, M. Lee, T. Lee, Y. Duan, W. Wang, O. Donini, P. Cieplak, J. Srinivasan, D. A. Case and T. E. Cheatham, *Acc. Chem. Res.*, 2000, **33**, 889.
47. A. Okur and C. Simmerling, in *Annnual Reports in Computational Chemistry*, ed. D. Spellmeyer, Elsevier, Maryland Heights, 1st edn, 2006, vol. 2, ch. 6, pp. 97–109.
48. A. Warshel, P. K. Sharma, M. Kato and W. W. Parson, *Biochim. Biophys. Acta*, 2006, **1764**, 1647–1676.
49. A. Warshel, *Chem. Phys. Lett.*, 1978, **55**, 454.
50. G. King and A. Warshel, *J. Chem. Phys.*, 1989, **91**, 3647.
51. W. Im, S. Bernèche and B. Roux, *J. Chem. Phys.*, 2001, **114**, 2924.
52. P. Schaefer, D. Ricciardi and Q. Cui, *J. Chem. Phys.*, 2005, **123**, 014905.
53. N. Rega, G. Brancato and V. Barone, *Chem. Phys. Lett.*, 2006, **422**, 367–371.

54. G. Brancato, N. Rega and V. Barone, *J. Chem. Phys.*, 2008, **128**, 144501.
55. V. Barone, R. Improta and N. Rega, *Acc. Chem. Res.*, 2008, **41**, 605–616.
56. J. Tomasi, *Wiley Interdiscip. Rev.: Comput. Mol. Sci.*, 2011, **1**, 855–867.
57. B. Mennucci, *Wiley Interdiscip. Rev.: Comput. Mol. Sci.*, 2012, **2**, 386–404.
58. A. Klamt, *Wiley Interdiscip. Rev.: Comput. Mol. Sci.*, 2011, **1**, 699–709.
59. M. Elstner, D. Porezag, G. Jungnickel, J. Elsner, M. Haugk, T. Frauenheim, S. Suhai and G. Seifert, *Phys. Rev. B: Condens. Matter Mater. Phys.*, 1998, **58**, 7260–7268.
60. Q. Cui and M. Elstner, in *Multi-scale Quantum Models for Biocatalysis*, ed. D. M. York and T.-S. Lee, Springer, Amsterdam, 1st edn, 2009, vol. 7, ch. 7, pp. 173–196.
61. M. Gaus, Q. Cui and M. Elstner, *J. Chem. Theory Comput.*, 2011, **7**, 931–948.
62. C. Ochsenfeld, J. Kussmann and D. S. Lambrecht, in *Reviews in Computational Chemistry*, ed. K. B. Lipkowitz and T. L. Cundari, VCH Publishers, New York, 1st edn, 2007, vol. 23, ch. 1, pp. 1–82.
63. F. Weigend, *Phys. Chem. Chem. Phys.*, 2002, **4**, 4285.
64. C. Hättig, *Phys. Chem. Chem. Phys.*, 2005, **7**, 59.
65. T. B. Pedersen, F. Aquilante and R. Lindh, *Theor. Chem. Acc.*, 2009, **124**, 1.
66. F. Weigend, M. Kattanek and R. Ahlrichs, *J. Chem. Phys.*, 2009, **130**, 164106.
67. S. Saebo and P. Pulay, *Annu. Rev. Phys. Chem.*, 1993, **44**, 213–236.
68. M. Schütz, G. Hetzer and J. H. J. Werner, *J. Chem. Phys.*, 1999, **111**, 5691–5705.
69. M. Schütz, *J. Chem. Phys.*, 2000, **113**, 9986–10001.
70. M. Schütz and H. J. Werner, *J. Chem. Phys.*, 2001, **114**, 661–681.
71. H. J. Werner, F. R. Manby and P. J. Knowles, *J. Chem. Phys.*, 2003, **118**, 8149–8160.
72. M. Schütz, H. J. Werner, R. Lindh and F. R. Manby, *J. Chem. Phys.*, 2004, **121**, 737–750.
73. W. Klopper, F. R. Manby, S. Ten-No and E. F. Valeev, *Int. Rev. Phys. Chem.*, 2006, **25**, 427.
74. C. Hättig, W. Klopper, A. Köhn and D. P. Tew, *Chem. Rev.*, 2012, **112**, 4–74.
75. T. B. Adler, H. J. Werner and F. R. Manby, *J. Chem. Phys.*, 2009, **130**, 054106.
76. R. Lonsdale, J. N. Harvey and A. J. Mulholland, *Chem. Soc. Rev.*, 2012, **41**, 3025–3038.
77. C. D. Sherrill, *J. Chem. Phys.*, 2010, **132**, 110902.
78. R. Zaleśny, M. G. Papadopoulos, P. Mezey and J. Leszczynski, *Linear-Scaling Techniques in Computational Chemistry and Physics Methods and Applications*, Springer, Berlin-Heidelberg, 2011.
79. T. B. Adler and H. J. Werner, *J. Chem. Phys.*, 2011, **135**, 144117.
80. H. J. Werner and M. Schütz, *J. Chem. Phys.*, 2011, **135**, 144116.

81. D. J. Cole, C. K. Skylaris, E. Rajendra, A. R. Venkitaraman and M. C. Payne, *EPL*, 2010, **91**, 37004.
82. A. D. Mackerell, *J. Comput. Chem.*, 2004, **25**, 1584–1604.
83. W. D. Cornell, P. Cieplak, C. I. Bayly, K. M. Gould, K. M. Merz, D. M. Ferguson, G. L. Seibel, U. C. Singh, P. K. Weiner and P. A. Kollmann, *J. Am. Chem. Soc.*, 1995, **117**, 5179.
84. D. A. Case, T. E. Cheatham, T. Darden, H. Gohlke, R. Luo, K. M. Merz, A. Onufriev, C. Simmerling, B. Wang and R. J. Woods, *J. Comput. Chem.*, 2005, **26**, 1668–1688.
85. A. Perez, I. C. P. Marchan, D. Svozil, J. Sponer, T. E. Cheatham, C. A. Laughton and M. Orozco, *Biophys. J.*, 2007, **92**, 3817–3829.
86. C. Oostenbrink, A. Villa, A. E. Mark and W. F. Van Gunsteren, *J. Comput. Chem.*, 2004, **25**, 1656–1676.
87. R. D. Lins and P. H. Hünenberger, *J. Comput. Chem.*, 2005, **26**, 1400–1412.
88. N. Schmid, A. Eichenberger, A. Choutko, S. Riniker, M. Winger, A. Mark and W. van Gunsteren, *Eur. Biophys. J.*, 2011, **40**, 843–856.
89. M. M. Reif, P. H. Hünenberger and C. Oostenbrink, *J. Chem. Theory Comput.*, 2012, **8**, 3705–3723.
90. A. D. MacKerell Jr., C. L. Brooks, L. Nilsson, B. Roux, Y. Won and M. Karplus, in *The Encyclopedia of Computational Chemistry*, ed. P. v. R. Schleyer, N. L. Allinger, T. Clark, J. Gasteiger, P. A. Kollman, H. F. Schaefer, P. R. Schreiner, Wiley, Chichester, 1st edn, 1998, vol. 1, pp. 271–277.
91. A. D. MacKerell, J. J. Wiorkiewicz-Kuczera and M. Karplus, *J. Am. Chem. Soc.*, 2000, **117**, 11946.
92. A. D. MacKerell, N. Banavali and N. Foloppe, *Biopolymers*, 2000, **56**, 257–265.
93. X. Zhu, P. E. M. Lopes and A. D. MacKerell, *Wiley Interdiscip. Rev.: Comput. Mol. Sci.*, 2012, **2**, 167–185.
94. A. D. Mackerell, M. Feig and C. L. Brooks, *J. Comput. Chem.*, 2004, **25**, 1400–1415.
95. W. L. Jorgensen, J. Chandrasekhar, J. D. Madura, R. W. Impey and M. L. Klein, *J. Chem. Phys.*, 1983, **79**, 926–935.
96. W. L. Jorgensen and J. D. Madura, *Mol. Phys.*, 1985, **56**, 1381–1392.
97. K. Kiyohara, K. E. Gubbins and A. Z. Panagiotopoulos, *Mol. Phys.*, 1998, **94**, 803–808.
98. H. J. C. Berendsen, J. R. Grigera and T. P. Straatsma, *J. Phys. Chem.*, 1987, **91**, 6269–6271.
99. K. E. Shaw, C. J. Woods and A. J. Mulholland, *J. Phys. Chem. Lett.*, 2010, **1**, 219–223.
100. J. Gao, *J. Comput. Chem.*, 1997, **18**, 1061–1071.
101. F. Lipparini and V. Barone, *J. Chem. Theory Comput.*, 2011, **7**, 3711–3724.
102. B. T. Thole, *Chem. Phys.*, 1981, **59**, 341–350.
103. G. Lamoureux and B. Roux, *J. Chem. Phys.*, 2003, **119**, 3025–3039.

104. S. W. Rick, S. J. Stuart and B. J. Berne, *J. Chem. Phys.*, 1994, **101**, 6141–6156.
105. P. Liu, E. Harder and B. J. Berne, *J. Phys. Chem. B*, 2005, **109**, 2949–2955.
106. T. Yan, S. Li, W. Jiang, X. Gao, B. Xiang and G. A. Voth, *J. Phys. Chem. B*, 2006, **110**, 1800–1806.
107. D. Bakowies and W. Thiel, *J. Phys. Chem.*, 1996, **100**, 10580.
108. J. P. Perdew, A. Ruzsinszky, J. Tao, V. N. Staroverov, G. E. Scuseria and G. A. Csonka, *J. Chem. Phys.*, 2005, **123**, 062201.
109. G. E. Scuseria and V. N. Staroverov, in *Theory and Applications of Computational Chemistry*, ed. C. E. Dykstra, G. Frenking, K. S. Kim and G. E. Scuseria, Elsevier, Amsterdam, 1st edn., 2005, ch. 24, pp. 669–724.
110. Y. Zhao, N. E. Schultz and D. G. Truhlar, *J. Chem. Theory Comput.*, 2006, **2**, 364–382.
111. Y. Zhao and D. G. Truhlar, *Acc. Chem. Res.*, 2008, **41**, 157–167.
112. G. I. Csonka, J. P. Perdew and A. Ruzsinszky, *J. Chem. Theory Comput.*, 2010, **6**, 3688–3703.
113. K. Burke, *J. Chem. Phys.*, 2012, **136**, 150901.
114. R. Car and M. Parrinello, *Phys. Rev. Lett.*, 1985, **55**, 2471–2474.
115. D. Marx and J. Hutter, *Ab initio Molecular Dynamics: Basic Theory and Advanced Methods*, Cambridge University Press, Cambridge, 2009.
116. ed. W. Andreoni, D. Marx and M. Sprik, *ChemPhysChem*, 2005, **6**, Special issue: Parrinello Festschrift, Wiley-VCH.
117. J. Hutter, *Wiley Interdiscip. Rev.: Comput. Mol. Sci.*, 2012, **2**, 604–612.
118. A. Hassanali, M. K. Prakash, H. Eshet and M. Parrinello, *Proc. Natl. Acad. Sci. U. S. A.*, 2011, **108**, 20410–20415.
119. A. Laio, J. VandeVondele and U. Rothlisberger, *J. Chem. Phys.*, 2002, **116**, 6941–6947.
120. E. Ippoliti, J. Dreyer, P. Carloni and U. Rothlisberger, in *Hierarchical Methods for Dynamics in Complex Molecular Systems*, ed. J. Grotendorst, G. Sutmann, G. Gompper and D. Marx, Forschungszentrum Jülich, 1st edn, 2012, ch. 7, pp. 163–182.
121. W. R. P. Scott, P. H. Hünenberger, I. G. Tironi, A. E. Mark, S. R. Billeter, J. Fennen, A. E. Torda, T. Huber, P. Krüger and W. F. van Gunsteren, *J. Phys. Chem. A*, 1999, **103**, 3596.
122. T. Laino, F. Mohamed, A. Laio and M. Parrinello, *J. Chem. Theory Comput.*, 2005, **1**, 1176–1184.
123. T. Laino, F. Mohamed, A. Laio and M. Parrinello, *J. Chem. Theory Comput.*, 2006, **2**, 1370–1378.
124. J. Hutter, M. Iannuzzi, F. Schiffmann and J. VandeVondele, *WIREs Comput. Mol. Sci*, 2014, **4**, 15–25.
125. M. Born and J. R. Oppenheimer, *Ann. Phys.*, 1927, **84**, 864.
126. P. Hohenberg and W. Kohn, *Phys. Rev.*, 1964, **136**, 864–871.
127. T. A. Wesolowski, *J. Phys. Chem.*, 1993, **97**, 8050.
128. S. G. Louie, S. Froyen and M. L. Cohen, *Phys. Rev. B: Condens. Matter Mater. Phys.*, 1982, **26**, 1738.

129. D. Sebastiani and U. Rothlisberger, in *Medicinal Quantum Chemistry*, ed. P. Carloni and F. Alber, Wiley, Weinheim, 1st edn, 2003, ch. 1, pp. 5–39.
130. W. Kohn and L. J. Sham, *Phys. Rev.*, 1965, **140**, 1133–1138.
131. O. A. von Lilienfeld, I. Tavernelli, U. Rothlisberger and D. Sebastiani, *J. Chem. Phys.*, 2005, **122**, 014113.
132. A. Laio, J. VandeVondele and U. Rothlisberger, *J. Phys. Chem. B*, 2002, **106**, 7300–7307.
133. A. Laio, F. Gervasio, M. Sulpizi and U. Rothlisberger, *J. Phys. Chem.*, 2004, **108**, 7983.
134. D. Yarne, M. E. Tuckerman and G. J. Martyna, *J. Chem. Phys.*, 2001, **115**, 3531.
135. R. W. Hockney, in *Methods in Computational Physics*, 1st edn, Academic Press, New York, 1970, vol. 9, ch. 4, 135–211.
136. G. J. Martyna and M. E. Tuckerman, *J. Chem. Phys.*, 1999, **110**, 2810–2821.
137. P. Hünenberger, *J. Chem. Phys.*, 2000, **113**, 10464.
138. R. Rajamani, K. J. Naidoo and J. L. Gao, *J. Comput. Chem.*, 2003, **24**, 1775.
139. T. K. Woo, P. Margl, P. E. Bloechl and T. Ziegler, *J. Phys. Chem. A*, 2002, **106**, 1173.
140. A. Laio and M. Parrinello, *Proc. Natl. Acad. Sci. U. S. A.*, 2002, **99**, 12562–12566.
141. J. VandeVondele and U. Rothlisberger, *J. Chem. Phys.*, 2000, **113**, 4863.
142. J. VandeVondele and U. Rothlisberger, *J. Am. Chem. Soc.*, 2002, **124**, 8163.
143. L. Guidoni and U. Rothlisberger, *J. Chem. Theory Comput.*, 2005, **1**, 554.
144. S. Grimme, *Wiley Interdiscip. Rev.: Comput. Mol. Sci.*, 2011, **1**, 211–228.
145. O. A. von Lilienfeld, I. Tavernelli, D. Sebastiani and U. Rothlisberger, *Phys. Rev. Lett.*, 2004, **93**, 15300.
146. I. C. Lin, M. D. Coutinho-Neto, C. Felsenheimer, O. A. von Lilienfeld, I. Tavernelli and U. Rothlisberger, *Phys. Rev. B: Condens. Matter Mater. Phys.*, 2007, **75**, 205131.
147. I. C. Lin, A. P. Seitsonen, I. Tavernelli and U. Rothlisberger, *J. Chem. Theory Comput.*, 2012, **8**, 3902–3910.
148. V. Weber, C. Bekas, T. Laino, A. Curioni, A. Bertsch and S. Futral, 28th IEEE International Symposium on Parallel and Distributed Processing, 2014, 735–744.
149. A. D. Becke, *J. Chem. Phys.*, 1993, **98**, 5648.
150. C. T. Lee, W. T. Yang and R. G. Parr, *Phys. Rev. B: Condens. Matter Mater. Phys.*, 1988, **37**, 785.
151. S. H. Vosko, L. Wilk and M. Nusair, *Can. J. Phys.*, 1980, **58**, 1200–1211.
152. P. J. Stephens, F. J. Devlin, C. F. Chabalowski and M. J. Frisch, *J. Phys. Chem.*, 1994, **98**, 11623.
153. R. Peverati and D. G. Truhlar, *J. Phys. Chem. Lett.*, 2011, **2**, 2810.

154. A. Dal Corso, S. Baroni, R. Resta and S. de Gironcoli, *Phys. Rev. B: Condens. Matter Mater. Phys.*, 1993, **47**, 3588–3592.
155. C. Hartwigsen, S. Goedecker and J. Hutter, *Phys. Rev. B: Condens. Matter Mater. Phys.*, 1998, **58**, 3641.
156. X. Assfeld and J. L. Rivail, *Chem. Phys. Lett.*, 1996, **263**, 100.
157. X. Biarnés, A. Ardèvol, A. Planas, C. Rovira, A. Laio and M. Parrinello, *J. Am. Chem. Soc.*, 2007, **129**, 10686.
158. M. Dal Peraro, P. Ruggerone, S. Raugei, F. L. Gervasio and P. Carloni, *Curr. Opin. Struct. Biol.*, 2007, **17**, 149–156.
159. M. Dal Peraro, A. J. Vila, P. Carloni and M. L. Klein, *J. Am. Chem. Soc.*, 2007, **129**, 2808–2816.
160. P. Vidossich, M. Alfonso-Prieto, X. Carpena, I. Fita, P. C. Loewen and C. Rovira, *Arch. Biochem. Biophys.*, 2010, **500**, 37.
161. P. Vidossich, G. Fiorin, M. Alfonso-Prieto, E. Derat, S. Shaik and C. Rovira, *J. Phys. Chem. B*, 2010, **114**, 5161–5169.
162. M. Dal Peraro, A. J. Vila and P. Carloni, in *Quantum Biochemistry*, ed. C. F. Matta, Wiley, Verlag GmbH & Co. KGaA, Weinheim, 1st edn, 2010, ch. 21, pp. 605–622.
163. J. Sgrignani, A. Magistrato, M. Dal Peraro, A. Vila, P. Carloni and R. Pierattelli, *J. Comput. -Aided Mol. Des.*, 2012, **26**, 425–435.
164. G. Palermo, A. Cavalli, M. L. Klein, M. Alfonso-Prieto, M. D. Peraro and M. D. Vivo, *Acc. Chem. Res.*, 2015, **48**, 220–228.
165. L. Guidoni and P. Carloni, *J. Recept. Signal Transduct.*, 2002, **22**, 315–331.
166. D. Bucher, S. Raugei, L. Guidoni, M. Dal Peraro, U. Rothlisberger, P. Carloni and M. L. Klein, *Biophys. Chem.*, 2006, **124**, 292–301.
167. D. Bucher, L. Guidoni, P. Carloni and U. Rothlisberger, *Biophys. J.*, 2010, **98**, 47–49.
168. V. Calandrini, J. Dreyer, E. Ippoliti and P. Carloni, *J. Chem. Phys.*, 2014, 141.
169. E. Brunk, N. Ashari, P. Athri, P. Campomanes, F. F. de Carvalho, B. F. E. Curchod, P. Diamantis, M. Doemer, J. Garrec, A. Laktionov, M. Micciarelli, M. Neri, G. Palermo, T. J. Penfold, S. Vanni, I. Tavernelli and U. Rothlisberger, *Chimia*, 2011, **65**, 667–671.
170. F. L. Gervasio, A. Laio, M. Iannuzzi and M. Parrinello, *Chem. Eur. - J.*, 2004, **10**, 4846.
171. M. Boero, F. L. Gervasio and M. Parrinello, *Mol. Simul.*, 2007, **33**, 57.
172. Y. A. Mantz, F. L. Gervasio, T. Laino and M. Parrinello, *Phys. Rev. Lett.*, 2007, **99**, 058104.
173. E. Brunk, J. S. Arey and U. Rothlisberger, *J. Am. Chem. Soc.*, 2012, **134**, 8608–8616.
174. B. Mollwitz, E. Brunk, S. Schmitt, F. Pojer, M. Bannwarth, M. Schiltz, U. Rothlisberger and K. Johnsson, *Biochemistry*, 2012, **51**, 986–994.
175. E. Brunk, B. Mollwitz and U. Rothlisberger, *ChemBioChem*, 2013, **14**, 703–710.
176. C. Gossens, I. Tavernelli and U. Rothlisberger, *Chimia*, 2005, **59**, 81–84.

177. A. Magistrato, P. Ruggerone, K. Spiegel, P. Carloni and J. Reedijk, *J. Phys. Chem. B*, 2006, **110**, 3604–3613.
178. K. Spiegel and A. Magistrato, *Org. Biomol. Chem.*, 2006, **4**, 2507–2517.
179. K. Spiegel, A. Magistrato, P. Maurer, P. Ruggerone, U. Rothlisberger, P. Carloni, J. Reedijk and M. L. Klein, *J. Comput. Chem.*, 2008, **29**, 38–49.
180. C. Gossens, I. Tavernelli and U. Rothlisberger, *J. Phys. Chem. A*, 2009, **113**, 11888–11897.
181. T. H. Nguyen, F. Arnesano, S. Scintilla, G. Rossetti, E. Ippoliti, P. Carloni and G. Natile, *J. Chem. Theory Comput.*, 2012, **8**, 2912–2920.
182. V. Calandrini, F. Arnesano, A. Galliani, T. H. Nguyen, E. Ippoliti, P. Carloni and G. Natile, *Dalton Trans.*, 2014, **43**, 12085–12094.
183. V. Calandrini, T. H. Nguyen, F. Arnesano, A. Galliani, E. Ippoliti, P. Carloni and G. Natile, *Chem. Eur. – J.*, 2014, **20**, 11719–11725.
184. A. Lodola and M. D. Vivo, in *Structural and Mechanistic EnzymologyBringing Together Experiments and Computing*, ed. C. Christov and T. Karabencheva-Christova, Academic Press, 1st edn, 2012, vol. 87, pp. 337–362.
185. A. Khandelwal, V. Lukacova, D. Comez, D. M. Kroll, S. Raha and S. Balaz, *J. Med. Chem.*, 2005, **48**, 5437–5447.
186. F. Gräter, S. M. Schwarzl, A. Dejaegere, S. Fischer and J. C. Smith, *J. Phys. Chem. B*, 2005, **109**, 10474–10483.
187. K. Raha, J. Kenneth and M. Merz, *J. Am. Chem. Soc.*, 2004, **126**, 1020–1021.
188. Z. Yang, Y. Liu, Z. Chen, Z. Xu, J. Shi, K. Chen and W. Zhu, *J. Mol. Model.*, 2015, 21.
189. V. Gogonea, D. Suárez, A. van der Vaart and K. M. M. Merz Jr, *Curr. Opin. Struct. Biol.*, 2001, **11**, 217–223.
190. J. Khandogin and D. M. York, *Proteins: Struct., Funct., Bioinf.*, 2004, **56**, 724–737.
191. A. Isvoran, D. Craciun, V. Martiny, O. Sperandio and M. Miteva, *BMC Pharmacol. Toxicol.*, 2013, **14**, 31.
192. M. Dal Peraro, A. Vila and P. Carloni, *J. Biol. Inorg. Chem.*, 2002, **7**, 704–712.
193. M. Dal Peraro, A. J. Vila and P. Carloni, *lnorg. Chem.*, 2003, **42**, 4245–4247.
194. A. Cavalli, M. De Vivo and M. Recanatini, *Chem. Commun.*, 2003, 1308–1309.
195. A. Cavalli, G. Bottegoni, C. Raco, M. D. Vivo and M. Recanatini, *J. Med. Chem.*, 2004, **47**, 3991–3999.
196. A. Cavalli, A. E. Prota, T. Stehle, T. S. Dermody, M. Recanatini, G. Folkers and L. Scapozza, *Biophys. J.*, 2004, **86**, 3423–3431.
197. M. De Vivo, A. Cavalli, P. Carloni and M. Recanatini, *Chem. Eur. – J.*, 2007, **13**, 8437–8444.
198. M. De Vivo, B. Ensing and M. L. Klein, *J. Am. Chem. Soc.*, 2005, **127**, 11226–11227.

199. M.-H. Ho, M. D. Vivo, M. D. Peraro and M. L. Klein, *J. Am. Chem. Soc.*, 2010, **132**, 13702–13712.
200. G. Palermo, D. Branduardi, M. Masetti, A. Lodola, M. Mor, D. Piomelli, A. Cavalli and M. D. Vivo, *J. Med. Chem.*, 2011, **54**, 6612–6623.
201. G. Palermo, U. Rothlisberger, A. Cavalli and M. D. Vivo, *Eur. J. Med. Chem.*, 2015, **91**, 15–26.
202. V. L. Schramm, *Arch. Biochem. Biophys.*, 2005, **433**, 13–26.
203. V. L. Schramm, *Annu. Rev. Biochem.*, 2011, **80**, 703–732.
204. J. P. Davis, G. A. Cain, W. J. Pitts, R. L. Magolda and R. A. Copeland, *Biochemistry*, 1996, **35**, 1270–1273.
205. M. De Vivo, M. Dal Peraro and M. L. Klein, *J. Am. Chem. Soc.*, 2008, **130**, 10955–10962.
206. T. A. Steitz and J. A. Steitz, *Proc. Natl. Acad. Sci. U. S. A.*, 1993, **90**, 6498–6502.
207. G. Ciccotti, M. Ferrario, J. T. Hynes and R. Kapral, *Chem. Phys.*, 1989, **129**, 241–251.
208. M. C. Starnes and Y. C. Cheng, *J. Biol. Chem.*, 1989, **264**, 7073–7077.
209. E. R. Goedken and S. Marqusee, *J. Mol. Biol.*, 2001, **314**, 863–871.
210. M. M. De Vivo, B. Ensing, M. Dal Peraro, G. A. Gomez, D. W. Christianson and M. L. Klein, *J. Am. Chem. Soc.*, 2007, **129**, 387–394.
211. M. F. Moghaddam, D. F. Grant, J. M. Cheek, J. F. Greene, K. C. Williamson and B. D. Hammock, *Nat. Med.*, 1997, **3**, 562–566.
212. K. Node, Y. Huo, X. Ruan, B. Yang, M. Spiecker, K. Ley, D. C. Zeldin and J. K. Liao, *Science*, 1999, **285**, 1276–1279.
213. K. R. Schmelzer, L. Kubala, J. W. Newman, I.-H. Kim, J. P. Eiserich and B. D. Hammock, *Proc. Natl. Acad. Sci. U. S. A.*, 2005, **102**, 9772–9777.
214. C. Morisseau and B. D. Hammock, *Annu. Rev. Pharmacol. Toxicol.*, 2013, **53**, 37–58.
215. G. Palermo, M. Stenta, A. Cavalli, M. D. Peraro and M. D. Vivo, *J. Chem. Theory Comput.*, 2013, **9**, 857–862.
216. B. H. Schmidt, A. B. Burgin, J. E. Deweese, N. Osheroff and J. M. Berger, *Nature*, 2010, **465**, 641–644.
217. M. Nowotny, S. A. Gaidamakov, R. J. Crouch and W. Yang, *Cell*, 2005, **121**, 1005–1016.
218. V. Genna, R. Gaspari, M. Dal Peraro and M. De Vivo, *Nucleic Acids Res.*, 2016, **44**, 2827–2836.
219. G. Palermo, I. Bauer, P. Campomanes, A. Cavalli, A. Armirotti, S. Girotto, U. Rothlisberger and M. De Vivo, *PLoS Comput. Biol.*, 2015, **11**, e1004231.
220. S. H. Ackerman and D. L. Gatti, *PLoS One*, 2013, **8**, e55136.
221. M. Dal Peraro, L. I. Llarrull, U. Rothlisberger, A. J. Vila and P. Carloni, *J. Am. Chem. Soc.*, 2004, **126**, 12661–12668.
222. G. Palermo, P. Campomanes, M. Neri, D. Piomelli, A. Cavalli, U. Rothlisberger and M. De Vivo, *J. Chem. Theory Comput.*, 2013, **9**, 1202–1213.

223. G. Palermo, P. Campomanes, A. Cavalli, U. Rothlisberger and M. De Vivo, *J. Phys. Chem. B*, 2015, **119**, 789–801.
224. H. L. Woodcock, M. Hodoscek, A. T. B. Gilbert, P. M. W. Gill, H. F. Schaefer and B. R. Brooks, *J. Comput. Chem.*, 2007, **28**, 1485.
225. M. J. Loferer, H. H. Loeffler and K. R. Liedl, *J. Comput. Chem.*, 2003, **24**, 1240.
226. H. Hu and W. Yang, *Annu. Rev. Phys. Chem.*, 2008, **59**, 573–601.
227. J. D. McGeagh and A. J. Mulholland, in *Kinetics and Dynamics*, ed. P. Paneth and A. Dybala-Defratyka, Springer, Netherlands, 1st edn, 2010, vol. 12, ch. 13, pp. 375–395.
228. S. Ferrer, J. Ruiz-Pernía, S. Martí, V. Moliner, I. T. Tuñón, J. Bertrán and J. Andrés, *Adv. Protein Chem. Struct. Biol.*, 2011, **85**, 81–142.
229. D. R. Glowacki, J. N. Harvey and A. J. Mulholland, *Biochem. Soc. Trans.*, 2012, **40**, 515–521.
230. D. R. Glowacki, J. N. Harvey and A. J. Mulholland, *Nat. Chem.*, 2012, **4**, 169–176.
231. F. Claeyssens, J. N. Harvey, F. R. Manby, R. A. Mata, A. J. Mulholland, K. E. Ranaghan, M. Schütz, S. Thiel, W. Thiel and H. J. Werner, *Angew. Chem.*, 2006, **118**, 7010–7013.
232. R. A. Mata, H. J. Werner, S. Thiel and W. Thiel, *J. Chem. Phys.*, 2008, **128**, 025104.
233. M. W. van der Kamp, J. Żurek, F. R. Manby, J. N. Harvey and A. J. Mulholland, *J. Phys. Chem. B*, 2010, **114**, 11303–11314.
234. R. Lonsdale, J. N. Harvey and A. J. Mulholland, *J. Chem. Theory Comput.*, 2012, **8**, 4637–4645.
235. R. Lonsdale, K. T. Houghton, J. Zurek, C. M. Bathelt, N. Foloppe, M. J. de Groot, J. N. Harvey and A. J. Mulholland, *J. Am. Chem. Soc.*, 2013, **135**, 8001–8015.
236. R. Lonsdale, S. Hoyle, D. T. Grey, L. Ridder and A. J. Mulholland, *Biochemistry*, 2012, **51**, 1774–1786.
237. R. Lonsdale, J. N. Harvey and A. J. Mulholland, *J. Phys. Chem. Lett.*, 2010, **1**, 3232–3237.
238. S. Grimme, J. Antony, S. Ehrlich and H. Krieg, *J. Chem. Phys.*, 2010, **132**, 154104.
239. S. Ehrlich, J. Moellmann and S. Grimme, *Acc. Chem. Res.*, 2012, **46**, 916–926.
240. A. D. Becke and E. R. J. Johnson, *J. Chem. Phys.*, 2007, **127**, 124108.
241. V. Barone, *Computational Strategy for Spectroscopy: From Small Molecules to Nano Systems*, Wiley, New York, 2011.
242. U. F. Röhrig, L. Guidoni, A. Laio, I. Frank and U. Rothlisberger, *J. Am. Chem. Soc.*, 2004, **126**, 15328–15329.
243. A. M. Conte, E. Ippoliti, R. Del Sole, P. Carloni and O. Pulci, *Phys. Status Solidi B*, 2010, **247**, 1920–1924.
244. D. Jacquemin, B. Mennucci and C. Adamo, *Phys. Chem. Chem. Phys.*, 2011, **13**, 16987–16998.
245. M. Cascella, M. A. Cuendet, I. Tavernelli and U. Rothlisberger, *J. Phys. Chem. B*, 2007, **111**, 10248–10252.

246. E. Cannuccia, O. Pulci, R. D. Sole and M. Cascella, *Chem. Phys.*, 2011, **389**, 35–38.
247. Q. Sun, M. Doerr, Z. Li, S. C. Smith and W. Thiel, *Phys. Chem. Chem. Phys.*, 2010, **12**, 2450–2458.
248. Q. Sun, Z. Li, Z. Lan, C. Pfisterer, M. Doerr, S. Fischer, S. C. Smith and W. Thiel, *Phys. Chem. Chem. Phys.*, 2012, **14**, 11413–11424.
249. C. Filippi, F. Buda, L. Guidoni and A. Sinicropi, *J. Chem. Theory Comput.*, 2012, **8**, 112–124.
250. N. A. Murugan, J. Kongsted, Z. Rinkevicius and H. Ågren, *Phys. Chem. Chem. Phys.*, 2012, **14**, 1107–1112.
251. O. Valsson, P. Campomanes, I. Tavernelli, U. Rothlisberger and C. Filippi, *J. Chem. Theory Comput.*, 2013, **9**, 2441–2454.
252. M. B. Goncalves, J. Dreyer, P. Lupieri, C. Barrera-Patino, E. Ippoliti, M. R. Webb, J. E. T. Corrie and P. Carloni, *Phys. Chem. Chem. Phys.*, 2013, **15**, 2177–2183.
253. M. Olivucci, *Computational Photochemistry*, Elsevier, Amsterdam, 2005.
254. D. Roca-Sanjuán, F. Aquilante and R. Lindh, *Wiley Interdiscip. Rev.: Comput. Mol. Sci.*, 2011, **2**, 585–603.
255. I. Schapiro, F. Melaccio, E. N. Laricheva and M. Olivucci, *Photochem. Photobiol. Sci.*, 2011, **10**, 867–886.
256. M. Boggio-Pasqua, C. F. Burmeister, M. A. Robb and G. Groenhof, *Phys. Chem. Chem. Phys.*, 2012, **14**, 7912–7928.
257. G. Onida, L. Reining and A. Rubio, *Rev. Mod. Phys.*, 2002, **74**, 601–659.
258. Y. Ma, M. Rohlfing and C. Molteni, *J. Chem. Theory Comput.*, 2010, **6**, 257–265.
259. D. Rocca, D. Lu and G. Galli, *J. Chem. Phys.*, 2010, **133**, 164109.
260. L. González, D. Escudero and L. Serrano-Andrés, *ChemPhysChem*, 2012, **13**, 28–51.
261. M. Guglielmi, M. Doemer, I. Tavernelli and U. Rothlisberger, *Faraday Discuss.*, 2013, **163**, 189–203.
262. B. Lasorne, G. A. Worth and M. A. Robb, *Wiley Interdiscip. Rev.: Comput. Mol. Sci.*, 2011, **1**, 460–475.
263. U. F. Röhrig, I. Frank, J. Hutter and U. Rothlisberger, *ChemPhysChem*, 2003, **4**, 1177.
264. U. F. Röhrig, L. Guidoni and U. Rothlisberger, *ChemPhysChem*, 2005, **6**, 1836–1847.
265. E. Tapavicza, I. Tavernelli and U. Rothlisberger, *Phys. Rev. Lett.*, 2007, **98**, 023001.
266. I. Tavernelli, B. F. E. Curchod and U. Rothlisberger, *Chem. Phys.*, 2011, **391**, 101–109.
267. C. M. Isborn, A. W. Götz, M. A. Clark, R. C. Walker and T. J. Martínez, *J. Chem. Theory Comput.*, 2012, **8**, 5092–5106.
268. L. Spörkel, G. Cui and W. Thiel, *J. Phys. Chem. A*, 2013, **117**, 4574–4583.
269. E. A. Briggs, N. A. Besley and D. Robinson, *J. Phys. Chem. A*, 2013, **117**, 2644–2650.
270. M. E. Casida, *J. Mol. Struct.: THEOCHEM*, 2009, **914**, 3–18.

271. M. Sulpizi, P. Carloni, J. Hutter and U. Rothlisberger, *Phys. Chem. Chem. Phys.*, 2003, **5**, 4798–4805.
272. M. Sulpizi, U. F. Röhrig, J. Hutter and U. Rothlisberger, *Int. J. Quantum Chem.*, 2005, **101**, 671–682.
273. M. E. Moret, E. Tapavicza, L. Guidoni, U. F. Röhrig, M. Sulpizi, I. Tavernelli and U. Rothlisberger, *Chimia*, 2005, **59**, 493–498.
274. N. A. Murugan and H. Ågren, *J. Phys. Chem. A*, 2009, **113**, 2572–2577.
275. P. Lupieri, E. Ippoliti, P. Altoè, M. Garavelli, M. Mwalaba and P. Carloni, *J. Chem. Theory Comput.*, 2010, **6**, 3403–3409.
276. E. Tapavicza, I. Tavernelli, U. Rothlisberger, C. Filippi and M. E. Casida, *J. Chem. Phys.*, 2008, **129**, 124108.
277. M. Wohlgemuth, V. Bonačić-Koutecký and R. Mitrić, *J. Chem. Phys.*, 2011, **135**, 054105.
278. M. Guglielmi, M. Doemer, I. Tavernelli and U. Rothlisberger, *Faraday Discuss.*, 2013, **163**, 189.
279. B. F. E. Curchod, U. Rothlisberger and I. Tavernelli, *ChemPhysChem*, 2013, **14**, 1314–1340.
280. A. J. Cohen, P. Mori-Sánchez and W. Yang, *Chem. Rev.*, 2012, **112**, 289–320.
281. S. Grimme and M. Parac, *ChemPhysChem*, 2003, **4**, 292–295.
282. T. Ziegler and M. Krykunov, *J. Chem. Phys.*, 2010, **133**, 074104.
283. L. Goerigk and S. Grimme, *J. Chem. Phys.*, 2010, **132**, 184103.
284. J. D. Chai and M. Head-Gordon, *J. Chem. Phys.*, 2008, **128**, 084106.
285. R. Baer, E. Livshits and U. Salzner, *Annu. Rev. Phys. Chem.*, 2010, **61**, 85–109.
286. N. Kuritz, T. Stein, R. Baer and L. Kronik, *J. Chem. Theory Comput.*, 2011, **7**, 2408–2415.
287. Y.-S. Lin, G.-D. Li, S.-P. Mao and J.-D. Chai, *J. Chem. Theory Comput.*, 2013, **9**, 263–272.
288. B. O. Roos, in *Advances in Chemical Physics; Ab Initio Methods in Quantum Chemistry – II*, ed. K. P. Lawley, Wiley, Chichester, England, 1st edn, 1987, ch. 7, pp. 399–445.
289. B. O. Roos, *Acc. Chem. Res.*, 1999, **32**, 137–144.
290. P. Pulay, *Int. J. Quantum Chem.*, 2011, **111**, 3273–3279.
291. K. Andersson, P. Å. Malmqvist and B. O. Roos, *J. Chem. Phys.*, 1992, **96**, 1218–1226.
292. G. Groenhof, M. Bouxin-Cademartory, B. Hess, S. P. de Visser, H. J. C. Berendsen, M. Olivucci, A. E. Mark and M. A. Robb, *J. Am. Chem. Soc.*, 2004, **126**, 4228–4233.
293. A. Sinicropi, R. Basosi and M. Olivucci, *J. Phys.*, 2008, **101**, 012001.
294. I. Schapiro, M. N. Ryazantsev, W. J. Ding, M. M. Huntress, F. Melaccio, T. Andruniow and M. Olivucci, *Aust. J. Chem.*, 2010, **63**, 413–429.
295. I. Schapiro, M. N. Ryazantsev, L. M. Frutos, N. Ferré, R. Lindh and M. Olivucci, *J. Am. Chem. Soc.*, 2011, **133**, 3354–3364.
296. L. M. Frutos, T. Andruniów, F. Santoro, N. Ferré and M. Olivucci, *Proc. Natl. Acad. Sci. U. S. A.*, 2007, **104**, 7764–7769.

297. F. Melaccio, N. Ferre and M. Olivucci, *Phys. Chem. Chem. Phys.*, 2012, **14**, 12485–12495.
298. S. Gozem, I. Schapiro, N. Ferré and M. Olivucci, *Science*, 2012, **337**, 1225–1228.
299. L. V. Schaefer, G. Groenhof, M. Boggio-Pasqua, M. A. Robb and H. Grubmüller, *PLoS Comput. Biol.*, 2008, **4**, e1000034.
300. D. Yarkony and W. Domcke, *Conical Intersections: Electronic Structure Dynamics and Spectroscopy*, World Scientific, Singapore, 2004.
301. A. G. Worth, M. A. Robb and B. Lasorne, *Mol. Phys.*, 2008, **106**, 2077.
302. T. Takahashi, J. Sugiura and K. Nagayama, *J. Chem. Phys.*, 2002, **116**, 8232.
303. M. S. Lee, J. F. R. Salsbury and M. A. Olson, *J. Comput. Chem.*, 2004, **25**, 1967.
304. G. Brancato, A. Di Nola, V. Barone and A. Amadei, *J. Chem. Phys.*, 2005, **122**, 154109.
305. G. Brancato, N. Rega and V. Barone, *J. Chem. Phys.*, 2006, **124**, 214505.
306. G. Petraglio, M. Ceccarelli and M. Parrinello, *J. Chem. Phys.*, 2005, **123**, 044103.
307. V. Makarov, B. M. Pettitt and M. Feig, *Acc. Chem. Res.*, 2002, **35**, 376–384.
308. M. S. Cheung, A. E. García and J. N. Onuchic, *Proc. Natl. Acad. Sci. U. S. A.*, 2002, **99**, 685–690.
309. S. Chalmet and M. F. Ruiz-López, *J. Chem. Phys.*, 2001, **115**, 5220–5227.
310. J. Gao, F. J. Luque and M. Orozco, *J. Chem. Phys.*, 1993, **98**, 2975–2982.
311. J. Guillemoles, V. Barone, L. Joubert and C. Adamo, *J. Phys. Chem. A*, 2002, **106**, 11354–11360.
312. R. Improta and V. Barone, *J. Am. Chem. Soc.*, 2004, **126**, 14320–14321.
313. M. Pavone, G. Brancato, G. Morelli and V. Barone, *ChemPhysChem*, 2006, **7**, 148–156.
314. G. Brancato, N. Rega and V. Barone, *J. Chem. Phys.*, 2006, **125**, 164515.
315. C. Adamo and V. Barone, *J. Chem. Phys.*, 1999, **110**, 6158.
316. M. Ernzerhof and G. E. Scuseria, *J. Chem. Phys.*, 1999, **110**, 5029–5036.
317. P. E. Smith and B. M. Pettitt, *J. Chem. Phys.*, 1991, **95**, 8430.
318. P. E. Smith, H. D. Blatt and B. M. Pettitt, *J. Phys. Chem. B*, 1997, **101**, 3886.
319. P. H. Hunenberger and J. A. McCammon, *J. Chem. Phys.*, 1999, **110**, 1856.
320. M. Bergdorf, C. Peter and P. H. Hunenberger, *J. Chem. Phys.*, 2003, **119**, 9129.
321. S. Bogusz, T. E. Cheatham and B. R. Brooks, *J. Chem. Phys.*, 1998, **108**, 7070.
322. S. Sakane, H. S. Ashbaugh and R. H. Wood, *J. Phys. Chem. B*, 1998, **102**, 5673.
323. G. Hummer, L. R. Pratt and A. E. Garcìa, *J. Chem. Phys.*, 1997, **107**, 9275–9277.
324. M. A. Kastenholz and P. H. Hünenberger, *J. Chem. Phys.*, 2006, **124**, 124108.

325. A. Y. Lozovoi, A. Alavi, J. Kohanoff and R. M. Lynden-Bell, *J. Chem. Phys.*, 2001, **115**, 1661.
326. C. L. Brooks and M. Karplus, *J. Chem. Phys.*, 1983, **79**, 6312–6325.
327. D. Beglov and B. Roux, *J. Chem. Phys.*, 1994, **100**, 9050–9063.
328. J. W. Essex and W. L. Jorgensen, *J. Comput. Chem.*, 1995, **16**, 951–972.
329. P. Attard, *Mol. Phys.*, 2006, **104**, 1951–1960.
330. G. Brancato, N. Rega and V. Barone, *J. Am. Chem. Soc.*, 2007, **129**, 15380–15390.
331. G. Brancato, N. Rega and V. Barone, *Chem. Phys. Lett.*, 2010, **500**, 104–110.
332. G. Brancato, N. Rega and V. Barone, *Phys. Chem. Chem. Phys.*, 2010, **12**, 10736–10739.
333. G. Brancato, N. Rega and V. Barone, *Chem. Phys. Lett.*, 2008, **451**, 53–57.
334. G. Brancato, N. Rega, M. Causá and V. Barone, *Theor. Chem. Acc.*, 2008, **120**, 499–506.
335. N. Rega, G. Brancato, A. Petrone, P. Caruso and V. Barone, *J. Chem. Phys.*, 2011, **134**, 074504.
336. G. Brancato, N. Rega and V. Barone, *Chem. Phys. Lett.*, 2008, **453**, 202–206.
337. V. Barone, M. Biczysko and G. Brancato, in *Advances in Quantum Chemistry*, ed. R. S. John and E. Brändas, Academic Press, 1st edn, 2010, vol. 59, ch. 2, pp. 17–57.
338. G. Brancato and N. Rega, in *Computational Strategies for Spectroscopy*, ed. V. Barone, Wiley, 1st edn, 2011, ch. 11, pp. 517–547.
339. J. Tomasi, B. Mennucci and R. Cammi, *Chem. Rev.*, 2005, **105**, 2999–3094.
340. A. Ben-Naim, *Solvation Thermodynamics*, Plenum Press, New York, 1987.
341. F. M. Floris and J. Tomasi, *J. Comput. Chem.*, 1989, **10**, 616.
342. F. M. Floris, J. Tomasi and J. L. Pascual-Ahuir, *J. Comput. Chem.*, 1991, **12**, 784.
343. J. D. Thompson, C. J. Cramer and D. G. Truhlar, *Theor. Chem. Acc.*, 2005, **113**, 107.
344. C. P. Kelly, C. J. Cramer and D. G. Truhlar, *J. Chem. Theory Comput.*, 2005, **1**, 1133.
345. I. Tuñón, M. F. Ruiz-López, D. Rinaldi and J. Bertrán, *J. Comput. Chem.*, 1996, **17**, 148.
346. C. Curutchet, M. Orozco and F. J. Luque, *J. Comput. Chem.*, 2001, **22**, 1180.
347. C. Curutchet, A. Bidon-Chanal, I. Soteras, M. Orozco and F. J. Luque, *J. Phys. Chem. B*, 2005, **109**, 3565.
348. A. Klamt, V. Jonas, T. Bürger and J. C. W. Lohrenz, *J. Phys. Chem. A*, 1998, **102**, 5074.
349. S. Dapprich, I. Komaromi, K. S. Byun, K. Morokuma and M. J. Frisch, *J. Mol. Struct.: THEOCHEM*, 1999, **462**, 1.
350. T. Vreven, K. Morokuma, O. Farkas, H. B. Schlegel and M. J. Frisch, *J. Comput. Chem.*, 2003, **24**, 760.

351. N. Rega, S. S. Iyengar, G. A. Voth, H. B. Schlegel, T. Vreven and M. J. Frisch, *J. Phys. Chem. B*, 2004, **108**, 4210.
352. A. Klamt and G. Schürmann, *J. Chem. Soc., Perkin Trans. 2*, 1993, **2**, 799–805.
353. V. Barone and M. Cossi, *J. Phys. Chem. A*, 1998, **102**, 1995–2001.
354. M. Cossi, N. Rega, G. Scalmani and V. Barone, *J. Comput. Chem.*, 2003, **24**, 669–681.
355. H. B. Schlegel, J. M. Millam, S. S. Iyengar, G. A. Voth, A. D. Daniels, G. E. Scuseria and M. J. Frisch, *J. Chem. Phys.*, 2001, **114**, 9758–9763.
356. S. S. Iyengar, H. B. Schlegel, J. M. Millam, G. A. Voth, G. E. Scuseria and M. J. Frisch, *J. Chem. Phys.*, 2001, **115**, 10291.
357. H. B. Schlegel, S. S. Iyengar, X. Li, J. M. Millam, G. A. Voth, G. E. Scuseria and M. J. Frisch, *J. Chem. Phys.*, 2002, **117**, 8694–8704.
358. D. Marx and J. Hutter, in *Modern Methods and Algorithms of Quantum Chemistry*, ed. J. Grotendorst, John vonNeumann Institute for Computing, Julich, 1st edn, 2000, vol. 1, pp. 301–449.
359. C. Houriez, N. Ferrée, D. Siri and M. Masella, *J. Phys. Chem. B*, 2009, **113**, 15047–15056.
360. P. Cimino, A. Pedone, E. Stendardo and V. Barone, *Phys. Chem. Chem. Phys.*, 2010, **12**, 3741–3746.
361. S. Grubišić, G. Brancato, A. Pedone and V. Barone, *Phys. Chem. Chem. Phys.*, 2012, **14**, 15308–15320.
362. M. A. O'Neill and J. K. Barton, in *Charge Transfer in DNA: From Mechanism to Application*, ed. H. A. Wagenknecht, Wiley-VCH, New York, 1st edn, 2005, vol. 1, p. 27.
363. A. J. Bertinchamps, J. Hüttermann, W. Köhnlein and R. Teoule, *Effects of Ionizing Radiation on DNA: Physical, Chemical and Biological Aspects*, Springer-Verlag, Berlin, 1978.
364. C. von Sonntag, *The Chemical Basis of Radiation Biology*, Taylor and Francis, London, 1987.
365. D. K. Ghosh and D. H. Wiffen, *J. Chem. Soc.*, 1960, 1869–1873.
366. J. R. Morton, *J. Am. Chem. Soc.*, 1964, **86**, 2325.
367. P. Neta and R. W. Fessenden, *J. Phys. Chem.*, 1971, **75**, 738–748.
368. A. Sanderud and E. Sagstuen, *J. Phys. Chem. B*, 1998, **102**, 9353–9361.
369. N. Rega, M. Cossi and V. Barone, *J. Am. Chem. Soc.*, 1997, **119**, 12962–12967.
370. V. Barone, C. Adamo, A. Grand, Y. Brunel, M. Fontecave and R. Subra, *J. Am. Chem. Soc.*, 1995, **117**, 1083.
371. G. P. F. Wood, D. Moran, R. Jacob and L. Radom, *J. Phys. Chem. A*, 2005, **119**, 6318–6325.
372. S. Kaprzac, R. Reviakine and M. Kaupp, *J. Phys. Chem. B*, 2007, **111**, 811.
373. N. Rega, M. Cossi and V. Barone, *J. Am. Chem. Soc.*, 1998, **120**, 5723–5732.
374. E. Pauwels, V. V. Speybroeck and M. Waroquier, *J. Phys. Chem. A*, 2004, **108**, 11321–11332.

375. D. K. Ghosh and D. H. Whiffen, *Mol. Phys.*, 1959, **2**, 285.
376. R. l. Bachorz, A. W. Klopper, M. Gutowski, X. Li and K. H. Bowen, *J. Chem. Phys.*, 2008, **129**, 054309–054310.
377. M. Haranczyk, M. Gutowski and A. Warshel, *Phys. Chem. Chem. Phys.*, 2008, **10**, 4442–4448.
378. R. l. Bachorz, A. J. Rak and M. Gutowski, *Phys. Chem. Chem. Phys.*, 2005, **7**, 2116–2125.

CHAPTER 10

Nuclear Quantum Effects in Enzymatic Reactions

DAN THOMAS MAJOR,* REUVEN EITAN, SUSANT DAS, ANIL MHASHAL AND VIJAY SINGH

Department of Chemistry and the Lise Meitner-Minerva Center of Computational Quantum Chemistry and the Institute for Nanotechnology and Advanced Materials, Bar-Ilan University, Ramat-Gan 52900, Israel
*Email: majort@biu.ac.il

10.1 Introduction

10.1.1 Enzymes – the Par Excellence Catalysts of Nature

Enzymes are the par excellence catalysts known to mankind, and constitute the molecular factories in Nature. Indeed, life as we know it would not be possible without enzymes. Enzymes are present in all organisms, and perform an astonishing range of different chemistries, ranging from transfer of single atoms to multistep synthesis of highly complex molecules with numerous stereochemical centers.[1] These molecular machines enhance reaction rates by tens of orders of magnitude, when compared to the analogous reactions in aqueous solution, reaching reaction rates close to the bimolecular diffusion limit.[2,3] Indeed, some enzymes have evolved to a level of perfection, where non-chemical steps, such as conformational changes, are rate limiting and not bond shuffling.[4,5]

Enzymes draw on all available physical means in order to reduce the free energy barrier of chemical reactions. Perhaps the primary catalytic strategy adopted by enzymes is preorganisation of the active site.[6–9] According to this

RSC Theoretical and Computational Chemistry Series No. 9
Simulating Enzyme Reactivity: Computational Methods in Enzyme Catalysis
Edited by Iñaki Tuñón and Vicent Moliner

Published by the Royal Society of Chemistry, www.rsc.org

mechanism, the uncatalysed reaction in solution requires significant reorganisation of the solvent molecules in order to reach a reactive state from which chemistry can occur. Reaching such a state is a rare event, and hence is slow. In contrast, in an enzyme, the polar protein environment is preorganised in a reactive state, from which the reaction can occur, and hence is much faster than in the absence of the catalyst.[10] However, enzymes are multifaceted molecular constructs and may employ additional catalytic strategies, such as reactant destabilisation,[11] desolvation,[12] covalent bonding,[13] dynamics effects,[14,15] chemical control,[16,17] and even quantum mechanical effects.[18,19] This latter effect is the focus of this chapter. However, prior to embarking a description of such nuclear quantum effects (NQEs) and related methods, we will briefly describe a prerequisite for performing meaningful quantum simulations, namely an accurate potential energy surface (PES) and some basic simulation approaches.[20]

10.1.2 Enzyme Simulations using Hybrid PESs

The most basic requirement when modelling enzyme reactions is an accurate PES. An accurate PES should correctly describe the physical interactions in the biomolecular system, which typically includes the enzyme, substrates, cofactors, water molecules and ions. To effectively and accurately treat such large and complex systems, it is usually necessary to adopt a multiscale approach. Specifically, some reactive fragment within the active site, and possibly selected amino acid residues, are described by quantum mechanics (QM) to facilitate modelling of chemical events. The remaining enzyme, solvent, and ions are treated classically by molecular mechanics (MM), and hence we have a hybrid QM/MM PES (Figure 10.1).[21]

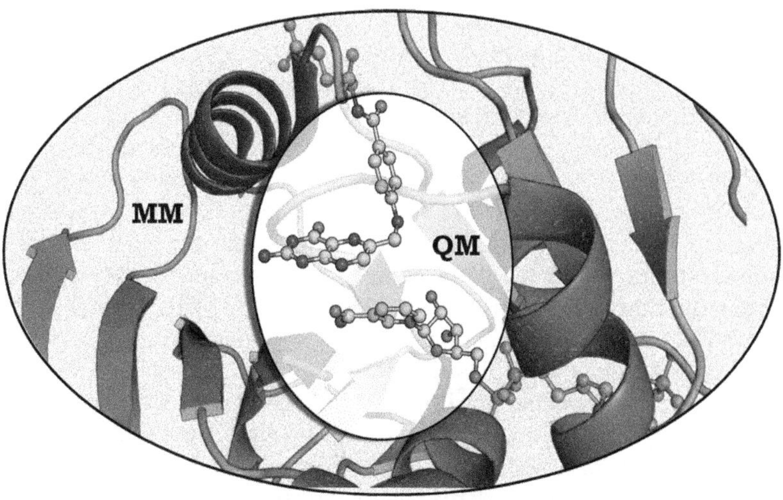

Figure 10.1 QM/MM partitioning for the enzyme DHFR.

This multiscale approach was introduced by Karplus, Warshel, and Levitt in the early 70's of the previous century,[21–23] and later adopted by numerous researchers.[24–26] QM/MM methods have been reviewed extensively in the literature (see Chapter 7 in this book).[27–31]

The accuracy of the simulated chemical reaction can in principle only be as accurate as the QM method describing the reactive fragment. It is therefore of great importance to employ a reliable QM method. This includes high-level *ab initio* (AI) methods, density functional theory (DFT), and in some carefully validated cases semi-empirical (SE) methods.[20] Unfortunately, AI and DFT methods are often too computationally demanding for enzyme simulations, and SE approaches are not accurate enough. Therefore, in many cases, it is necessary to adopt a more pragmatic approach, which entails fine-tuning the PES to the system at hand. This may be achieved for SE methods within the so-called specific reaction parameter (SRP) approach, introduced by Rossi and Truhlar[32] or in the empirical valence bond approach (EVB) of Warshel.[33,34] The latter method, which is based on the definition of two valence bond states, ϵ_1 and ϵ_2, has been particularly useful in modelling enzyme reactions (see Chapter 9 in this book). An additional approach includes a simple valence bond (SVB) correction to existing PES.[35]

In our group, we have employed DFT,[16,17] traditional SE methods,[36,37] as well as SE-SRP approaches,[18,38–43] to study solution and enzyme reactions. When choosing a PES the guiding principle should be to employ the cheapest possible method, without sacrificing the goal of high accuracy.

10.1.3 Classical Simulation Methods for Enzyme Modelling

Atomic-level molecular dynamics (MD) simulations of proteins date back to the mid-1970's,[44,45] and since these early days of computational biochemistry MD has become an essential tool for both theoreticians and experimentalists.[46] MD propagates the enzyme system in phase space (*i.e.* coordinates and momenta) according to Newton's classical equations of motion. This discloses a wealth of information regarding the dynamic behavior of the enzyme system, and powerful tools exist that can analyse MD trajectories.[47] However, chemical reactions are almost always rare events, and currently brute force MD simulations do not allow sufficient time for the reaction to be observed in such *in silico* experiments. Therefore, specialised techniques have been designed that facilitate barrier climbing within reasonable simulation times (see Chapter 5 in this book).[48–54] A widely used method is the umbrella sampling (US) technique,[55] which provides a simple way to model enzymatic reactions.[56,57] In US one applies an umbrella potential, which serves as a bias that ideally makes it seem like there is no barrier along the chemical coordinate of interest, hence facilitating proper sampling along this direction. The chemical coordinate is usually termed the reaction coordinate and will be discussed further below in the framework of calculation of rate constants. This bias is subsequently removed to obtain an unbiased probability distribution and free energy profile. To facilitate construction of complete free energy profiles for multistep enzyme reactions, one may employ the weighted histogram analysis method (WHAM).[58] Multidimensional versions of this method have been developed, and also combined with other techniques, such as the string method.[59–61] Many additional techniques exist, such as free energy perturbation,[62] thermodynamic integration,[63] and transition path sampling.[64]

10.1.4 Nuclear Quantum Effects in Enzymes

NQEs are a result of the QM behaviour of all particles. This non-classical conduct of nuclei is most pronounced for light atoms, such as hydrogen and its isotopes. However, in many cases considerably heavier atoms, such as lithium, carbon, nitrogen and oxygen, may have significant NQE, and such effects may be detected employing isotope effects.[65–67] The principle NQEs are quantum vibrations, and in particular zero point energy (ZPE), and tunnelling. Additional quantum effects, such as non-classical reflection, may also be important although less so than tunnelling; this is because reflection occurs at higher, and hence less populated, energy levels than tunnelling. Non-classical reflection is a scattering phenomenon and may both increase or reduce the rate constant.

NQE can result in significant differences between the quantum and the classical chemical rate constants, and hence when modelling enzymatic reactions it is often very important to include such effects.[68–70] We note that this does not necessarily mean that NQE accelerate the enzyme reaction relative to the uncatalysed analogous reaction,[68] although in certain cases tunnelling can have a small catalytic effect.[18] For instance, in the enzyme nitroalkane oxidase, tunnelling is responsible for a threefold acceleration in the rate relative to the uncatalysed reaction.[18]

10.1.4.1 Zero Point Energy (ZPE)

A fundamental property in QM – and its related Heisenberg uncertainty principle – is that the energy of a confined system can never equal zero, or in other words, there is nullpunktsenergie.[71] In contrast to Newtonian, classical mechanics – in which a particle can co-exist in a state with well-defined position and momentum – a confined quantum particle, even at its minimal energy state (*i.e.* the ground state), has a distribution 'width' that inherently results in a non-zero energy state. In physical systems the ZPE phenomena influences the nature of covalent bonds,[72] as well as H-bonds.[73–75] Interestingly, in condensed phases, one may observe that confined water molecules on the surface of a protein may have greater ZPE than bulk water, and hence greater kinetic energy, in clear 'violation' of the classical equipartition theorem.[76–78] The physical source of this quantum confinement is that it increases the curvature of the nuclei's wavefunction, with a concomitant increase in nuclear kinetic energy. Vibrational energy may also be computed in enzymes, and since the vibrational energy is sensitive to its quantum confinement, it is a useful reporter on changes in bonding and the local molecular environment. Isotope effects is an invaluable tool in elucidating this local environment, and in particular binding isotope effects (BIEs), equilibrium isotope effects (EIEs) and kinetic isotope effects (KIEs),[65] and these will be discussed later in this chapter (see also Chapter 4 in this book).

10.1.4.2 Tunnelling

Tunnelling is a QM phenomenon entirely absent in a classical world. In classical mechanics a particle can only cross a potential energy barrier if it has total energy greater than the height of the barrier. However, within a QM description, there is a finite probability for a particle to tunnel through a barrier, due to the wave behavior of the particle, and this tunnelling ability is highly sensitive to the mass of the particle and the nature of the barrier.

The effects of tunnelling in many chemical reactions have been studied extensively, and it is well established that tunnelling can be significant at low temperatures,[79–81] including remote locations such as interstellar clouds.[82] Tunnelling has also been indicated in enzyme catalysed reactions based on temperature-dependent KIE experiments, and has been rationalised within a generalised Marcus-like description.[19,83–85] According to a Marcus-like theory for hydrogen transfer reactions, one may define a collective reaction coordinate for hydrogen transfer. This reaction coordinate describes the collective equilibrium motion of the environment (*i.e.* everything in the system but the hydrogen) as the system undergoes a chemical transformation from the reactant state (RS), *via* the transition state (TS), to the product state (PS) (Figure 10.2). Thus, one may view such a fluctuating environmental potential as a three state scenario. Orthogonal to this collective reaction coordinate one may envision a hydrogen transfer coordinate, which also has three states (Figure 10.2): (a) RS where the hydrogen wavefunction is located near the donor atom. (b) TS where the hydrogen wavefunction is delocalised along the donor–acceptor axis. (c) PS where the hydrogen wavefunction is located in proximity to the acceptor atom. Optimal hydrogen tunnelling occurs when the RS and PS are degenerate, and the extent of tunnelling depends on the height and width of the barrier at the TS. The question of whether tunnelling can be catalytic has also been raised, and to date very few such cases have been identified.[18,86,87]

10.1.5 The Classical and Quantum Rate Constants

The calculation of the classical rate constant, k_C, for chemical reactions is typically based on transition state theory (TST). Within this formalism, the rate constant of a unimolecular reaction may be defined as:[88–90]

$$k_C^{\mathrm{TST}} = (\beta h)^{-1} \mathrm{e}^{-\beta \Delta G_C^{\ddagger}} \tag{10.1}$$

Here, $\Delta G_C^{\ddagger}$ is the classical free energy barrier, h is Planck's constant, $\beta = (k_\mathrm{B}T)^{-1}$, k_B Boltzmann's constant, and T is the temperature. The TST rate

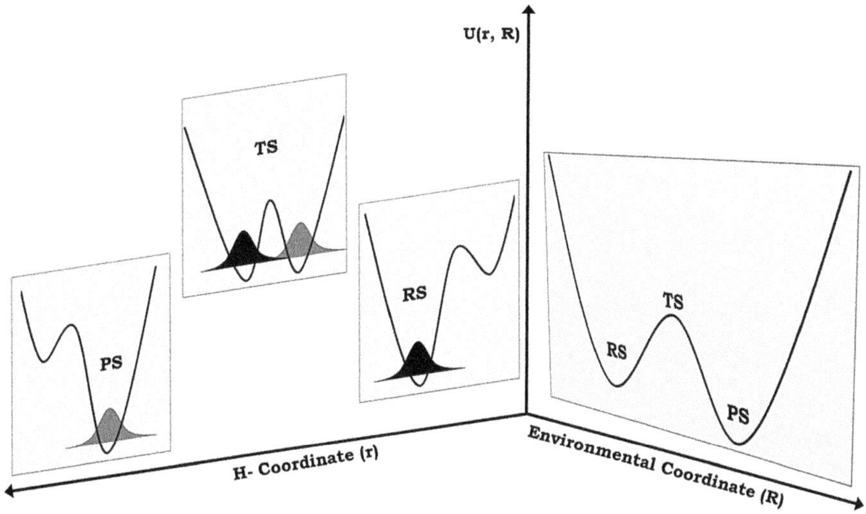

Figure 10.2 Depiction of a hypothetical symmetric H^+/H^- transfer reaction as a function of two coordinates: environmental and hydrogen transfer coordinates. The environmental coordinate changes from RS to the PS *via* the TS, in concert with the H^+/H^- transfer. In the RS well of the environmental coordinate, the corresponding H^+/H^- transfer potential favours localisation of the hydrogen wavefunction in the RS well. In the PS well of the environmental coordinate, the corresponding H^+/H^- transfer potential favours localisation of the hydrogen wavefunction in the PS well. At the TS, the adiabatic (*i.e.* ground state) H^+/H^- levels are degenerate and tunnelling may occur.

constant may be reformulated into a form more amendable to condensed phase simulations:[91]

$$k_C^{\mathrm{TST}} = \Omega \frac{\mathrm{P}_C(\zeta^{\ddagger})}{\int_{-\infty}^{\zeta^{\ddagger}} \mathrm{P}_C(\zeta)\mathrm{d}\zeta} \tag{10.2}$$

Here ζ is a chemical progress coordinate, termed the reaction coordinate, and $\mathrm{P}_C(\zeta)$ is the classical probability density defined as:

$$\mathrm{P}_C(\zeta) = \mathrm{e}^{-\beta W_C(\zeta)} \tag{10.3}$$

and $W_C(\zeta)$ is the classical potential of mean force (PMF). Hence, in eqn (10.2) the nominator is the probability density of being at the TS, whereas the denominator describes the probability of being in the RS well. The prefactor, Ω, is defined as:[84,92]

$$\Omega = \left\langle \left(\frac{Z_\zeta}{2\pi\beta} \right)^{1/2} \right\rangle_{\zeta^{\ddagger}} \tag{10.4}$$

where $\langle \cdots \rangle_{\zeta\ddagger}$ is a configurational average performed at the TS, and Z_ζ is the Jacobian factor defined as:

$$Z_\zeta \equiv \sum_{i=1}^{3N} \frac{1}{m_i}\left(\frac{\partial \zeta}{\partial r_i}\right)^2 \tag{10.5}$$

Here, r_i is the coordinates of degree of freedom i and m_i the mass associated with that degree of freedom. The activation free energy barrier in eqn (10.1) may be computed from the PMF in conjunction with US,[55] or other enhanced sampling methods (see Chapter 5 in this book):[89,93]

$$\Delta G_C^\ddagger = W^\ddagger - W^{RS} - G_\zeta^{RS} \tag{10.6}$$

In the above equation,G_ζ^{RS} is the free energy of the reaction coordinate motion of the RS, assuming the reaction coordinate is separable from the remaining degrees of freedom.

The classical TST may also be defined in terms of the TS and RS partition functions, Q, as follows:[94]

$$k_C^{TST} = \Omega \frac{Q_C^\ddagger}{Q_C^{RS}} \tag{10.7}$$

Here the RS and TS partition functions contain the same number of degrees of freedom. One may also define the prefactor, Ω, in the following slightly different manner:[95]

$$\Omega = \frac{\omega_S}{2\pi} \tag{10.8}$$

In this form of the TST rate constant, the reaction coordinate is a complete solvent coordinate (*e.g.* energy gap coordinate, $\zeta \equiv \Delta\epsilon = \epsilon_1 - \epsilon_2$), and the factor ω_S is the harmonic frequency of the oscillation of the solvent coordinate in the reactant well (*i.e.* assuming a quadratic dependence of the energy on the reaction coordinate). Such a prefactor is valid for cases where there is relatively large solvent reorganisation along the chemical progress coordinate.

To go beyond TST, while remaining within the realms of classical mechanics, one may account for recrossing of the TS hypersurface (Figure 10.3), as well as non-equilibrium solvation effects along the reaction coordinate (*i.e.* in cases the environment does not have time to fully relax as the reaction progresses). Classical recrossing may be accounted for by activated dynamics methods, and is included *via* a prefactor, κ_C $(0 \leq \kappa_C \leq 1)$:[56,57]

$$k_C = \kappa_C k_C^{TST} \tag{10.9}$$

Recrossing accounts for the fact that not all trajectories that reach the dividing surface actually end up in the product well. This is a multidimensional

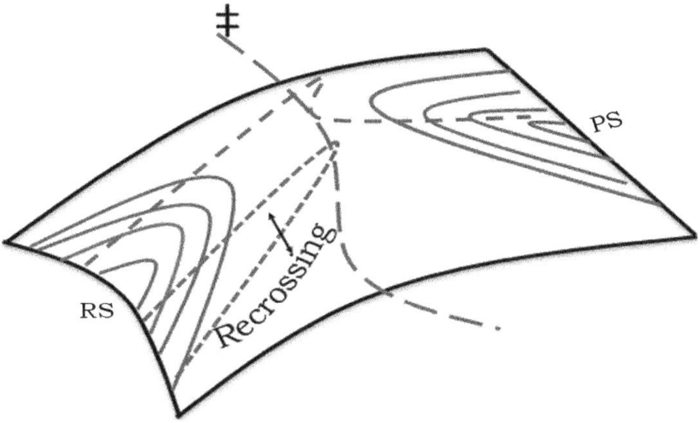

Figure 10.3 Example of crossing and recrossing trajectories across the dividing surface (*i.e.* transition state) of a reaction.

friction effect,[83] and classically always results in a rate lower than, or equal to, that predicted by TST. Non-equilibrium solvation effects may also be accounted for.[96]

A natural extension to classical TST is quantum TST (QTST).[94,97] Within this formalism we have:

$$k_Q^{\mathrm{TST}} = \Omega \frac{Q_Q^{\ddagger}}{Q_Q^{\mathrm{RS}}} \tag{10.10}$$

The difference between the quantum and classical partition functions may be defined as:

$$\delta(\zeta) = Q_Q(\zeta)/Q_C(\zeta) \tag{10.11}$$

Hence, the quantum rate, within QTST, may be calculated directly using eqn (10.10) or indirectly using eqn (10.7) and (10.11). In the former approach, one needs to perform quantum simulations directly, whereas in the latter approach one may run classical simulations, followed by quantum corrections. As we will see below, estimation of the QTST rate constants may be realised using a range of techniques, including wavefunction (WF) based approaches,[70] path integrals (PIs),[66,98–100] and semiclassical (SC) methods.[83] One may also go beyond the QTST approach,[97] and account for recrossing. This may be done by adding κ_C or alternatively, one may also add the effects of recrossing using quantum methods,[101] to obtain a purely quantum rate constant:

$$k_Q = \kappa_Q k_Q^{\mathrm{TST}} \tag{10.12}$$

For a more extensive discussion of rate theories for enzymatic reactions, see Chapter 2 in this book.

10.1.6 Kinetic, Equilibrium and Binding Isotope Effects

The kinetics of a chemical reaction, and therefore its rate constant, might be rather sensitive to isotopic substitutions. This phenomenon is known as the kinetic isotope effect (KIE) (see Chapter 4 in this book).[65] The KIE is defined to be the ratio between the light (k_L) and the heavy (k_H) isotope rate constants (here we use k as a generic symbol for the rate constant, regardless of theoretical framework):

$$\mathrm{KIE} = \frac{k_L}{k_H} \tag{10.13}$$

KIE arises mainly from ZPE and tunnelling effects. KIE is an important experimental tool in elucidating mechanisms and the nature of TSs.[102] KIEs also provide a direct way to compare theoretical models with experiments, and hence an avenue for validating computational results.[66]

The KIE can be defined as either primary or secondary. The former KIE describes an isotopic substitution of atoms taking part in the chemical reaction, whereas the latter type describes an isotopic substitution of an atom not directly involved in the bond breaking or forming. An important property of KIE is that it may be used to identify tunnelling in reactions. For instance, as a rule of thumb a $k_{\mathrm{H}}/k_{\mathrm{D}} > 7$ is an indication of tunnelling, although a value smaller than that cannot rule out a tunnelling mechanism. Another way to detect tunnelling is *via* temperature-dependent KIE studies, as tunnelling is highly sensitive to the temperature.[65]

Another related phenomenon is the difference in the fractionation of reactants and products at equilibrium, which is known as the equilibrium isotope effect (EIE).[65] The definition is:

$$\mathrm{EIE} = \frac{K_L}{K_H} \tag{10.14}$$

where K_L and K_H are the light and heavy isotope equilibrium constants. The EIE principly reflects the difference in vibrational energy distribution in the initial and final states of the reaction. Primary and secondary EIE may be defined in an analogous manner to that for KIE.

One may also define the BIEs,[65,103,104] which we here define as an equilibrium effect on binding (as opposed to a kinetic BIE). Thus, a BIE reflects the difference in vibrational energy in the unbound and bound states of a substrate or product. Binding of a ligand to a protein usually involves weak, non-covalent interactions, and hence BIEs are typically small in magnitude. Nonetheless, the BIE is a potent reported on the environment encapsulating a ligand. To date, BIE have been somewhat scarcely studied and their role in kinetic isotope complexity is not well understood.

Summary

This section introduces some basic concepts of computational enzymology: PES, hybrid QM/MM approaches and enhanced simulation methods for modelling enzyme reactions. We also introduced some basic concepts of NQE, namely bound vibrations, ZPE, and tunnelling. Subsequently, we discussed classical and quantum rate expressions typically used in enzyme modelling. Finally, we mentioned various kinds of isotope effects, which are a direct consequence of NQE.

10.2 How Can We Include NQE in Enzyme Modelling?

In previous sections we highlighted the importance of including NQE when studying enzyme reactions, in particular when modelling hydrogen transfers or isotope effects. In this section we will discuss some of the main methods that facilitate incorporating NQE in enzyme modelling. Possibly the most basic approach for capturing some of the NQE is *via* normal mode analysis within the harmonic approximation. This was realised early on,[105,106] and is a rather straightforward extension of gas phase and continuum solution phase normal mode analysis and will not be discussed further here. Here we will focus on more specialised methods that were developed with enzyme simulations in mind. Furthermore, the methods described below were all coupled with hybrid QM/MM PESs. Thus, the methods described herein quantise the electrons in the QM region to allow chemistry to occur (*i.e.* QM/MM), while including quantisation of nuclei *via* specialised NQE techniques. A common feature of current methods is to treat the motion of only a subset of nuclei by quantum dynamics, whereas the motion of the remaining nuclear degrees of freedom classically. In the following, $\mathbf{r}$ represents the nuclei treated as QM, whereas $\mathbf{R}$ denotes the classical nuclear coordinates.

We will commence with the ensemble-averaged semiclassical approach of Truhlar, Gao, and coworkers, followed by a description of the wavefunction-based method of Hammes-Schiffer and coworkers, and finally describe various PI-based methods. We will focus more extensively on the latter category, as this is the type of methods developed in our group and with which we are most familiar.

10.2.1 Semiclassical Approach to Enzyme Modelling

In this section we will describe a widely used computational method due to Truhlar and coworkers for calculating reaction rate constants for enzymatic reactions. This method is based on variational TST with multidimensional tunnelling (VTST/MT), which has been applied to a large number of reactions in the gas phase and solution.[107] VTST/MT is implemented in the Polyrate code, and is interfaced with many popular quantum chemistry codes, such as Gamess, Gaussian, Jaguar, Mopac, NWChem, and more.[108] The method was generalised to treat enzymatic reactions in collaboration with the Gao group.[93] The enzyme version of the method is an ensemble-averaged (EA) extension, and is termed EA-VTST/MT, and the method has been reviewed extensively.[69,83,109] The method takes into account quantum vibrations, tunnelling effects, dynamical recrossing, and some non-equilibrium solvation effects. The quantum rate constant is written as follows:

$$k_Q^{\text{EA-VTST/MT}} = (\beta h)^{-1} \cdot \gamma(\mathbf{r};\, \mathbf{R}) \cdot \mathrm{e}^{-\beta \Delta G^{\ddagger}_{\text{QV}}(\mathbf{r},\mathbf{R})} \tag{10.15}$$

In this rate expression, $\Delta G^{\ddagger}_{\text{QV}}$ represents a so-called quasi-classical free energy barrier, which constitutes the classical free energy barrier with corrections for quantised vibrations (QV) for degrees of freedom orthogonal to the reaction coordinate, as well as quantisation of the motion along the reaction coordinate in the RS. γ is a correction coefficient defined as $\Gamma \cdot \kappa \cdot g$, which accounts for tunnelling effects (Γ), dynamical recrossing (κ), and non-equilibrium solvation (g). The latter factor is usually set to unity, and the remaining factors are computed as an ensemble average:

$$\gamma(\mathbf{r};\mathbf{R}) = \frac{1}{M}\sum_{j=1}^{M} \Gamma_j(\mathbf{r};\mathbf{R})\kappa_j(\mathbf{r};\mathbf{R}) \tag{10.16}$$

Here M represents the number of configurations included in the ensemble averaging process. Here, **r** represents the nuclei treated as QM (*i.e.* the primary zone), whereas **R** denotes the classical nuclear coordinates (*i.e.* the secondary zone). The primary zone is usually the reactive fragment and possibly some additional atoms, whereas the secondary zone contains the remaining parts of the enzyme system. This division into zones is helpful in reducing the computational cost of the method, as the quantisation occurs only within the primary zone. In practice, the EA-VTST/MT rate constant is calculated in three separate stages, where at each stage additional quantum effects are included.[93]

Stage 1 has two separate steps: in the first step, one computes the classical PMF, using US, FEP or any other suitable method, to obtain the free energy and the classical rate constant within TST (eqn (10.1) and (10.6)). In the second step of stage 1, one adds the first quantum correction term, by including vibrational quantisation effects to the classical free energies that were calculated in first step. This entails computing the difference in quantum and classical vibrational free energy, the latter being implicitly included in evaluation of the classical PMF for all degrees of freedom other than the reaction coordinate. To this end, one performs a normal mode analysis of an ensemble of instantaneous configurations drawn from the PMF simulations. As these configurations are not at local minima, the method may best be defined as an instantaneous normal mode analysis technique.[110] In practice one can view this as incorporating some degree of anharmonicity to what is a standard harmonic approximation, as the instantaneous normal mode vibrational frequencies are typically slightly lower than vibrational frequencies computed at true local minima. The Hessian is computed numerically, and as mentioned above in order to reduce computational cost, one defines a subset of flexible atoms, typically in the active site region (*i.e.* primary zone 1 with N1 atoms). The dynamic primary atoms are embedded in a static field of the remaining atoms of the system, which constitute the secondary zone. In practice, one need to quantise 3N1 − 6 modes of the RS and only 3N1 − 7 modes at the TS, as well as the reaction coordinate mode in the RS. This first stage then provides the Boltzmann factor in eqn (10.15), $e^{-\beta\Delta G^{\ddagger}_{QV}}$.

In stage 2, the recrossing (Figure 10.3) and tunnelling correction effects (Figure 10.4) are included in the rate constant that was calculated in the previous stage. An ensemble averaging procedure over M samples is performed (*i.e.* eqn (10.16)), to yield the EA-VTST/MT rate constant (eqn (10.15)). Practically, for the stage 2 calculations one selects a flexible primary zone with N2 atoms in the static field of the remaining atoms in the secondary zone. First, the recrossing factor is estimated by computing an ensemble of variational TSs (*i.e.* that have no recrossing). Subsequently, the ratios of the Boltzmann factors of the variational ($s^{\ddagger}$) and non-variational TS ($\zeta^{\ddagger}$) are computed as the transmission coefficients for configurations j, *i.e.* κ_j:[93]

$$\kappa_j = e^{-\beta\{\Delta G_j(s^{\ddagger}) - \Delta G_j(\zeta^{\ddagger})\}} \tag{10.17}$$

Subsequently, the zero, small and large curvature (LC) tunnelling pathways are computed (Figure 10.4), and the tunnelling mechanism that maximises the rate is selected to yield the transmission coefficients, Γ_j, for each configuration, j. Hence, one may then obtain the EA composite transmission coefficient, γ (eqn (10.16)).

In the second stage calculations, thermal energy and entropy of the atoms in the secondary zone are not included when computing the transmission coefficient, γ. These contributions are added in the third stage calculations. In stage 3, one takes into account effects on the transmission coefficients due to the breakdown of the frozen bath assumption (*i.e.* fixed secondary zone) that was imposed in stage 2. Practically, one allows the secondary zone atoms to relax, and then estimates recrossing and tunnelling effects.

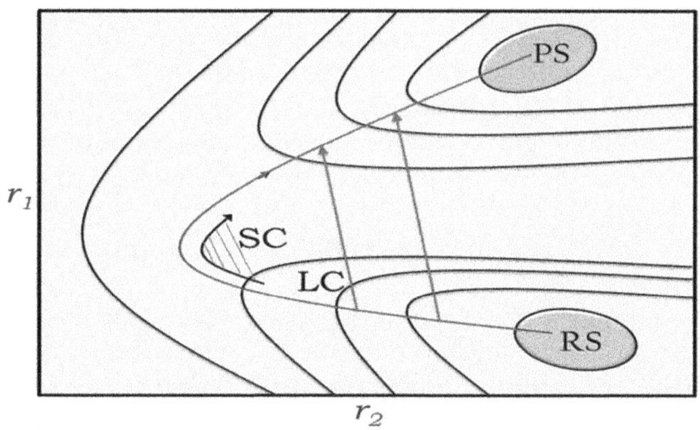

Figure 10.4 Various tunnelling paths on a hypothetical potential energy surface. Small curvature (SC) and large curvature (LC) paths are marked.

The third stage assumes there is sufficient time for the environment to equilibrate as the chemical event occurs, whereas in the second stage only a subset of the system equilibrates along the reaction path (*i.e.* the primary zone atoms). In practice, this last stage is often not included.

The EA-VTST/MT method is implemented in the CHARMM software package[111,112] *via* an interface between CHARMM and Polyrate (*i.e.* CHARMMRATE),[108] as well as in Dynamo.[113] Important advantages of the method are that it includes all main quantum effects in a systematic and identifiable way, and tunnelling effects are treated in a multidimensional manner. A potential disadvantage is that it assumes separability of degrees of freedom and assumes that the classical trajectories sampled are similar to the true quantum trajectories. There are also several technical subtleties, such as how to best generate the RS, TS and PS triplet needed when forming the isoinertial minimum energy pathways for tunnelling calculations.

10.2.2 Vibrational Wave Function Approach to Enzyme Modelling

Another method for inclusion of NQE is the hybrid quantum-classical (QC) approach due to Hammes-Schiffer and coworkers,[70,101] which was developed in order to simulate proton or hydride transfer reactions in solution and enzymes. In this hybrid QC method, a single nucleus is treated as a three-dimensional wavefunction embedded in a classical bath of the remaining atoms. The transferring hydrogen nucleus wavefunctions (*i.e.* ground and excited states) are obtained by solving the time-independent nuclear Schrödinger equation at each simulation step.

$$[T_H + V(\mathbf{r}, \mathbf{R})]\Phi_j(\mathbf{r}; \mathbf{R}) = E_j(\mathbf{R})\Phi_j(\mathbf{r}; \mathbf{R}) \tag{10.18}$$

Here, $\mathbf{r}$ represents the hydrogen nuclear coordinate, $\mathbf{R}$ denotes the remaining classical nuclear coordinates, T_H is the kinetic energy operator of the hydrogen nucleus, and $V(\mathbf{r}, \mathbf{R})$ is the potential energy of the electronic ground state (EVB, vide infra). In practice, the wavefunctions are solved on a grid with a Fourier grid Hamiltonian (FGH) method.[114,115]

To generate mixed QC trajectories, the classical nuclei are propagated according to the Hellmann–Feynman forces, which account for the classical particles and the wavefunction nature of the transferring hydrogen.

$$\mathbf{F} = -\nabla_{\mathbf{R}} \mathrm{E}_k(\mathbf{R}) = -\langle \Phi_k(\mathbf{r}; \mathbf{R}) | \nabla_{\mathbf{R}} V(\mathbf{r}, \mathbf{R}) | \Phi_k(\mathbf{r}; \mathbf{R}) \rangle_{\mathbf{r}} \tag{10.19}$$

The quantum particle feels the classical particles by solving the nuclear time-independent Schrödinger's equation in the presence of the hybrid QM/MM potential, hence ensuring a proper feedback between the classical and quantum regions. The classical nuclei are propagated in space and time, whereas the wavefunction is solved adiabatically at each time step. Based on this hybrid approach, one may then calculate the free energy surface as a function of a collective reaction coordinate, including vibrationally adiabatic NQE (*i.e.* ground vibrational state). In order to facilitate H^+/H^- transfer within a practical time, a mapping potential is employed which drives the system from RS to PS along an adiabatic reaction coordinate. Hence, no dynamical behavior of the hydrogen nucleus is available from these free energy simulations. Moreover, the free energy EVB-FGH method is inherently adiabatic, so no thermal NQE are included. The QC QTST rate constant, k_Q^{TST}, may be calculate according to eqn (10.10).[84,101]

In order to account for the dynamic behavior of the hydrogen nucleus, one needs to account for non-adiabatic dynamics. To this end, Hammes-Schiffer and coworkers adopted a surface hopping method, which allows the quantum particle to occupy excited vibrational states as well. This non-adiabatic QC simulation method is termed MD with quantum transitions (MDQT).[101] The basic principle of the MDQT method is to propagate an ensemble of trajectories, and each trajectory moves classically on a single adiabatic surface as described above, except for instantaneous transitions between the adiabatic vibrational states. For each trajectory, the time-independent Schrödinger equation is solved using the FGH method; however to facilitate hopping between vibrational states, a time-dependent quantum amplitude factor is computed by integrating the time-dependent Schrödinger equation. A fewest switches algorithm is adopted, which correctly apportions trajectories among the adiabatic states according to the time-dependent quantum transition amplitude factor. The MDQT method may be combined with a specialised quantum reactive flux method to estimate dynamical recrossings, and hence the recrossing factor, κ_Q. The combined quantum rate constant, k_Q, may then be computed according to eqn (10.12).

In the implementation of Hammes-Schiffer and coworkers, the PES is described by an empirical QM/MM approach, namely the EVB method of Warshel and coworkers.[33,34] Within this treatment, the reaction coordinate is described as the difference between the energy of two valence bond states ϵ_1 and ϵ_2, *i.e.* energy gap, or collective reaction coordinate (*i.e.* $\zeta \equiv \Delta\epsilon$). In the QC version of the theory (EVB-QC), the collective reaction coordinate, $\Delta\epsilon$, is computed as an expectation value over the ground vibrational state of the hydrogen atom. This hybrid EVB-QC method allows fast generation of potential values and gradients due to the low cost of EVB calculations, and therefore is highly suited for both adiabatic quantum free energy simulations, as well as non-adiabatic dynamic propagation implemented in the MDQT method. However, in principle other PES could be used. The EVB, FGH, and MDQT methods have been implemented in an in-house version of the DLPROTEIN code.[116]

The advantage of the EVB-QC wavefunction method is that it is a true quantum dynamics approach, and not a scheme for correcting classical trajectories. The disadvantage is that the method is currently limited to a single quantised particle (*e.g.* transferring H^+ or H^-). Hence, some multi-dimensional NQE might not be captured. Moreover, since the method is in practice designed for hydrogenic atom transfer reactions, it leaves much of chemical space beyond reach. In spite of these limitations, it is possibly the most elegant theoretical method accounting for NQE in enzymes.

10.2.3 Path Integral Methods

10.2.3.1 Closed Path Integral Methods for Enzyme Modelling

An additional approach for treating NQE is the Feynman PI formalism to QM.[117] In the PI approach to QM, one replaces the wavefunction picture with one of paths (Figure 10.5). Thus, according to Feynman, ensembles of paths, weighed by the classical action for each path, capture the delocalised, wave behavior of particles inherent to Schrödinger's description. A great advantage of PIs is that it may be generalised to multi-particle systems in a straightforward manner, and the method naturally includes thermal NQE. Additionally, one may adopt standard classical simulation methods, such as MD or Monte Carlo (MC), to sample the Feynman paths.

The QTST rate constant may be computed within PI theory (PI-QTST) using the thermodynamic formulation of the method.[94] A dynamical formulation also exists, which facilitates calculating the full quantum dynamics rate constant, although this approach is less used due to its computational cost and implementational complexity.

Assume we have a system composed of a set of QM atoms embedded in a bath of classical atoms in thermodynamic equilibrium. The QM partition

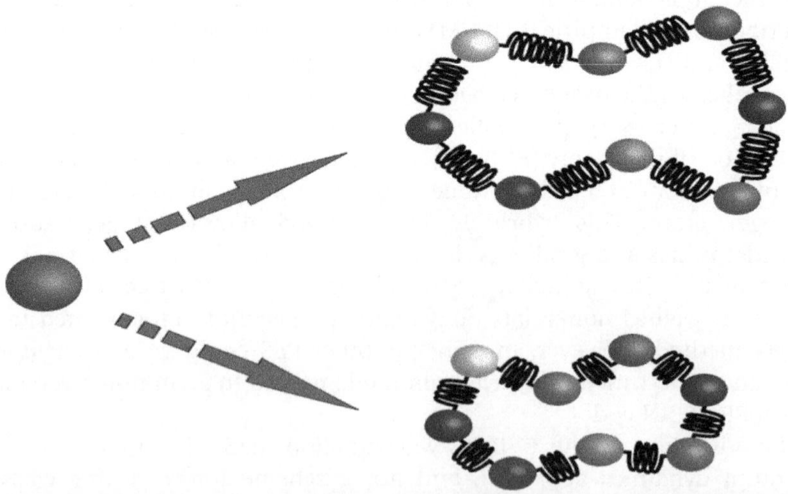

Figure 10.5 Path-integral description, where particles are described by ring polymers, which possess quantum delocalisation. Large ring corresponds to light isotope, while contracted ring corresponds to heavy isotope.

function, Q_Q, for this system may be written as the trace of the thermal density matrix (DM) ρ:

$$Q_Q = \mathrm{Tr}(\rho) = \int d\mathbf{R} \int d\mathbf{r} \rho(\mathbf{r}, \mathbf{r}; \mathbf{R}, \beta) = \int d\mathbf{R} \int d\mathbf{r} \langle \mathbf{r}; \mathbf{R} | \, e^{-\beta \hat{H}} \, | \mathbf{r}; \mathbf{R} \rangle \tag{10.20}$$

Here $\hat{H}$ is the system Hamiltonian, which is a sum of the kinetic ($\hat{K}$) and potential ($\hat{V}$) energy operators. $|\mathbf{r}; \mathbf{R}\rangle$ are the Dirac spatial basis functions that depend on the positions of the quantum particles, $\mathbf{r}$, and parametrically on the classical particles, $\mathbf{R}$. According to the PI formalism the integral in eqn (10.20) can be decomposed into so-called higher temperature thermal DM multiplication ($\tau = \beta/P$):

$$Q_Q = \int d\mathbf{R} \int d\mathbf{r} \langle \mathbf{r} | e^{-\tau H} e^{-\tau H} \ldots e^{-\tau H} | \mathbf{r} \rangle = \int d\mathbf{R} \int d\mathbf{r}_1 \ldots d\mathbf{r}_P \prod_{j=1}^{P} \rho(\mathbf{r}_j, \mathbf{r}_{j+1}; \mathbf{R}, \tau) \tag{10.21}$$

P is usually taken to be an integer, which represents the number of quasi-particles or beads. For a closed ring-polymer chain PI, the boundary condition $\mathbf{r}_1 = \mathbf{r}_{P+1}$ is taken (Figure 10.5). In the high temperature regime (*i.e.* in the limit $\tau \rightarrow 0$ and $P \rightarrow \infty$), one can approximate the DM according to the semiclassical primitive approximation (PA):[118]

$$\rho_{\mathrm{PA}}(\mathbf{r}_j, \mathbf{r}_{j+1}; \mathbf{R}, \tau) = \left(\frac{m}{2\pi\tau\hbar^2} \right)^{D/2} e^{-\left(\frac{m}{2\tau\hbar^2} (\mathbf{r}_{j+1} - \mathbf{r}_j)^2 + \frac{\tau}{2} \left(V(\mathbf{r}_{j+1}; \mathbf{R}) + V(\mathbf{r}_j; \mathbf{R}) \right) \right)} \tag{10.22}$$

where V is the system potential (*i.e.* a QM/MM potential), m is the particle mass, and D is the dimension of the system (the dimension of $\mathbf{r}$). The ρ_{PA} DM is sometimes also referred as the high temperature density matrix (HTDM), since when $\tau \rightarrow 0$, $T \rightarrow \infty$. The description above is isomorphic to a classical ring of beads system, where the beads are connected *via* springs (*i.e.* harmonic potential) (Figure 10.5).

One may define the rate constant according to PI theory, yielding PI-QTST:[94]

$$k_Q^{\mathrm{PI-TST}} = \Omega_{\mathrm{FP}} \frac{Q_Q^{\ddagger}}{Q_Q^{\mathrm{RS}}} \tag{10.23}$$

In this case, the prefactor Ω_{FP} was defined above (eqn (10.4)), although within PI terminology it is called the free-particle (FP) term.

Within the PI approach one may compute the rate constant using eqn (10.9) and (10.11), where the quantum correction term to the classical rate (eqn (10.11)) is calculated using PI:

$$\delta^{\ddagger} = \frac{\left(Q_Q^{\ddagger} / Q_C^{\ddagger} \right)}{\left(Q_Q^{\mathrm{RS}} / Q_C^{\mathrm{RS}} \right)} \tag{10.24}$$

where $Q_Q^{\ddagger}$, $Q_C^{\ddagger}$, Q_Q^{RS}, and Q_C^{RS} are the quantum and classical partition functions of the TS (‡) and RS, respectively. To calculate the quantum to classical ratio of the partition function, one can use the quantised classical path (QCP) approach derived by Warshel and coworkers:[98,99]

$$\delta = \frac{Q_{\mathrm{QM}}}{Q_{\mathrm{CM}}} = \left\langle \left\langle \mathrm{e}^{-\tau\left(\sum_{j=1}^{P}(V(\mathbf{r}_i;\mathbf{R})-V(\mathbf{r}_c;\mathbf{R}))\right)} \right\rangle_{\mathrm{FP},\mathbf{r}_c} \right\rangle_{V(\mathbf{r}_c;\mathbf{R})} \tag{10.25}$$

where $\mathbf{r}_c$ is the centroid coordinate. The external average represents a classical average, while the delocalised QM description comes from the inner average – the so-called FP average. In this approach, the classical and quantum simulations are performed separately, providing a practical and efficient framework for quantum simulations in enzymes. We note that the δ as defined in eqn (10.24) and (10.25) includes both vibrational and tunnelling effects, and mirrors (but is not identical to) the factor $\Gamma \mathrm{e}^{-\beta(\Delta G_{\mathrm{QV}}^{\ddagger}-\Delta G_{\mathrm{C}}^{\ddagger})}$ in EA-VTST/MT.

A well-known problem with PI simulations is the difficulty in sampling the polymer ring due the harmonic coupling between the beads.[118] Our initial attempts to implement the QCP approach clearly demonstrated this issue, and we turned our focus to combining QCP with enhanced sampling methods.[119] Thus, we combined QCP with the bisection sampling algorithm due to Ceperley,[119–122] and later we also coupled the method with the staging algorithm.[123,124]

In spite of the greatly improved convergence of the QCP approach with the bisection or staging algorithms, the approach was not accurate enough to predict KIEs. KIEs are due to minute differences in free energy, and specialised techniques are required to obtain quantitative results. We therefore developed a mass-perturbation technique, termed PI-FEP, wherein the free energy difference between the different isotopes is computed directly.[125] This greatly improved the accuracy of the computed KIE, and has been used extensively since.[36,37,40,124,126–129] Mass-perturbation techniques for PI KIE have also been adopted by others in the study of small molecules.[130,131]

To further enhance the efficiency of the PI simulations, one may use higher-order factorisations of the DM operator. Thus, following the work of Chin[132–134] we generated a higher-order version of the QCP method, which converges with a considerably smaller number of beads, at the cost of computing the potential gradient in addition to the potential.[124] This approach may also be combined with the mass-perturbation approach, to efficiently compute the KIE.[36] The above-mentioned methods have all been implemented in the CHARMM software platform.[111,112]

Additional PI flavors have been adopted in enzyme studies, such as a hybrid quantum-classical PIMC method implemented in DLPROTEIN,[116,135] ring-polymer MD PI,[136] and the instanton method.[137]

10.2.3.2 Open Path Integral Methods for Enzyme Modelling

It is possible to obtain information regarding particle tunnelling by inspecting the particle momentum distribution. For instance, the hydrogen momentum distribution contains important information regarding the PES the hydrogen experiences in condensed phase environments. Hence, inspection of the population of different regions of the momentum spectrum reveals valuable details regarding the system at hand. For instance, low momentum regions correspond to rotational motion, librational motion, and low-frequency vibrational modes, while the high momentum regions correspond to bond stretching modes. Experimentally, the momentum distribution may be determined from neutron Compton scattering experiments. In such experiments high-energy neutrons are scattered by the hydrogen atoms in the system, yielding a scattering cross-section that may be related to the momentum distribution.[138,139] Momentum distribution has been employed as a reporter for quantum behavior in systems such as ferroelectric potassium dihydrogen phosphate,[140] water in various phases,[141–143] water confined in silica pores,[144] superprotonic conductors,[145] and the hydration shell of globular proteins.[146] A peculiar feature in the momentum distribution in some of these systems is the presence of secondary peaks, or shoulders, in regions of elevated momenta. These features have been attributed to underlying nodes in the momentum distribution, and studies of model systems have suggested that these are related to ground state tunnelling. Numerous molecular simulations have also computed the momentum distribution.[147–155]

To inspect tunnelling effects in enzyme catalysis we have therefore implemented an open chain PI code within the CHARMM program.[154] Whereas closed-chain PI simulations only sample diagonal elements of the DM, open chain PI simulations also sample off-diagonal DM elements. These elements are required to compute the momentum distribution. The current version of the code can sample both the anisotropic and isotropic momentum distribution of a transferring $H^+/H^-/H^\bullet$ during a reaction. Thus far, we have employed open PIs to study NQE in several enzymes, as will be described below.

10.2.3.3 Path Integral Eigenvalue Methods

Another approach for calculating the quantum partition function using the PI-HTDM, is the PI eigenvalue (PIEV) method.[100,156,157] In the PIEV method, one takes advantage of the fact that the system energies and eigenstates are temperature independent. It is straightforward to show that the eigenstates of the HTDM are the system eigenstates, $\Phi_j(\mathbf{r};\, \mathbf{R})$, and that the system energies $E_j(\mathbf{R})$ are related to the eigenvalues of the HTDM $\eta_j(\mathbf{R})$ by:

$$E_j(\mathbf{R}) = -\frac{\ln(\eta_j(\mathbf{R}) \cdot \Delta\mathbf{r})}{\tau} \tag{10.26}$$

where $\Delta\mathbf{r}$ is the grid step size. Therefore, by diagonalising the HTDM one can calculate the system energies and eigenstates. Having them, one can directly calculate the quantum partition function at any temperature:

$$Q_Q = \sum_j e^{-\beta E_j(\mathbf{R})} \Phi_j(\mathbf{r};\mathbf{R}) \Phi_j^*(\mathbf{r};\mathbf{R}) \tag{10.27}$$

In fact, one can directly diagonalise the Hamiltonian matrix to calculate the system energies and eigenstates, and not make use of the DM. However, due to the fast decaying nature of the exponential in eqn (10.20), the HTDM is sparse. This sparse nature of the HTDM makes it computationally more efficient to diagonalise than the Hamiltonian.

The rate constant can then be calculated by the QM correction term (eqn (10.11)) or directly using the Miller *et al.*[158] flux–flux autocorrelation formalism, which is equivalent to the full quantum transition rate constant within QTST. The quantum rate constant is the plateau value at long time, t, of the amplitude summation over all the flux contributions according to the Boltzmann distributions (normalised with respect to the RS partition function):

$$k_{\mathrm{QM}} = \lim_{t\to\infty} k = \lim_{t\to\infty} \left(\frac{1}{Q_Q^{\mathrm{RS}}} \sum_{i,j} e^{-\frac{\beta}{2}(E_i+E_j)} \frac{\sin(\omega_{ij})t}{\omega_{ij}} |F_\zeta|^2 \right) \tag{10.28}$$

Here $\omega_{ij} = (E_i + E_j)/\hbar$, F_ζ is the flux operator along the reaction coordinate, ζ, and Q^{RS} is the reactant partition function per unit volume.

Summary

We have reviewed several of the state-of-the-art methods available for quantum simulations in enzymes. These methods include semiclassical methods, wavefunction-based methods, as well methods based on Feynman PIs. The methods covered include both direct quantum dynamics methods, as well as methods for correction of classical simulations for NQEs. These methods facilitate inclusion of both quantum vibrational effects, as well as tunnelling effects.

10.3 Applying NQE Methods to Enzymes: Dihydrofolate Reductase (DHFR) – the Gold Standard in Enzymology

10.3.1 NQE in Enzyme Reactions

The importance of NQE in general, and isotope effects in particular, in elucidating mechanisms and physical aspects of enzyme reactions has been suggested by the experimental work of Cleland,[159] Klinman and Kohen,[19]

Romesberg and Schowen,[160] Schramm[161,162] and others. In particular, the role of tunnelling in enzyme reactions has been proposed by Klinman and Kohen,[19] Nagel and Kliman[163] and others, and has been the source of much debate.[87] The most significant experimental marker for tunnelling is the temperature dependence of the KIE. A temperature independent KIE has been interpreted as indicative of tunnelling,[19,163] and has been rationalised as environmentally coupled tunnelling within the framework of an extended Marcus model.[163]

It is widely appreciated that even if NQE are not catalytic (*i.e.* are similar in solution and enzyme reactions), it is essential to include such effects in simulations to accurately predict chemical rates.[69,70,83] Hence, it is important to include both the effects of bound vibrations, and in particular ZPE, as well as tunnelling effects. Moreover, since hydrogenic transfers may be dominated by tunnelling (although tunnelling might not be catalytic), it is essential to include such effects from a basic science point of view.[84]

In the following, we will give an overview on some experimental work performed on *E. Coli* DHFR (ec-DHFR) prior to entering a discussion of some theoretical studies of this enzyme. Many independent research groups have studied DHFR, and importantly, many different computational NQE approaches have been applied to DHFR, allowing careful comparison between them.

10.3.2 DHFR – Background

ec-DHFR is a flexible, small, monomeric protein with no metals or SS bonds (Figure 10.6). The enzyme catalyses a simple chemical transformation (C–H→C): the reduction of 7,8-dihydrofolate (H2folate) to 5,6,7,8-tetrahydrofolate (H4folate) with the stereospecific transfer of a hydride from the pro-R C4 position of the nicotinamide ring of nicotinamide adenine dinucleotide phosphate hydride (NADPH) to the si face of the C6 of the pterin ring (Scheme 10.1 and Figure 10.6). H4folate is an important cofactor in many biochemical processes including biosynthesis of nucleotides, and thus DHFR is a target for various antibiotic and chemotherapeutic drugs. DHFR has served as a platform for many experimental[164–171] and theoretical studies,[43,60,136,154,172–184] and its protein dynamics-function relations have been a matter of on-going debate.[129,179–182,185–188] DHFR is thus an ideal model system for exploring basic physical features in enzymology.

10.3.2.1 Experimental Background

A wealth of experimental data is available for this reaction, and KIEs have been measured for the wild type (WT) enzyme, many mutants, and numerous temperature-dependent studies have also been performed.[189] The experimental WT enzyme KIE values at 25 °C range from 2.8 to 3.5, depending on the protocols employed. As a prelude to the discussion on the experimentally determined data, let us briefly mention some of the methods

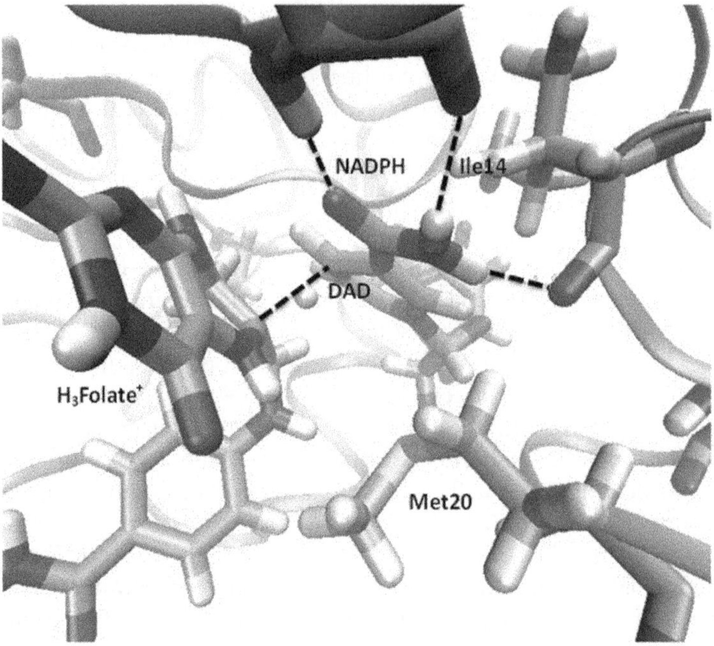

Figure 10.6 Transition state configuration for the hydride transfer in DHFR obtained from a transition path simulation.

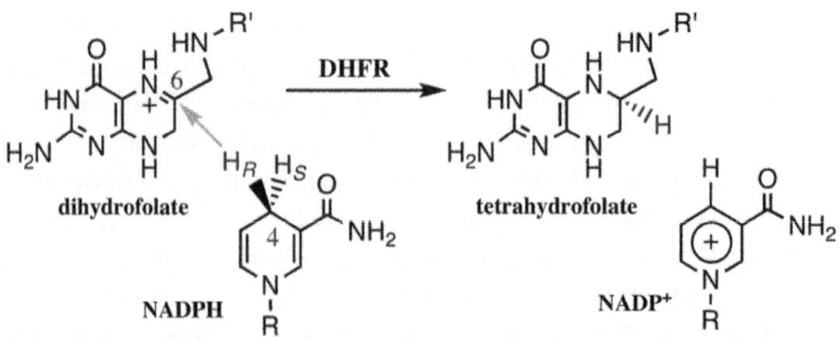

Scheme 10.1 The hydride transfer reaction catalysed by DHFR. R: adenine dinucleotide 2′-phosphate; R′: *p*-aminobenzoyl-glutamate.

applied in determining KIEs in DHFR.[1] A method that has been widely used is a pre-steady-state kinetics approach, where one follows the fluorescence resonance energy transfer (FRET) burst during stopped-flow experiments. Under such pre-steady-state conditions, the chemical step is expressed in the loss of fluorescence as the NADH moiety is oxidised to NAD^+. The advantage of this method is that kinetic masking due to potential rate-limiting steps, such as product release, are minimised. An alternative approach is to conduct the kinetics measurements at steady-state conditions, by measuring the initial rates with constant enzyme concentration as a function of different substrate or cofactor concentrations. This approach may be coupled with

Northrop's method to obtain intrinsic KIEs, hence reducing the amount of kinetic complexity.

Benkovic *et al.* measured the pre-steady-state KIE using stopped-flow FRET burst experiments (pH 7), as well as steady-state experiments (pH 9), for this enzyme, obtaining values of 2.8–3.0.[164,165] Similar results were obtained by Allemann and coworkers using similar pre-steady-state conditions (pH 7),[190,191] whereas Francis *et al.*[192] extracted an intrinsic KIE (*i.e.* for the chemical step) of 3.5 using the Northrop method from steady-state kinetics at pH 9.[166] Recently, Wang *et al.*[193] obtained a value of 3.2 using pre-steady-state stopped-flow FRET burst experiments (pH 7). Additionally, the Kohen group measured a secondary KIE of 1.13.[166] Regarding the temperature dependency of the KIE, there is fundamental variability, depending on experimental techniques and conditions, such as pH. For instance, Sikorski *et al.* found that the H/D and H/T primary KIEs are temperature independent in the range 5–45 °C at pH 9 using steady-state conditions and Northrop's method.[166] Moreover, these authors consider this basic pH to mimic the chemically relevant one, as at this pH the chemical step is fully rate limiting. However, Loveridge and Allemann obtained temperature-dependent KIEs at pH 7 using pre-steady-state kinetics, which they claim best represent the biological cell-like conditions.[191] Wang *et al.* obtained temperature independent KIEs at pH 7, also using pre-steady-state kinetics. Based on these data, Kohen and coworkers suggest that there are environmentally coupled quantum effects, whereas Allemann and coworkers and Schramm, Schwartz, and coworkers maintain that there is no coupling between the enzyme environment and the chemical step. Therefore, although the absolute value of the KIE for WT DHFR at 25 °C is *ca.* 3, there is considerable disagreement regarding the temperature dependency of the KIEs, and their physical interpretation. To complicate matters, a large body of KIE data is also available for mutant forms of DHFR, which display complex temperature dependency patterns.[167,171,179,194–197] Additional studies have addressed heavy isotope version of the enzyme.[180,181,193,197]

10.3.2.2 Computational Background

Historically, the first multiscale study of the DHFR reaction was performed by Castillo *et al.*, and the focus in this early work was the influence of the environment on the TS structure.[172] The TS structure was found to be considerably different in the enzyme compared to that obtained *in vacuo*, hence emphasising the importance of including the protein environment in *in silico* enzyme studies. The role of NQE in DHFR was addressed *via* normal mode analysis and the harmonic approximation.

Subsequent multiscale studies by various researchers estimated the NQE on the DHFR catalysed reaction and good agreement was obtained among the various methods. Below we will address the effect of NQE on several properties, such as effective free energy barrier, tunnelling, donor–acceptor distance (DAD) at the TS, and KIE.

10.3.3 NQE Effects in DHFR

10.3.3.1 NQE on the Effective Free Energy Barrier

The combined effect of quantum vibrations and tunnelling effects is to lower the effective free energy barrier. The main source of the vibrational effect is the loss of a bound degree of freedom at the TS and hence loss of vibrational energy, resulting in an effectively lower barrier. The tunnelling effect is due to the wavelike behavior of light particles that results in a finite probability of crossing the barrier at energies below the barrier height.

In ec-DHFR most NQE approaches find that the combined result of NQE on the hydride transfer step is to reduce the effective free energy barrier by 1.8–3.5 kcal mol^{-1}.[43,173,198] Agarwal *et al.* employed the EVB-FGH approach in their studies of the hydride transfer step in DHFR and with this method the free energy barrier was reduced by 2.2–2.4 kcal mol^{-1} due to NQE.[135,198] It is possible that this value is a slight underestimation, as only the transferring hydrogen nucleus is treated as QM, and that only the ground state wavefunction is populated (*i.e.* no thermal effects). In a subsequent study by Garcia-Viloca, the EA-VTST/MT method was employed to elucidate the rate and the importance of NQE on the reaction.[173] In this work, the PES was described by a hybrid QM(AM1-SVB)/MM potential. The total contribution of NQE to barrier reduction was 3.1 kcal mol^{-1}.[173,177] The vibrational contribution to the barrier reduction was found to be 2.5 kcal mol^{-1}. Luk *et al.* also employed EA-VTST/MT,[180] in conjunction a QM(AM1-SRP)/MM potential developed by Doron *et al.*[43] In this work, the total magnitude of barrier reduction due to NQE was 1.8 kcal mol^{-1}, and of this 1.2 kcal mol^{-1} was ascribed to vibrational effects. The considerable difference in vibrational contribution compared to the work of Garcia-Viloca *et al.* may be ascribed to inaccuracies in the underlying AM1 PES,[43] which has been corrected in the AM1-SRP Hamiltonian.[43] Several studies employing different PI methods have also tackled the DHFR reaction. Doron *et al.* found using a QM(AM1-SRP)/MM potential that inclusion of NQE reduces the barrier by 2.1 kcal mol^{-1} within the centroid approximation,[43] whereas Boekelheide *et al.* found that quantisation reduces the barrier by 3.5 kcal mol^{-1} using EVB-RPMD. An EVB-QC-PIMC by Wang and Hammes-Schiffer found a barrier reduction of 3.5 kcal mol^{-1} due to NQE.[135] The difference between this latter value and the one obtained by the same group using the EVB-FGH method (2.2 kcal mol^{-1}), was carefully analysed by the authors and found to be due to the vibrationally adiabatic nature of the EVB-FGH method. Indeed, the PI approach naturally includes thermal excitations of vibrational states, and readily captures differences in the density of vibrational states in the RS and at the TS.

10.3.3.2 Effect of Tunnelling in DHFR

Wavefunction and PI methods do not facilitate a direct separation of the NQE into vibrational and tunnelling contributions. Indeed, to scrutinise the effect of tunnelling on chemical rates, the EA-VTST/MT method is invaluable. Multidimensional tunnelling calculations with EA-VTST/MT and QM(AM1-SVB)/MM suggested that on average, tunnelling reduces the free energy barrier by *ca.* 0.6 kcal mol^{-1}.[173,177] Similar values were obtained by Luk *et al.* using the same NQE method and QM(AM1-SRP)/MM.[180] The dominant tunnelling mechanism in ec-DHFR is the so-called small curvature (SC) tunnelling (Figure 10.4). Together, these results suggest that there is no extensive, deep tunnelling at play in DHFR. Rather, there is limited tunnelling at the top of the reaction barrier, with minor deviations from the classical reaction path.

The EVB-FGH-MDQT wavefunction approach of Agarwal *et al.* afforded profound insight into the nature of the hydride transfer.[198] The ensemble of TS configurations obtained corresponded to a near symmetric hydride double well potential. Inspection of the three-dimensional hydride wavefunctions showed bilobal functions delocalised between the carbon donor and acceptor atoms. Similar results were obtained using the PIEV method (Figure 10.7). Further analysis of the tunnelling splitting between the ground and excited vibrational states indicated that tunnelling plays a moderate role in the hydride transfer reaction.

Engel *et al.* employed the momentum distribution computed from open PI simulations with QM(AM1-SRP)/MM PES to recover fingerprints of tunnelling or quantum delocalisation at the TS in ec-DHFR (Figure 10.8).[154] In this work secondary features were observed in the high-energy tail of the momentum distribution and were used as indicators of quantum tunnelling or delocalisation. However, the tunnelling fingerprints were weak, and hence suggest only moderate tunnelling effects.

10.3.3.3 NQE on the Donor–Acceptor Distance

The thermal de Broglie wavelength for a hydrogen nucleus at ambient biological temperatures is comparable to the distance over which the hydrogen must move in a typical chemical reaction, hinting that tunnelling could play

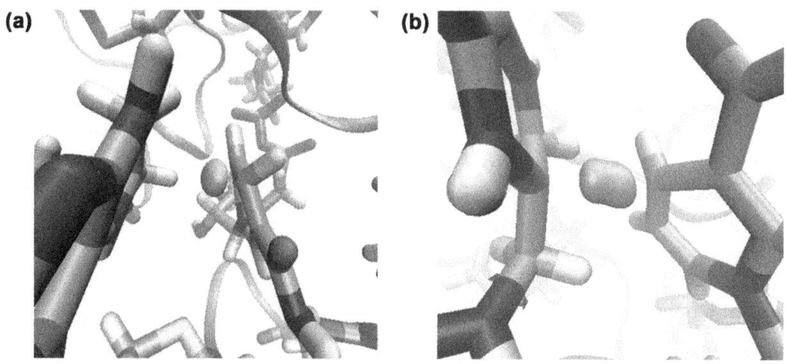

Figure 10.7 Ground state nuclear wavefunctions obtained for DHFR with the path-integral eigenvalue method. (a) Reactant state. (b) Transition state.

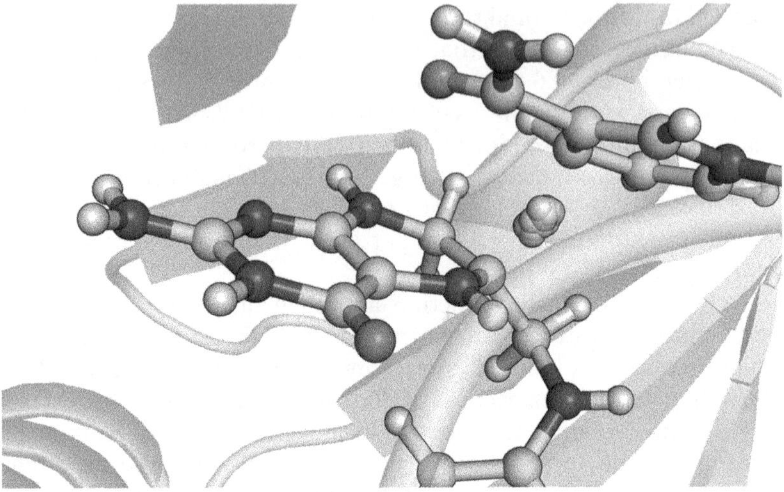

Figure 10.8 Open path-integral description of the transferring hydride at the transition state for DHFR.

a role. Moreover, the extent of tunnelling is highly sensitive to this distance, and it has been suggested that enzymes have a unique ability to reach so-called tunnelling-ready states (TRS).[19,163,188] Therefore, the optimal DAD at the TS is a useful metric to compute from enzyme simulations.

Garcia-Viloca *et al.* obtained a DAD of 2.68 Å at the TS using QM(AM1-SVB)/MM in conjunction with classical US simulations, whereas Fan *et al.* obtained a value of 2.71 Å with the same methodology. Interestingly, the latter study found a DAD of 2.70 Å at the TS for the M42W/G121V double mutant, which had been predicted by experiments to have rather different temperature-dependent barrier crossing behavior than the WT enzyme.[166,167] Moreover, a phenomenological model predicted DAD at the TRS of 3.06 and 3.34 Å for the WT and M42W/G121V double mutant, respectively,[199] in disagreement with atomistic simulations. We further note that the DAD at the TS for a model DHFR reaction in the gas phase is 2.76 Å using the AM1 Hamiltonian.[43] Hence, the DAD observed in these classical QM/MM enzyme simulations with an SVB-corrected AM1 method are very similar to the intrinsic gas-phase value for a model reaction.

Doron *et al.* studied the WT and several active site mutants in the temperature range 5–45 °C using QM(AM1-SRP)/MM and classical US simulations, and found that the DAD at the TS is 2.63–2.65 Å, irrespective of the DHFR variants and temperature. The above-mentioned phenomenological model predicted DAD at the TRS ranging from 3.06 to 3.31 Å for the WT and various active site mutants.[199] We again note that the DAD at the TS for a model DHFR reaction in the gas phase is 2.63 Å using the AM1-SRP Hamiltonian,[43] suggesting that the enzyme has little influence on the computed DAD at the TS. In the above mentioned work of Doron *et al.*, the DAD were obtained from classical US simulations; including NQE in the donor–acceptor distance increases the DAD by *ca.* 0.02 Å for WT and all mutant forms. Luk and coworkers obtained a DAD of 2.63 Å at the TS using the same QM(AM1-SRP)/MM Hamiltonian and classical simulations, but with a different computational protocol and software.[180] In a follow-up study, Ruiz-Pernía studied the double mutant N23PP/S148A, and concluded that in this kinetically impaired mutant there is greater coupling between enzyme motion and the chemical step.[181] Nevertheless, the DAD at TS was 2.66 Å, which is identical to the WT enzyme within the reported uncertainty.

Using the EVB-FGH approach, which is a direct quantum simulation method, Hammes-Schiffer and coworkers obtained TS DAD values of 2.73[198] for the WT enzyme, while a value of 2.74 Å for the G121V mutant form.[200] We note that the EVB potential employed in these studies is based partially on HF/6-31G** results, and gas-phase model calculations at this *ab initio* level resulted in a DAD of 2.69 Å (no NQE included). Miller and coworkers reported a value of 2.76 Å based on direct quantum EVB-RPMD simulations, while classical EVB simulations (*i.e.* no NQE) yielded a DAD of 2.72 Å.[136] Warshel and coworkers reported values in the range 2.6 to > 3 Å.[175,176,201]

Thus, we may conclude that atomistic QM/MM simulations do note predict a great sensitivity of the DAD on the enzyme environment, in contrast to experimental inferrals. We note that transition path simulations suggest that the biased simulation methods might slightly underestimate the DAD at the TS for such reactions, but the general conclusions reached using standard biased simulation techniques are nonetheless fully valid.[187] We further note that the donor-hydride-acceptor angles at the TS are very similar in model gas-phase reactions and in the enzyme.[43,198]

10.3.3.4 Kinetic Isotope Effects in DHFR

Agarwal *et al.* combined their EVB-FGH-MDQT approach to compute a KIE of 3.2,[198] which is in good agreement with the experimental estimates mentioned above. EA-VTST/MT studies yielded a KIE of 2.8, while a secondary KIE of 1.13 was predicted.[173] This secondary KIE value was later confirmed by experiments.[166] Importantly, it was suggested that the primary KIE is mainly a result of changes in bound vibrations, while the secondary KIE is entirely due to effects on the reaction coordinate, such as tunnelling. PI simulations have resulted in KIE values of 3.0–3.5, depending on the sampling scheme.[43]

A considerably more challenging test for multiscale simulation methods is prediction of the temperature dependence of KIE. This has seen some success,[176,177,202] although no consensus has yet emerged regarding the ability of modern methods to reproduce and explain experimental results. An initial attempt to reproduce the temperature independent KIE for WT DHFR, was due to Pu *et al.* who employed EA-VTST/MT to compute free energy profiles and KIEs for the hydride transfer reaction.[177] These calculations managed to reproduce the temperature independent, or weakly temperature-dependent, KIE. The authors pointed to two main reasons for this weak temperature dependence: (1) A change in the position of the variationally optimised TS along the predefined reaction coordinate, which resulted in gradually varying contribution of the quantum vibrations to the reaction rate. (2) A slight increase in the symmetry in the barrier as the temperature increased. This latter effect was then used to explain a slight increase in tunnelling. Subsequent studies using a similar, yet more accurate potential, was unable to reproduce these trends.[203] Fan *et al.* also computed the temperature dependence for DHFR and several mutants, and also obtained temperature independent KIEs for the WT enzyme, although a rational for this was not provided.[128] Warshel *et al.* also computed the temperature dependence of the KIE and concluded that this is largely due to variations in the DAD at different temperatures and for different mutants.[176] Specifically, these authors suggested that a longer DAD increases the KIE. Moreover, the varying temperature dependence in the KIE for WT and mutant forms of DHFR is due to the temperature dependence of the DAD. However, these conclusions were not based on direct simulations results, but rather the authors showed that by varying the DAD one could obtain KIE in agreement with experiments. Indeed, the authors concluded by stating "... it is extremely challenging to evaluate the temperature dependence by computer simulations that do not involve any special parametrisation. In fact, we are not aware of any successful attempts to reproduce the temperature dependence of KIE of an enzymatic reaction in a quantitative way by a first principle simulation, despite attempts to do so ...".[176]

Summary

We have presented some of the main conclusions arising from experimental and computational studies of NQE in the enzyme DHFR. From an experimental perspective, there is widespread agreement regarding the value of the KIE at room temperature, but greater fundamental variability regarding the temperature-dependent behavior of the KIE. From a computational perspective, there is remarkable agreement regarding many aspects of the hydride transfer reaction in DHFR. All methods suggest that: (i) the rate enhancement due to NQE is similar, with slight variations most likely due to differences in the underlying PES; (ii) there is only moderate tunnelling involved in the reaction; (iii) the DAD is rather insensitive to the enzyme environment; (iv) the predicted KIE is in reasonable agreement with experiment. Finally, predicting temperature-dependent KIE remains a challenge from first-principles methods.

10.4 Concluding Words

This chapter attempted to provide an overview of the different methods available for calculation of NQE in enzyme systems. A variety of different approaches were covered, and their application to DHFR, a widely studied enzyme, was presented. It is encouraging to see that very different approaches that are implemented in different software packages give fairly consistent results. Firstly, the prescriptions for inclusion of NQE, ranging from semiclassical to fully quantum wavefunction treatments, *via* Feynman PI methods, all give rather similar results. Secondly, the different studies mentioned above used very different PES treatments for the QM region within the QM/MM formalism. Finally, these studies were performed by many different research groups and have been tested and retested over the years. Yet, challenges still persist, particularly in the area of temperature-dependent KIEs, which are particularly difficult to predict accurately.

References

1. P. A. Frey and A. D. Hegeman, *Enzymatic Reaction Mechanisms*, Oxford University Press, New York, 2007.
2. R. Wolfenden and M. J. Snider, *Acc. Chem. Res.*, 2001, **34**, 938–945.
3. R. A. Alberty and G. G. Hammes, *J. Phys. Chem.*, 1958, **62**, 154–159.
4. G. G. Hammes, *Biochemistry*, 2002, **41**, 8221–8228.
5. E. D. Watt, H. Shimada, E. L. Kovrigin and J. P. Loria, *Proc. Natl. Acad. Sci. U. S. A.*, 2007, **104**, 11981–11986.
6. L. Pauling, *Chem. Eng. News*, 1946, **24**, 1375–1377.
7. A. Warshel, *Proc. Natl. Acad. Sci. U. S. A.*, 1978, **75**, 5250–5254.
8. R. L. Schowen, *Transition States of Biochemical Processes*, Plenum, New York, 1978.
9. A. Warshel, P. K. Sharma, M. Kato, Y. Xiang, H. Liu and M. H. M. Olsson, *Chem. Rev.*, 2006, **106**, 3210–3235.
10. A. Warshel, *Proc. Natl. Acad. Sci. U. S. A.*, 1984, **81**, 444–448.
11. N. Wu, Y. Mo, J. Gao and E. F. Pai, *Proc. Natl. Acad. Sci. U. S. A.*, 2000, **97**, 2017–2022.
12. M. J. S. Dewar and D. M. Storch, *Proc. Natl. Acad. Sci. U. S. A.*, 1985, **82**, 2225–2229.
13. X. Zhang and K. N. Houk, *Acc. Chem. Res.*, 2005, **38**, 379–385.
14. S. D. Schwartz and V. L. Schramm, *Nat. Chem. Biol.*, 2009, **5**, 551–558.
15. S. Hay and N. S. Scrutton, *Nat. Chem.*, 2012, **4**, 161–168.
16. M. Weitman and D. T. Major, *J. Am. Chem. Soc.*, 2010, **132**, 6349–6360.
17. D. T. Major and M. Weitman, *J. Am. Chem. Soc.*, 2012, **134**, 19454–19462.
18. D. T. Major, A. Heroux, A. M. Orville, M. P. Valley, P. F. Fitzpatrick and J. Gao, *Proc. Natl. Acad. Sci. U. S. A.*, 2009, **106**, 20734–20739.
19. J. P. Klinman and A. Kohen, *Annu. Rev. Biochem.*, 2013, **82**, 471–496.
20. C. J. Cramer, *Essentials of Computational Chemistry: Theories and Models*, John Wiley & Sons, Ltd, Wiltshire, England, 2004.
21. A. Warshel and M. Levitt, *J. Mol. Biol.*, 1976, **103**, 227–249.
22. A. Warshel and A. Bromberg, *J. Chem. Phys.*, 1970, **52**, 1262–1269.
23. A. Warshel and M. Karplus, *J. Am. Chem. Soc.*, 1972, **94**, 5612–5625.
24. U. C. Singh and P. A. Kollman, *J. Comput. Chem.*, 1986, **7**, 718–730.
25. M. J. Field, P. A. Bash and M. Karplus, *J. Comput. Chem.*, 1990, **11**, 700–733.

26. J. Gao, *J. Phys. Chem.*, 1992, **96**, 537–540.
27. J. Gao, in *Reviews in Computational Chemistry*, ed. K. B. Lipkowitz and D. B. Boyd, VCH Publishers, New York, 1995, vol. 7, pp. 119–185.
28. H. M. Senn and W. Thiel, *Angew. Chem., Int. Ed.*, 2009, **48**, 1198–1229.
29. H. Lin and D. G. Truhlar, *Theor. Chem. Acc.*, 2007, **117**, 185–199.
30. K. Meier, A. Choutko, J. Dolenc, A. P. Eichenberger, S. Riniker and W. F. V. Gunsteren, *Angew. Chem., Int. Ed.*, 2013, **52**, 2820–2834.
31. A. T. P. Carvalho, A. Barrozo, D. Doron, A. V. Kilshtain, D. T. Major and S. C. L. Kamerlin, *J. Mol. Graphics Modell.*, 2014, **54**, 62–79.
32. I. Rossi and D. G. Truhlar, *Chem. Phys. Lett.*, 1995, **233**, 231–236.
33. A. Warshel and R. M. Weiss, *J. Am. Chem. Soc.*, 1980, **102**, 6218–6226.
34. A. Warshel, *Computer Modeling of Chemical Reactions in Enzymes and Solutions*, John Wiley & Sons, New York, 1991.
35. L. S. Devi-Kesavan, M. Garcia-Viloca and J. Gao, *Theor. Chem. Acc.*, 2003, **109**, 133–139.
36. A. Vardi-Kilshtain, A. Azuri and D. T. Major, *J. Comput. Chem*, 2012, **33**, 435–441.
37. A. Vardi-Kilshtain, D. Doron and D. T. Major, *Biochemistry*, 2013, **52**, 4382–4390.
38. D. T. Major, D. M. York and J. Gao, *J. Am. Chem. Soc.*, 2005, **127**, 16374–16375.
39. D. T. Major and J. L. Gao, *J. Am. Chem. Soc.*, 2006, **128**, 16345–16357.
40. A. Rubinstein and D. T. Major, *J. Am. Chem. Soc.*, 2009, **131**, 8513–8521.
41. A. Rubinstein and D. T. Major, *Biochemistry*, 2010, **49**, 3957–3964.
42. A. Vardi-Kilshtain, D. T. Major, A. Kohen and D. Doron, *J. Chem. Theory Comput.*, 2012, **8**, 4786–4796.
43. D. Doron, D. T. Major, A. Kohen, W. Thiel and X. Wu, *J. Chem. Theory Comput.*, 2011, **7**, 3420–3437.
44. M. Levitt and A. Warshel, *Nature*, 1975, **253**, 694–698.
45. M. Levitt, *J. Mol. Biol.*, 1976, **104**, 59–107.
46. M. Karplus and J. A. McCammon, *Nat. Struct. Biol.*, 2002, **9**, 646–652.
47. M. P. Allen and D. J. Tildesley, *Computer Simulation of Liquids*, Oxford University Press, Oxford, 1989.
48. G. N. Patey and J. P. Valleau, *J. Chem. Phys.*, 1975, **63**, 2334–2339.
49. P. A. Bash, U. C. Singh, F. K. Brown, R. Langridge and P. A. Kollman, *Science*, 1987, **235**, 574–576.
50. E. A. Carter, G. Ciccotti, J. T. Hynes and R. Kapral, *Chem. Phys. Lett.*, 1989, **156**, 472–477.
51. A. Ferrenberg and R. Swendsen, *Phys. Rev. Lett.*, 1989, **63**, 1195–1198.
52. H. Grubmüller, *Phys. Rev. E: Stat., Nonlinear, Soft Matter Phys.*, 1995, **52**, 2893–2906.
53. C. Jarzynski, *Phys. Rev. Lett.*, 1997, **78**, 2690–2693.
54. C. Jarzynski, *Phys. Rev. E*, 1997, **56**, 5018–5035.
55. G. M. Torrie and J. P. Valleau, *J. Comp. Phys*, 1977, **23**, 187–199.
56. J. A. McCammon and M. Karplus, *Proc. Natl. Acad. Sci. U. S. A.*, 1979, **76**, 3585–3589.

57. S. H. Northrup, M. R. Pear, C. Y. Lee, J. A. McCammon and M. Karplus, *Proc. Natl. Acad. Sci. U. S. A.*, 1982, **79**, 4035–4039.
58. S. Kumar, D. Bouzida, R. H. Swendsen, P. A. Kollman and J. M. Rosenberg, *J. Comput. Chem.*, 1992, **13**, 1011–1021.
59. R. Rajamani, K. J. Naidoo and J. Gao, *J. Comput. Chem.*, 2003, **24**, 1775–1781.
60. D. Doron, A. Kohen and D. T. Major, *J. Chem. Theory Comput.*, 2012, **8**, 2484–2496.
61. W. E. W. Ren and E. Vanden-Eijnden, *Phys. Rev. B*, 2002, **66**, 052301.
62. R. W. Zwanzig, *J. Chem. Phys.*, 1954, **22**, 1420–1426.
63. J. G. Kirkwood, *J. Chem. Phys.*, 1935, **3**, 300–313.
64. P. G. Bolhuis, C. Dellago, D. Chandler and P. Geissler, *Annu. Rev. Phys. Chem.*, 2002, **53**, 291–318.
65. A. Kohen and H. H. Limbach, *Isotope Effects in Chemistry and Biology*, Taylor and Francis Group, CRC Press, New York, 2006.
66. A. Vardi-Kilshtain, N. Nitoker and D. T. Major, *Arch. Biochem. Biophys.*, 2015, **582**, 18–27.
67. K. Y. Wong, Y. Q. Xu and L. Xu, *Biochim. Biophys. Acta*, 2015, **1854**, 1782–1794.
68. J.-K. Hwang and A. Warshel, *J. Am. Chem. Soc.*, 1996, **118**, 11745–11751.
69. D. G. Truhlar, J. Gao, C. Alhambra, M. Garcia-Viloca, J. Corchado, M. L. Sanchez and J. Villa, *Acc. Chem. Res.*, 2002, **35**, 341–349.
70. S. Hammes-Schiffer, *Curr. Opin. Struct. Biol.*, 2004, **2004**, 192–201.
71. H. Kragh, *Arch. Hist. Exact Sci.*, 2012, **66**, 199–240.
72. L. Pauling, *The Nature of the Chemical Bond and the Structure of Molecules and Crystals: An Introduction to Modern Structural Chemistry*, Cornell University Press, Ithaca, New York, 1960.
73. X.-Z. Li, B. Walker and A. Michaelides, *Proc. Natl. Acad. Sci. U. S. A.*, 2011, **108**, 6369–6373.
74. M. Ceriotti, J. Cuny, M. Parinello and D. E. Manolopoulos, *Proc. Natl. Acad. Sci. U. S. A.*, 2013, **110**, 15591–15596.
75. L. Wang, S. D. Fried, S. G. Boxer and T. E. Markland, *Proc. Natl. Acad. Sci. U. S. A.*, 2015, **111**, 18454–18459.
76. G. F. Reiter, R. Senesi and J. Mayers, *Phys. Rev. Lett.*, 2010, **105**, 148101–148104.
77. G. F. Reiter, A. I. Kolesnikov, S. J. Paddison, P. M. Platzman, A. P. Moravsky, M. A. Adams and J. Mayers, *Phys. Rev. B*, 2012, **85**, 045403–045405.
78. R. Senesi, G. Romanelli, M. A. Adams and C. Andreani, *Chem. Phys.*, 2013, **427**, 111–116.
79. R. P. Bell, *The Tunnel Effect in Chemistry*, Chapman and Hall, London, 1980.
80. R. J. McMahon, *Science*, 2003, **299**, 833–834.
81. D. Ley, D. Gerbig and P. R. Schreiner, *Org. Biomol. Chem.*, 2012, **10**, 3781–3790.

82. R. J. Shannon, M. A. Blitz, A. Goddard and D. E. Heard, *Nat. Chem.*, 2013, **5**, 745–749.
83. J. Z. Pu, J. L. Gao and D. G. Truhlar, *Chem. Rev.*, 2006, **106**, 3140–3169.
84. J. P. Layfield and S. Hammes-Schiffer, *Chem. Rev.*, 2014, **114**, 3466–3494.
85. P. M. Kiefer and J. T. Hynes, *J. Phys. Org. Chem.*, 2010, **23**, 632–646.
86. K. M. Doll, B. R. Bender and R. G. Finke, *J. Am. Chem. Soc.*, 2003, **125**, 10877–10884.
87. P. Ball, *Nature*, 2004, **431**, 396–397.
88. H. Eyring, *J. Chem. Phys.*, 1935, **3**, 107–115.
89. G. K. Schenter, B. C. Garret and D. G. Truhlar, *J. Chem. Phys.*, 2003, **119**, 5828–5833.
90. P. Hanggi, P. Talkner and M. Borkovec, *Rev. Mod. Phys.*, 1990, **62**, 251–341.
91. S. P. Webb and S. Hammes-Schiffer, *J. Chem. Phys.*, 2000, **113**, 5214–5227.
92. K. Hinsen and B. Roux, *J. Chem. Phys.*, 1997, **106**, 3567–3577.
93. C. Alhambra, J. Corchado, M. L. Sanchez, M. Garcia-Viloca, J. Gao and D. G. Truhlar, *J. Phys. Chem. B*, 2001, **105**, 11326–11340.
94. G. A. Voth, *J. Phys. Chem.*, 1993, **97**, 8365–8377.
95. A. Staib, D. Borgis and J. T. Hynes, *J. Chem. Phys.*, 1995, **102**, 2487–2505.
96. J. Villa and A. Warshel, *J. Phys. Chem. B*, 2001, **105**, 7887–7907.
97. W. H. Miller, *Acc. Chem. Res.*, 1993, **26**, 174–181.
98. J. K. Hwang, Z. T. Chu, A. Yadav and A. Warshel, *J. Phys. Chem.*, 1991, **95**, 8445–8448.
99. J. K. Hwang and A. Warshel, *J. Phys. Chem.*, 1993, **97**, 10053–10058.
100. D. Doron, M. Weitman, A. Vardi-Kilshtain, A. Azuri, H. Engel and D. T. Major, *Isr. J. Chem.*, 2014, **54**, 1108–1117.
101. S. R. Billeter, S. P. Webb, T. Iordanov, P. K. Agarwal and S. Hammes-Schiffer, *J. Chem. Phys.*, 2001, **114**, 6925–6936.
102. V. L. Schramm, *Acc. Chem. Res.*, 2015, **48**, 1032–1039.
103. K. Swiderek and P. Paneth, *Chem. Rev.*, 2013, **113**, 7851–7879.
104. B. E. Lewis and V. L. Schramm, in *Isotope Effects in Chemistry and Biology*, ed. A. Kohen and H. H. Limbach, Taylor and Francis Group, CRC Press, New York, 2006.
105. J. A. Barnes and I. H. Williams, *Chem. Commun.*, 1996, 193–194.
106. J. A. Barnes and I. H. Williams, *Biochem. Soc. Trans.*, 1996, **24**, 263–268.
107. A. Fernandez-Ramos, B. A. Ellingson, B. C. Garret and D. G. Truhlar, in *Reviews in Computational Chemistry*, ed. K. B. Lipkowitz and T. R. Cundari, Wiley-VCH, Hoboken, NJ, 2007, vol. 23, pp. 125–232.
108. J. C. Corchado, Y.-Y. Chuang, P. L. Fast, W.-P. Hu, Y.-P. Liu, G. C. Lynch, K. A. Nguyen, C. F. Jackels, A. F. Ramos, B. A. Ellingson, B. J. Lynch, V. S. Melissas, J. Villa, I. Rossi, E. L. Costino, J. Pu, T. V. Albu, R. Steckler, B. C. Garrett, A. D. Isaacson and D. G. Truhlar, POLYRATE 9.3.1, (2005), University of Minnesota: Minneapolis.

109. L. Masgrau and D. G. Truhlar, *Acc. Chem. Res.*, 2015, **48**, 431–438.
110. R. M. Stratt, *Acc. Chem. Res.*, 1995, **28**, 201–207.
111. B. R. Brooks, R. E. Bruccoleri, B. D. Olafson, D. J. States, S. Swaminathan and M. Karplus, *J. Comput. Chem.*, 1983, **4**, 187–217.
112. B. R. Brooks, C. L. Brooks III, A. D. MacKerell Jr., L. Nilsson, R. J. Petrella, B. Roux, Y. Won, G. Archontis, C. Bartels, S. Boresch, A. Caflisch, L. Caves, Q. Cui, A. R. Dinner, M. Feig, S. Fischer, J. Gao, M. Hodoscek, W. Im, K. Kuczera, T. Lazaridis, J. Ma, V. Ovchinnikov, E. Paci, R. W. Pastor, C. B. Post, J. Z. Pu, M. Schaefer, B. Tidor, R. M. Venable, H. L. Woodcock, X. Wu, W. Yang, D. M. York and M. Karplus, *J. Comput. Chem.*, 2009, **30**, 1545–1614.
113. M. J. Field, M. Albe, C. Bret, F. Proust-De Martin and A. Thomas, *J. Comput. Chem.*, 2000, **21**, 1088–1100.
114. T. Iordanov, S. R. Billeter, S. P. Webb and S. Hammes-Schiffer, *Chem. Phys. Lett.*, 2001, **338**, 389–397.
115. J. B. Watney, A. V. Soudackov, K. F. Wong and S. Hammes-Schiffer, *Chem. Phys. Lett.*, 2006, **418**, 268–271.
116. S. Melchionna and S. Cozzini, DLPROTEIN, (2001), INFM UDr SISSA, Triste, Italy.
117. R. P. Feynman and A. R. Hibbs, *Quantum Mechanics and Path Integrals*, McGraw-Hill, New York, 1965.
118. B. J. Berne and D. Thirumalai, *Annu. Rev. Phys. Chem.*, 1986, **37**, 401–424.
119. D. T. Major and J. L. Gao, *J. Mol. Graphics Modell.*, 2005, **24**, 121–127.
120. E. L. Pollock and D. M. Ceperley, *Phys. Rev. B*, 1984, **30**, 2555–2568.
121. D. M. Ceperley, *Rev. Mod. Phys.*, 1995, **67**, 279–355.
122. D. T. Major, M. Garcia-Viloca and J. L. Gao, *J. Chem. Theory Comput.*, 2006, **2**, 236–245.
123. M. Sprik, M. L. Klein and D. Chandler, *Phys. Rev. B*, 1985, **31**, 4234–4244.
124. A. Azuri, H. Engel, D. Doron and D. T. Major, *J. Chem. Theory. Comput.*, 2011, **7**, 1273–1286.
125. D. T. Major and J. L. Gao, *J. Chem. Theory Comput.*, 2007, **3**, 949–960.
126. J. Gao, D. T. Major, Y. Fan, Y.-l. Lin, S. Ma and K. Y. Wong, *Molecular Modeling of Proteins*, Springer Verlag, 2008.
127. J. Gao, K.-Y. Wong, D. T. Major, A. Cembran, L. Song, Y.-l. Lin, Y. Fan and S. Ma, in *Quantum Tunnelling in Enzyme-Catalysed Reactions*, The Royal Society of Chemistry, 2009, pp. 105–134.
128. Y. Fan, A. Cembran, S. Ma and J. Gao, *Biochemistry*, 2013, **52**, 2036–2049.
129. A. J. Adamczyk, J. Cao, S. C. L. Kamerlin and A. Warshel, *Proc. Natl. Acad. Sci. U. S. A.*, 2011, **108**, 14115–14120.
130. J. Vaníček and W. H. Miller, *J. Chem. Phys.*, 2007, **127**, 114309–114318.
131. M. Ceriotti and T. E. Markland, *J. Chem. Phys.*, 2013, **138**, 014112.
132. S. A. Chin, *Phys. Lett. A*, 1997, **226**, 344–348.
133. S. A. Chin and C. R. Chen, *J. Chem. Phys.*, 2002, **117**, 1409–1415.

134. S. A. Chin, *Phys. Rev. E*, 2004, **69**, 046118–046125.
135. Q. Wang and S. Hammes-Schiffer, *J. Chem. Phys.*, 2006, **125**, 184102–184111.
136. N. Boekelheide, R. Salomon-Ferrer and T. F. Miller III, *Proc. Natl. Acad. Sci. U. S. A.*, 2011, **108**, 16159–16163.
137. J. B. Rommel, Y. Liu, H.-J. Werner and J. Kästner, *J. Phys. Chem. B*, 2012, **116**, 13682–13689.
138. J. Mayers, G. F. Reiter and P. Platzman, *J. Mol. Struct.*, 2002, **615**, 275–282.
139. G. F. Reiter, J. Mayers and J. Noreland, *Phys. Rev. B: Condens. Matter Mater. Phys.*, 2002, **65**, 104305.
140. G. F. Reiter, J. Mayers and P. Platzman, *Phys. Rev. Lett.*, 2002, **89**, 135505–135508.
141. C. J. Burnham, G. F. Reiter, J. Mayers, T. Abdul-Redah, H. Reichert and H. Dosch, *Phys. Chem. Chem. Phys.*, 2006, **8**, 3966–3977.
142. C. Pantalei, A. Pietropaolo, R. Senesi, S. Imberti, C. Andreani, J. Mayers, C. Burnham and G. Reiter, *Phys. Rev. Lett.*, 2008, **100**, 177801–177804.
143. A. Pietropaolo, R. Senesi, C. Andreani, A. Botti, M. A. Ricci and F. Bruni, *Phys. Rev. Lett.*, 2008, **100**, 127802–127805.
144. V. Garbuio, C. Andreani, S. Imberti, A. Pietropaolo, G. F. Reiter, R. Senesi and M. A. Ricci, *J. Chem. Phys.*, 2007, **127**, 154501–154511.
145. D. Homouz, G. Reiter, J. Eckert, J. Mayers and R. Blinc, *Phys. Rev. Lett.*, 2007, **98**, 115502–115505.
146. R. Senesi, A. Pietropaolo, A. Bocedi, S. E. Pagnotta and F. Bruni, *Phys. Rev. Lett.*, 2007, **98**, 138102–138105.
147. J. A. Morrone, V. Srinivasan, D. Sebastiani and R. Car, *J. Chem. Phys.*, 2007, **126**, 234504–234513.
148. J. A. Morrone and R. Car, *Phys. Rev. Lett.*, 2008, **101**, 017801–017804.
149. J. A. Morrone, L. Lin and R. Car, *J. Chem. Phys.*, 2009, **130**, 204511–204524.
150. L. Lin, J. A. Morrone, R. Car and M. Parrinello, *Phys. Rev. Lett.*, 2010, **105**, 110602–110606.
151. L. Lin, J. A. Morrone and R. Car, *J. Stat. Phys.*, 2011, **145**, 365–384.
152. L. Lin, J. A. Morrone, R. Car and M. Parrinello, *Phys. Rev. B*, 2011, **83**, 220302–220305.
153. A. Perez and M. E. Tuckerman, *J. Chem. Phys.*, 2011, **135**, 064104–064120.
154. H. Engel, D. Doron, A. Kohen and D. T. Major, *J. Chem. Theory Comput.*, 2012, **8**, 1223–1234.
155. O. Schutt and D. Sebastiani, *J. Comp. Chem.*, 2013, **34**, 827–835.
156. A. Sethia, S. Sanyal and Y. Singh, *J. Chem. Phys.*, 1990, **93**, 7268–7276.
157. M. Dixit, H. Engel, R. Eitan, D. Aurbach, M. Levi, M. Kosa and D. T. Major, *J. Phys. Chem. C*, 2015, **119**, 15801–15809.
158. W. H. Miller, S. D. Schwartz and J. W. Tromp, *J. Chem. Phys.*, 1983, **79**, 4889–4898.

159. W. W. Cleland, *Arch. Biochem. Biophys.*, 2005, **433**, 2–12.
160. F. E. Romesberg and R. L. Schowen, *Adv. Phys. Org. Chem.*, 2004, **39**, 27–77.
161. V. L. Schramm, *Curr. Opin. Struct. Biol.*, 2001, **11**, 657–665.
162. V. L. Schramm, *Curr. Opin. Chem. Biol.*, 2007, **11**, 529–536.
163. Z. D. Nagel and J. P. Klinman, *Nat. Chem. Biol.*, 2009, **5**, 543–550.
164. C. A. Fierke, K. A. Johnson and S. J. Benkovic, *Biochemistry*, 1987, **26**, 4085–4092.
165. P. Rajagopalan, S. Lutz and S. Benkovic, *Biochemistry*, 2002, **41**, 12618–12628.
166. R. S. Sikorski, L. Wang, K. A. Markham, P. T. R. Rajagopalan, S. J. Benkovic and A. Kohen, *J. Am. Chem. Soc.*, 2004, **126**, 4778–4779.
167. L. Wang, N. M. Goodey, S. J. Benkovic and A. Kohen, *Proc. Natl. Acad. Sci. U. S. A.*, 2006, **103**, 15753–15758.
168. L. Wang, N. M. Goodey, S. J. Benkovic and A. Kohen, *Philos. Trans. R. Soc., B*, 2006, **361**, 1307–1315.
169. L. Wang, S. Tharp, T. Selzer, S. J. Benkovic and A. Kohen, *Biochemistry*, 2006, **45**, 1383–1392.
170. V. Stojković, L. L. Perissinotti, J. Lee, S. J. Benkovic and A. Kohen, *Chem. Commun.*, 2010, **46**, 8974–8976.
171. V. Stojković, L. L. Perissinotti, D. Willmer, S. J. Benkovic and A. Kohen, *J. Am. Chem. Soc.*, 2012, **134**, 1738–1745.
172. R. Castillo, J. Andres and V. Moliner, *J. Am. Chem. Soc.*, 1999, **121**, 12140–12147.
173. M. Garcia-Viloca, D. G. Truhlar and J. Gao, *Biochemistry*, 2003, **42**, 13558–13575.
174. I. F. Thorpe and C. L. Brooks, *J. Phys. Chem. B*, 2003, **107**, 14042–14051.
175. H. Liu and A. Warshel, *Biochemistry*, 2007, **46**, 6011–6025.
176. H. Liu and A. Warshel, *J. Phys. Chem. B*, 2007, **111**, 7852–7861.
177. J. Pu, S. Ma, J. Gao and D. G. Truhlar, *J. Phys. Chem. B*, 2005, **109**, 8551–8556.
178. J. Pu, S. Ma, M. Garcia-Viloca, J. Gao, D. G. Truhlar and A. Kohen, *J. Am. Chem. Soc.*, 2005, **127**, 14879–14886.
179. E. J. Loveridge, E. M. Behiry, J. Guo and R. K. Allemann, *Nat. Chem*, 2012, **4**, 292–297.
180. L. Y. P. Luk, J. J. Ruiz-Pernía, W. M. Dawson, M. Roca, E. J. Loveridge, D. R. Glowacki, J. N. Harvey, A. J. Mulholland, I. Tuñón, V. Moliner and R. K. Allemann, *Proc. Natl. Acad. Sci. U. S. A.*, 2013, **110**, 16344–16349.
181. J. J. Ruiz-Pernia, L. Y. P. Luk, R. García-Meseguer, S. Martí, E. J. Loveridge, I. A. Tuñón, V. Moliner and R. K. Allemann, *J. Am. Chem. Soc.*, 2013, **135**, 18689–18696.
182. L. Y. P. Luk, J. J. Ruiz-Pernía, W. M. Dawson, E. J. Loveridge, I. a. Tuñón, V. Moliner and R. K. Allemann, *J. Am. Chem. Soc.*, 2014, **136**, 17317–17323.

183. R. S. Swanwick, G. Maglia, L. Tey and R. K. Allemann, *Biochem. J.*, 2006, **394**, 259–265.
184. D. Roston, A. Kohen, D. Doron and D. T. Major, *J. Comput. Chem.*, 2014, **35**, 1411–1417.
185. D. D. Boehr, D. McElheny, H. J. Dyson and P. E. Wright, *Science*, 2006, **313**, 1638–1642.
186. G. Bhabha, J. Lee, D. C. Ekiert, J. Gam, I. A. Wilson, H. J. Dyson, S. J. Benkovic and P. E. Wright, *Science*, 2011, **332**, 234–238.
187. D. Doron, A. Kohen, K. Nam and D. T. Major, *J. Chem. Theor. Comput.*, 2014, **10**, 1863–1871.
188. D. B. Northrop, *Biochemistry*, 1975, **14**, 2644–2651.
189. K. Francis and A. Kohen, *Curr. Opin. Chem. Biol.*, 2014, **21**, 19–24.
190. P. Singh, A. Sen, K. Francis and A. Kohen, *J. Am. Chem. Soc.*, 2014, **136**, 2575–2582.
191. E. J. Loveridge and R. K. Allemann, *ChemBioChem*, 2011, **12**, 1258–1262.
192. K. Francis, V. Stojković and A. Kohen, *J. Biol. Chem.*, 2013, **288**, 35961–35968.
193. Z. Wang, D. Antoniou, S. D. Schwartz and V. L. Schramm, *Biochemistry*, 2016, **55**, 157–166.
194. A. Kohen, *Acc. Chem. Res.*, 2015, **48**, 466–473.
195. L. Y. P. Luk, J. J. Ruiz-Pernía, A. S. Adesina, E. J. Loveridge, I. Tuñón, V. Moliner and R. K. Allemann, *Angew. Chem., Int. Ed.*, 2015, **54**, 9016–9020.
196. C. T. Liu, K. Francis, J. P. Layfield, X. Huang, S. Hammes-Schiffer, A. Kohen and S. J. Benkovic, *Proc. Natl. Acad. Sci. U. S. A.*, 2014, **111**, 18231–18236.
197. M. Dametto, D. Antoniou and S. D. Schwartz, *Mol. Phys.*, 2012, **110**, 531–536.
198. P. K. Agarwal, S. R. Billeter and S. Hammes-Schiffer, *J. Phys. Chem. B*, 2002, **106**, 3283–3293.
199. Z. Wang, P. Singh, C. M. Czekster, A. Kohen and V. L. Schramm, *J. Am. Chem. Soc.*, 2014, **136**, 8333–8341.
200. D. Roston, C. M. Cheatum and A. Kohen, *Biochemistry*, 2012, **51**, 6860–6870.
201. J. B. Watney, P. K. Agarwal and S. Hammes-Schiffer, *J. Am. Chem. Soc.*, 2003, **125**, 3745–3750.
202. D. R. Glowacki, J. N. Harvey and A. J. Mulholland, *Nat. Chem*, 2012, **4**, 169–176.
203. D. Doron, V. Stojkovic, L. Gakhar, A. Kohen and D. T. Major, *J. Phys. Chem. B*, 2015, **119**, 906–916.

Section III: Applications

CHAPTER 11

QM/MM Methods for Simulating Enzyme Reactions

KARA E. RANAGHAN AND ADRIAN J. MULHOLLAND*

Centre for Computational Chemistry, University of Bristol, Clifton, Bristol, BS8 1TS, UK
*Email: Adrian.Mulholland@bristol.ac.uk

11.1 Introduction

The application of combined quantum mechanics/molecular mechanics (QM/MM) methods to the study of biological systems is a thriving and growing area, with advances in computer hardware and software opening up exciting new avenues of research. QM/MM methods have proved particularly useful in studies of enzyme-catalysed reaction mechanisms, but are increasingly also finding application in other areas such as biomolecular spectroscopy and drug design. The award of the 2013 Nobel Prize for Chemistry to Martin Karplus, Michael Levitt and Arieh Warshel recognised the development of QM/MM methods and the important role these methods can play in understanding biological systems.[1–4] Figure 11.1 shows a plot of the results of a search of the ISI Web of Science (v5.21: http://wok.mimas.ac.uk) using the criteria: Topic = (qm/mm or mm/qm or qm-mm or mm-qm or qm:mm or mm:qm or quantum mechanics/molecular mechanics or quantum mechanical/molecular mechanical) and Topic = (enzyme) sorted by publication year. The number of papers describing QM/MM studies of enzymes has grown dramatically over the past two decades, reaching almost 200 studies in 2014. The number of citations (for the same search criteria) has grown from only a few tens of papers per year to thousands of citations per year over the

RSC Theoretical and Computational Chemistry Series No. 9
Simulating Enzyme Reactivity: Computational Methods in Enzyme Catalysis
Edited by Iñaki Tuñón and Vicent Moliner

Published by the Royal Society of Chemistry, www.rsc.org

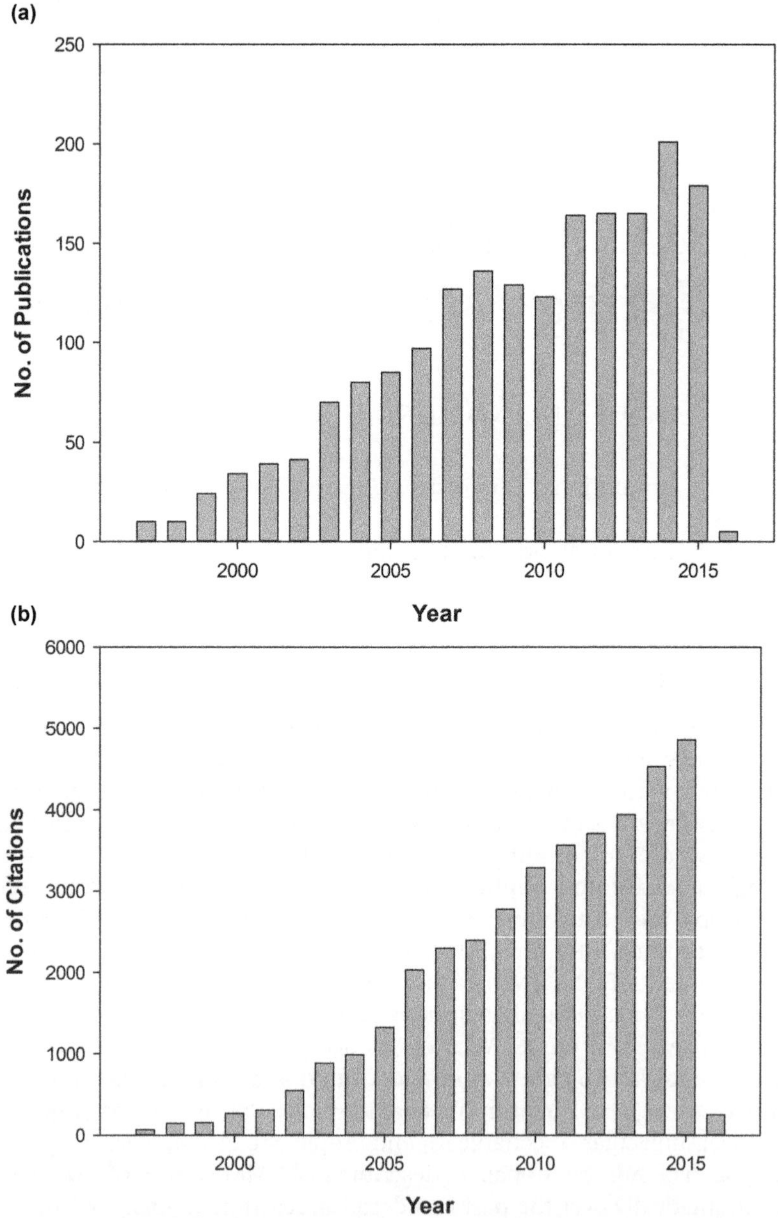

Figure 11.1 (a) Growth in publications and citations of work applying QM/MM methods to enzymes over the last 20 years: results of a search (February 2016) on ISI Web of Science (v5.21: http://wok.mimas.ac.uk/) using the criteria: Topic = (qm/mm or mm/qm or qm-mm or mm-qm or qm:mm or mm:qm or quantum mechanics/molecular mechanics or quantum mechanical/molecular mechanical) and Topic = (enzyme) sorted by publication year. (b) The citations in each year for the same search criteria.

time period. The whole field of QM/MM modelling of enzyme reactions is now too large to reasonably cover all QM/MM studies of enzymes in recent years in a single review. Progress in the field has been discussed in detail in many useful reviews.[5–9] Practical aspects of setting up and running QM/MM calculations have also been described in, for example, the Tutorial Review of Lonsdale *et al.*[10] Many organisations and networks provide practical training in QM/MM methods, for example CCP-BioSim (the UK Collaborative Computational Project on Biomolecular Simulation, see ccpbiosim.ac.uk).

Despite the recognition that QM/MM methods and other molecular simulation techniques now receive, some experimentalists still remain sceptical about molecular modelling and simulation in general, and about QM/MM methods in particular. It is important to view the results of modelling with a critical eye, considering the capabilities of the methods used, their strengths and limitations, and ranges of application, in assessing results critically and realistically. Bond breaking and formation processes are often what one aims to understand using QM/MM methods (particularly in studies of enzyme-catalysed reactions, of course). One crucial factor is the accuracy of the QM method employed in a QM/MM calculation. It is now possible in some cases to approach 'chemical accuracy' (*i.e.* energy barriers accurate to $\sim$1 kcal mol^{-1}) for reaction barriers with QM/MM calculations using coupled cluster methods for the QM region,[11,12] but these methods are very computationally expensive, particularly for large QM regions. The spin-component-scaled (SCS) version of MP2, SCS-MP2[13] is a promising method for QM/MM applications: it has been shown to give results close to that of coupled cluster methods without any noticeable additional computational expense above standard MP2 calculations.[12,14] In general, the expense of such correlated *ab initio* methods currently precludes QM/MM geometry optimisations (energy profiles would typically be calculated from single point calculations on structures from lower levels of QM theory) or applications to large QM regions or those that contain transition metals. Density functional theory (DFT) nowadays provides the typical 'work horse' QM methods for QM/MM applications (at least those not involving extensive molecular dynamics simulations) because it generally provides a good compromise between accuracy and computational expense. Unlike *ab initio* methods, DFT is not systematically improvable and the best choice of functional for a particular application is not always obvious. B3LYP has become the de facto functional applied for many QM/MM studies, but mechanistic conclusions based on this level of theory alone may not always be reliable.[12,14,15] QM/MM MD simulations at the DFT QM level are possible, but the computational expense restricts simulations to relatively short timescales. Semiempirical QM methods such as SCC-DFTB[16] or PM3[17,18] are still extensively used in QM/MM calculations because their much more modest computational expense allows QM/MM MD simulations to be performed for longer timescales.[19–26] As with DFT-based QM methods, the choice of semiempirical method for a particular application may significantly influence the accuracy of the QM/MM calculation, and where possible, benchmarking against higher levels of theory should be performed. This chapter aims to highlight the performance of QM/MM methods in addressing some important questions in enzymology, emphasising the strengths of using a combined experimental and modelling approach to investigating enzyme-catalysed reactions, and illustrating the insight that QM/MM calculations can provide.

11.2 Applications of QM/MM Methods

11.2.1 A Catalytic Role for Methionine Revealed by Computation and Experiment

Identification and analysis of catalytic functions is one area in which the combination of mechanistic modelling and experiments is proving increasingly important in enzymology.[6,27–32] This combined approach was central to the discovery of a catalytic role for methionine in phosphite dehydrogenase (PTDH), where the catalytic importance of Met53 in the PTDH reaction was initially identified from QM/MM simulations, and subsequently verified by experiments.[33] Kinetics experiments confirmed a catalytic role of Met53 (shown by a 50-fold reduction in kcat for M53A, 200-fold reduction in k_{cat} for M53N; K_M was not affected by these mutations). Mutagenesis experiments using the non-natural amino acids norleucine and selenomethionine indicated that the sulfur atom in Met53 specifically is important for catalysis (again k_{cat} is reduced in these mutants), but the experimental data did not identify the cause of this catalytic effect, *e.g.* whether the effect of Met53 mutation is due to a disruption of hydrogen bonds, a change in the steric environment of the active site, or some other interaction. Methionine can act as a ligand for metal ions in proteins,[34] and form stabilising (including non-hydrogen bond) interactions with other residues.[35] Methionine residues can be oxidised to more reactive methionine sulfoxide or sulfone forms, but no clear catalytic function for methionine had previously been identified until experiments were carried out on PTDH. The identification of the catalytic role for methionine in PTDH came from QM/MM simulations, and was confirmed be experiments. As well as identifying this intriguing catalytic role, the QM/MM simulations analysed its molecular nature.

PTDH catalyses the oxidation of phosphite (the widely used name for hydrogen phosphonate) to phosphate with the concurrent reduction of nicotinamide adenine dinucleotide (NAD^+) to NADH (see Figure 11.2).[36] This process has applications in biotechnology and biocatalysis, because it allows the regeneration of NADH.[37] The reaction is unusual because it involves nucleophilic attack by water (or hydroxide) on phosphite, displacing hydride.[38] Despite much experimental effort,[39–41] a detailed picture of the reaction mechanism is yet to be obtained. His292 is a likely candidate for the catalytic base in the reaction. However, the pH rate profile suggests that this residue is protonated and thus not basic. Another interpretation of these results involves a reverse protonation mechanism, in which phosphite is protonated and His292 is neutral. In order to investigate this reverse protonation mechanism,

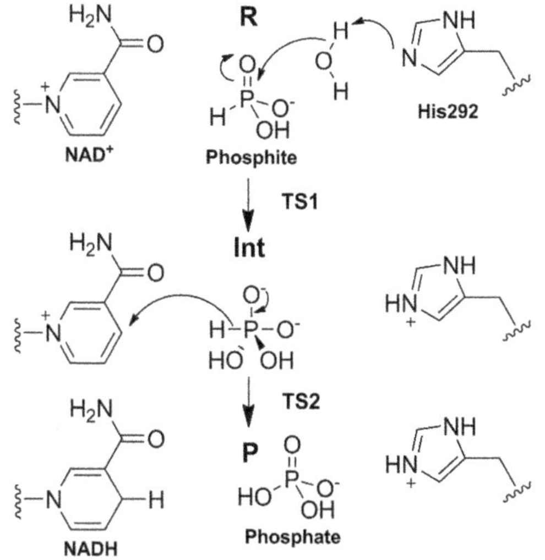

Figure 11.2 The mechanism of phosphite oxidation to form phosphate catalysed by PTDH (33). QM/MM umbrella sampling techniques (AM1/CHARMM27) have been used to model the reaction as an associative process, involving a pentacoordinated phosphorus intermediate (INT). Analysis of the simulations showed that the side chain of Met53 stabilises the positively charged His292 at TS2. This catalytic role was confirmed by kinetics experiments that used the non-natural amino acids norleucine and selenomethionine to show that the sulfur (chalcogen) atom is necessary for catalysis. This is the first case where a clearly defined, catalytic role for methionine has been identified.

a computational model of the system was built from the co-crystal structure of TS-PTDH with NAD^+ and the inhibitor sulphite.[42] The QM region was defined as phosphite, the nicotinamide ring of NAD^+, a (crystallographic) water molecule (Wat61) involved in the reaction, and the imidazole ring of (neutral) His292 to act as the base.

The dynamics and conformational behaviour of the ternary (Michaelis) complex of PTDH was explored through two separate 1 ns QM/MM MD simulations at the AM1/CHARMM27 level of theory, initiated from different structures of the model taken from MM MD simulations. Hydrogen bond analysis of the MD trajectories identified a hydrogen bond, where the protonated oxygen of phosphite (O3) donates a hydrogen bond to the sulfur atom of Met53 in the majority of structures in one trajectory (86%) but only 40% of structures in the second trajectory, due to rotation of the hydroxyl group of phosphite and also some flexibility in the position of the sidechain of Met53.

The reaction was modelled as an associative process (Figure 11.2) using umbrella sampling MD techniques, again at the AM1/CHARMM27 level of QM/MM theory. The first step, attack on phosphorus by the water/hydroxide nucleophile ($Z_{\text{step1}} = d(\text{OH2–H2T}) - d(\text{H2T–NE1}) - d(\text{OH2–P})/\text{Å}$), was calculated to have a free energy barrier of 28.8 kcal mol^{-1}, forming a pentacoordinated phosphorus intermediate (Int, Figure 11.2) with an energy of 4.6 kcal mol^{-1} above the reactant complex (R). The calculated free energy barrier to the second step ($Z_{step2} = d$(P1–H1) – d(H1–C4N)/Å) is 20.7 kcal mol^{-1} (relative to Int). The apparent barrier (from the experimental k_{cat}) to reaction is 16.4 kcal mol^{-1}, significantly lower than the barrier calculated from simulations: the QM/MM barriers are too high, which is often true for calculations with semiempirical methods such as AM1,[43–45] and also particularly here because of the known limitations of the AM1 method for modelling phosphorus.[46]

Accurate calculation of the reaction barrier for a chemical (*e.g.* rate limiting) step is clearly useful, both to identify the mechanism and to validate and demonstrate computational methods. Typically, however, predicting the rate of a natural enzyme-catalysed reaction is not an important biochemical question, as the rate may already have been measured. Rather, the questions of interest are usually in identifying the mechanism, understanding the origins of catalysis, and practical problems such as predicting the activity of mutant enzymes and reactivity of alternative substrates. QM/MM calculations have an important role in all of these questions. Simulations can model transition states and unstable intermediates in enzymes, and analyse how they are formed and stabilised. In phosphite dehydrogenase, the simulations identified an unprecedented catalytic interaction that exists only in the transition state. In the simulations, the Met53 sidechain changes position as the system changes from the intermediate structure, Int, to the transition state, TS2. It moves from a position close to the phosphite to one near the His292 sidechain (Figure 11.3). Decomposition of the energies from QM/MM calculations into different components provides further insight into the nature of the interactions. The electrostatic component of the QM/MM interaction energy is particularly useful for identifying residues involved in catalysis.[20,22,43,47] Here, the QM/MM electrostatic interaction is defined as the difference in the total QM/MM electrostatic interaction energy and the QM/MM electrostatic energy between the QM region and the chosen residue (at the MM level) when the MM atomic partial charges have been set to zero.

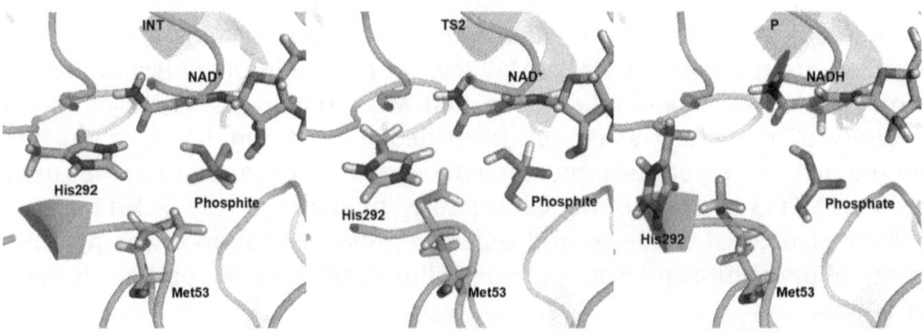

Figure 11.3 QM/MM modelling of the reaction in the enzyme phosphite dehydrogenase. Left: the ternary complex TS–PTDH–NAD$^+$–Phosphite at INT, showing the side chain of Met53 oriented towards phosphite. Middle: the complex at TS2 showing reorientation of the Met53 side chain towards the positively charged His292. Right: the product of hydride transfer to NAD$^+$ (33).

Decomposition of the QM/MM electrostatic energy in PTDH show that Met53 stabilises the positively charged histidine in TS2. The average QM/MM electrostatic interaction between Met53 and the QM region is slightly destabilising in the first half of the reaction but becomes significantly stabilising at TS2. The average QM/MM electrostatic energy between Met53 and the QM region is -7.9 kcal mol^{-1} in TS2 (relative to R), the largest contribution of any neutral residue. The QM/MM van der Waals interaction energy remains relatively constant along the path at ~-3 kcal mol^{-1}; the fact that this QM/MM interaction energy does not change significantly during the reaction shows that van der Waals interactions with the protein environment do not contribute to lowering the reaction barrier. Further decomposition of the interaction between Met53 and the QM region in TS2 showed that the majority of this stabilising interaction is due to interaction between the QM region and the sidechain of Met53, rather than the backbone. A C–H···O hydrogen bond is present between the methyl group of Met53 and the amide oxygen of the NAD^+ cofactor (Figure 11.3), but this is a weak interaction, which does not make a significant contribution to stabilisation: *e.g.* in optimisation of complexes in the gas phase, this interaction is lost. Rotation of the methionine methyl group also means that this interaction it is not present in all structures. These interactions were not present in structures of R or Int. Therefore, the QM/MM simulations showed that the sidechain of this methionine residue stabilises the crucial transition state structure.

These interactions were identified by relatively low-level QM/MM simulations, which use QM methods with known and important limitations, as noted above. To test the findings of transition state stabilisation, correlated *ab initio* calculations (at the MP2/aug-cc-pVTZ//MP2/6-31+G(d) level) were carried out on small models of His292 and Met53 using the Gaussian09[48] and Molpro programs.[49] A 4-methyl-imizadole/imidazolium fragment was used to represent neutral and charged histidine, respectively, and methylthioethane to represent methionine. One complex was a face-on interaction between the imidazolium group and methylthioethane, similar to the interaction seen in the PTDH transition state structure (Figure 11.4). It is encouraging that, despite the well-known limitations of the AM1 method, QM/MM methods were able to identify this face-on interaction in good agreement with *ab initio* methods, in terms of both geometry and energy. In the AM1/CHARMM27 TS2 structure (Figure 11.4), the sulfur of the methylthioethane fragment sits quite centrally over the centre of the imidazolium ring, with sulfur to nitrogen distances d2 and d3 of 3.36 and 3.64 Å, respectively, and a Nd (N1)–H–S angle of 57°. When this structure was optimised in the gas phase at the MP2/6-31G(d) level of *ab initio* QM theory, a similar face-on complex was obtained with distances d2 and d3 of 3.63 Å and 3.65 Å and a Nd (N1)–H–S angle of 64° (Figure 11.4). The interaction energy calculated for this complex at the MP2[50]/aug-cc-pVTZ and SCS-MP2[13]/aug-cc-VTZ levels of theory (including counterpoise correction for basis set superposition error) is -10.4 and -9.0 kcal mol^{-1} at the MP2 and SCS-MP2 levels, respectively. These values are in quite good agreement with the average QM/MM electrostatic energy (see discussion

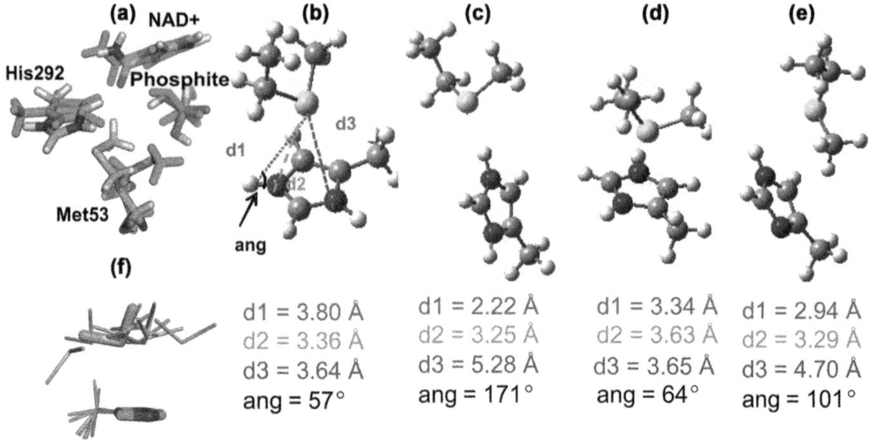

Figure 11.4 (a) Representative structures of the interaction between Met53 and the QM region at the intermediate (Int, green atoms) and at the transition state for the hydride transfer step TS2 (atoms coloured by type) from AM1/CHARMM27 QM/MM calculations. (b)–(e) Geometries of the complexes of 4-methyl imidazole/imidazolium and methylthioethane fragments (representing His and Met) (optimised at the MP2/6-31+G(d) level unless otherwise stated). (b) The average AM1/CHARMM27 (QM/MM) geometry of the 4-methyl imidazolium and methylthioethane fragments (averaged over 110 snapshots at the reaction coordinate window for TS2). (c) Geometry of a hydrogen-bonded complex between 4-methylimidazolium and methylthioethane. (d) Geometry of a face-on complex between 4-methylimidazolium and methylthioethane. (e) Geometry of a neutral complex between 4-methylimidazole and methylthioethane. (f) The structural template for the His-Met interaction search shown in sticks (green carbons) superimposed (on His atoms only) with the 10 hits from the Protein Data Bank with the lowest RMSD (grey carbons). As indicated in (b) d1 is the S–H separation, d2 is the S–N1 (Nd) distance, d3 is the S–N2 (Ne) distance and ang is the (Nd) N1–H–S angle.
Reproduced from ref. 33 with permission from The Royal Society of Chemistry.

above) of -6.9 kcal mol^{-1} for Met53 at TS2 in the QM/MM simulations. The interaction energy of a neutral complex of 4-methylimidazole and methylthioethane, constrained to the geometry of the face-on complex, was significantly smaller (only -0.57 kcal mol^{-1}). The smaller interaction energy for the neutral complex with the same geometry is consistent with the interaction between the histidine and methionine being formed in the second step of the reaction when the histidine is protonated. The 50 to 200-fold reduction in k_{cat} (with K_M unchanged) observed experimentally for the Met53 mutants studied is equivalent to a catalytic contribution of 2–3 kcal mol^{-1} for Met53. The interaction energies calculated here are larger, but are of the correct order of magnitude; it should be remembered that compensating interactions are likely to be formed in the mutant enzymes, so the comparison is not exact because the mutants were not modelled here. The interaction energies calculated here are not directly comparable to the experimental $\Delta\Delta^{\ddagger}G$ values for the mutations because they do not include the effects of dielectric shielding, for example. Analysis of protein and small-molecule structures in the Protein Databank[51] and Cambridge Structural Database indicates that face-on Met-His interactions of this type are relatively widespread, with potential importance for the structure and functioning (*e.g.* catalysis) of other enzymes (particularly those in which histidine acts as a base).

The experimental results do not allow identification of the reaction step in which Met53 contributes to catalysis. Whilst other mechanisms for the reaction cannot be ruled out, the experimentally observed catalytic contribution of Met53 provides indirect evidence in support of the modelled mechanism in which His292 acts as the base in an associative process. In a dissociative mechanism, protonation of His292 would occur after the rate-limiting hydride transfer event, and hence Met53 would not be expected to stabilise the transition state that governs kcat. If another residue acted as a catalytic base, then according to the pH rate profile, His292 would be expected to be protonated throughout the reaction. The protonated His292 and Met53 might then be expected to form this type of interaction throughout the reaction and thus it would not be catalytic. The results therefore support the mechanism as modelled.

11.2.2 QM/MM Simulations as an Assay for Carbapenemase Activity in Class A β-Lactamases

Antimicrobial resistance is a growing global concern. One particular current worry is the emergence and spread of resistance to carbapenem, particularly in carbapenem resistant Gram negative bacteria. Resistance to carbapenem antibiotics – currently the 'last resort' treatment against multidrug resistant Gram negative bacteria – has now spread to all regions of the world.[52] Much of the resistance in such bacteria can be attributed to β-lactamases, which break down the four-membered β-lactam ring, characteristic (and necessary for the antibacterial activity) of β-lactam antibiotics.[53] β-Lactamases capable of hydrolysing carbapenem antibiotics (carbapenemases) have been identified in classes A, B and D of β-lactamase enzymes.[54–56]

β-Lactamase hydrolysis involves the formation of an acyl–enzyme intermediate, coincident with the opening of the β-lactam ring.[57] A deacylation step then follows to release the inactivated lactam and complete the enzymatic turnover. The deacylation stage involved nucleophilic attack by a conserved water molecule that has been activated by an active site glutamate residue (Figure 11.5).[58] Carbapenem-inhibited class A β-lactamases are readily acylated but have a long lived acyl–enzyme.[59] In contrast, carbapenemases deacylate efficiently, allowing bacteria which carry these enzymes to become resistant to the carbapenem antibiotics. The causes of differences in carbapenemase activity in the different β-lactamases remains unclear, *e.g.* because the

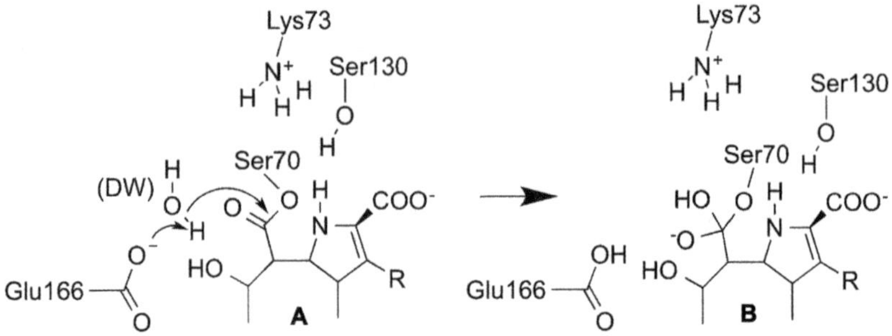

Figure 11.5 The first step of deacylation in the breakdown of carbapenem acyl–enzymes in class A β-lactamases: the acyl–enzyme (A) reacts to form a tetrahedral intermediate (B) (numbering for the TEM-1 enzyme). Reproduced from ref. 19 with permission from The Royal Society of Chemistry.

different enzymes have apparently very similar active site (and overall) structures (see, for example, references for crystal structures given in ref. 19).

Chudyk *et al.*[19] have developed a protocol to predict carbapenemase activity of class A β-lactamases using QM/MM simulations. This protocol is as follows: starting from the crystal structure of the enzyme, a model of the acyl–enzyme complex is generated, based on the conformation of meropenem in the crystallographically observed acyl–enzyme complex with SFC-1 (PDB ID:4EV4[54]). Using the AMBER[60] ff12SB force field for the protein, the TIP4P-Ew water model for the solvent and the General AMBER force field (GAFF) for meropenem (for the atoms not treated by QM), the model system is heated to 300K and allowed to equilibrate using QM/MM MD at the semiempirical SCC-DFTB[16] QM/MM level. Unrestrained QM/MM MD simulations are performed for the acyl–enzyme complex and starting structures for reaction modelling are selected based on two geometric criteria: the conformation of the 6α-1R-hydroxyethyl group of meropenem and the position of the deacylating water (DW). For the systems with meropenem, the QM region consisted of 41 atoms and three link atoms (Figure 11.6). The Glu166 side chain up to the CG atom and the DW molecule were treated QM. The acylated meropenem was included in the QM region from CB of Ser70 up to the S atom. The final stage is to perform QM/MM umbrella sampling simulations along two reaction coordinates, describing the proton transfer from DW to Glu166 $[r_x = d(\text{OE1}_{\text{Glu166}} - \text{H2}_{\text{DW}}) - d(\text{O}_{\text{DW}} - \text{H2}_{\text{DW}})]$ and the nucleophilic attack on meropenem $[r_y = d(\text{CC}_{\text{Mer}} - \text{O}_{\text{DW}})]$, to generate the two-dimensional free energy surface. The minimum energy path for the reaction is then identified from the free energy surface and the activation free energy obtained. The protocol was tested for eight different class A lactamases: KPC-2, NMC-A, SPC-1, SME-1, CTX-M-16, BlaC, TEM-1 and SHV-1. For some enzymes, the crystal structure of the acyl–enzyme intermediate was available, but for others the structure of the *apo* enzyme was used to construct a model of the acyl–enzyme.

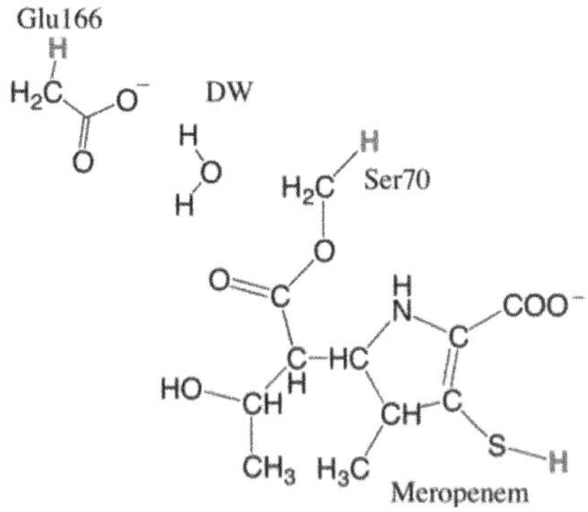

Figure 11.6 The QM region used in QM/MM calculations on the meropenem acyl–enzyme in class A β-lactamase enzymes; these atoms were treated at the QM (SCC-DFTB) level in QM/MM MD simulations, with the surrounding atoms in the acyl–enzyme complex treated by MM. The hydrogen atoms shown in red are 'link' atoms.
Reproduced from Figure S1 in ref. 19 with permission from the Royal Society of Chemistry.

Figure 11.7 shows that the predicted reaction barriers fall into two distinct groups: one group with reaction barriers around 10 kcal mol^{-1} and the second with activation free energies around 17–18 kcal mol^{-1}. The lower activation barrier of ∼10 kcal mol^{-1} indicates efficient deacylation *i.e.* carbapenemase activity, with the higher barrier indicating the long lived acyl–enzyme complex expected in carbapenem-inhibited cases. Comparison of the calculated barriers with experimental results for these enzymes shows that the method clearly distinguishes between the two types of enzymes, despite some underestimation of the barriers, as is common with the SCC-DFTB method. The results show that Chudyk *et al.*[19] have developed a robust QM/MM protocol that can reliably predict carbapenemase activity using the crystal structure of the apo form or the acyl–enzyme. Application of the protocol to new structures from clinical isolates and other classes of β-lactamase will further our understanding of β-lactam resistance and may aid the development of new strategies to combat this important threat to human health.

11.2.3 QM/MM Simulations Indicate That Asp185 is the Catalytic Base in HIV-1 Reverse Transcriptase

According to the World Health Organization (WHO), in 2014 1.2 million people died from HIV-related causes globally, with approximately 2 million people becoming newly infected. Significant advances have been made in the treatment of HIV/AIDS, but it remains one of the biggest social, economic and health challenges in the world.[61] HIV-1 reverse transcriptase (RT) is a member of the family of DNA polymerases, assisting DNA replication from the 5′ to the 3′ directions.[62] The mechanism involves deprotonation of

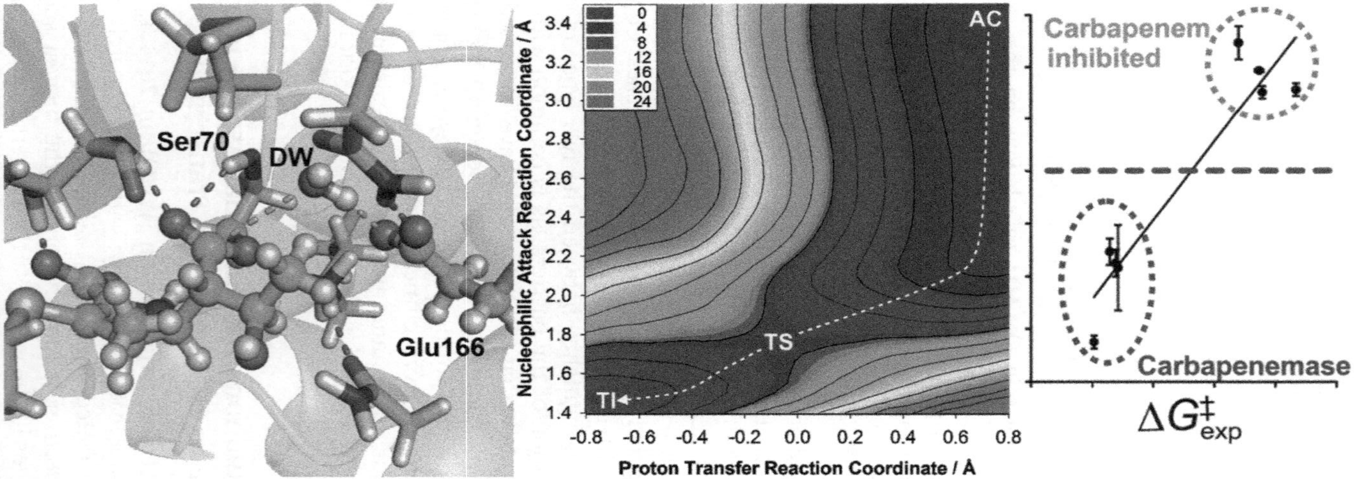

Figure 11.7 Left: the acyl–enzyme complex of meropenem in the SFC-1 β-lactamase. Middle: A representative QM/MM free energy surface for the deacylation reaction, indicating the minimum energy path. Right: A plot of calculated *vs.* experimental activation free energies for deacylation (which is probably rate limiting) in eight different class A β-lactamases. QM/MM simulations (at the SCC-DFTB level of QM theory), while underestimating the barriers in absolute terms, correctly predict the relative activities of the different enzymes. In particular, the QM/MM simulations distinguish between those enzymes with carbapenemse activity (those which acylate rapidly) from those that do not. This is an example of QM/MM simulations being used as a computational 'assay' for enzyme activity, in this case relevant to bacterial antimicrobial resistance.[19]

the 3′-primer terminus followed by nucleotide addition.[63] The pyrophosphate leaving group then accepts a proton from K220.[64] The general base for the initial protonation step is uncertain.[63] Rungrotmongkol *et al.*[65] performed QM/MM calculations to compare three different mechanisms for the deprotonation step in HIV-1 RT. Three different possible mechanisms were simulated and compared: either Asp185, or Asp186 or the 2′-deoxy-L-thymidine-5′-triphosphate (dTTP) substrate as the catalytic base accepting the proton, along the subsequent nucleotide addition step. They were modelled using umbrella sampling MD simulations at the PM3/CHARMM27[66,67] level of QM/MM theory. Their model of the RT-DNA-dTTP ternary complex was based of the PDB structure 1RTD used in previous work.[68] The QM region comprised dTTP, 2 Mg^{2+} ions, Asp110, Asp185, Asp186, Val111 and a water molecule (see Figure 11.8). Eight HQ link atoms[69] were required to terminate the QM boundary. For the deprotonation step, the reaction coordinate was defined as $R_x = d(\text{O3}'\text{-H3T}) - d(\text{H3T-O}_{base})$, where O_{base} was OD1 of Asp185, OD1 of Asp186 or O5′ of dTTP. For the polymerisation step the reaction coordinate

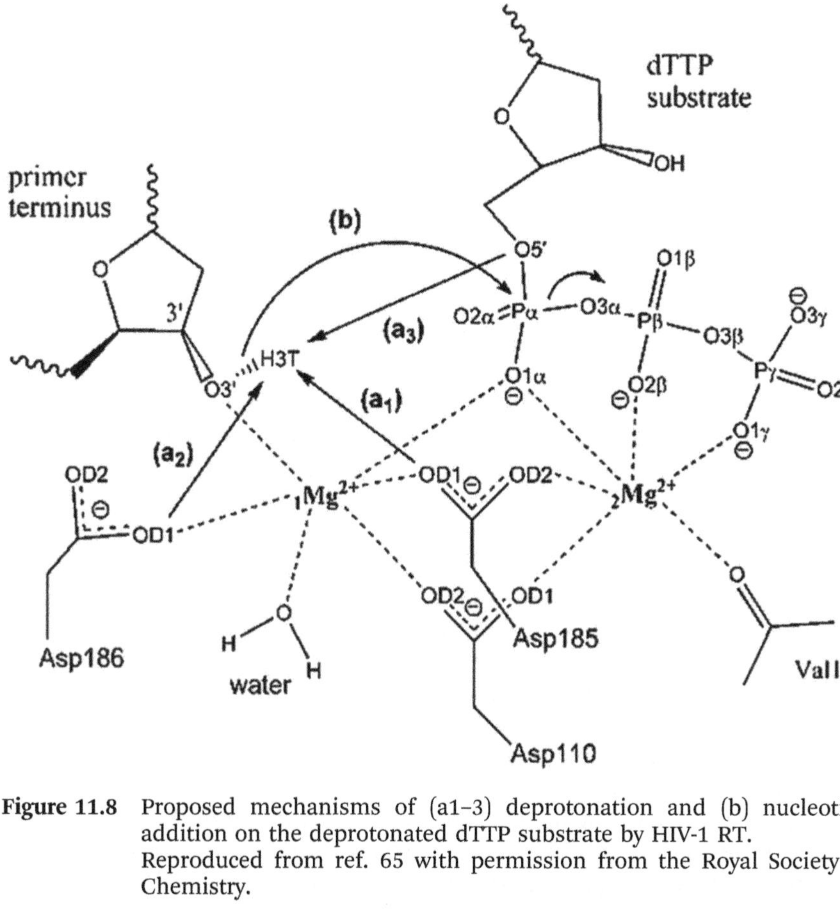

Figure 11.8 Proposed mechanisms of (a1–3) deprotonation and (b) nucleotide addition on the deprotonated dTTP substrate by HIV-1 RT. Reproduced from ref. 65 with permission from the Royal Society of Chemistry.

was defined as $R_y = d(P_\alpha\text{-}O3_\alpha) - d(O3'\text{-}P_\alpha)$. QM/MM MD was carried out for 30 ps at each step along the reaction coordinate at 0.1 intervals and free energy profiles were obtained using the weighted histogram analysis method (WHAM).[70,71]

Comparison of the free energy profiles for the different model systems indicated that dTTP is unlikely to act as base in the reaction. The reaction barrier for the deprotonation step calculated for this mechanism is significantly higher and the product is very unstable. Based on the free energy barriers for the deprotonation step only, reaction *via* either Asp185 or Asp186 would apparently be possible. However, when Asp186 acts as base, the conformation of the system after the deprotonation has occurred is not favourable for the subsequent polymerisation step. Overall, the results of the QM/MM calculations indicate that the dTTP substrate in its fully deprotonated form is more active and in a more suitable conformation for the DNA polymerisation when Asp185 acts as the base (see Figure 11.9). The atomic-level understanding resulting from these QM/MM simulations of the RT reaction mechanism may aid the development of new anti HIV/AIDS drugs that target HIV-1 RT.

11.2.4 The Origins of Catalysis in Chorismate Mutase Analysed by QM/MM Simulations

Catalysis in chorismate mutase (CM) has been at the centre of debate surrounding enzyme catalysis since the early 1980s[72,73] and computer simulation has played an important role in the debate (see Figure 11.10).[47,74,75] As a test case, it highlights the potential of modelling to contribute significantly to enzymology and no review of QM/MM methods would be complete without a discussion of CM. A search of the ISI Web of Science (v5.21: http://wok.mimas.ac.uk) using the criteria: Topic = (qm/mm or mm/qm or qm-mm or mm-qm or qm:mm or mm:qm or quantum mechanics/molecular mechanics or quantum mechanical/molecular mechanical) and Topic = (chorismate mutase), yields 122 results, showing that it is a popular test case both for the development and testing of QM/MM methods and analyses of enzyme catalysis. The simple, unimolecular rearrangement catalysed by CM makes it an ideal system for analysing the origins of catalysis, because the reaction does not involve any covalent interaction between the enzyme and the substrate, and also because the same reaction occurs in solution with the same reaction mechanism. CM catalyses the Claisen rearrangement of chorismate to prephenate (Figure 11.11). The experimental free energy barrier $\Delta^{\ddagger}G = 15.4$ kcal mol^{-1} ($\Delta^{\ddagger}H = 12.7$ kcal mol^{-1}) in *Bacillus subtilis* CM is significantly lower than that for the uncatalysed reaction in aqueous solution ($\Delta^{\ddagger}G = 24.5$ kcal mol^{-1}, $\Delta^{\ddagger}H = 20.7$ kcal mol^{-1}).[76] This translates to a rate acceleration of 10^6 by the enzyme ($\Delta\Delta^{\ddagger}G = 9.1$ kcal mol^{-1}). The reaction barrier in the enzyme can be reproduced to within chemical accuracy using 'state of the art' LCCSD(T0)/CHARMM27 QM/MM techniques.[11,77] However,

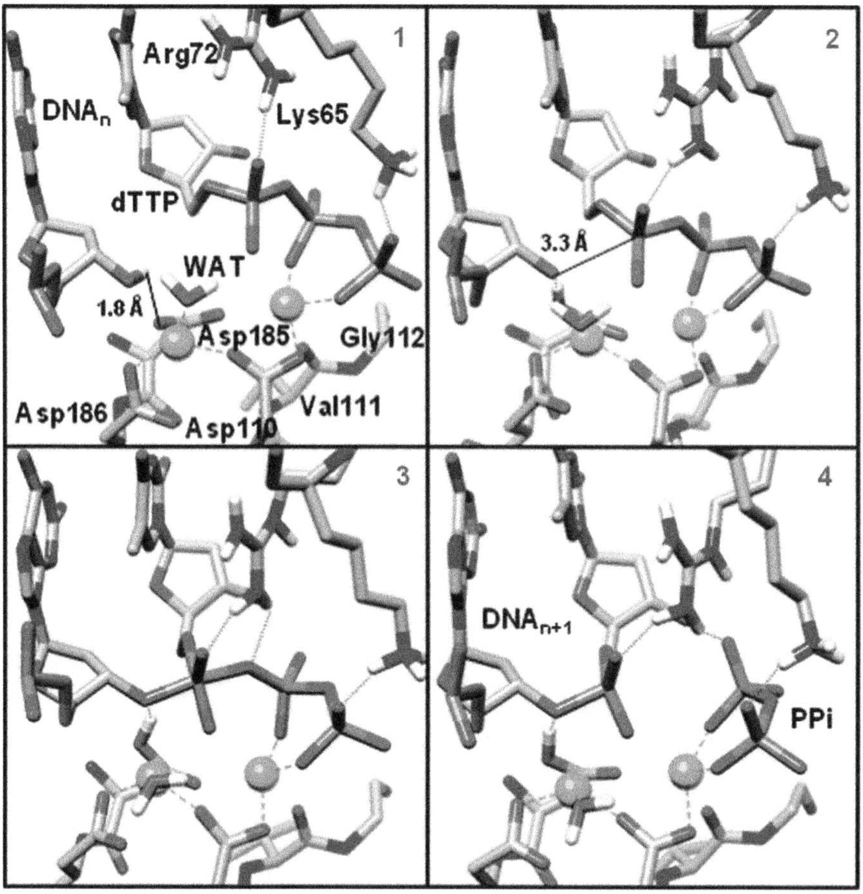

Figure 11.9 Representative active site structures of the full-deprotonated dTTP–DNA–HIV-1RT ternary complex taken from QM/MM MD simulations of the reaction mechanism with deprotonation by Asp185 and subsequent nucleotide addition: (1) reactant, (2) product of deprotonation, (3) pentahedral intermediate, and (4) 3′-5′ phosphodiester and PPi final products. Hydrogen bonds with Lys65 and Arg72 (green lines) and Mg^{2+} coordinations (dashed lines) are also shown.
Reproduced from ref. 65 with permission from the Royal Society of Chemistry.

these methods remain highly computationally expensive and mechanistic insight can be obtained from lower levels of QM/MM theory. QM/MM calculations (*e.g.* at the semiempirical AM1/CHARMM or *ab initio* QM level) have shown TS stabilisation by the enzyme to be the major contribution to catalysis.[21,47,78–82] The electrostatic nature of catalysis in this enzyme is also supported by mutagenesis experiments that show a significant decrease in catalytic activity when Arg90 is mutated to the isosteric, but neutral, citrulline.[83] The enzyme-bound conformation of chorismate is significantly different from that in solution, and more closely resembles the

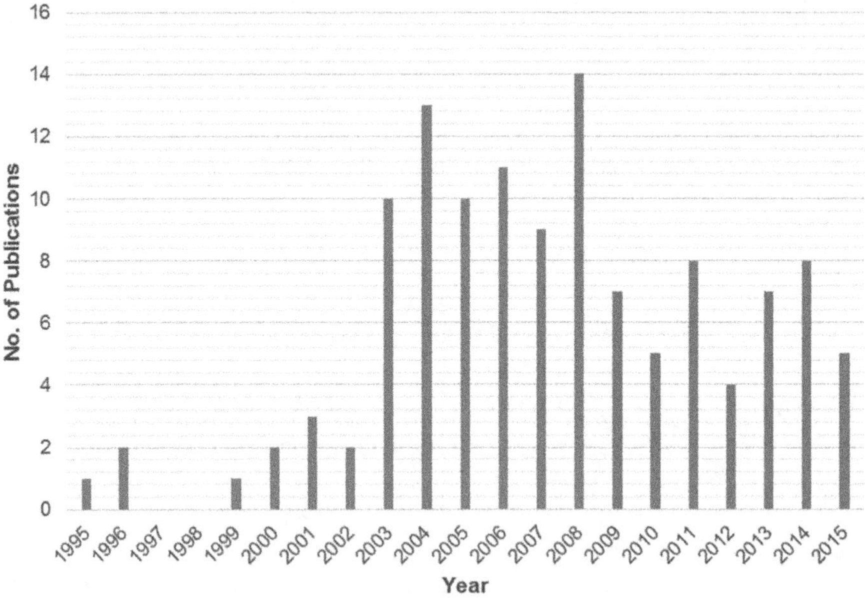

Figure 11.10 Chorismate mutase is one of the most widely studied enzymes by QM/MM techniques. QM/MM studies of chorismate mutase: a (February 2106) search on ISI Web of Science (v5.21: http://wok.mimas.ac.uk/) using the criteria: Topic = (qm/mm or mm/qm or qm-mm or mm-qm or qm:mm or mm:qm or quantum mechanics/molecular mechanics or quantum mechanical/molecular mechanical) and Topic = (chorismate mutase) sorted by publication year.

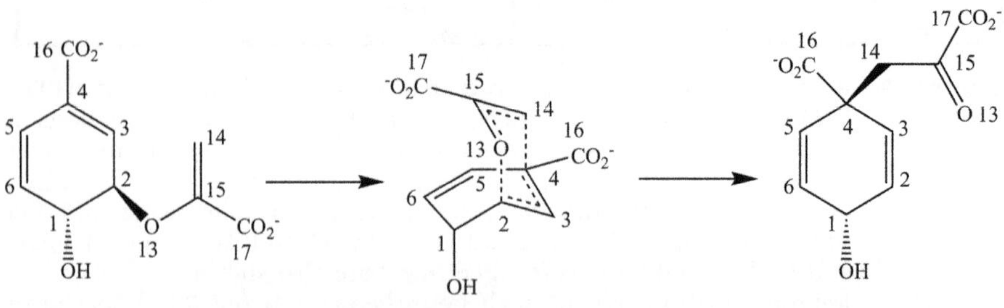

Figure 11.11 The Claisen rearrangement of chorismate to form prephenate (reproduced from ref. 87 with permission from the Royal Society of Chemistry). QM/MM methods at the B3LYP/6-31G(d)/CHARMM27 level of theory have been used to compare multiple reaction pathways for this simple unimolecular reaction in solution and in the *Bacillus subtilis* chorismate mutase. Electrostatic stabilisation of the transition state by the enzyme active site (particularly Arg90) is the major contribution to catalysis in this enzyme.

TS,[21,47,74,75,84–86] and this also contributes to the barrier lowering in the enzyme: this conformational effect also arises because the enzyme is complementary to the TS for the reaction (see below).

Claeyssens *et al.*[87] have investigated the origins of catalysis in CM through the calculation of multiple QM/MM adiabatic reaction pathways in the BsCM enzyme and in water at the B3LYP/CHARMM27 level of theory. The performance of the B3LYP method for QM/MM calculations on this system has been thoroughly tested and the results agree well with experiment. Catalysis by CM can be understood by comparison of potential energy barriers, because the small entropic, zero-point and thermal contributions to the barrier are probably similar in the two environments, given that the mechanism of reaction is the same in both.[76] The activation enthalpies for the reaction in the enzyme and in water at 25 °C are $\Delta^{\ddagger}H = 12.7 \pm 0.4$ kcal mol^{-1} and $\Delta^{\ddagger}H = 20.7 \pm 0.4$ kcal mol^{-1}, respectively, a reduction of $\Delta\Delta^{\ddagger}H = 8 \pm 0.4$ kcal mol^{-1} by the enzyme.[76] Claeyssens *et al.*[87] attribute the 6.1 kcal mol^{-1} average difference in potential barrier between the two environments to two contributions: (i) better TS stabilisation by the enzyme (6.3 kcal mol^{-1}); and (ii) a small effect (0.5 kcal mol^{-1}) due to compression of the distance between the C4 and C14 (the atoms which form a bond during the reaction) from ~3.5 Å in water to ~3.3 Å in the enzyme. This compression of the bond-forming distance can be considered as strain. These results generally agree with the conclusions of the first QM/MM investigation of CM,[47] although the higher level methods suggest a smaller role for substrate strain.

The more controversial aspect of catalysis in CM involves the initial conformation of the bound substrate, and its potential contribution to barrier lowering. The pseudo-diaxial form of chorismate binds to the enzyme but it is not the lowest energy conformation in solution. From experimental[88] and computational[89,90] studies, the contribution of the binding of the substrate in this form is estimated to be 0.9–3.6 kcal mol^{-1}. The lower bounds are the result of experimental studies and the upper limit from the MC/FEP study of chorismate in solution by Carlson and Jorgensen.[89] These values suggest that, at most, binding of the pseudo-diaxial conformation contributes ~26% catalysis in CM, with TS stabilisation playing the major role. These results are in line with previous calculations,[21,75] but not with the very approximate (and indirect) estimates reported by Bruice and coworkers.[91–95] who attributed ~90% or more of the total $\Delta\Delta^{\ddagger}G$ to the so-called 'near attack conformers' or 'NAC effect' in CMs from all the species studied. All the substrate conformations found in the pathways generated by Claeyssens *et al.*,[87] both in water and in the enzyme, qualify as NACs by all previous definitions. Therefore, the differences in reaction barrier found here are not due to the so-called (and now discredited) 'NAC effect'. It should also be pointed out that the binding of the pseudo-diaxial form, and compression of the substrate, are the result of the complementarity of the enzyme active site for the TS, thus they are consequences of the high affinity of the enzyme for the TS.

Analysis of the effects of the environment on the reaction pathways shows that both environments reduce the barrier, compared to the same reaction pathways in vacuum, by TS stabilisation. However, the enzyme environment is much better at stabilising the TS than water (on average TS stabilisation energy of 7.3 *versus* 1.0 kcal mol^{-1}). The TS stabilisation is linearly correlated with the barrier height for both environments (see Figure 11.12). In each environment, the reaction pathways are structurally similar (though not identical). The variation in barrier height in either environment is due to differences in the structure of the environment, and depends on the degree of TS stabilisation. When the effect of the environment is completely removed, the pathways for the reaction in the enzyme and in solution show similar intrinsic reaction barriers. However, the reaction pathway in the two environments is slightly different. This is exemplified by the larger distance

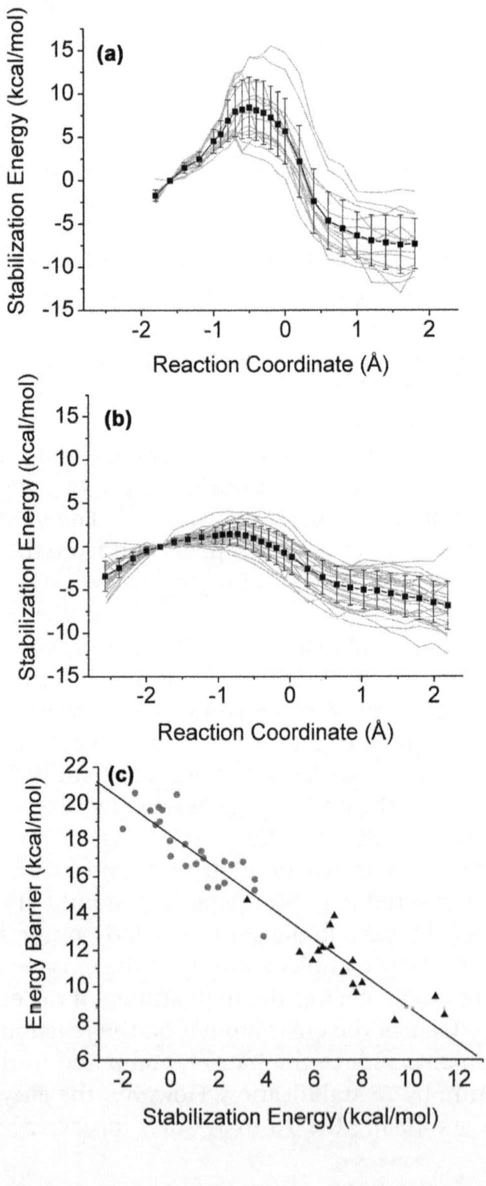

between the two carboxylate moieties of the substrate in water throughout the reaction. This indicates that electrostatic repulsion between the two carboxylate moieties is the origin of this 'strain' effect, as proposed previously.[75,86] The compressed conformation is stabilised in the enzyme by positively charged groups in the active site (Arg90, Arg7, Arg63). This compression equates to a destabilisation (strain) of the reactant in the enzyme compared to the water environment but this destabilisation remains constant during the reaction, and does not contribute significantly to catalysis. It should be noted also that the product conformation is also compressed (strained) compared to that in solution. This may assist the overall catalytic process by reducing the binding affinity for the product somewhat, thus speeding product release.

CM has been the subject of a significant amount of research effort over the years. It is heartening that the key findings of TS stabilisation by the enzyme, and the central role of Arg90 in TS stabilisation, have been found consistently at a variety of different levels of theory, by a number of different groups. QM/MM methods are able to model the essential physical processes involved in the reaction and can be used to analyse the features giving rise to catalysis (comparing the enzyme-catalysed reaction to the reaction in solution, for example). QM/MM simulations have been particularly important in showing how the transition state is stabilised within the enzyme.[47,75,78–80,96–99]

11.3 Conclusions

QM/MM methods are now an established approach for the study of enzyme-catalysed reactions, and are increasingly important in this area. Highly accurate QM methods can now be applied in QM/MM calculations on enzymes. Software developments are making it increasingly easy for 'non-expert' modellers to carry out QM/MM simulations and to gain useful insight into enzyme reactions from them. This will undoubtedly lead to even more

Figure 11.12 QM/MM modelling of the Claisen rearrangement of chorismate, forming prephenate, in water and in the *Bacillus subtilis* chorismate mutase enzyme. Stabilisation energy along the reaction coordinate (relative to the reactant (which is found at a reaction coordinate value of $r = -1.8$ Å)) (a) in the enzyme and (b) in water, the interaction energies for the individual pathways are shown in grey, while the average (with standard deviations as error bars), is shown in black. This is the interaction energy between the reacting (QM) system and the MM protein/water environment. A positive value indicates stabilisation of the reacting system by the environment (*e.g.* the stabilisation energy reaches a maximum for the (TS) in the enzyme. In (c) the reaction barrier height *versus* stabilisation energy at the TS (at $r = -0.5$ Å) for both environments (at the B3LYP/6-31G(d)/CHARMM27 level of QM/MM theory) is shown (grey circles: reaction in water environment; black triangles: reaction in the enzyme).
Reproduced from ref. 87 with permission from the Royal Society of Chemistry.

publications involving QM/MM methods in the future. With this increasing volume of work, it will become even more important to assess the quality and reliability of the results. Testing against high-level *ab initio* QM/MM calculations is useful in this respect. In many cases, the accuracy of QM/MM simulations of enzymes is limited more by the level of QM/MM theory used than by the errors introduced by QM/MM partitioning. For example, DFT calculations can provide misleading energetics, which can lead to qualitatively wrong mechanistic conclusions. Obtaining sufficient sampling also remains a challenge for simulations of many enzyme-catalysed reactions. Experimental data is obviously important for testing QM/MM results. Beyond this, QM/MM simulations and modelling are demonstrating their value in analysing and interpreting experimental data. Many multidisciplinary approaches involving both experiment and simulation have already provided insight into issues such as antibiotic resistance, and identified a role for methionine sidechains in catalysis, as examples above illustrate. QM/MM simulations and modelling are providing detailed and valuable understanding of enzymes and enzyme-catalysed reaction mechanisms, and complement experimental studies of these fascinating biological catalysts.

References

1. http://www.nobelprize.org/nobel_prizes/chemistry/laureates/2013/advanced-chemistryprize2013.pdf, 2013.
2. M. Karplus, Development of Multiscale Models for Complex Chemical Systems: From H + H-2 to Biomolecules (Nobel Lecture), *Angew. Chem., Int. Ed.*, 2014, **53**(38), 9992–10005.
3. M. Levitt and L. Michael, Birth and Future of Multiscale Modeling for Macromolecular Systems (Nobel Lecture), *Angew. Chem., Int. Ed.*, 2014, **53**(38), 10006–10018.
4. A. Warshel, Multiscale Modeling of Biological Functions: From Enzymes to Molecular Machines (Nobel Lecture), *Angew. Chem., Int. Ed.*, 2014, **53**(38), 10020–10031.
5. M. W. van der Kamp and A. J. Mulholland, Combined Quantum Mechanics/Molecular Mechanics (QM/MM) Methods in Computational Enzymology, *Biochemistry*, 2013, **52**(16), 2708–2728.
6. H. M. Senn and W. Thiel, QM/MM Methods for Biomolecular Systems, *Angew Chem., Int. Ed.*, 2009, **48**(7), 1198–1229.
7. A. T. P. Carvalho, A. Barrozo, D. Doron, A. V. Kilshtain, D. T. Major and S. C. L. Kamerlin, Challenges in computational studies of enzyme structure, function and dynamics, *J. Mol. Graphics Modell.*, 2014, **54**, 62–79.
8. R. Lonsdale and A. J. Mulholland, QM/MM Modelling of Drug-Metabolizing Enzymes, *Curr. Top. Med. Chem.*, 2014, **14**(11), 1339–1347.
9. S. F. Sousa, P. A. Fernandes and M. J. Ramos, Computational enzymatic catalysis – clarifying enzymatic mechanisms with the help of computers, *Phys. Chem. Chem. Phys.*, 2012, **14**(36), 12431–12441.

10. R. Lonsdale, J. N. Harvey and A. J. Mulholland, A practical guide to modelling enzyme-catalysed reactions, *Chem. Soc. Rev.*, 2012, **41**(8), 3025–3038.
11. F. Claeyssens, J. N. Harvey, F. R. Manby, R. A. Mata, A. J. Mulholland, K. E. Ranaghan *et al.*, High-Accuracy Computation of Reaction Barriers in Enzymes, *Angew. Chem., Int. Ed.*, 2006, **45**, 6856–6859.
12. M. W. van der Kamp, J. Zurek, F. R. Manby, J. N. Harvey and A. J. Mulholland, Testing High-Level QM/MM Methods for Modeling Enzyme Reactions: Acetyl-CoA Deprotonation in Citrate Synthase, *J. Phys. Chem. B*, 2010, **114**(34), 11303–11314.
13. S. Grimme, Improved second-order Møller–Plesset perturbation theory by separate scaling of parallel- and antiparallel-spin pair correlation energies, *J. Chem. Phys.*, 2003, **118**, 9095.
14. N. Lawan, K. E. Ranaghan, F. R. Manby and A. J. Mulholland, Comparison of DFT and ab initio QM/MM methods for modelling reaction in chorismate synthase, *Chem. Phys. Lett.*, 2014, **608**, 380–385.
15. N. Kaiyawet, R. Lonsdale, T. Rungrotmongkol, A. J. Mulholland and S. Hannongbua, High-Level QM/MM Calculations Support the Concerted Mechanism for Michael Addition and Covalent Complex Formation in Thymidylate Synthase, *J. Chem. Theory Comput.*, 2015, **11**(2), 713–722.
16. M. Elstner, T. Frauenheim, E. Kaxiras, G. Seifert and S. Suhai, A self-consistent charge density-functional based tight-binding scheme for large biomolecules, *Phys. Status Solidi B*, 2000, **217**(1), 357–376.
17. J. J. P. Stewart, Optimization of parameters for semiempirical methods I. Method, *J. Comput. Phys.*, 1989, **10**, 209–220.
18. J. J. P. Stewart, Optimization of Parameters for Semiempirical Methods .3. Extension of Pm3 to Be, Mg, Zn, Ga, Ge, as, Se, Cd, in, Sn, Sb, Te, Hg, Tl, Pb, and Bi, *J. Comput. Chem.*, 1991, **12**(3), 320–341.
19. E. I. Chudyk, M. A. L. Limb, C. Jones, J. Spencer, M. W. van der Kamp and A. J. Mulholland, QM/MM simulations as an assay for carbapenemase activity in class A beta-lactamases, *Chem. Commun.*, 2014, **50**(94), 14736–14739.
20. K. E. Ranaghan, L. Masgrau, N. S. Scrutton, M. J. Sutcliffe and A. J. Mulholland, Analysis of classical and quantum paths for deprotonation of methylamine by methylamine dehydrogenase, *Chem. Phys. Chem.*, 2007, **8**(12), 1816–1835.
21. K. E. Ranaghan and A. J. Mulholland, Conformational effects in enzyme catalysis: QM/MM free energy calculation of the 'NAC' contribution in chorismate mutase, *Chem. Commun.*, 2004, **10**, 1238–1239.
22. L. Masgrau, K. E. Ranaghan, N. S. Scrutton, A. J. Mulholland and M. J. Sutcliffe, Tunneling and classical paths for proton transfer in an enzyme reaction dominated by tunneling: Oxidation of tryptamine by aromatic amine dehydrogenase, *J. Phys. Chem. B*, 2007, **111**(11), 3032–3047.
23. J. Jitonnom, M. A. L. Limb and A. J. Mulholland, QM/MM Free-Energy Simulations of Reaction in Serratia marcescens Chitinase B Reveal the

Protonation State of Asp142 and the Critical Role of Tyr214, *J. Phys. Chem. B*, 2014, **118**(18), 4771–4783.

24. L. Capoferri, A. Lodola, S. Rivara and M. Mor, Quantum Mechanics/Molecular Mechanics Modeling of Covalent Addition between EGFR-Cysteine 797 and N-(4-Anilinoquinazolin-6-yl) Acrylamide, *J. Chem. Inf. Model.*, 2015, **55**(3), 589–599.
25. N. Nitoker and D. T. Major, Understanding the Reaction Mechanism and Intermediate Stabilization in Mammalian Serine Racemase Using Multiscale Quantum-Classical Simulations, *Biochemistry*, 2015, **54**(2), 516–527.
26. R. Lonsdale, S. Hoyle, D. T. Grey, L. Ridder and A. J. Mulholland, Determinants of Reactivity and Selectivity in Soluble Epoxide Hydrolase from Quantum Mechanics/Molecular Mechanics Modeling, *Biochemistry*, 2012, **51**(8), 1774–1786.
27. R. Lonsdale, K. E. Ranaghan and A. J. Mulholland, Computational Enzymology, *Chem. Commun.*, 2010, **46**, 2354–2372.
28. S. C. L. Kamerlin and A. Warshel, At the dawn of the 21st century: Is dynamics the missing link for understanding enzyme catalysis? *Proteins: Struct., Funct., Bioinf.*, 2010, **78**(6), 1339–1375.
29. K. Meier, A. Choutko, J. Dolenc, A. P. Eichenberger, S. Riniker and W. F. van Gunsteren, Multi-Resolution Simulation of Biomolecular Systems: A Review of Methodological Issues, *Angew Chem., Int. Ed.*, 2013, **52**(10), 2820–2834.
30. L. Masgrau, A. Roujeinikova, L. O. Johannissen, P. Hothi, J. Basran, K. E. Ranaghan *et al.*, Atomic description of an enzyme reaction dominated by proton tunneling, *Science*, 2006, **312**(5771), 237–241.
31. L. Y. P. Luk, J. J. Ruiz-Pernia, W. M. Dawson, M. Roca, E. J. Loveridge, D. R. Glowacki *et al.*, Unraveling the role of protein dynamics in dihydrofolate reductase catalysis, *Proc. Natl. Acad. Sci. U. S. A.*, 2013, **110**(41), 16344–16349.
32. D. H. Min, H. R. Josephine, H. Z. Li, C. Lakner, I. S. MacPherson, G. J. P. Naylor *et al.*, An enzymatic atavist revealed in dual pathways for water activation, *PLoS Biol.*, 2008, **6**(8), 1802–1810.
33. K. E. Ranaghan, J. E. Hung, G. J. Bartlett, T. J. Mooibroek, J. N. Harvey, D. N. Woolfson *et al.*, A catalytic role for methionine revealed by a combination of computation and experiments on phosphite dehydrogenase, *Chem. Sci.*, 2014, **5**(6), 2191–2199.
34. J. Reedijk, Plasticity in the copper-thioether bond: Manifestation in blue Cu proteins and in synthetic analogs, *J. Inorg. Biochem.*, 2012, **115**, 182–185.
35. D. Pal and P. Chakrabarti, Non-hydrogen bond interactions involving the methionine sulfur atom, *J. Biomol. Struct. Dyn.*, 2001, **19**(1), 115–128.
36. A. M. G. Costas, A. K. White and W. W. Metcalf, Purification and characterization of a novel phosphorus-oxidizing enzyme from Pseudomonas stutzeri WM88, *J. Biol. Chem.*, 2001, **276**(20), 17429–17436.
37. J. M. Vrtis, A. K. White, W. W. Metcalf and W. A. van der Donk, Phosphite dehydrogenase: A versatile cofactor-regeneration enzyme, *Angew Chem., Int. Ed.*, 2002, **41**(17), 3257–3259.

38. J. M. Vrtis, A. K. White, W. W. Metcalf and W. A. van der Donk, Phosphite dehydrogenase: An unusual phosphoryl transfer reaction, *J. Am. Chem. Soc.*, 2001, **123**(11), 2672–2673.
39. R. Woodyer, W. A. van der Donk and H. M. Zhao, Relaxing the nicotinamide cofactor specificity of phosphite dehydrogenase by rational design, *Biochemistry*, 2003, **42**(40), 11604–11614.
40. R. Woodyer, J. L. Wheatley, H. A. Relyea, S. Rimkus and W. A. van der Donk, Site-directed mutagenesis of active site residues of phosphite dehydrogenase, *Biochemistry*, 2005, **44**(12), 4765–4774.
41. R. Woodyer, H. M. Zhao and W. A. van der Donk, Mechanistic investigation of a highly active phosphite dehydrogenase mutant and its application for NADPH regeneration, *FEBS J.*, 2005, **272**(15), 3816–3827.
42. Y. Zou, H. Zhang, J. S. Brunzelle, T. W. Johannes, R. Woodyer, J. E. Hung *et al.*, Crystal structures of phosphite dehydrogenase provide insights into nicotinamide cofactor regeneration, *Biochemistry*, 2012, **51**, 4263–4270.
43. K. Ranaghan, L. Ridder, B. Szefczyk, W. Sokalski, J. Hermann and A. Mulholland, Insights into enzyme catalysis from QM/MM modelling: transition state stabilization in chorismate mutase, *Mol. Phys.*, 2003, **101**(17), 2695–2714.
44. L. Ridder, A. J. Mulholland, I. Rietjens and J. Vervoort, A quantum mechanical/molecular mechanical study of the hydroxylation of phenol and halogenated derivatives by phenol hydroxylase, *J. Am. Chem. Soc.*, 2000, **122**(36), 8728–8738.
45. L. Ridder, B. A. Palfey, J. Vervoort and I. Rietjens, Modelling flavin and substrate substituent effects on the activation barrier and rate of oxygen transfer by p-hydroxybenzoate hydroxylase, *FEBS Lett.*, 2000, **478**(1–2), 197–201.
46. J. Zurek, A. L. Bowman, W. A. Sokalski and A. J. Mulholland, MM and QM/MM modeling of threonyl-tRNA synthetase: Model testing and simulations, *Struct. Chem.*, 2004, **15**(5), 405–414.
47. P. D. Lyne, A. J. Mulholland and W. G. Richards, Insights into chorismate mutase catalysis from a combined QM/MM simulation of the enzyme reaction, *J. Am. Chem. Soc.*, 1995, **117**(45), 11345–11350.
48. M. J. Frisch, G. W. Trucks, H. B. Schlegel, G. E. Scuseria, M. A. Robb and J. R. Cheeseman *et al.*, Gaussian. 09, Revision A.1 ed. Wallingford CT: Gaussian, Inc., 2009.
49. H.-J. Werner, P. J. Knowles, R. Lindh, F. R. Manby, M. Schütz and P. Celani *et al.*, Molpro. www.molpro.net, 2006.
50. C. Møller and M. S. Plesset, Note on an Approximation Treatment for Many-Electron Systems, *Phys. Rev.*, 1934, **46**(7), 618–622.
51. H. M. Berman, J. Westbrook, Z. Feng, G. Gilliland, T. N. Bhat, H. Weissig *et al.*, The Protein Data Bank, *Nucleic Acids Res.*, 2000, **28**(1), 235–242.
52. World Health Organziation (WHO): Antimicrobial resistance: global report on surveilance, 2014.

53. J. F. Fisher, S. O. Meroueh and S. Mobashery, Bacterial resistance to beta-lactam antibiotics: Compelling opportunism, compelling opportunity, *Chem. Rev.*, 2005, **105**(2), 395–424.
54. F. Fonseca, E. I. Chudyk, M. W. van der Kamp, A. Correia, A. J. Mulholland and J. Spencer, The Basis for Carbapenem Hydrolysis by Class A beta-Lactamases: A Combined Investigation using Crystallography and Simulations, *J. Am. Chem. Soc.*, 2012, **134**(44), 18275–18285.
55. P. Nordmann, G. Cuzon and T. Naas, The real threat of Klebsiella pneumoniae carbapenemase-producing bacteria, *Lancet Infect. Dis.*, 2009, **9**(4), 228–236.
56. T. R. Walsh, Emerging carbapenemases: a global perspective, *Int. J. Antimicrob. Agents*, 2010, **36**, S8–S14.
57. J. C. Hermann, J. Pradon, J. N. Harvey and A. J. Mulholland, High Level QM/MM Modeling of the Formation of the Tetrahedral Intermediate in the Acylation of Wild Type and K73A Mutant TEM-1 Class A beta-Lactamase, *J. Phys. Chem. A*, 2009, **113**(43), 11984–11994.
58. J. C. Hermann, L. Ridder, H. D. Hotje and A. J. Mulholland, Molecular mechanisms of antibiotic resistance: QM/MM modelling of deacylation in a class A beta-lactamase, *Org. Biomol. Chem.*, 2006, **4**(2), 206–210.
59. L. Mourey, K. Miyashita, P. Swaren, A. Bulychev, J. P. Samama and S. Mobashery, Inhibition of the NMC-A beta-lactamase by a penicillanic acid derivative and the structural bases for the increase in substrate profile of this antibiotic resistance enzyme, *J. Am. Chem. Soc.*, 1998, **120**(36), 9382–9383.
60. D. A. Case, T. A. Darden, T. E. Cheatham, C. L. Simmerling, J. Wang and R. E. Duke *et al.*, AMBER 12. University of California, San Francisco, 2012.
61. World Health Organization (WHO): Global update on the health sector response to HIV, 2014.
62. E. M. Wondrak, J. Lower and R. Kurth, Functional purification and enzymatic characterization of the RNA-dependent DNA-polymerase of Human-Immunodeficiency-Virus, *J. Gen. Virol.*, 1986, **67**, 2791–2797.
63. C. Castro, E. Smidansky, K. R. Maksimchuk, J. J. Arnold, V. S. Korneeva, M. Gotte *et al.*, Two proton transfers in the transition state for nucleotidyl transfer catalyzed by RNA- and DNA-dependent RNA and DNA polyrnerases, *Proc. Natl. Acad. Sci. U. S. A.*, 2007, **104**(11), 4267–4272.
64. C. Castro, E. D. Smidansky, J. J. Arnold, K. R. Maksimchuk, I. Moustafa, A. Uchida *et al.*, Nucleic acid polymerases use a general acid for nucleotidyl transfer, *Nat. Struct. Mol. Biol.*, 2009, **16**(2), 212–218.
65. T. Rungrotmongkol, A. J. Mulholland and S. Hannongbua, QM/MM simulations indicate that Asp185 is the likely catalytic base in the enzymatic reaction of HIV-1 reverse transcriptase, *MedChemComm*, 2014, **5**(5), 593–596.
66. N. Foloppe and A. D. MacKerell, All-atom empirical force field for nucleic acids: I. Parameter optimization based on small molecule and

condensed phase macromolecular target data, *J. Comput. Chem.*, 2000, **21**(2), 86–104.

67. A. D. MacKerell, D. Bashford, M. Bellott, R. L. Dunbrack, J. D. Evanseck, M. J. Field *et al.*, All-atom empirical potential for molecular modeling and dynamics studies of proteins, *J. Phys. Chem. B*, 1998, **102**(18), 3586–3616.
68. T. Rungrotmongkol, A. J. Mulholland and S. Hannongbua, Active site dynamics and combined quantum mechanics/molecular mechanics (QM/MM) modelling of a HIV-1 reverse transcriptase/DNA/dTTP complex, *J. Mol. Graphics Modell.*, 2007, **26**(1), 1–13.
69. M. J. Field, P. A. Bash and M. Karplus, A combined quantum-mechanical and molecular mechanical potential for molecular-dynamics simulations, *J. Comput. Chem.*, 1990, **11**(6), 700–733.
70. C. Bartels and M. Karplus, Multidimensional adaptive umbrella sampling: Applications to main chain and side chain peptide conformations, *J. Comput. Chem.*, 1997, **18**(12), 1450–1462.
71. S. Kumar, J. M. Rosenberg, D. Bouzida, R. H. Swendsen and P. A. Kollman, Multidimensional free-energy calculations using the weighted histogram analysis method, *J. Comput. Chem.*, 1995, **16**(11), 1339–1350.
72. L. Addadi, E. K. Jaffe and J. R. Knowles, Secondary tritium isotope effects as probes of the enzymic and non-enzymic conversion of chorismate to prephenate, *Biochemistry*, 1983, **22**(19), 4494–4501.
73. S. G. Sogo, T. S. Widlanski, J. H. Hoare, C. E. Grimshaw, G. A. Berchtold and J. R. Knowles, Stereochemistry of the rearrangement of chorismate to prephenate - chorismate mutase involves a chair transition-state, *J. Am. Chem. Soc.*, 1984, **106**(9), 2701–2703.
74. S. Marti, M. Roca, J. Andres, V. Moliner, E. Silla, I. Tunon *et al.*, Theoretical insights in enzyme catalysis, *Chem. Soc. Rev.*, 2004, **33**(2), 98–107.
75. M. Štrajbl, A. Shurki, M. Kato and A. Warshel, Apparent NAC effect in chorismate mutase reflects electrostatic transition state stabilisation, *J. Am. Chem. Soc.*, 2003, **125**, 10228–10237.
76. P. Kast, M. Asif-Ullah and D. Hilvert, Is Chorismate Mutase a Prototypic Entropy Trap? *Tetrahedron Lett.*, 1996, **37**(16), 2691–2694.
77. M. W. van der Kamp, J. Żurek, F. R. Manby, J. N. Harvey and A. J. Mulholland, Testing High-Level QM/MM Methods for Modeling Enzyme Reactions: Acetyl-CoA Deprotonation in Citrate Synthase, *J. Phys. Chem. B*, 2010, **114**(34), 1303–1114.
78. K. E. Ranaghan, L. Ridder, B. Szefczyk, W. A. Sokalski, J. C. Hermann and A. J. Mulholland, Insights into enzyme catalysis for QM/MM modelling: transition state stabilisation in chorismate mutase, *Mol. Phys.*, 2003, **101**, 2695–2714.
79. K. E. Ranaghan, L. Ridder, B. Szefczyk, W. A. Sokalski, J. C. Hermann and A. J. Mulholland, Transition state stabilization and substrate strain in enzyme catalysis: ab initio QM/MM modelling of the chorismate mutase reaction, *Org. Biomol. Chem.*, 2004, **2**, 968–980.

80. S. Martí, J. Andrés, V. Moliner, E. Silla, I. Tuñón and J. Bertrán, Transition structure selectivity in enzyme catalysis: a QM/MM study of chorismate mutase, *Theory, Chem. Acc.*, 2001, **105**, 207–212.
81. A. Shurki, M. Štrajbl, J. Villà and A. Warshel, How much do enzymes really gain by restraining their reaction fragments? *J. Am. Chem. Soc.*, 2002, **124**, 4097–4107.
82. F. Claeyssens, K. Ranaghan, F. Manby, J. Harvey and A. Mulholland, Multiple high-level QM/MM reaction paths demonstrate transition-state stabilization in chorismate mutase: correlation of barrier height with transition-state stabilization, *Chem. Commun.*, 2005, **40**, 5068–5070.
83. A. Kienhofer, P. Kast and D. Hilvert, Selective stabilization of the chorismate mutase transition state by a positively charged hydrogen bond donor, *J. Am. Chem. Soc.*, 2003, **125**(11), 3206–3207.
84. H. Guo, Q. Cui, W. N. Lipscomb and M. Karplus, Substrate conformational transitions in the active site of chorismate mutase: Their role in the catalytic mechanism, *Proc. Natl. Acad. Sci. U. S. A.*, 2001, **98**(16), 9032–9037.
85. S. Marti, J. Andres, V. Moliner, E. Silla, I. Tunon and J. Bertran, Preorganization and reorganization as related factors in enzyme catalysis: The chorismate mutase case, *Chem. – A Eur. J.*, 2003, **9**(4), 984–991.
86. C. R. W. Guimaraes, M. P. Repasky, J. Chandrasekhar, J. Tirado-Rives and W. L. Jorgensen, Contributions of conformational compression and preferential transition state stabilization to the rate enhancement by chorismate mutase, *J. Am. Chem. Soc.*, 2003, **125**(23), 6892–6899.
87. F. Claeyssens, K. E. Ranaghan, N. Lawan, S. J. Macrae, F. R. Manby, J. N. Harvey *et al.*, Analysis of chorismate mutase catalysis by QM/MM modelling of enzyme-catalysed and uncatalysed reactions, *Org. Biomol. Chem.*, 2011, **9**(5), 1578–1590.
88. S. D. Copley and J. R. Knowles, The conformational equilibrium of chorismate in solution: Implications for the mechanism of the non-enzymatic and the enzyme catalysed rearrangement of chorismate to prephenate, *J. Am. Chem. Soc.*, 1987, **109**, 5008–5013.
89. H. A. Carlson and W. L. Jorgensen, Monte Carlo investigations of solvent effects on the chorismate to prephenate rearrangement, *J. Am. Chem. Soc.*, 1996, **118**, 8475–8484.
90. S. Martí, J. Andrés, V. Moliner, E. Silla, I. Tuñón and J. Bertrán, Conformational equilibrium of chorismate. A QM/MM theoretical study combining statistical simulations and geometry optimisations in gas phase and in aqueous solution, *J. Mol. Struct.: THEOCHEM*, 2003, **632**, 197–206.
91. S. Hur and T. C. Bruice, Comparison of formation of reactive conformers (NACs) for the Claisen rearrangement of chorismate to prephenate in water and in the E-coli mutase: The efficiency of the enzyme catalysis, *J. Am. Chem. Soc.*, 2003, **125**(19), 5964–5972.
92. S. Hur and T. C. Bruice, Just a near attack conformer for catalysis (chorismate to prephenate rearrangements in water, antibody, enzymes, and their mutants), *J. Am. Chem. Soc.*, 2003, **125**(35), 10540–10542.

93. S. Hur and T. C. Bruice, The near attack conformation approach to the study of the chorismate to prephenate reaction, *Proc. Natl. Acad. Sci. U. S. A.*, 2003, **100**(21), 12015–12020.
94. X. D. Zhang, X. H. Zhang and T. C. Bruice, A definitive mechanism for chorismate mutase, *Biochemistry*, 2005, **44**(31), 10443–10448.
95. X. H. Zhang and T. C. Bruice, The proficiency of a thermophilic chorismate mutase enzyme is solely through an entropic advantage in the enzyme reaction, *Proc. Natl. Acad. Sci. U. S. A.*, 2005, **102**(51), 18356–18360.
96. S. Martí, J. Andrés, V. Moliner, E. Silla, I. Tuñón and J. Bertrán, A QM/MM Study of the conformational equilibria in the chorismate mutase active site. The role of enzymatic deformation energy contribution, *J. Phys. Chem. B*, 2000, **104**, 11308–11315.
97. B. Szefczyk, W. A. Sokalski, K. E. Ranaghan and A. J. Mulholland, Differential transition-state stabilization in enzyme catalysis: Quantum chemical analysis of interactions in the chorismate mutase reaction and prediction of the optimal catalytic field, *J. Am. Chem. Soc.*, 2004, **126**(49), 16148–16159.
98. T. Ishida, Effects of Point Mutation on Enzymatic Activity: Correlation between Protein Electronic Structure and Motion in Chorismate Mutase Reaction, *J. Am. Chem. Soc.*, 2010, **132**(20), 7104–7118.
99. C. R. W. Guimaraes, M. Udier-Blagovic, I. Tubert-Brohman and W. L. Jorgensen, Effects of Arg90 neutralization on the enzyme-catalyzed rearrangement of chorismate to prephenate, *J. Chem. Theory Comput.*, 2005, **1**(4), 617–625.

CHAPTER 12

Ribozymes

J. BERTRAN* AND A. OLIVA

Universitat Autònoma de Barcelona, Department of Chemistry, Edifici Cn, 08193 Bellaterra (Cerdanyola del Vallès), Spain
*Email: joan.bertran@uab.cat

12.1 Introduction

The concept of the ribozyme is quite recent. In fact, until the early 1980s RNA was considered just a copy of the DNA instructions and its role was to be a messenger to direct protein synthesis. However, in 1982, the research group of Cech found the first example of an RNA molecule – the self-splicing intron of the ciliated protozoan Tetrahymena – with enzyme-like properties.[1] This unexpected result led to the coining of the word 'ribozyme' to describe this kind of RNA with enzyme-like activity, even if only one example was known at that moment. But only one year later, the group of Altman discovered that the RNA component of a ribonuclease enzyme – the RNase P – was the first case of a multiple turnover enzyme using RNA-based catalysis.[2,3] The importance of this finding, which showed that RNA acts sometimes as an active participant in the chemistry of life, led Cech and Altman to be awarded with the Nobel Prize in Chemistry in 1989.

12.1.1 Natural Ribozymes

Since its initial discovery, a large number of naturally occurring ribozymes have been found.[4–7] Reactions catalysed in nature include phosphoryl and aminoacyl transfer reactions, the phosphoryl transfer being able to proceed *via* transesterification or hydrolysis.

RSC Theoretical and Computational Chemistry Series No. 9
Simulating Enzyme Reactivity: Computational Methods in Enzyme Catalysis
Edited by Iñaki Tuñón and Vicent Moliner

Published by the Royal Society of Chemistry, www.rsc.org

According to their size, natural ribozymes can be divided into three families: small self-cleaving RNAs (less than 200 nucleotides), medium-sized self-splicing introns, and larger catalytic ribonuclear–protein (RNP) complexes. Some authors[8] assemble the two last families in the same one. The class of small ribozymes include the hammerhead, hairpin, hepatitis delta virus (HDV), Varkud satellite (VS), glmS and twister. All of them catalyse a site-specific RNA backbone cleavage reaction as well as the reverse ligation process. As shown in Figure 12.1A cleavage is achieved through an S_N2 reaction mechanism in which the 2′-hydroxyl (2′-OH) of the upstream nucleotide is activated for nucleophilic attack leading to a 2′,3′-cyclic-phosphate and a 5′-OH termini. The process can take place in the absence of Mg^{2+}, albeit slower, if enough high concentrations of monovalent cations are present to allow the folding of RNA.

In contrast, larger ribozymes (more than 200 nucleotides) are critically dependent on Mg^{2+} ions for catalytic activity, thus implying that they are metalloenzymes. Introns are non-coding segments that interrupt the protein-coding sequences of most genes. There are two groups of self-splicing introns, called I and II. They are of intermediate complexity as some of them require protein cofactors and others do not. Group I introns catalyse the self-excision of the non-coding sequences through two separate phosphoryl-transfer reactions. Figure 12.1B shows that, in the first step, the 5′-exon-intron junction is cleaved following attack by the 3′-oxygen of an exogenous guanosine cofactor. In contrast, in group II introns, the initial nucleophilic attack on the 5′-exon junction is performed by either an endogenous 2′-OH group or a water molecule.

The family of largest ribozymes includes spliceosome, ribonuclease P (RNase P) and ribosome. Spliceosome is a dynamic ribonucleoprotein (RNP) that usually comprises more than 100 proteins and five small nuclear RNAs. It carries out the removal of the introns in eukaryotic cells through two successive transesterification steps. The spliceosome is in essence a protein-directed ribozyme, the protein components being essential for the delivery of RNA molecules into close proximity at the right time.[9] The structure of its active site is similar to that of the group II introns, thus suggesting that they both evolved from a common ancestor.[10] RNase P, found in all domains of life (Bacteria, Archea and Eukarya), catalyses site-specific hydrolysis of precursor RNA substrates, mainly tRNA, to generate its mature 5′ end. Finally, the ribosome is a complex molecular machine that catalyses the peptide bond formation in living organisms. An important advance in the study of this machine was made when crystal structures of the two ribosomal subunits were obtained.[11–13] This lead Steitz, Yonath and Ramakrishnan to be awarded the Nobel Prize in Chemistry in 2009. The ribosome has three tRNA-binding sites: A, P and E. Its largest subunit contains the peptidyl transferase centre (PTC), which catalyses the formation of peptide bonds by the nucleophilic attack of an aminoacyl-tRNA in the A-site on ester carbon of the peptidyl-tRNA in the P-site (see Figure 12.1C). It is worth mentioning the two main differences between ribosome and spliceosome in eukaryotes.

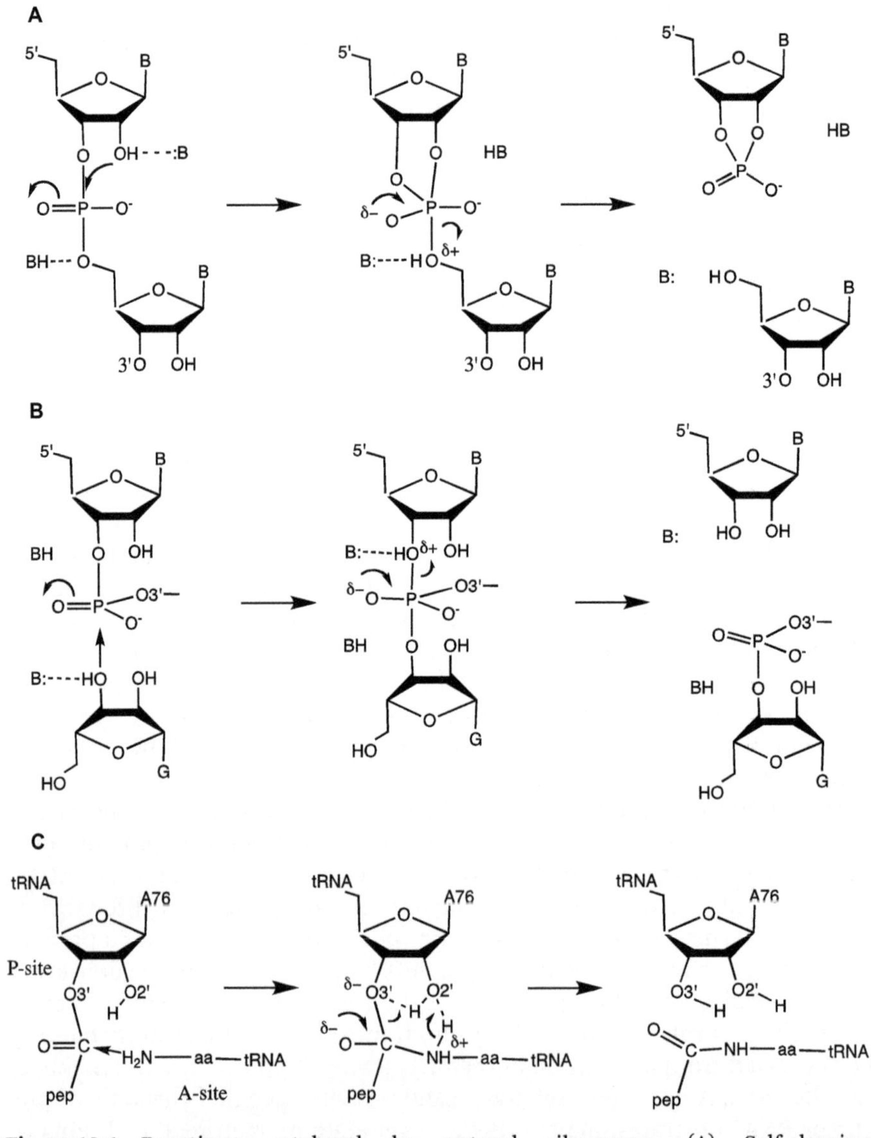

Figure 12.1 Reactions catalysed by natural ribozymes. (A) Self-cleaving reactions catalysed by small ribozymes. (B) Self-splicing reactions catalysed by group I introns. (C) Peptide bond formation catalysed by ribosomes.

On one side, the protein components and the RNA molecules are in a mass ratio of approximately 1 : 1 in ribosome while the corresponding ratio is 10 : 1 in spliceosome. On the other side, the protein components of the ribosome are mainly located on the exterior, away from the catalytic centre, while they surround and support the RNA-based catalytic centre in the spliceosome.[14]

12.1.2 Artificial Ribozymes

As described above, only two reactions are observed to be catalysed by RNA in nature: phosphoryl transfer and peptide bond formation. However, these naturally occurring ribozymes use a wide range of catalytic strategies that could be applied to other reactions.[15] The discovery of catalytic RNA showed that ribozymes could have had a key role in the evolution of life on our planet. There is clearly a 'chicken-or-egg' problem of which came first: DNA or protein. In modern cells, DNA codes for proteins, but proteins are needed to copy DNA in a process called 'replication' that occurs when cells divide. Because RNA appears to be chemically capable of serving simultaneously as an information carrier as well as a self-replicator, it is likely that primitive cellular life was RNA-based. Such an RNA world would imply that RNA is a very versatile molecule that could, in principle, catalyse the myriad of reactions necessary to sustain life.[16,17] For example, ribozymes generated by *in vitro* selection can catalyse the relatively difficult formation of carbon–carbon bonds, as in Michael addition,[18] aldol condensation[19] and Diels–Adler reactions.[20] Still unidentified ribozymes are likely to be low-cost alternatives to proteins, while maintaining high catalytic activity. The versatility of RNA indicates that a wider range of ribozymes fitting this description may be possible.[15]

12.1.3 Origin of Catalysis in Ribozymes

The chemical structure of proteins has evolved to provide a highly adaptable catalytic framework with a broad repertoire of functional groups. It is based on an electrically neutral backbone with side chains that introduce a wide variety of chemistries: carboxylic acids, amines, hydroxyl and thiol groups, as well as aliphatic or aromatic hydrophobic side chains. In contrast, RNA comprises four nucleotide bases of similar chemical nature, connected by an electrically-charged ribose-phosphate backbone. So, what resources are available to RNA that can be used to build a catalyst?[21]

First, RNA might exploit its structure. Substrate binding can result in acceleration of reaction rate owing to proximity and orientation of reactants. Second, RNA can create specific metal-ion binding sites that can activate nucleophiles or provide electrostatic stabilisation of negative charge such as a phosphorane TS. Third, the nucleobases have donor and acceptor hydrogen bonds that can be used to bind the substrate and, potentially, to stabilise a transition state. They could also act as general acids and bases, although, at first sight, their pK_a values are not ideally suited to this role at neutral pH values. However, the situation can be improved because environment can change the pK_a significantly.[21]

Ribozymes were initially classified as metalloenzymes, because metal ions were thought to be essential for their efficient catalysis and for the stabilisation of their structure.[22] In fact, metal ions could play several different direct roles in catalysis. First, water molecules contained within the inner sphere of metal ion coordination can, in principle, participate in Brønsted acid–base catalysis. Second, a metal ion could form an inner-sphere complex with RNA, acting as a Lewis acid to assist deprotonation of the attacking nucleophile or stabilisation of the charged leaving group. Third, a metal ion could stabilise the charged transition state electrostatically, possibly by binding to one of the non-bridging oxygen atoms. Several protein enzymes use a two-metal-ion strategy that involves binding one metal ion to the attacking nucleophile while the other is bound to the leaving group.[23] Crystal structure of a self-spliced group II intron suggests that a two-metal ion mechanism is also used in the case of ribozymes.[24]

In order to establish the origin of catalysis in each ribozyme, it is necessary to clarify its particular mechanism. For this purpose dynamic molecular simulations play a fundamental role. This is just the main topic of the present chapter.

12.2 Methodological Aspects

There are two widely used nucleic acid force fields, CHARMM[25] and AMBER,[26] which share similar functional form but differ in parametrisation. Both have been successfully used for describing the folding of proteins. In the case of RNA, they present several problems that are related to the conformation versatility due to the geometrical parameters shown in Figure 12.2. First, the torsional angles (α, β, γ, δ, ε, ζ) give a large flexibility to the main chain and their parametrisation is not easy since they are strongly correlated. Second, the sugar puckering and the glycosidic torsion angle, χ, also increase the number of conformational possibilities. Finally, the 2′OH group, not existing in DNA, makes RNA chemically reactive and, which is most important, promotes key hydrogen bond donor/acceptor

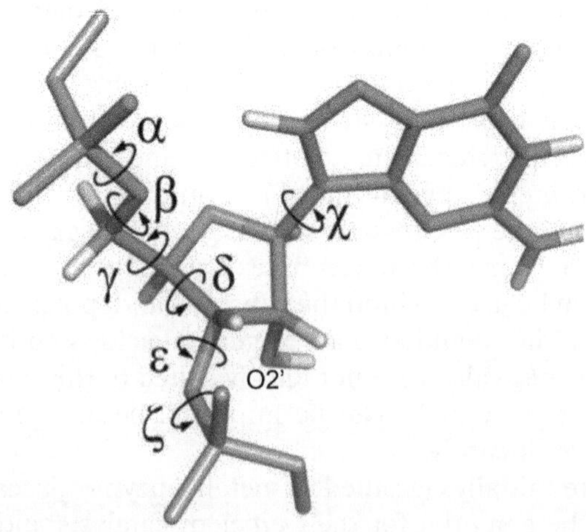

Figure 12.2 Structure of RNA nucleotides with depicted torsion angles and identification of the O2′ atom.

interactions, leading to a dramatic increase of the conformational versatility of RNA.

In the case of CHARMM, force field parameters specifically optimised for the residues important in the study of RNA catalysis were derived from density-functional calculations consistent with the CHARMM27 all-atom empirical force fields.[27] More recently, it has been reparameterised to CHARMM36 by correcting the 2′OH ribose torsional potential in an attempt to suppress rather rapidly developing instabilities in base-paired regions.[28]

The first parametrisation of AMBER, in 1995, was later modified to improve the sugar puckering[29] and some torsional angles.[30] The new version was called AMBER99. In 2007, the bsc0 correction reparameterised the α/γ dihedrals to prevent collapse of the B-DNA due to accumulation of irreversible non-native γ-*trans* backbone dihedral states.[31] In 2010, several reparameterisation of the torsional parameters for the glicosidic dihedral angle, χ, were carried out in order to remove destabilisation of the anti region and thus prevent the formation of spurious ladder-like structural distortions in RNA simulations.[32–34] They also improve the description of the *syn* region and the *syn*-anti balance. The same groups have also carried out additional reparameterisations, derived from quantum-mechanical calculations, of ε, ζ and other torsional parameters.[35,36]

Another aspect of AMBER which is worth mentioning is the parameterisation of the electrostatic term by fitting the charges to reproduce the electrostatic potential around the nucleic acid building blocks. This fixed charge model well describes the electrostatic component of base stacking, base pairing and the interaction involving the 2′-OH group of ribose, but gives a worse description of the backbone, which is generally flexible, anionic and populates diverse conformational classes.[37] In addition, the performance of pairwise additive force fields is obviously unsatisfactory for inclusion of divalent cations, because their interactions lead to major polarisation and charge-transfer effects. Neglect of polarisation also limits the description accuracy of other contributions, such as interactions with monovalent cations, conformational flexibility, hydrogen bonding and solute polarisation by the solvent.[38]

12.3 Mechanisms in Natural Ribozymes

Large ribozymes that catalyse phosphoryl-transfer reactions (spliceosome, RNase P and group I and II introns) are metalloenzymes. It was therefore widely believed that all ribozymes would be metalloenzymes. However, in 1998, Murray and coworkers[39] showed that the small nucleolytic ribozymes function in the absence of divalent cations. As a result, the paradigm was revised to a model in which the nucleolytic ribozymes do not use metal ions to catalyse their cleavage. For this reason, we will focus the first part of this section to the study of the mechanism of the self-cleaving reaction catalysed by the six smallest ribozymes. The second part will be devoted to the peptide bond formation catalysed by the ribosome.

12.3.1 Self-cleaving Reaction

Ribonuclease A (RNase A), a protein enzyme, catalyses the same chemical reaction as the self-cleaving ribozymes and provides a clear example of concerted general acid–base catalysis.[40,41] In this mechanism (see Figure 12.3), His12 serves as a general base catalyst to remove a proton from the attacking 2′-oxygen nucleophile, His119 acts as a general acid catalyst to protonate the 5′-oxygen leaving group and Lys41 is involved in stabilising the negatively charged transition state. The imidazole side chain of histidine ionises with a pK_a value near 6, making histidine residues particularly well suited to accept and donate protons at neutral pH. So, RNase A uses four basic strategies to promote the self-cleaving reaction: geometric constraints, activation of the nucleophile, transition state stabilisation, and leaving group protonation. The six small self-cleaving ribozymes can be split into two groups, based on whether their active site is formed by an irregular helix (hammerhead, hairpin and VS) or by a double pseudoknot (PK) structure (HDV, glmS and twister). In the following sections we will discuss if the mechanism of the six self-cleaving ribozymes is similar or not to that of the RNase A.

12.3.1.1 Hammerhead Ribozyme

In 1987, Olke Ukklenbeck first reported the existence of a small self-cleaving RNA known as the hammerhead ribozyme, but its catalytic mechanism remains still elusive. Early studies focused on a minimal ribozyme design, whose crystal structure was the first riboxyme to be characterised by X-ray crystallography. Surprisingly, this structure did not seem to correspond to a catalytically active conformation, since it was not supported by the extensive biochemical data. Other crystal structures of the hammerhead ribozyme

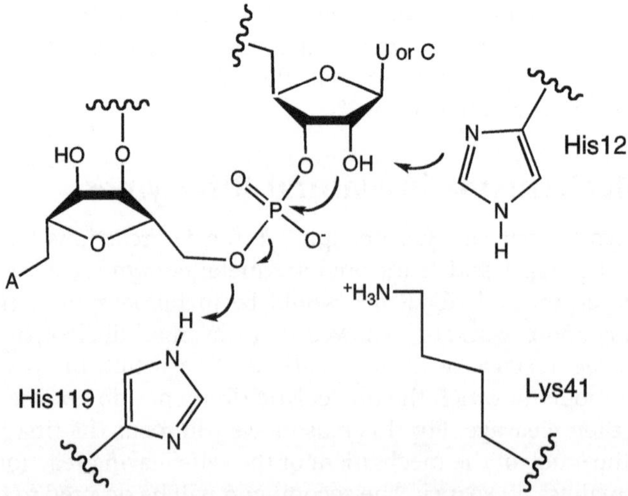

Figure 12.3 Self-cleaving reaction catalysed by RNase A.

have been solved. They have been the basis of molecular dynamics (MD) simulations on this system and only one showed the spontaneous rearrangement to give a near in-line attack conformation. Starting from this conformation, a quantum-mechanical study of a reduced model was carried out by Torres *et al.*,[42] using a $[Mg^{2+}(OH)^{-}(H_2O)_4]^{+}$ cluster which is bound to a non-bridging oxygen atom of phosphate. In a first step, the hydroxyl anion removes a proton from the O2′ attacking nucleophile and, in a second step, one of the water molecules, activated by the magnesium cation, gives a proton to the O5′ leaving group. Later on, Leclerc and Karplus[43] obtained a more favourable reaction path using two magnesium cations.

An important change was produced in 2006, since the determination of a full-length structure of the hammerhead ribozyme revealed the role played by tertiary interactions in the 1000-fold catalysis enhancement. This allowed explaining the previously irreconcilable sets of experimental results.[44] It is also worth mentioning that the guanosines G-8 and G-12 are now well positioned for their suggested roles in acid–base catalysis (see Figure 12.4). The elucidation of the full-length structure of the hammerhead ribozyme has lead to a substantial controversy about the mechanistic role played by metal ions in the catalytic process. At one extreme, it has been suggested that Mg^{2+} plays a purely structural role and is not a required participant in the chemical mechanism of catalysis. At the other extreme, some proposals argue that Mg^{2+} ions directly participate in the transition state of the

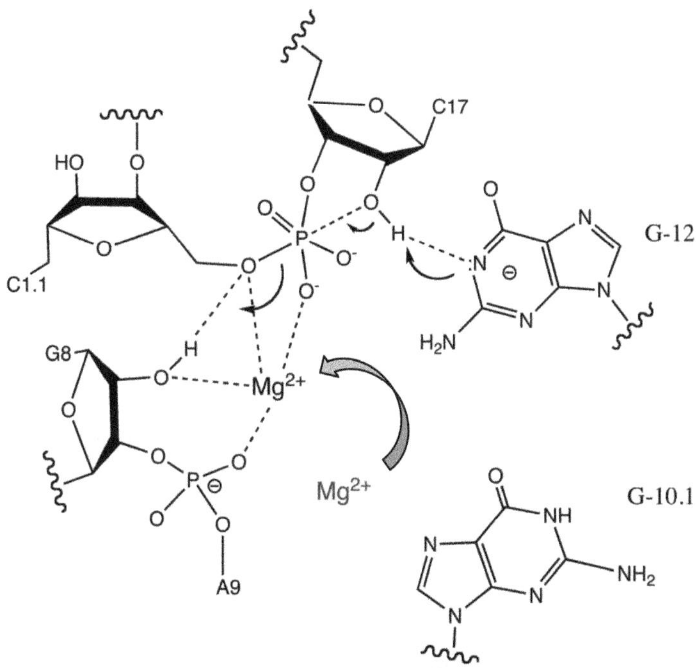

Figure 12.4 Catalytic role of Mg^{2+} in the full-length hammerhead ribozyme.

hammerhead self-cleavage reaction. As an excess of negative charge is predicted to accumulate in the transition state and the proximity of the A-9 phosphate (see Figure 12.4), one would think that its stabilisation through the presence of cations would likely contribute to catalytic enhancement. Although no Mg^{2+} ions were observed in the crystal structure, a single Mn^{2+} is observed to bind directly to the A9 phosphate in the active site, thus increasing the reaction rate.[44,45]

The role of Mg^{2+} in hammerhead ribozyme catalysis has been also studied by means of MD simulations[46,47] based on the above mentioned full-length crystal structure. The simulation results show that the deprotonation of the nucleophile is correlated with the migration of the Mg^{2+} ion, thus leading to a catalytically active conformation with the Mg^{2+} ion bridging the A9 and scissile phosphates. Once in the bridging position, Mg^{2+} interacts, as a Lewis acid, with the O5′ leaving group and the 2′OH of G8. On one side, it stabilises directly the accumulating negative charge on the O5′ leaving group and, at the same time, it increases the pK_a on the 2′OH to facilitate the proton transfer from the O2′ of G8 to O5′ (see Figure 12.4). Additional QM/MM simulations suggest that the barrier for the general acid proton-transfer step may be sufficiently low so as to occur on the nanosecond time scale. This mechanism has been confirmed through mutant simulations at the C3, G8 and G5 positions, thus suggesting that the Watson–Crick base pair interaction between G8 and C3, the hydrogen bond network between C17 and G5, and the base stacking interactions between G8 and C1.1 collectively contribute to the enhancement of the catalytic activity.[48] The role of monovalent (Na^+) and divalent (Mg^{2+}) metal ion binding to stabilise repulsions between negatively charged groups in the reactant and in the activated precursor has been also the object of MD simulations. An important conclusion is that Mg^{2+}, in addition to stabilising active conformations, plays a specific chemical role in the catalytic mechanism.[49] Finally, QM/MM free-energy simulations, assuming that the 2′OH group of C17 is already deprotonated, have been carried out.[50] The potential of mean force profiles reported in this study have allowed to show that phosphoryl transfer and general acid catalysis are stepwise and that the phosphoryl transfer is the rate-determining step with a free-energy barrier of about 24.4 $kcal\,mol^{-1}$, in a quite good agreement with the experimental rate. They also show that both steps are sensitive to the Mg^{2+} binding mode and that the position of this ion follows the negative charge along the phosphoryl-transfer reaction coordinate in order to provide electrostatic stabilisation, the change of position being more pronounced in the initial and final stages when the nucleophile and the leaving group have the greatest negative charge.

12.3.1.2 HDV Ribozyme

The HDV ribozyme was initially discovered in the human pathogen HDV. It has been found in all kingdoms of life, including the human genome. So, understanding these molecules has implications for human health and general RNA function *in vivo*. Recent crystallographic studies have demonstrated that the HDV ribozyme binds a Mg^{2+} ion within its active site.[51] Mutational analyses suggest a more clear participation of this ion in the catalysis by HDV than in the case of the hammerhead ribozyme. In the genomic RNA strand as well as in its antigenomic counterpart generated during virus replication there are two sequences with ribozyme activities. These ribozymes, the genomic and anti-genomic types, are required for self-cleavage of polymeric RNA transcripts.[52] Both forms adopt an overall similar secondary structure and likely employ similar catalytic mechanisms. Several crystal structures have been solved and this has allowed to find two different structures of the ribozyme, named *cis*-acting and *trans*-acting, depending on whether the cleavage site and the active site are on the same (*cis*) or different (*trans*) strands.

Crystal structures of the precursor and product of the *cis*-acting genomic HDV ribozyme and of the *trans*-acting precursor have confirmed that a cytosine (C75 in the genomic ribozyme and C76 in the antigenomic form) plays an important role in the catalysis. Two different mechanisms have been proposed (see Figure 12.5). In the first mechanism (Figure 12.5A), the cytosine acts as a general base accepting a proton from the 2′-hydroxyl while a water molecule, coordinated to the Mg^{2+} ion, protonates the 5′-oxygen. In the second mechanism (Figure 12.5B) a protonated cytosine acts as a general acid mediating proton transfer to the 5′-oxygen leaving group, while a hydroxyl anion, coordinated to the Mg^{2+} ion abstracts the proton from the 2′-OH. The introduction of mutations at C75 gave support to the first mechanism.[53] This conclusion was confirmed by another paper through the study of several crystal structures of the HDV ribozyme in its pre-cleaved

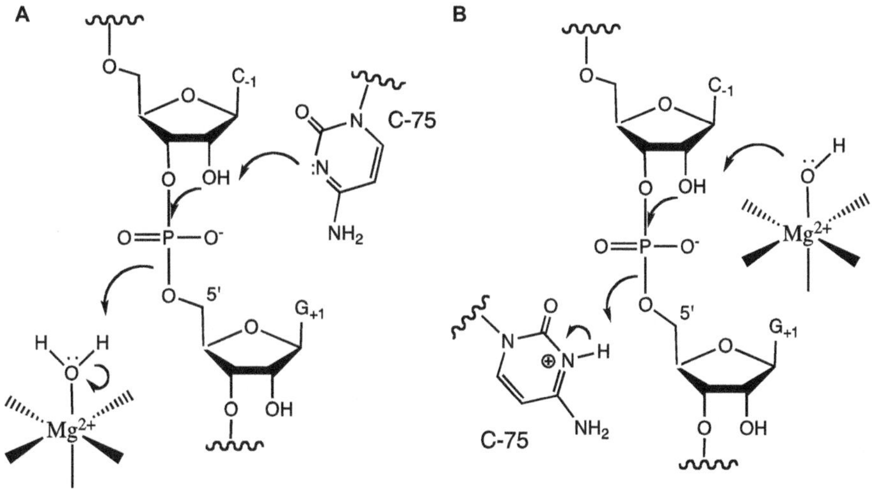

Figure 12.5 Roles proposed for the active-site cytosine (C75) in the HDV ribozyme reaction. (A) Cytosine acts as a general base. (B) Protonated cytosine acts as a general acid.

state, as it showed that cytidine is positioned to activate the 2′-OH nucleophile in the precursor structure.[54] In contrast, reactivity–pH profiles, as well as distance to the leaving group oxygen gave support to the second mechanism,[55] a conclusion that was also reached by other authors[56] through the use of a hyperactivated RNA substrate bearing a 5′-phosphorothiolate, a much better leaving group that suppresses the deleterious effects of cytosine mutations. More information about the role of metal ions in the mechanism was obtained through the analysis of the precursor crystal structure.[51,57] It was shown that the Mg^{2+} ion interacts with the G25 · U20 reverse wobble and that it is positioned to directly coordinate to the 2′-OH, thus acting as a Lewis acid to facilitate its deprotonation. Overall, these studies support a model in which the partially hydrated Mg^{2+} positioned at the G25 · U20 reverse wobble is catalytic and could serve as a Lewis acid, a Brønsted base, or both. Finally, a recent study suggests that the HDV ribozyme transition state resembles the cleavage product in the degree of proton transfer from C75 to the leaving group.[58]

Let us now consider the theoretical studies that have been carried out on the mechanism of the HDV ribozyme. The first of them investigated the cleavage reaction of fully hydrated RNA enzymes by means of Car–Parrinello calculations, using one or two Mg^{2+} ions. The results obtained showed that the double-metal-ion reaction mechanism in the presence of a hydroxyl anion is favoured with respect to single-metal-ion mechanisms.[59]

MD simulations have also been carried out. Two of these simulations[60,61] found that the first mechanism (the cytosine acting as a general base) is the most favourable. A Mg^{2+} ion binds stably, by both inner- and outer-sphere contacts. In the case of unprotonated C75, the simulation indicates the formation of a hydrogen bond between the 2′OH and the N3 atom of cytosine, this hydrogen bond being required for C75 to act as a general base. When the C75 is protonated, the hydrogen bond between $C75H^+$ and the O5′ atom, which would be expected if cytosine acted as a general acid, is not observed. More recent simulations[57,62–65] give support to the second mechanism (the cytosine acting as a general acid). The starting point of all of them is the X-ray structure obtained by Chen *et al.*[51] Three of this recent simulations[57,62,63] underline that the G25 · U20 reverse wobble interacts with the Mg^{2+} or Na^+ ions, but while Mg^{2+} is positioned to directly coordinate to the 2′-OH to act as a Lewis acid, this does not occur in the case of Na^+. Through electrostatic calculations, they also show that the pK_a of C75 is shifted towards neutrality. The group of York[64] has carried out MD simulations at different stages along the reaction path and, in this way, they support the role of protonated C75 as the general acid in the transition state and permit to make experimentally testable predictions regarding the roles of key active residues that are not obvious from any available crystal structure. Finally, the simulation of Sripathi *et al.*[65] shows the presence of two wobble GU pairs flanking the active site and that the mutation of either one results in decreased catalytic activity. They also focus on linking the mutations to specific structural effects on catalytic fitness.

Let us now consider the QM/MM calculations. To our knowledge, the first of these studies[66] examined the two opposite mechanisms in which the cytosine residue can act as a general base or as a general acid. Their results supported the C75-acid mechanism, which has an unfavourable starting geometry (in agreement with the crystallographic data) but a significantly lower energy barrier as compared to the C75-base mechanism. A second QM/MM study[67] used several starting configurations differing in magnesium ion placement to build potential energy surfaces that permitted to explore plausible reaction paths. It was found that C75 is readily capable of acting as a general base while the hydrated magnesium ion acts as an acid. The calculated energy barrier (about 20 kcal mol^{-1}) lowers around 15 kcal mol^{-1} the reaction barrier of the uncatalysed reaction. More recent studies[68,69] support a mechanism in which C75 acts as a general acid while Mg^{2+} ion acts as a Lewis acid. The calculations indicate that the self-cleavage reaction is concerted when a divalent ion is bound to the catalytic site while it is sequential when a monovalent ion, such as Na^+, is at this site. Furthermore, electrostatic potential calculations suggest that the divalent metal ion lowers the pK_a of C75. It has also been shown that the metal ion contributes to catalysis by interacting simultaneously with a non-bridging oxygen atom and with the nucleophilic 2′-hydroxyl, a result that is in good agreement with the experimental findings. Finally, four free-energy simulations[70–73] have been carried out, making possible a quantitative comparison between theoretical and experimental results. The group of Hammes-Schiffer has used QM/MM free-energy simulations to investigate three different reaction mechanisms with a Mg^{2+} ion bound at the catalytic site.[70,71] They found that the free-energy barrier along the concerted pathway is about 13 kcal mol^{-1}, which is consistent with experimental results. They have also studied the reaction with a Na^+ ion instead of a Mg^{2+} ion at the catalytic site, the self-cleaving reaction being sequential in this case. The free-energy barriers along this sequential path are much lower (about 3.5 and 2 kcal mol^{-1}), although the reaction rate is ~25-fold slower since the authors indicate that the activation of the O2′ nucleophile by deprotonation and orientation is disfavoured in the case of Na^+. The third free-energy simulation[72] showed that the deprotonation and nucleophile attack are sequential and that the nucleophile attack occurs concurrently with the second proton transfer from the protonated $C75H^+$ to the O5′ leaving group. They have investigated eight reaction pathways, six of them giving a feasible reaction barrier ranging from 14.2 to 21.9 kcal mol^{-1}. This range of energy barriers indicates that the specific position and coordination of Mg^{2+} ions have a significant impact on the self-cleavage reaction. Finally, the recent free-energy simulation of Radak *et al.*[73] is consistent with a mechanistic model whereby an active site metal Mg^{2+} ion facilitates nucleophile activation and C75 acts as a general acid catalyst. However, results for a similar pathway in the absence of Mg^{2+} are not consistent with experimental data, suggesting that a structural model in which the crystallographically determined Mg^{2+} is simply replaced with Na^+ is likely incorrect.

12.3.1.3 Hairpin, VS and Twister Ribozymes

Small nucleolytic ribozymes can be divided into two distinct classes based on their ability to react in the presence of $[Co(NH_3)_6]^{3+}$, a trivalent ion with nonexchangable ligands that mimic hydrated Mg^{2+} ions but cannot directly coordinate to RNA ligands or participate in proton-transfer reactions because the pK_a of the amine ligands is much higher than that of water. Hairpin, VS, twister and glmS ribozymes can react in the presence of $[Co(NH_3)_6]^{3+}$ alone. Consistent with this lack of dependence on Mg^{2+}, none of these four ribozymes appears to use metal ion-mediated catalysis in the reaction mechanism and, therefore, nucleic acid moieties alone must be responsible for increasing the cleavage reaction rate. In contrast, hammerhead and HDV ribozymes do not have high activity in the presence of $[Co(NH_3)_6]^{3+}$ alone, since this ion strongly inhibits the Mg^{2+} catalysed reaction.[74]

The hairpin ribozyme catalysed phosphate cleavage presents a rate enhancement of about 10^6-fold relative to the nonenzymatic process in aqueous solution. This ribozyme also catalyses the reverse ligation reaction at a rate up to an order of magnitude faster than the forward cleavage process. There is experimental evidence that an adenine (A38) and a guanine (G8) have an important role in the catalytic mechanism. Vanadate has been used as a transition state mimic of protein enzymes that catalyse the same reaction. The comparison of the structure of a vanadate-hairpin ribozyme complex with structures of precursor and product complexes reveals that the catalytic role is essentially due to the relative stabilisation of the transition state through the formation of more hydrogen bonds than in the precursor or in the product.[75] Another study of the vanadate-hairpin ribozyme structures shows the presence of two water molecules near the non-bridging positions of the oxyphosphorane mimic in a manner that suggests an electrostatic stabilisation of the transition state.[76] Alternatively, it has been proposed that A38 and G8 act in a general acid–base catalysis (see Figure 12.6).

In a recent work,[77] it has been shown that a large reduction in cleavage rate is produced when A38 is substituted by purine and that this effect can be reversed by replacement of the 5′-oxygen atom at the scissile phosphate by sulphur, which is a much better leaving group. This is consistent with A38 acting as a general acid. Experimental measures of pK_a values show that the A38 N1 imino group of a hairpin ribozyme in distinct conformational states shifts towards neutrality, thus confirming the acid role of A38.[78] Evidence suggesting the role of G8 as a general base is somewhat less compelling. It has been shown[79] that most of G8 is protonated at neutral pH. This does not exclude a role in proton transfer, but a simple interpretation is that the protonated form can act by donating hydrogen bonds.

To gain insight into the catalytic mechanism of hairpin ribozyme self-cleavage, MD simulations of protonated and unprotonated A38 and G8 have been carried out. It has been shown that N1 of unprotonated A38 moves into close proximity of the active site 2′OH, indicating that A38 may act as a general base during cleavage, a role that had not been previously considered

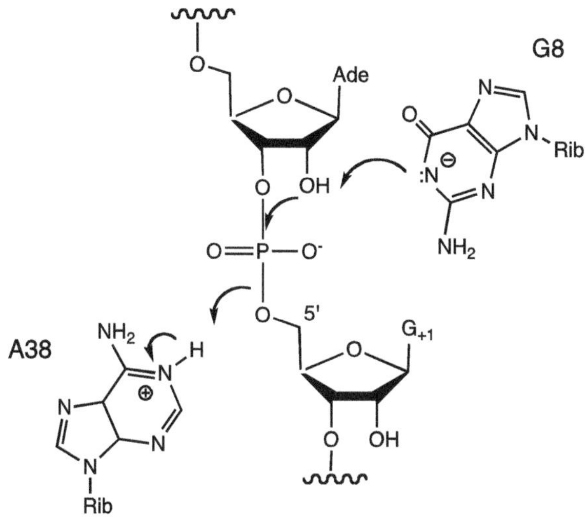

Figure 12.6 Mechanism of self-cleaving reaction catalysed by the hairpin ribozyme in which A38 and G8 act in a general acid–base catalysis.

due to the long distance observed in crystal structures. By contrast, simulations of protonated A38 indicate that N1 is placed near to the 5′-oxygen leaving group, thus confirming that A38 serves as a general acid. These results allow to suggest a plausible mechanism in which A38 acts bifunctionally and shuttles a proton directly from the 2′OH to the 5′ oxygen.[80] Furthermore, MD simulations have also shown that a $A38H^+$ is more consistent with the crystallography data than the unprotonated A38 and that a canonical G8 agrees well with the crystal structure while a deprotonated G8 ($G8^-$) profoundly distorts the active site, thus confirming that $G8^-$ does not straightforward participate in catalysis.[81] More recent MD simulations, predict that G8 donates two hydrogen bonds to the nucleophile, and this may facilitate its activation and positioning for in-line attack.[82]

MD simulations using a combined QM/MM potential allow getting a deeper insight into the mechanism of catalysis by ribozymes. Several possible mechanisms for the self-cleaving reaction catalysed by the hairpin ribozyme have been studied carrying out one-dimensional (1-D) and two-dimensional (2-D) free-energy profiles.[83,84] Of special interest is the mechanism whereby the reaction proceeds without explicit chemical participation by conserved nucleobases in the active site, since the proton is transferred from O2′ to O5′ through the non-bridging oxygen atoms of phosphate. The ribozyme lowers the overall free-energy barrier by up to 16 kcal mol^{-1}, accounting for the majority of the observed rate enhancement. The barrier reduction is achieved mainly by the electrostatic environment provided by the ribozyme.[83] QM/MM simulations have also been used to suggest the feasibility of multiple competing mechanisms.[85] In all of them, the initial nucleophilic attack is predicted to be rate-limiting and the calculated activation barriers are in good agreement with those from experiment (20–21 kcal mol^{-1}). In particular, the general acid–base mechanism, with the participation of G8^{-} and A38H^{+} yields an activation barrier of 20.4 kcal mol^{-1}. Finally, it is worth mentioning that *ab initio* and semi-empirical QM/MM approaches provide acceptable activation barriers, but predict significantly different reaction pathways.[86]

The VS ribozyme is the largest of the nucleolytic ribozymes, its crystal structure having been determined very recently.[87] Its proposed active site has common features with the one of the hairpin ribozyme and these similarities could extend to their catalytic mechanisms. The evidence for an important contribution from general acid–base catalysis in both ribozymes is strong. In the case of the VS ribozyme, there is good evidence that the cleavage reaction is catalysed by the guanine (G638), acting as a general base, and the adenine (A756), as general acid.[88] Like any functional RNA species, the VS ribozyme requires the presence of metal ions to fold into its active conformation. But the ribozyme is active in high concentrations of monovalent metal ions,[38] so that the direct participation of a site-bound metal ion as a Lewis acid, or in general acid–base catalysis, is unlikely. In fact, it has been suggested[89] that the ion-specific rate enhancement can be explained solely by the effect of the ions on nucleobase pK_a. MD simulations in the presence of monovalent and divalent ions have permitted to confirm that Mg^{2+} plays an important role in the formation of the adequate tertiary structure.[90]

Quite recently, a new nucleolytic ribozyme were discovered,[91] and soon thereafter, crystal structures were published.[92–95] This ribozyme was called twister because its conserved secondary structure resembles the ancient Egyptian hieroglyph ‘twisted flax’. Representatives of the twister ribozyme are found in all domains of life, but its biological role has yet to be determined. It shows one of the fastest known catalytic rates of any naturally occurring small self-cleaving ribozyme. The crystal structure obtained from *Oryza sativa* has permitted to suggest that a guanosine deprotonates the 2′OH, acting as a general base, while an adenosine, instead of acting as a general acid, neutralises the negative charge on a non-bridging phosphate oxygen atom.[93] A second crystal structure, the env22 twister ribozyme, shows that both an invariant guanosine and a Mg^{2+} ion are directly coordinated to the non-bridging phosphate oxygens at the uracyl-adenine cleavage state, thus indicating that the divalent metal ion plays a key role in the catalytic mechanism.[94,95]

12.3.1.4 glmS Ribozyme

The glmS ribozyme is the first natural self-cleaving ribozyme known to require a cofactor for activity. Binding of the cognate cofactor, D-glucosamine-6-phosphate (GlcN6P), initiates ribozyme cleavage, which triggers the degradation of the mRNA. The crystal structure of the glmS ribozyme allowed proposing a possible catalytic mechanism which is schematically depicted in Figure 12.7.[96] First, the amine of GlcN6P could serve as a general base.

Although the 2′-OH nucleophile is sterically inaccessible to GlcN6P, the crystal structure reveals a buried water molecule that hydrogen bonds to the nucleophile and a second water molecule that would bridge the amine of the cofactor with the first buried water. Thus the metabolite may activate the nucleophile through a proton relay. Alternatively, the N1 of G40, which is located 3.2 Å from the nucleophile, could play the role of a general base (see Figure 12.7A). Once the amine becomes positively charged, it is ideally placed to stabilise the increased negative charge of the pentacoordinate phosphorous transition state (see Figure 12.7B). Finally, the ion resulting from protonation of the amino group of GlcN6P could act as a general acid, donating a proton to the 5′-oxo leaving group. Two experimental studies are in agreement with this acid behaviour. In one of them,[97] Raman crystallography has shown that binding to the RNA lowers the pK_a of the amine of the cofactor. The second study[98] has found that there is an strong correlation between the pH dependence of the cleavage reaction and the intrinsic acidity of the cofactor. A third study[99] has questioned the role of guanine in general base catalysis from the relationship between its pK_a values and the pH dependence of cleavage kinetics. Instead, this study proposes that guanine in its neutral protonated form donates a hydrogen bond to stabilise the transition state.

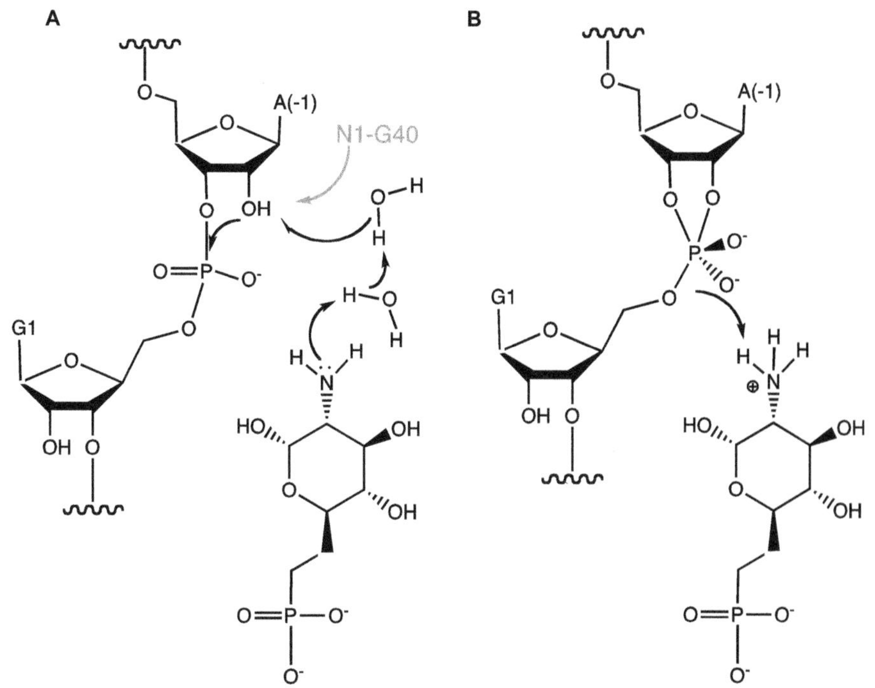

Figure 12.7 Mechanism of the self-cleaving reaction catalysed by glmS ribozyme. (A) First, the amine of GlcN6P acts as a general base through a chain of two water molecules. (B) Second, the protonated amine acts as a general acid, transferring a proton to the leaving group.

Because the X-ray crystallography is not able to identify proton positions, the MD simulations can be employed to assign a protonation state, which corresponds to the state reflected in the X-ray structure. Three MD studies[100–102] have investigated the catalytic mechanism of the glmS ribozyme. The aim of the first study[100] was to decipher the role of the cofactor through the determination of the preferred protonation state of the amine group in the wild type and in an inactive G40A mutant using MD simulations and free-energy calculations. It was shown that the binding of GlcN6P to wild type glmS decreases by about 2.2 the pK_a of the amine moiety (8.2 in solution). In contrast, the pK_a slightly increases to 8.4 for the inactive mutant. These results suggest that GlcN6P acts as a general acid and that the ability to easily release its proton diminishes when G40 is replaced by adenine. In the second study, Banás *et al.*[101] suggest that the canonical form of G40 plays a structural role by stabilising an in-line attack conformation of the cleavage site 2′-OH nucleophile rather than a more direct chemical role. They propose that 2′-OH is activated as a nucleophile by a non-bridging oxygen atom of the scissile phosphate, whereas the ammonium group of GlcN6P acts as the general acid protonating the O5′ leaving group. In the third study, Zhang *et al.*[102] investigated two different mechanisms. In the first one, the reaction is initiated by deprotonation of the N1 of G40 by an external base. The self-cleavage reaction continues as a proton transfers from the cofactor amine group to O5′ and the P–O bond making/breaking is completed in a concerted yet asynchronous reaction. The overall free-energy barrier for this mechanism was calculated to be about 23 kcal mol^{-1} in excellent agreement with the experimental one (22.6 ± 0.2 kcal mol^{-1}). The second mechanism does not require initial deprotonation by an external base and it is sequential. In the first step, there is a proton transfer from O2′ to a non-bridging oxygen atom of the phosphate, in such a way that the proton transferred stabilises the phosphorane intermediate. In this case, the calculated free-energy barrier is ~30 kcal mol^{-1}. This value is noticeably higher than the experimental one.

The glmS ribozyme is the only self-cleaving ribozyme that is also a riboswitch, since it regulates the cellular production of GlcN6P. Riboswitches, discovered in 2002, are a class of RNAs that function as regulators of gene expression. They are comprised of two domains, a sensing domain (also called aptamer domain) and a regulating domain (called expression platform). Binding of a metabolite to the aptamer domain results in a conformational change within the expression platform thereby affecting the expression of the corresponding gene. Riboswitches are important for developing antibiotics, designing new molecular sensors and being integrated into synthetic circuits. Unlike other riboswitch ligands that induce a conformational change in the RNA upon binding, the role of GlcN6P in the case of glmS appears to be entirely chemical. Solving a series of crystal structures of the glmS ribozyme, it has been found that the active site is preorganised in the absence of GlcN6P.[103]

The fusion of aptamers and ribozymes generates aptazymes, with the catalytic activity of the ribozyme being allosterically regulated by the aptamer.[104] The activity enhancement upon ligand activation of an aptazyme is typically between 10 and 10^3, the higher end of this range being more than sufficient for practical biosensor assays. The first example of an aptazyme was developed by the Breaker's group.[105] They fused an ATP aptamer to a minimal hammerhead ribozyme by rational design. Depending on the linker sequence between aptamer and ribozyme (the so called communication module), they observed that hammerhead cleavage could be induced or inhibited allosterically.

12.3.2 Peptide Bond Formation Catalysed by the Ribosome

The ribosome is a complex molecular machine which implements the genetic code by reading instructions encoded in mRNA and producing the corresponding protein. Among the different processes that take place in the ribosome, we will focus in the formation of peptide bonds by the nucleophilic attack of an aminoacyl-tRNA in the A-site on ester carbon of the peptidyl-tRNA in the P-site. This reaction takes place in the PTC of the largest subunit, the ribosome acting as a ribozyme. It is important to understand if this ancient enzyme exploits non-standard catalytic factors.

This Section will be divided in two parts, devoted to explain, respectively, the experimental works and the theoretical studies on the mechanism of this reaction.

12.3.2.1 Experimental Studies

Crystal structures, at a resolution of 2.6 Å, of the large ribosomal subunit complexed with minimal substrate, transition state or product analogues suggest[106] that the peptidyl transferase reaction proceeds through a tetrahedral zwitterionic intermediate (see Figure 12.8A), its formation being the rate-determining step of the process. The oxyanion of the tetrahedral intermediate interacts with a water molecule that is positioned by two nucleotides. Then, the intermediate rearranges and decomposes to yield deacylated tRNA in the P-site and peptidyl-tRNA that is prolonged by one aminoacid in the A-site. However, a more recent paper[107] argues from experimental data that the stabilisation of the transition structure due to the water molecule is quite small.

The transformation of this intermediate into products requires the protonation of O3′ and the deprotonation of the amino group. It has been proposed that the shuttle of the proton through one group of the ribosome could also take place through a group of the substrate itself. In particular, the A76 2′-OH of the peptidyl-tRNA is hydrogen bonded to the α-amino group and could facilitate peptide bond formation by acting as a proton shuttle between the α-amino group and the A76 3′-hydroxyl of the peptidyl-tRNA. Weinger *et al.*[108] reported experimental evidence of this participation of the substrate itself in the catalytic process. They showed that substitution of the P-site tRNA A76 2′-OH with 2′-H or 2′-F results in at least a 10^6-fold

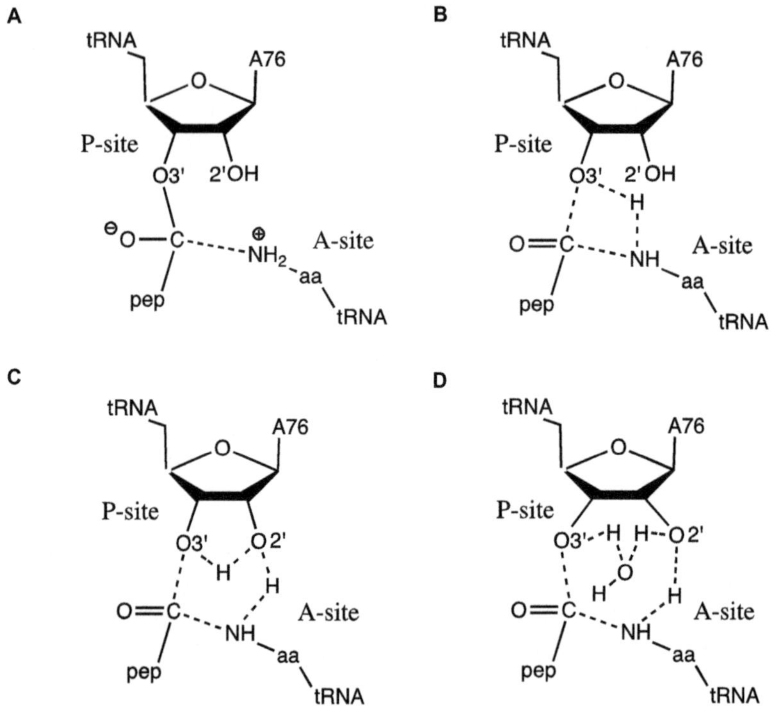

Figure 12.8 Mechanisms of the peptide bond formation, through a zwitterionic intermediate, catalysed by the ribosome. (A) Zwitterionic intermediate. (B) Four-membered cycle TS. (C) Six-membered cycle TS. (D) Eight-membered cycle TS.

reduction in the rate of peptide bond formation, but does not affect binding of the modified substrates. So, they conclude that substrate-assisted catalysis, although quite uncommon among modern protein enzymes, is essential for the evolution of enzymatic function and they suggest that substrate assistance has been retained as a catalytic strategy during the evolution of the prebiotic PTC into the modern ribosome. This proton shuttle mechanism has been questioned by Sprinzl's lab.[109,110] In a more recent paper, Zaher *et al.*[111] reinvestigated the kinetic contribution of the 2′-OH group to the catalysis and found that it is significantly smaller (about 100-fold). These new findings reconcile the conflict in the literature and support a model where interactions between active site residues and the 2′-OH are pivotal in orienting substrates in the active site for optimal catalysis.

So, according to Schmeing *et al.*,[106] several catalytic pathways can be envisaged. The α-amino group could nucleophilically attack the ester carbon, forming a zwitterion intermediate, which could break down by the direct transfer of a hydrogen atom from the amino group to O3′ through a four-membered cycle (Figure 12.8B) or by a concerted mechanism in which the hydrogen atom is transferred to the 2′ hydroxyl while its hydrogen passes to the 3′-hydroxyl leaving group, a six-membered cycle being formed in this case (Figure 12.8C). Finally, the proton shuttle from the attacking α-amino group via the 2′-hydroxyl to the 3′-hydroxyl could include a second water molecule, which was christallographically detected[106] and which is simultaneously hydrogen bonded to the 2′-hydroxyl as an acceptor and the 3′-hydroxyl oxygen as a donor through the formation of an eight-membered ring (Figure 12.8D).

Although the above mentioned crystal structures have greatly advanced the understanding of ribosomal protein synthesis, they nevertheless have restricted the interpretation of the available experimental data to the context of the isolated large ribosomal subunit. Ramakrishnan's group[112] has obtained crystal structures, at a resolution of 3.6 Å, of the 70S ribosome in complex with A- and P-site tRNAs. These structures show that the PTC is very similar in the 50S subunit and in the whole ribosome and they also reveal interactions between the protein L27 and the tRNA substrates. More recently, Steitz's group[113] has proposed an alternative mechanism based on new X-ray diffraction studies of 70S ribosome at a resolution of 2.6 Å. This structure has allowed to detect the intervention of a third water molecule that could be part of a proton wire, along which three proton transfers would take place in a concerted rate-limiting formation of the tetrahedral intermediate. In this proton wire (see Figure 12.9A), the proton from NH2 is transferred to the O2′ hydroxyl of A76, from this O2′ a proton jumps to the hydroxyl of A2451, and from this hydroxyl the proton is transferred to the third water molecule that is activated by a phosphate group and a terminal NH2 group of an alanine residue of L27. The tetrahedral intermediate presents a negative charge since the proton has been lost through the described proton network.

Brønsted linear free-energy relationships with slopes close to zero, for the peptidyl transfer reaction between the α-amino nucleophile with a series of puromycin derivatives, indicate that in the ribosome-catalysed reaction the nucleophile is neutral at the transition state,[114] in contrast to the substantial positive charge reported for typical uncatalysed aminolysis reactions. This suggests that the ribosomal transition state involves deprotonation to a degree commensurate with nitrogen–carbon bond formation. The observation of a normal effect for ^{15}N substitution of the incoming nucleophile and the fact that it does not change as a function of pH suggests[115,116] that the nitrogen is being deprotonated simultaneously with the formation of the C–N bond, thus confirming that the ribosome promotes peptide bond formation by a mechanism that differs in its details from an uncatalysed aminolysis reaction in solution. The same conclusion is reached when kinetic isotopes effects of several atoms are measured for the uncatalysed[117] and the ribosome catalysed[116] reaction. Finally, solvent isotope effects and proton inventories indicate[118] that the rate-limiting step is the formation of three hydrogen bonds with about equal contributions, this fact being consistent with a concerted eight-membered proton shuttle in the transition state and also with the recent mechanism proposed by the Steitz's group.[113]

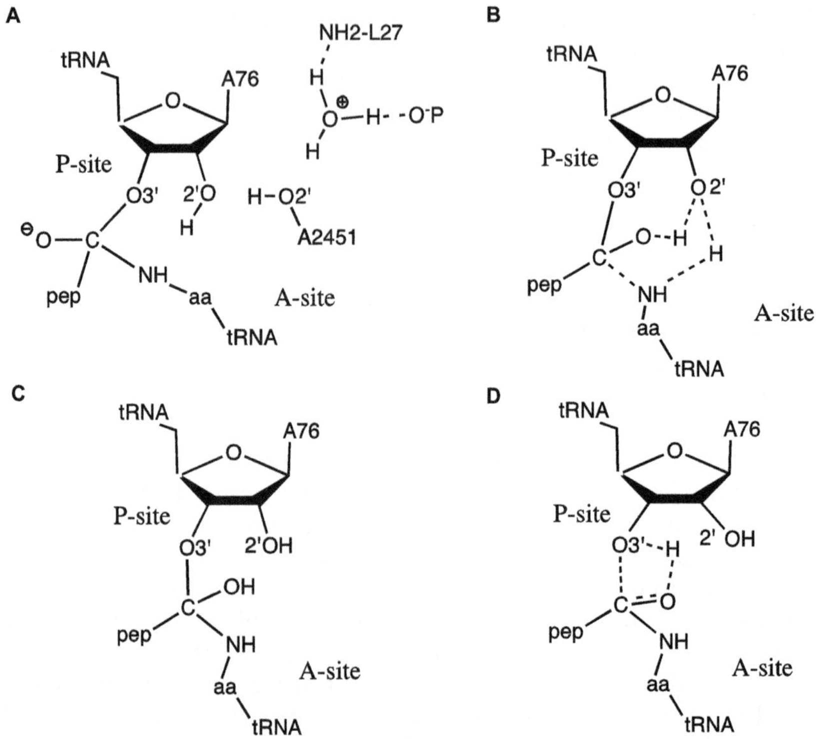

Figure 12.9 Mechanisms of the peptide bond formation, through an anionic or neutral intermediate, catalysed by the ribosome. (A) Anionic intermediate. (B) First TS. (C) Neutral intermediate. (D) Second TS.

To determine the effectiveness of the ribosome as a catalyst, several kinetic studies have been carried out.[119,120] They have shown that, in contrast with most protein enzymes, the enthalpy of activation is slightly less favourable on the ribosome than in solution. The $2{\times}10^7$-fold rate enhancement produced by the ribosome is achieved entirely by lowering the entropy of activation. This suggests that the ribosome promotes the reaction of the amino acid condensation by properly orienting the reaction substrates.[121] So, the ribosome acts as an entropic trap[122] that draws its catalytic power from the ability to orient and position substrates.

It is important to underline that kinetic studies, along with kinetic isotope effects and Brønsted linear free-energy relationships clearly suggest that the mechanism of the ribosome-catalysed reaction is quite different from that of the uncatalysed reaction in solution.

12.3.2.2 Theoretical Studies

Several QM studies, using the cluster model, have been devoted to investigate the mechanism of the peptide bond formation catalysed by the ribosome.[123–127] They have explored the mechanisms involving the formation of cycles with 4, 6 or 8 members. In all cases, the Intrinsic Reaction Coordinate (IRC) from the localised transition states leads to the reactants and products complexes, but no zwitterionic intermediate is found. In contrast, a swallow zwitterionic intermediate was found in the mechanism in solution through a four-membered ring, the effect of the solvent being introduced using a continuum model.[127] The energy barriers for the mechanisms through the formation of four- or six-membered rings are very similar, thus indicating that the mechanism involving a cycle of six members is not specially favourable and that substrate-assisted catalysis is not confirmed by theoretical calculations. These facts were confirmed by Sharma *et al.*,[128] who found that the 2′-OH of the A76 ribose, which is associated with a large rate acceleration in the ribosome reaction, does not catalyse peptide bond formation in water. In contrast, the addition of a second water molecule, leading to an eight-membered cycle, lowers the activation barrier by about 8 kcal mol^{-1} with respect to the ring of six members.[126,127]

Alternatively, other quantum mechanics calculations with cluster models have suggested a two-step mechanism, but through a neutral intermediate where the proton of the α-amino group was already transferred to the carbonyl oxygen atom, O1.[127,129–131] Wang *et al.*[129] have carried out a B3LYP study of this two-step mechanism (see Figure 12.9) in which the A76 2′-OH group transfers one proton to the carbonyl oxygen atom while, simultaneously, it receives one proton from the α-NH_2 group (Figure 12.9B). Protonation of the carbonyl oxygen neutralises the negative charge of the oxyanion, thus leading to a neutral intermediate (Figure 12.9C). In a second step, the proton is transferred to the O3′ leaving group (Figure 12.9D). Similar two-step mechanisms through different cycles in the two steps have also been considered.[127,130,131]

To explore some other options for the ribosomal peptidyl transfer reaction, several authors[128,132,133] have carried out extensive MD free-energy calculations by using the Warshel's empirical valence bond (EVB) method. Due to the use of a two-states model, a two-step mechanism through a zwitterionic intermediate was explored in solution and in the ribosome environment. They found that a significant part of the observed activation entropy of the reference solution reaction is due to solvation entropy, and that the proximity effect is smaller than previously thought. Trobro and Aqvist[132,133] stated that catalysis is mainly achieved through a stable network of hydrogen bonds to the reactants that reduces the reorganisation energy and the activation entropy of the reaction. This predicted hydrogen bonding network, including positions of key solvent molecules and the stereochemical route of the peptidyl transfer reaction, was verified in crystal structures with transition state analogues.[106]

Two *ab initio* QM/MM free-energy studies have been recently carried out to study the peptide bond formation reaction in the peptidyl transferase centre of the ribosome.[134,135] In one of them,[134] the QM part of the system was treated by density-functional theory at the B3LYP/6-31G* level, while the MM part, including the solvent and the RNA environment, was described by molecular force field. The calculated two-dimensional free-energy surfaces for the two mechanisms, the six- and eight-membered ring ones, exhibited large energy differences, the mechanism through the eight-membered cycle being the most favourable. This work has two important limitations. On one side, the Michaelis complexes are not in the calculated 2D free-energy surfaces, since they have been reduced in order to keep the computation time into reasonable limits. Furthermore, the two components of the reaction coordinate that the authors use are adequate for the case of a four-membered ring mechanism, but not in the other two cases. In fact, the authors are aware of this problem and they comment in their paper that their TS structures can be considered as 'quasi-transition states'. The second study follows a quite different strategy, since it uses the so called free-energy perturbation (FEP) method. It implies the sampling of the environment (usually the MM region) along an IRC previously traced from a TS located at QM/MM level. This strategy is cheaper and higher levels Hamiltonians can be used to describe the chemical system (the QM region). Moreover, keeping in mind that the sampling is performed along the IRC, the free-energy profile is obtained along a more realistic reaction coordinate. Nevertheless, since there is no sampling on the chemical system, the result could be biased by the fact that just one TS structure is used. In this second study, the mechanism through the eight-membered ring is again the most favourable. One important conclusion of this work is that a zwitterionic structure, similar to the one found for the reaction in solution,[136,137] appears along the reaction coordinate. However, while this zwitterion corresponds to a minimum in the free-energy profile in solution, the zwitterionic structure obtained when the reaction takes place in the ribosome is not a minimum. So, the mechanisms in solution and in the ribosome are not so different, the main difference being due to the effect of both environments. In solution, the path from the reactant complex to the TS needs a big reorganisation of the solvent. In the ribosome, on the contrary, the environment is already preorganised, this fact increasing the catalytic effect. It is also worth mentioning that this second study finds that the presence of a second magnesium ion near to the active centre leads to a new decrease of the free-energy barrier. Finally, it is important to emphasise that the stabilisation which explains the whole catalytic effect of this fossil enzyme which is the ribosome already appears in the formation of the Michaelis complex.

12.4 Conclusions, Challenges and Perspectives

As mentioned above, proteins and RNA are very different biopolimers. While proteins have a neutral backbone and side chains with a variety of functional groups which can have a catalytic role, RNA has a negative backbone and only four nucleotide bases with a similar nature and with the pK_a of their heteroatoms quite far from neutrality. Along this chapter we have seen that the catalytic effect of the ribozymes is not due to factors not present in enzymes, but to a different weight of some of these factors. In fact, the main difference comes from the role played by the ions acting as counterions. While in the traditional enzymes there are a few ions to neutralise the scarce charged groups in the side chains, in the ribozymes they play a central role since they have to neutralise the strongly negative charged backbone. Two limiting roles can be distinguished: on one side, some cations may act as a diffuse atmosphere, interacting with RNA *via* weak but numerous long-range electrostatic interactions; on the other side, site-bound cations interact strongly with the RNA at short distances, their mobility being low.[138] These interactions lead to the folding of RNA to attain a particular structure which is able to have a catalytic function. Apart from the folding, we have seen that site-bound cations may also play, in several ways, a direct role in the chemistry of the catalytic action. One possible way is to change the pK_a of the nucleic bases in such a way that these bases could participate in acid–base catalysis. They can also change the pK_a of a water molecule in their first coordination shell or of the 2′ OH of ribose to facilitate that they can act as a Brønsted acid–base catalyst. A second way is when the metal cation forms an inner-shell complex with RNA atoms and operates as a Lewis acid to facilitate the activation of the attacking nucleophile and the stabilisation of both the charged leaving group and of the increase of negative charge in the non-bridging oxygen atoms of phosphate in the TS. Examples of all these roles of the cations have been found in the mechanisms of self-cleaving reactions catalysed by the six smallest ribozymes. In contrast, in the ribosome there are no cations in the active centre and so, the catalytic effect is directly related to the formation of a structure which favours the adequate orientation of the reactants. In fact, the catalytic effect in ribosome is mostly reached in the formation of the Michaelis complex. The main conclusion of this chapter is just that not all the ribozymes can be considered as metalloenzymes, but that dications play an important catalytic role in many of them.

The main challenge in the MM or QM/MM simulations of ribozymes lies in the limitations of the force fields, especially in the parametrisation of the dications. As charge transfer and polarisation are not additives, the inclusion of both important terms in additive pair potentials is a big problem. For this reason, it is convenient to include the cations in the quantum subsystem, but the increase of the size of the QM subsystem implies to reduce the sampling and/or the level of calculation. We have already seen that the reduction of the quality level of the QM methodology might predict significantly different reaction pathways.[86]

This chapter has been mainly devoted to the natural ribozymes. However, the future perspectives in this field are clearly related to the synthesis of artificial ribozymes. As it has been mentioned above (see Section 12.1.2), the discovery of RNA catalysts (ribozymes) for accelerating chemical reactions in modern biological systems has lead to postulation of 'the RNA world' hypothesis, in which RNA molecules undertook two critical functions in primordial life: biological catalysts and genetic information carriers. Convincing evidence to support the RNA world hypothesis has been provided by the development of Darwinian-like *in vitro* evolution and selection approaches to obtaining ribozymes from artificial pools in laboratories. A critical event in the origin of life is thought to have been the emergence of an RNA molecule capable of replicating a primordial RNA 'genome'. Although the ancestral replicase appears to have been lost, modern ribozymes, such as the R18 RNA polymerase ribozyme, were generated by *in vitro* selection.[139] It is limited to transcribing stretches of RNA up to 14 nucleotides long. Starting from R18, new RNA polymerase ribozymes, capable to synthesise RNAs up to 95 nucleotides, have been obtained through both, RNA evolution and engineering.[140] It is also important to mention that, although no natural DNA catalysts have been reported, since 1994 many artificial DNA enzymes (deoxyribozymes) have been obtained, thus enlarging the future perspectives in the field.[141]

As an example of artificial ribozyme let us consider the above mentioned RNA Diels–Alderase which increases the rate and the selectivity of a specific Diels–Alder reaction.[20] It is well known that Diels–Alder reactions are a cornerstone in the organic synthesis. The crystal structure of this ribozyme shows[142] that it has a preformed hydrophobic pocket which is precisely complementary in shape to the reaction product. Atomic mutagenesis, in combination with various analytical techniques, has permitted to detect[143] the presence of three hydrogen bonds which stabilise the TS and which are also responsible for the tight binding of the product by the ribozyme. MD simulations in both crystalline and aqueous solution environments show[144] that the catalytic pocket is highly dynamic. At low Mg^{2+} ion concentrations, inactive states with the catalytic pocket closed dominate, a high concentration being necessary to stabilise the open state and thus obtaining the catalytic structure of the ribozyme. So, the catalytic activity depends on the Mg^{2+} concentration, but the possibility of a direct intervention of this ion remains an unsolved question. The ability of biopolymers to discriminate between optical isomers is vital for living systems. It has been shown that a simple strategy to selectively synthesise both stereoisomers in one catalytic pocket is by controlling access to the active site through opposite 'doors'.[145] This strategy has been confirmed through MD simulations which have shown that the dienophile is the first reaction partner to enter the hydrophobic pocket and, in a second step, the diene can enter trough the narrower back door or through the wider front door, one different enantiomer being obtained in each case.[146] Finally, it is worth mentioning that high enantioselectivities have been obtained *via* hybrid catalysis by combining a DNA-based catalyst (deoxyribozomes) and a Cu(II) complex. The role of DNA is mainly to be a chiral scaffold.[147,148]

As mentioned above, a first strategy to synthesise artificial ribozymes is by directed evolution although a limitation is that a minimum threshold of activity is required to start the cycles of mutations and selection. During the last years, a second strategy, the so called rational design, has been proposed. Rational design refers to the introduction of direct mutations of selected residues on specific positions of an already existing enzyme.

These mutations are driven by the analysis of the data obtained from different sources, ranging from X-ray to theoretical simulations. In particular, when computational analysis of the enzyme activity is used to guide the mutations, one usually refers to computational design.[149] The number of simulations carried out for artificial ribozymes is unfortunately very low. One of the main goals of this chapter is to encourage theoretical scientist to fulfil this void in order to accomplish the dream of obtaining low-cost enzymes in a very near future.

Acknowledgements

The authors gratefully acknowledge Prof. Vicent Moliner for his help in the preparation of the Figures of the present chapter, and Prof. Vicenç Branchadell for his careful reading and his valuable suggestions.

References

1. K. Kruger, P. J. Grabowski, J. Sands, D. E. Gottschling and T. R. Cech, *Cell*, 1982, **31**, 147.
2. C. Guerrier-Takada, K. Gardiner, T. Marsh, N. Pace and S. Altman, *Cell*, 1983, **35**, 849.
3. C. Guerrier-Takada and S. Altman, *Science*, 1984, **223**, 285.
4. J. A. Doudna and T. R. Cech, *Nature*, 2002, **418**, 222.
5. *Ribozymes and RNA Catalysis*, ed. D. M. Lilley and F. Eckstein, RSC Publishing, London, 2007.
6. *Ribozymes, Methods and Protocols*, ed. J. S. Hartig, Humana Press, New York, 2012.
7. W. L. Ward, K. Plakos and V. J. DeRose, *Chem. Rev.*, 2014, **114**, 4318.
8. B. L. Golden, *Biochemistry*, 2011, **50**, 9424.
9. J. Hang, R. Wan, C. Yan and Y. Shi, *Science*, 2015, **349**, 1191.
10. T. H. D. Nguyen, W. P. Galej, X. Bai, C. G. Savva, A. J. Newman, S. H. W. Scheres and K. Nagai, *Nature*, 2015, **523**, 47.
11. N. Ban, P. Nissen, J. Hansen, P. B. Moore and T. A. Steitz, *Science*, 2000, **289**, 905.
12. F. Schluenzen, A. Tocilj, R. Zarivach, J. Harms, M. Gluehmann, D. Janell, A. Bashan, H. Bartels, I. Agmon, F. Franceschi and A. Yonath, *Cell*, 2000, **102**, 615.
13. B. T. Wimberly, D. E. Brodersen, W. M. Clemons, R. J. Morgan-Warren, A. P. Carter, C. Vonrhein, T. Hartsch and V. Ramakrishnan, *Nature*, 2000, **407**, 327.
14. C. Yan, J. Hang, R. Wan, M. Huang, C. C. L. Wong and Y. Shi, *Science*, 2015, **349**, 1182.
15. D. A. Hiller and S. A. Strobel, *Philos. Trans. R. Soc. B*, 2011, **366**, 2929.
16. T. R. Cech, *Cell*, 2009, **136**, 599.
17. J. Sponer, J. E. Sponer, A. I. Petrov and N. B. Leontis, *J. Phys. Chem. B*, 2010, **114**, 15723.

18. G. Sengle, A. Eisenführ, P. S. Arora, J. S. Nowick and M. Famulok, *Chem. Biol.*, 2001, **8**, 459.
19. M. Oberhuber and G. F. Joyce, *Angew. Chem.*, 2005, **44**, 7580.
20. B. Seelig and A. Jäschke, *Chem. Biol.*, 1999, **6**, 167.
21. D. M. J. Lilley, *Philos. Trans. R. Soc. B*, 2011, **366**, 2910.
22. A. M. Pyle, *Science*, 1993, **261**, 709.
23. D. M. J. Lilley, *Trends Biochem. Sci.*, 2003, **28**, 495.
24. N. Toor, K. S. Keating, S. D. Taylor and A. M. Pyle, *Science*, 2008, **320**, 77.
25. N. Foloppe and A. D. Mackerell Jr., *J. Comput. Chem.*, 2000, **21**, 86.
26. W. D. Cornell, P. Cieplak, C. I. Bayly, I. R. Gould, K. M. Merz Jr., D. M. Ferguson, D. C. Spellmeyer, T. Fox, J. W. Caldwell and P. A. Kollman, *J. Am. Chem. Soc.*, 1995, **117**, 5179.
27. E. Mayaan, A. Moser, A. D. Mackerrel Jr. and D. M. York, *J. Comput. Chem.*, 2007, **28**, 495.
28. E. J. Denning, U. D. Priyakumar, L. Nilsson and A. D. Mackerrel, *J. Comput. Chem.*, 2011, **32**, 1929.
29. T. E. Cheatham III, P. Cieplak and P. A. Kollman, *J. Biomol. Struct. Dyn.*, 1999, **16**, 845.
30. J. Wang, P. Cieplak and P. A. Kollman, *J. Comput. Chem.*, 2000, **21**, 1049.
31. A. Pérez, I. Marchán, D. Svozil, J. Sponer, T. E. Cheatham III, C. A. Laughton and M. Orozco, *Biophys. J.*, 2007, **92**, 3817.
32. I. Yildirim, H. A. Stern, S. D. Kennedy, J. D. Tubbs and D. H. Turner, *J. Chem. Theory Comput.*, 2010, **6**, 1520.
33. M. Zgarbová, M. Otyepka, J. Sponer, A. Mládek, P. Banás, T. E. Cheatham III and P. Jurecka, *J. Chem. Theory Comput.*, 2011, **7**, 2886.
34. M. Krepl, M. Zgarbová, P. Stadlbauer, M. Otyepka, P. Banás, J. Koca, T. E. Cheatham III, P. Jurecka and J. Sponer, *J. Chem. Theory Comput.*, 2012, **8**, 2506.
35. I. Yildirim, S. D. Kennedy, H. A. Stern, J. M. Hart, R. Kierzek and D. H. Turner, *J. Chem. Theory Comput.*, 2012, **8**, 172.
36. M. Zgarbová, F. J. Luque, J. Sponer, T. E. Cheatham III, M. Otyepka and P. Jurecka, *J. Chem. Theory Comput.*, 2013, **9**, 2339.
37. J. Sponer, P. Banás, P. Jurecka, M. Zgarbová, P. Kührová, M. Havrila, M. Krepl, P. Stadlbauer and M. Otyepka, *J. Phys. Chem. Lett.*, 2014, **5**, 1771.
38. M. A. Ditzler, M. Otyepka, J. Sponer and N. G. Walter, *Acc. Chem. Res.*, 2010, **43**, 40.
39. J. B. Murray, A. E. Seyhan, N. G. Walter, J. M. Burke and W. G. Scott, *Chem. Biol.*, 1998, **5**, 587.
40. J. C. Cochrane and S. A. Strobel, *Acc. Chem. Res.*, 2008, **41**, 1027.
41. M. J. Fedor, *Annu. Rev. Biophys.*, 2009, **38**, 271.
42. R. A. Torres, F. Himo, T. C. Bruice, L. Noodleman and T. Lovell, *J. Am. Chem. Soc.*, 2003, **125**, 9861.
43. F. Leclerc and M. Karplus, *J. Phys. Chem. B*, 2006, **110**, 3395.

44. M. Martick and W. G. Scott, *Cell*, 2006, **126**, 309.
45. M. Martick, T. S. Lee, D. M. York and W. G. Scott, *Chem. Biol.*, 2008, **15**, 332.
46. T. S. Lee, C. Silva-López, M. Martick, W. G. Scott and D. M. York, *J. Chem. Theory Comput.*, 2007, **3**, 325.
47. T. S. Lee, C. Silva-López, G. M. Giambasu, M. Martick, W. G. Scott and D. M. York, *J. Am. Chem. Soc.*, 2008, **130**, 3053.
48. T. S. Lee and D. M. York, *J. Am. Chem. Soc.*, 2010, **132**, 13505.
49. T. S. Lee, G. M. Giambasu, C. P. Sosa, M. Martick, W. G. Scott and D. M. York, *J. Mol. Biol.*, 2009, **388**, 195.
50. K. Y. Wong, T. S. Lee and D. M. York, *J. Chem. Theory Comput.*, 2011, **7**, 1.
51. J. H. Chen, R. Yajima, D. M. Chadalavada, E. Chase, P. C. Bevilacqua and B. L. Golden, *Biochemistry*, 2010, **49**, 6508.
52. J. Wrzesinski, M. Legiewicz, B. Smólska and J. Ciesiolka, *Nucleic Acids Res.*, 2001, **29**, 4482.
53. A. T. Perrotta, I. H. Shih and M. D. Been, *Science*, 1999, **286**, 123.
54. A. Ke, K. Zhou, F. Ding, J. H. D. Cate and J. A. Doudna, *Nature*, 2004, **429**, 201.
55. S. I. Nakano, D. M. Chadalavada and P. C. Bevilacqua, *Science*, 2000, **287**, 1493.
56. S. R. Das and J. A. Piccirilli, *Nat. Chem. Biol.*, 2005, **1**, 45.
57. J. Chen, A. Ganguly, Z. Miswan, S. Hammes-Schiffer, P. C. Bevilacqua and B. L. Golden, *Biochemistry*, 2013, **52**, 557.
58. S. C. Koo, J. Lu, N. S. Li, E. Leung, S. R. Das, M. E. Harris and J. A. Piccirilli, *J. Am. Chem. Soc.*, 2015, **137**, 8973.
59. M. Boero, M. Tateno, K. Terakura and A. Oshiyama, *J. Chem. Theory Comput.*, 2005, **1**, 925.
60. M. V. Krasovska, J. Sefcikova, N. Spacková, J. Sponer and N. G. Walter, *J. Mol. Biol.*, 2005, **351**, 731.
61. M. V. Krasovska, J. Sefcikova, K. Réblová, B. Schneider, N. G. Walter and J. Sponer, *Biophys. J.*, 2006, **91**, 626.
62. N. Veeraraghavan, A. Ganguly, J. H. Chen, P. C. Bevilacqua, S. Hammes-Schiffer and B. L. Golden, *Biochemistry*, 2011, **50**, 2672.
63. N. Veeraraghavan, A. Ganguly, B. L. Golden, P. C. Bevilacqua and S. Hammes-Schiffer, *J. Phys. Chem. B*, 2011, **115**, 8346.
64. T. S. Lee, G. M. Giambasu, M. E. Harris and D. M. York, *J. Phys. Chem. Lett.*, 2011, **2**, 2538.
65. K. N. Sripathi, P. Banás, K. Réblová, J. Sponer, M. Otyepka and N. G. Walter, *Phys. Chem. Chem. Phys.*, 2015, **17**, 5887.
66. K. Wei, L. Liu, Y. H. Cheng, Y. Fu and Q. X. Guo, *J. Phys. Chem. B*, 2007, **111**, 1514.
67. P. Banás, L. Rulísek, V. Hánosová, D. Svozil, N. G. Walter, J. Sponer and M. Otyepka, *J. Phys. Chem. B*, 2008, **112**, 11177.
68. A. Ganguly, P. Bevilacqua and S. Hammes-Schiffer, *J. Phys. Chem. Lett.*, 2011, **2**, 2906.

69. P. Thaplyal, A. Ganguly, B. L. Golden, S. Hammes-Schiffer and P. C. Bevilacqua, *Biochemistry*, 2013, **52**, 6499.
70. A. Ganguly, P. Thaplyal, E. Rosta, P. C. Bevilacqua and S. Hammes-Schiffer, *J. Am. Chem. Soc.*, 2014, **136**, 1483.
71. P. Thaplyal, A. Ganguly, S. Hammes-Schiffer and P. C. Bevilacqua, *Biochemistry*, 2015, **54**, 2160.
72. V. Mlynsky, N. G. Walter, J. Sponer, M. Otyepka and P. Banás, *Phys. Chem. Chem. Phys.*, 2015, **17**, 670.
73. B. K. Radak, T. S. Lee, M. E. Harris and D. M. York, *RNA*, 2015, **21**, 1566.
74. A. Mir, J. Chen, K. Robinson, E. Lendy, J. Goodman, D. Neau and B. L. Golden, *Biochemistry*, 2015, **54**, 6369.
75. P. B. Rupert, A. P. Massey, S. T. Sigurdsson and A. R. Ferré-D'Amaré, *Science*, 2002, **298**, 1421.
76. A. T. Torelli, J. Krucinska and J. E. Wedekind, *RNA*, 2007, **13**, 1052.
77. S. Kath-Schorr, T. J. Wilson, N. S. Li, J. Lu, J. A. Piccirilli and D. M. J. Lilley, *J. Am. Chem. Soc.*, 2012, **134**, 16717.
78. J. A. Liberman, M. Guo, J. L. Jenkins, J. Krucinska, Y. Chen, P. R. Carey and J. E. Wedekind, *J. Am. Chem. Soc.*, 2012, **134**, 16933.
79. L. Liu, J. W. Cottrell, L. G. Scott and M. J. Fedor, *Nat. Chem. Biol.*, 2009, **5**, 351.
80. M. A. Ditzler, J. Sponer and N. G. Walter, *RNA*, 2009, **15**, 560.
81. V. Mlynsky, P. Banás, D. Hollas, K. Réblová, N. G. Walter, J. Sponer and M. Otyepka, *J. Phys. Chem. B*, 2010, **114**, 6642.
82. H. Heldenbrand, P. A. Janowski, G. Giambasu, T. J. Giese, J. E. Wedekind and D. M. York, *J. Am. Chem. Soc.*, 2014, **136**, 7789.
83. K. Nam, J. Gao and D. M. York, *RNA*, 2008, **14**, 1501.
84. K. Nam, J. Gao and D. M. York, *J. Am. Chem. Soc.*, 2008, **130**, 4680.
85. V. Mlynsky, P. Banás, N. G. Walter, J. Sponer and M. Otyepka, *J. Phys. Chem. B*, 2011, **115**, 13911.
86. V. Mlynsky, P. Banás, J. Sponer, M. W. van der Kamp, A. J. Mulholland and M. Otyepka, *J. Chem. Theory Comput.*, 2014, **10**, 1608.
87. N. B. Suslov, S. DasGupta, H. Huang, J. R. Fuller, D. M. J. Lilley, P. A. Rice and J. A. Piccirilli, *Nat. Chem. Biol.*, 2015, **11**, 840.
88. T. J. Wilson and D. M. J. Lilley, *RNA*, 2011, **17**, 213.
89. M. D. Smith, R. Mehdizadeh, J. E. Olive and R. A. Collins, *RNA*, 2008, **14**, 1942.
90. C. Bergonzo, K. B. Hall and T. E. Cheatham III, *J. Phys. Chem. B*, 2015, **119**, 12355.
91. A. Roth, Z. Weinberg, A. G. Y. Chen, P. B. Kim, T. D. Ames and R. R. Breaker, *Nat. Chem. Biol.*, 2014, **10**, 56.
92. Y. Liu, T. J. Wilson, S. A. McPhee and D. M. J. Lilley, *Nat. Chem. Biol.*, 2014, **10**, 739.
93. D. Eiler, J. Wang and T. A. Steitz, *Proc. Natl. Acad. Sci. U. S. A.*, 2014, **111**, 13028.
94. A. Ren, M. Kosutic, K. P. Rajashankar, M. Frener, T. Santner, E. Westhof, R. Micura and D. J. Patel, *Nat. Commun.*, 2014, **5**, 5534.

95. M. Kosutic, S. Neuner, A. Ren, S. Flür, C. Wunderlich, E. Mairhofer, N. Vusurovic, J. Seikowski, K. Breuker, C. Höbartner, D. J. Patel, C. Kreutz and R. Micura, *Angew. Chem., Int. Ed.*, 2015, **54**, 1.
96. D. J. Klein and A. R. Ferré-D'Amaré, *Science*, 2006, **313**, 1752.
97. B. Gong, D. J. Klein, A. R. Ferré-D'Amaré and P. R. Carey, *J. Am. Chem. Soc.*, 2011, **133**, 14188.
98. J. Viladoms and M. J. Fedor, *J. Am. Chem. Soc.*, 2012, **134**, 19043.
99. J. Viladoms, L. G. Scott and M. J. Fedor, *J. Am. Chem. Soc.*, 2011, **133**, 18388.
100. Y. Xin and D. Hamelberg, *RNA*, 2010, **16**, 2455.
101. P. Banás, N. G. Walter, J. Sponer and M. Otyepka, *J. Phys. Chem. B*, 2010, **114**, 8701.
102. S. Zhang, A. Ganguly, P. Goyal, J. L. Bingaman, P. C. Bevilacqua and S. Hammes-Schiffer, *J. Am. Chem. Soc.*, 2015, **137**, 784.
103. J. C. Cochrane, S. V. Lipchock, K. D. Smith and S. A. Strobel, *Biochemistry*, 2009, **48**, 3239.
104. A. Wittmann and B. Suess, *FEBS Lett.*, 2012, **586**, 2076.
105. J. Tang and R. R. Breaker, *Chem. Biol.*, 1997, **4**, 453.
106. T. M. Schmeing, K. S. Huang, D. E. Kitchen, S. A. Strobel and T. A. Steitz, *Mol. Cell*, 2005, **20**, 437.
107. N. Carrasco, D. A. Hiller and S. A. Strobel, *Biochemistry*, 2011, **50**, 10491.
108. J. S. Weinger, K. M. Parnell, S. Dorner, R. Green and S. A. Strobel, *Nat. Struct. Mol. Biol.*, 2004, **11**, 1101.
109. M. Koch, Y. Huang and M. Sprinzl, *Angew. Chem., Int. Ed.*, 2008, **47**, 7242.
110. Y. Huang and M. Sprinzl, *Angew. Chem., Int. Ed.*, 2011, **50**, 7287.
111. H. S. Zaher, J. J. Shaw, S. A. Strobel and R. Green, *EMBO J.*, 2011, **30**, 2445.
112. R. M. Voorhees, A. Weixlbaumer, D. Loakes, A. C. Kelley and V. Ramakrishnan, *Nat. Struct. Mol. Biol.*, 2009, **16**, 528.
113. Y. S. Polikanov, T. A. Steitz and C. A. Innis, *Nat. Struct. Mol. Biol.*, 2014, **21**, 787.
114. D. A. Kingery and S. A. Strobel, *Acc. Chem. Res.*, 2012, **45**, 495.
115. A. C. Seila, K. Okuda, S. Núñez, A. F. Seila and S. A. Strobel, *Biochemistry*, 2005, **44**, 4018.
116. D. A. Hiller, V. Singh, M. Zhong and S. A. Strobel, *Nature*, 2011, **476**, 236.
117. D. A. Hiller, M. Zhong, V. Singh and S. A. Strobel, *Biochemistry*, 2010, **49**, 3868.
118. S. Kuhlenkoetter, W. Wintermeyer and M. V. Rodnina, *Nature*, 2011, **476**, 351.
119. A. Sievers, M. Beringer, M. V. Rodnina and R. Wolfenden, *Proc. Natl. Acad. Sci. U. S. A.*, 2004, **101**, 7897.
120. G. K. Schroeder and R. Wolfenden, *Biochemistry*, 2007, **46**, 4037.
121. T. M. Schmeing, K. S. Huang, S. A. Strobel and T. A. Steitz, *Nature*, 2005, **438**, 520.
122. M. I. Page and W. P. Jencks, *Proc. Natl. Acad. Sci. U. S. A.*, 1971, **68**, 1678.

123. A. Gindulyte, A. Bashan, I. Agmon, L. Massa, A. Yonath and J. Karle, *Proc. Natl. Acad. Sci. U. S. A.*, 2006, **103**, 13327.
124. L. Massa, C. F. Matta, A. Yonath and J. Karle, in *Quantum Biochemistry*, ed. C. F. Matta, Wiley, Weinheim, 2010, p. 501.
125. K. Thirumoorthy and N. Nandi, *J. Phys. Chem. B*, 2008, **112**, 9187.
126. G. Wallin and J. Aqvist, *Proc. Natl. Acad. Sci. U. S. A.*, 2010, **107**, 1888.
127. C. Acosta-Silva, J. Bertran, V. Branchadell and A. Oliva, *J. Am. Chem. Soc.*, 2012, **134**, 5817.
128. P. K. Sharma, Y. Xiang, M. Kato and A. Warshel, *Biochemistry*, 2005, **44**, 11307.
129. Q. Wang, J. Gao, Y. Liu and C. Liu, *Chem. Phys. Lett.*, 2010, **501**, 113.
130. B. J. Byun and Y. K. Kang, *Phys. Chem. Chem. Phys.*, 2013, **15**, 14931.
131. Q. Wang, J. Gao, D. Zhang and C. Liu, *Chem. Phys.*, 2015, **450–451**, 1.
132. S. Trobro and J. Aqvist, *Proc. Natl. Acad. Sci. U. S. A.*, 2005, **102**, 12395.
133. S. Trobro and J. Aqvist, *Biochemistry*, 2006, **45**, 7049.
134. J. Xu, J. Z. H. Zhang and Y. Xiang, *J. Am. Chem. Soc.*, 2012, **134**, 16424.
135. K. Swiderek, S. Martí, I. Tuñón, V. Moliner and J. Bertran, *J. Am. Chem. Soc.*, 2015, **137**, 12024.
136. K. Swiderek, I. Tuñón, S. Martí, V. Moliner and J. Bertran, *J. Am. Chem. Soc.*, 2013, **135**, 8708.
137. K. Swiderek, I. Tuñón, S. Martí, V. Moliner and J. Bertran, *Chem. Commun.*, 2012, **48**, 11253.
138. J. C. Bowman, T. K. Lenz, N. V. Hud and L. D. Williams, *Curr. Opin. Struct. Biol.*, 2012, **22**, 262.
139. W. K. Johnston, P. J. Unrau, M. S. Lawrence, M. E. Glasner and D. P. Bartel, *Science*, 2001, **292**, 1319.
140. A. Wochner, J. Attwater, A. Coulson and P. Holliger, *Science*, 2011, **332**, 209.
141. S. K. Silverman, *Chem. Commun.*, 2008, 3467.
142. A. Serganov, S. Keiper, L. Malinina, V. Tereshko, E. Skripkin, C. Höbartner, A. Polonskaia, A. T. Phan, R. Wombacher, R. Micura, Z. Dauter, A. Jäschke and D. J. Patel, *Nat. Struct. Mol. Biol.*, 2005, **12**, 218.
143. S. Kraut, D. Bebenroth, A. Nierth, A. Y. Kobitski, G. U. Nienhaus and A. Jäschke, *Nucleic Acids Res.*, 2012, **40**, 1318.
144. T. Berezniak, M. Zahran, P. Imhof, A. Jäschke and J. C. Smith, *J. Am. Chem. Soc.*, 2010, **132**, 12587.
145. R. Wombacher, S. Keiper, S. Suhm, A. Serganov, D. J. Patel and A. Jäschke, *Angew. Chem., Int. Ed.*, 2006, **45**, 2469.
146. T. Berezniak, A. Jäschke, J. C. Smith and P. Imhof, *J. Comput. Chem.*, 2012, **33**, 1603.

147. G. Roelfes, A. J. Boersma and B. L. Feringa, *Chem. Commun.*, 2006, 635.
148. F. Rosati, A. J. Boersma, J. E. Klijn, A. Meetsma, B. L. Feringa and G. Roelfes, *Chem. - Eur. J.*, 2009, **15**, 9596.
149. K. Swiderek, I. Tuñón, V. Moliner and J. Bertran, *Arch. Biochem. Biophys.*, 2015, **582**, 68.

CHAPTER 13

Effects of Water and Non-aqueous Solvents on Enzyme Activity

EVA PLUHAŘOVÁ, NICOLAS CHÉRON AND DAMIEN LAAGE*

Ecole Normale Supérieure – PSL Research University, Chemistry Department, Sorbonne Universities – UPMC Univ Paris 06, CNRS UMR 8640 PASTEUR, 24 rue Lhomond, 75005 Paris, France
*Email: damien.laage@ens.fr

13.1 Introduction

Water is largely considered to be an indispensable ingredient to life. Living organisms can adapt to a surprisingly broad range of harsh conditions, including for example the glacial temperatures found in the Himalayas, the very hot (up to 400 °C) hydrothermal vents at the bottom of oceans, the very acidic or very basic conditions found in geysers and volcanic environments, and the high salinity of the Dead Sea.[1] However, despite the harshness of these extreme conditions, the presence of liquid water seems to remain an essential requirement. Even for organisms like seeds, which may survive in a dormant state in dry conditions, water is necessary to grow and develop. One may then ask what is so special about water. What are the molecular properties that make it so important for the functioning of living organisms? Is it for example its high polarity, or rather its ability to form an extended hydrogen-bonded network, or the lability of this network?

The layer of water molecules encasing a biomolecule is considered to play a large number of roles in biochemical systems, all essential to the proper

RSC Theoretical and Computational Chemistry Series No. 9
Simulating Enzyme Reactivity: Computational Methods in Enzyme Catalysis
Edited by Iñaki Tuñón and Vicent Moliner

Published by the Royal Society of Chemistry, www.rsc.org

functioning of biomolecules.[2–4] The presence of water is for example important for the structural integrity of proteins and DNA, as illustrated by the conformational change undergone by DNA at low hydration levels from the canonical B-form of the double helix to the A conformation.[5] The hydration shell is also a protection against any excess of energy that could denature the fragile scaffolding of biomolecules, held together by weak hydrogen bonds. Such an excess of energy due to, for example, a photon absorption can be quickly dissipated in the numerous vibrational modes available in the hydration shell,[6] which acts as a thermostat.

Many experimental and theoretical studies have investigated the properties of biomolecular hydration layers (see, for example, ref. 7 and 8). One key point has been to determine how much the hydration shell differs from bulk water, and especially to what extent its dynamics is perturbed by the presence of the biomolecular interface. In the limit of a very slow and viscous hydration shell, this could hinder the protein conformational motions. While some aspects are still debated, there is now a growing consensus that the water rotational and hydrogen-bond dynamics in most of the hydration shell is only moderately retarded with respect to the bulk, by a factor of 2 to 5, with some water molecules being more slowed[9–11] (Figure 13.1). This picture is supported both by NMR experiments[9,12] and by molecular dynamics simulations.[10,11] (We refer the interested reader to recent reviews on this topic.[12,13]) However, even though the perturbation induced by the biomolecule on its hydration shell appears to be limited, the presence of this hydration shell has a major impact on the functioning of the biomolecule.

In this chapter, we will focus on the effects of the hydration shell on the properties of proteins, and more specifically on the catalytic activity of enzymes. We will first review the widespread image that considers water as a lubricant of the protein motions necessary for catalysis: we will describe its

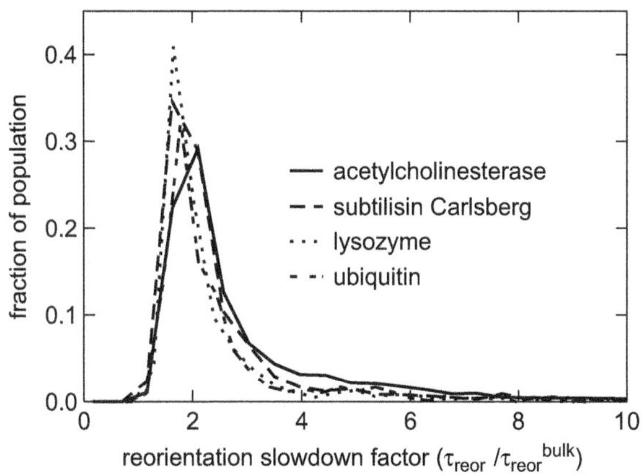

Figure 13.1 Distribution of retardation factors for the reorientation dynamics of water molecules in the first hydration shell of a series of proteins with respect to the bulk.[11]

origin and some recent experimental results that question this model. We will then discuss how some enzymes can retain a catalytic activity in non-aqueous conditions, thus opening a tremendous range of potential applications for biocatalysis. We will in particular review how changing the solvent can be used to tune the activity and the specificity of enzymes. We will then examine the different molecular pictures that have been proposed to rationalise these solvent effects on enzyme catalysis, and we will close with some concluding remarks.

13.2 Traditional Picture: Water Lubricates the Protein Motions

A widespread[2,3,14] picture sees water as a lubricant of the protein dynamics. The water molecules would act as a plasticiser that facilitates the protein conformational rearrangements necessary for the catalysis (we note that this image of water as a lubricant is also employed outside of the biochemical context, for synthetic objects including, for example, molecular machines[15]). The hydration layer would thus help proteins maintain the delicate balance between rigidity – without which the protein would lose its specificity – and flexibility – that is required to undergo conformational changes.

Protein motions take place on a rough high-dimensional free energy surface characterised by a succession of local minima corresponding to different conformations.[16] On this energy landscape, the different conformations are separated by barriers which range from low barriers for local rearrangements, as is the case for example for a simple rotation of a methyl group, up to very high barriers for large-amplitude, collective motions, including, for example, domain motions (Figure 13.2). This leads to a very broad spectrum of protein conformational motions, occurring on timescales ranging from picoseconds to milliseconds (see, for example, ref. 17).

During catalysis, the enzyme needs to rearrange from a conformation adapted to the reactant state to a conformation stabilising preferentially the reaction transition state, and in the traditional picture, water is considered to facilitate these conformational transitions.

13.2.1 Hydration, Protein Flexibility and Enzymatic Activity

Starting in the late 1970s, pioneering experimental studies investigated the effect of decreasing hydration levels on the activity of enzymes in hydrated powders (see ref. 18 for a review). For a series of enzymes including chymotrypsin, lysozyme, subtilisin Carlsberg, urease and amylase, a minimum amount of water was found to be necessary to detect a catalytic activity (see Figure 13.3). This critical hydration level was measured to range from 0.1 to 0.3 g of water per g of protein. For lysozyme, this amounts to close to 300 water molecules,[19] *i.e.* approximately the quantity of water necessary to form a monolayer around the protein. Above this minimum hydration, enzymatic

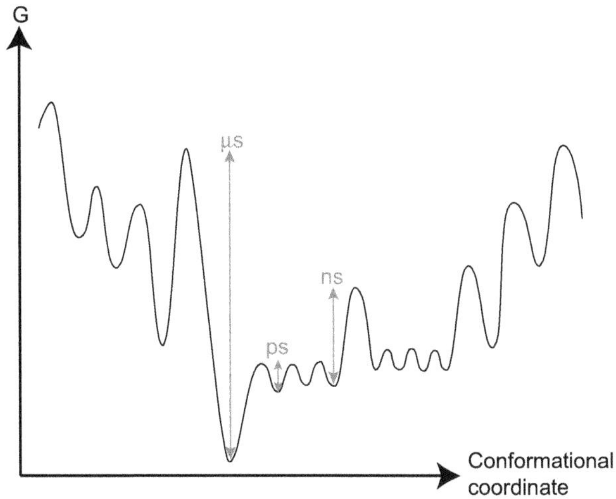

Figure 13.2 Schematic protein free energy landscape with different barrier heights, leading to a broad range of conformational rearrangement timescales.

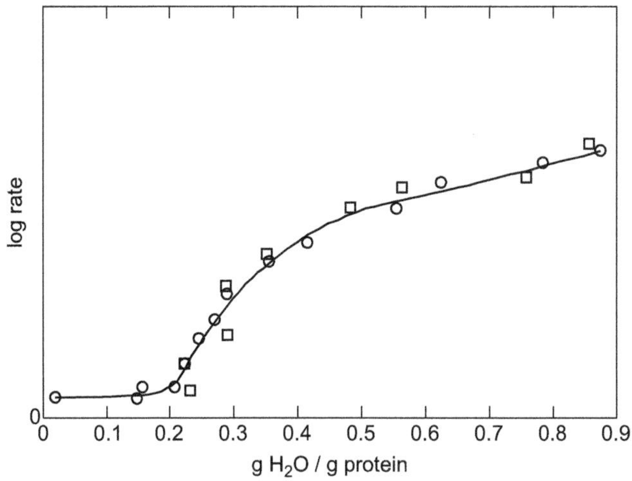

Figure 13.3 Hydration dependence of the lysozyme enzymatic activity (squares) and protein relaxation dynamics measured by ESR (circles) (data from ref. 19; the line is shown as a guide to the eye).

activity rises abruptly to reach a plateau value at high hydration levels (see Figure 13.3).

In addition, for systems like lysozyme,[18] the increase in the protein dynamics (measured here by the reorientation relaxation time of an ESR spin probe[18,19]) was measured to closely parallel the sigmoidal shape of the increase in the enzymatic activity (see Figure 13.3).

This correlation thus led to a picture[18] where the hydration shell gives the enzyme the mobility that is required to undergo the conformational rearrangements necessary to bind the substrate and convert it to a product (however, we stress that the experiments only showed a correlation between these properties, and did not establish an actual causality). In this image, the progressive addition of water is considered to enhance the protein flexibility for example by replacing the protein–protein hydrogen bonds with protein–water hydrogen bonds that can be broken and formed more easily, and by partly screening the electrostatic interactions between charged sites in the protein. According to this picture, a minimum amount of water is necessary to induce the protein flexibility, and once the protein is covered by approximately a full layer of water, the activity reaches a plateau value (Figure 13.3).

The relationship between protein hydration and protein dynamics was examined with a broad range of techniques. In addition to the above-mentioned ESR measurements, these include for example Mössbauer spectroscopy that showed that myoglobin does not exhibit significant motions at low hydration levels and ambient temperature, while a marked increase in the dynamical amplitudes is measured at a higher relative humidity.[20] The mean-square displacement of protein atoms was also measured by neutron scattering[21,22] (Figure 13.4). In dry protein powders, the amplitude of the protein motions remains very limited at every temperature. In contrast, for hydrated powder samples, a sudden increase in the amplitude of the protein motions is measured as the temperature is increased above approximately 200 K. This has been suggested to correspond to a dynamical transition[21] and to the onset of larger amplitude anharmonic protein motions, while at lower temperature the motions would be limited to harmonic fluctuations around a single conformation.

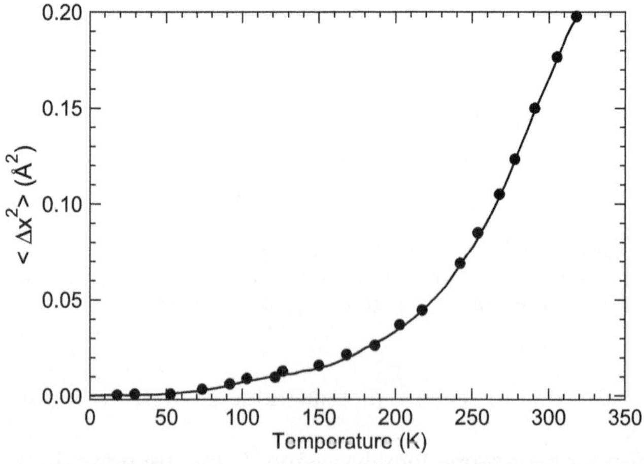

Figure 13.4 Mean-square displacements of hydrogen atoms in myoglobin measured by neutron scattering as a function of temperature (data from ref. 21).

13.2.2 Inconsistencies

Although the simplicity of this picture connecting the hydration level, the protein flexibility and the enzymatic activity is very appealing, subsequent experimental results revealed some inconsistencies, which suggest that a more elaborate description is required to understand the molecular-level effect of water on enzyme catalysis.

Some concerns were raised[23] about the experimental procedure that is employed to prepare the samples and measure the activity (see a description and references in ref. 18). In these experiments, the enzyme and the substrate are mixed quickly together in solution and immediately frozen and lyophilised to stop the reaction. The powder is then rehydrated at a controlled hydration level and the enzyme activity is measured at ambient temperature.[18] However, if the freeze-drying step is not fast enough with respect to the reaction time, there is a possibility that what is actually measured is the product release rather than the catalytic step.[23]

To overcome these difficulties, different experimental conditions can be employed, that do not require the freezing and unfreezing steps for the enzyme–substrate mixture, during which reactions can occur. They involve either enzymes in non-aqueous organic solvents with trace amounts of water, or enzymes with gaseous substrates where the relative humidity can be varied. Experiments performed on the pig liver esterase enzyme and the *Candida rugosa* lipase B in the vapor phase showed that an enzymatic activity can still be detected at very low hydration levels,[24] below 0.03 g of water per g of protein, *i.e.* with typically less than 30 water molecules per protein (see Figure 13.5). No sign of a minimum hydration level below which the enzymatic activity ceases completely was found in these experiments. This is further supported by very recent experiments, where myoglobin and a lipase enzyme were found to remain active in solvent-free environments when a polymer corona is attached to the biomolecule and replaces the solvent.[25,26] This series of experimental results at very low hydration levels therefore question the existence of a minimum hydration threshold for activity.

The connection between catalytic activity and protein flexibility may also need to be revised. At the very low hydration levels used in a study of the pig liver esterase enzyme and of the *Candida rugosa* lipase B,[24] the water molecules can cover only on the order of 10% of the protein exposed surface (we further note that this assumes that water molecules do not form droplets, which would lead to an even smaller coverage of the protein interface). Therefore, they cannot have a significant effect on its flexibility. In addition, measurements at temperatures below the protein dynamical transition have shown that while the protein anharmonic structural dynamics are arrested, a residual catalytic activity can still be detected.[27,28] This shows that although protein flexibility may facilitate the catalysis, it is not indispensable to the catalytic activity.[29]

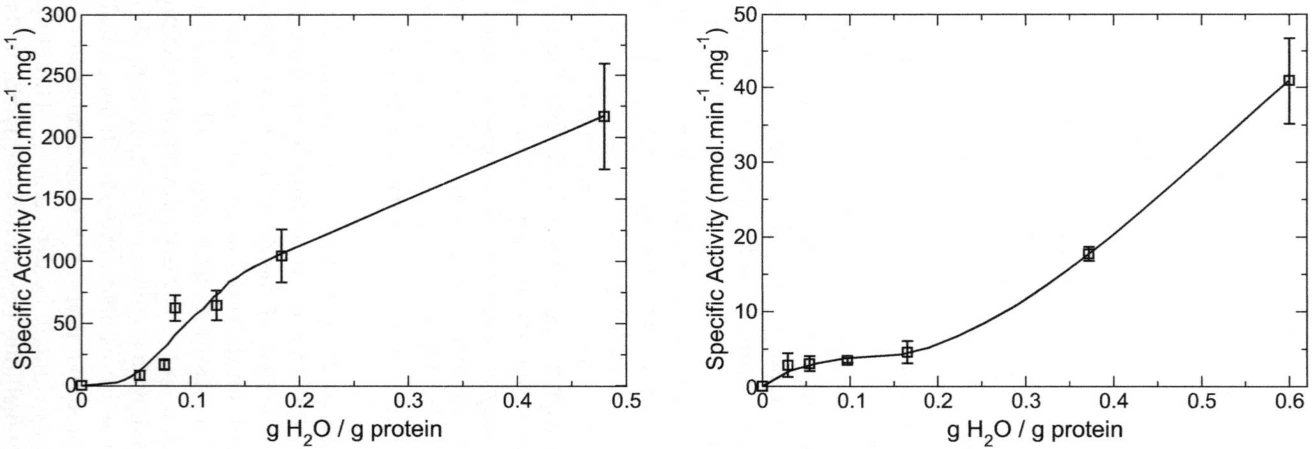

Figure 13.5 Changes in the vapour-phase catalytic activities of *Candida rugosa* lipase B (left) and of pig liver esterase (right) with hydration level (data from ref. 24).

13.3 Enzyme Catalysis in Non-aqueous Organic Solvents

13.3.1 Overview

Although the presence of water may not be a prerequisite for an enzyme to be catalytically active, the presence and the nature of the solvent surrounding the protein both have a strong effect on the functioning of the enzyme. And the range of solvent conditions in which enzymes can catalyse reactions is surprisingly broad![30]

This was realised a few decades ago when enzymes were found to remain active in non-aqueous organic solvents,[31–35] which paved the way for exciting new developments in biocatalysis. Indeed, enzymes are remarkably efficient catalysts, perfectly suited to supporting life, but usually not so well suited for technology. In their natural aqueous environment, exploiting their excellent catalytic power for synthetic chemistry and to catalyse 'non-natural' reactions is greatly limited by several obstacles, including the low water solubility of most compounds and the presence of unwanted degradation side reactions such as hydrolysis.[34,36] Upon transfer to non-aqueous media, it was long believed that enzymes' folded structures would not survive. However, recent studies have firmly established that a large range of enzymes, including lipases, esterases and proteases, not only retain their structural integrity[34,37] but also display a catalytic activity in the absence of bulk water, for example in organic solvents[1,34] and ionic liquids.[1] While the unfolded state may be more stable than the folded structure (because of favourable interactions between the hydrophobic protein core and the organic solvent), these proteins are kinetically trapped in their folded form, and their melting temperature can be even higher than in aqueous solution.[38] (However, we note that some organic solvents including, for example, acetonitrile tend to denature the proteins by 'stripping off' the residual water molecules at the protein interface.[39]) This therefore opened up a tremendous range of potentialities in chemistry, to catalyse non-natural reactions. New reactions that were suppressed in water can be catalysed, and the enzyme activity and selectivity can be finely tuned through the choice of organic solvent.[1,34] For example, hydrolase enzymes that catalyse a bond cleavage by hydrolysis in aqueous conditions can then catalyse transesterification reactions when used in organic solvents[31] (Figure 13.6). Enzyme catalysis in organic solvents is now widely used by the chemical and pharmaceutical industries, as illustrated for example by the synthesis of derivatives of the widely used antitumoral drug taxol that are more soluble in water than the parent molecule[40,41] (Figure 13.7).

13.3.2 Solvent Effects on Enzyme Activity and Specificity

Over recent decades, a broad range of experimental studies have established that for a given catalysed reaction, the enzymatic activity, *i.e.* the catalytic rate constant, is very sensitive to the nature of the solvent conditions.[34,42]

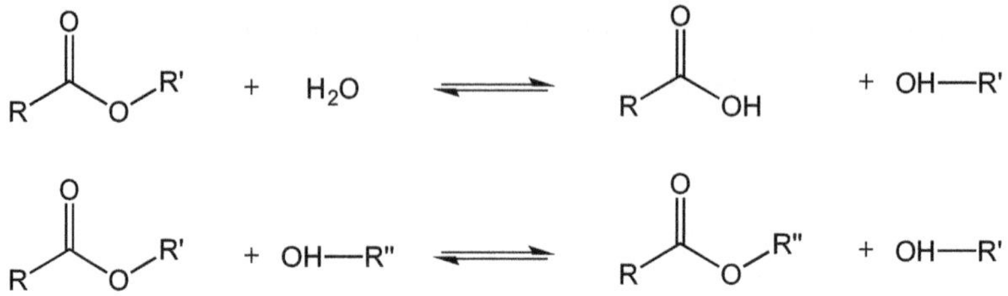

Figure 13.6 Mechanisms of hydrolysis (top) and transesterification (bottom) reactions of a carboxylic acid ester.

This has been a topic of major interest and a key challenge for the practical application of enzymes in non-aqueous conditions, because the enzymatic activity is usually greatly reduced in these solvents compared to that in aqueous conditions (up to a 10^4 decrease).[1,34] A series of empirical strategies have then been developed to activate the catalysis, including for example the addition of small amounts of water,[43] ions[44] or water mimics such as glycerol or ethylene glycol.[1] This trial and error approach has successfully recovered activities similar to those in water in some cases,[1] but a molecular-level understanding of the mechanisms underlying these rate enhancements has remained elusive, which precludes a systematic optimisation of the solvent conditions to maximise the catalytic activity.

Even if a residual catalytic activity remains at very low hydration levels, the addition of trace amounts of water in the non-aqueous organic solvent (*e.g.* iso-octane) can enhance the catalytic rate constant by more than one order of magnitude[43] (see Figure 13.8). It has also been found empirically that the addition of some ions such as sodium fluoride (NaF) or sodium acetate ($NaOOCCH_3$) can lead to a large catalytic enhancement, while some other ions (like sodium iodide NaI) only have a limited effect on the catalytic efficiency.[45]

In addition to the catalytic activity, the specificity of a given enzyme vis-à-vis two different substrates has been shown to be affected by the nature of the solvent conditions.[34] While in an aqueous solvent, such a change in the specificity would require a full site-directed mutagenesis study, in non-aqueous solvents it can be tuned by altering the solvent conditions. Examples include changes in the chemoselectivity between two chemically distinct substrates,[46] but also changes in the regioselectivity and in the enantioselectivity.[34] The latter is particularly interesting for applications in synthetic chemistry, in order to distinguish chiral molecules which otherwise have identical chemical and physical properties in an achiral environment. While an (over-)simplified approach could consider that the enzymatic catalytic rates for two enantiomers would be affected by a solvent change in the same way, thus keeping the enantiomeric ratio unchanged,[47] experimental measurements showed a pronounced solvent dependence of the enantioselectivity.[48] However, while the enantiomeric ratio exhibits some correlation with solvent parameters including for example the solvent dipole moment, its relative permittivity and its hydrophobicity,[48] recent

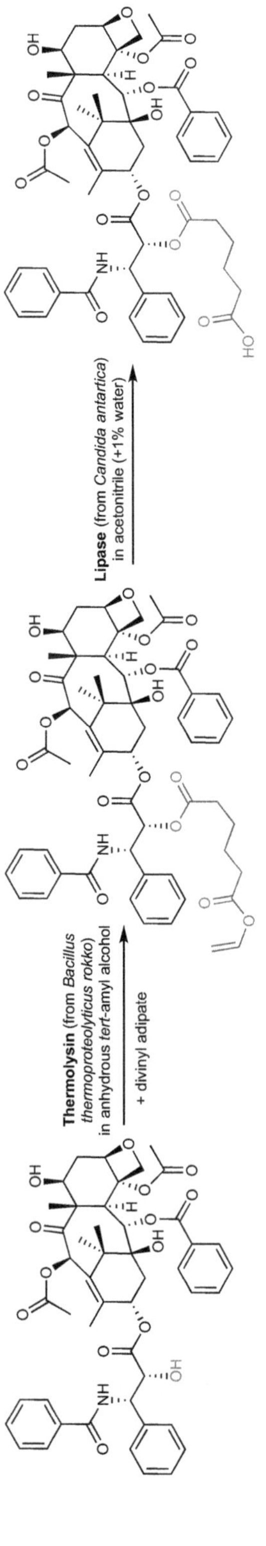

Figure 13.7 Biocatalytic synthesis of paclitaxel (taxol) derivatives.[41]

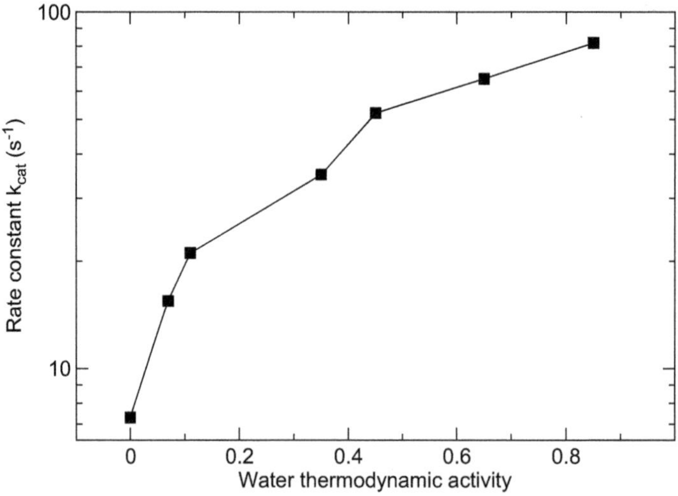

Figure 13.8 Catalytic rate constant for the transesterification reaction catalysed by Subtilisin Carlsberg in isooctane with increasing quantities of water (data from ref. 43).

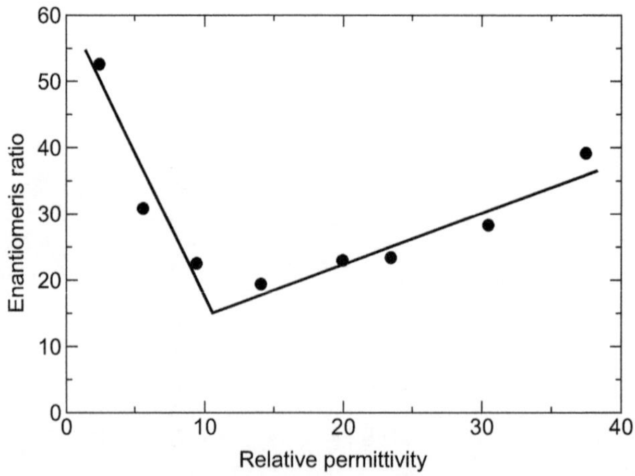

Figure 13.9 Ratio of the k_{cat}/K_M ratios for the *R* and *S* enantiomers for the acetylation of an alcohol catalysed by the *Candida antarctica* Lipase B enzyme for a series of solvent mixtures with different dielectric permittivities (data from ref. 49).

experiments revealed a puzzling non-monotonic dependence of the enantiomeric ratio with the composition of a mixed solvent for the resolution of primary and secondary alcohols by the Candida Antarctica Lipase B[49] (see Figure 13.9). Whether the solvent could affect the enzyme enantioselectivity by changing the protein flexibility is still unclear, since experimental results have lead to different, sometimes seemingly contradictory explanations.[50,51]

A decrease in the protein flexibility may be argued either to increase the enantioselectivity by providing a better discrimination of the two enantiomers, or on the contrary to decrease the enantioselectivity by hindering the accommodation of the fast reacting enantiomer more than that of the slow reacting one.

Therefore, here again, while some first studies have started to investigate the solvent effects on the enantiospecificity,[52] a full molecular-level understanding of the key effects at play is still missing.

13.4 Towards a Molecular Picture of Solvent Effects on Catalysis

We now review several possible molecular mechanisms that have been suggested to explain the effect of water on the enzymatic activity and specificity, together with some experimental results that are in agreement or in contradiction with these pictures.

As a preliminary remark, we stress that while increasing the fraction of water in the solvent usually leads to an increase in the enzymatic rate constant (as found, for example, for proteases in apolar organic solvents,[43] for an esterase in the gas phase[53] and for a reductase in water–cosolvent mixtures[54]), the opposite behavior has also been observed. It can be due to the 'stripping off' of the residual water molecules at the protein interface and the denaturation of the protein,[55,56] but the rate constant of the chemical step in the catalysis has also been measured to decrease for a structurally intact thermophilic reductase when the water fraction is increased.[57]

13.4.1 Solvent Polarity

Gradually replacing the aqueous solvent with another less polar solvent leads to a decrease in the solvent dielectric constant. Since the charge distributions of the reactant and transition states differ, this could affect differently the two states and modify the reaction free energy barrier. For example, a series of experiments performed on different DHFR homologues from mesophilic and thermophilic organisms in various water–cosolvent mixtures[54,58] were shown to yield a good correlation between the measured catalytic rate constant and the effective solvent dielectric constant (see Figure 13.10a). However, it is not fully clear whether a simple dielectric continuum picture that ignores the molecular structure of the active site is applicable to these systems.

13.4.2 Lubrication Picture

Another popular picture involves the lubricating effect due to the solvent.[34,36] This idea was already mentioned in Section 13.2.1 in connection with protein dynamics, and we focus here on its effect on the catalytic rate

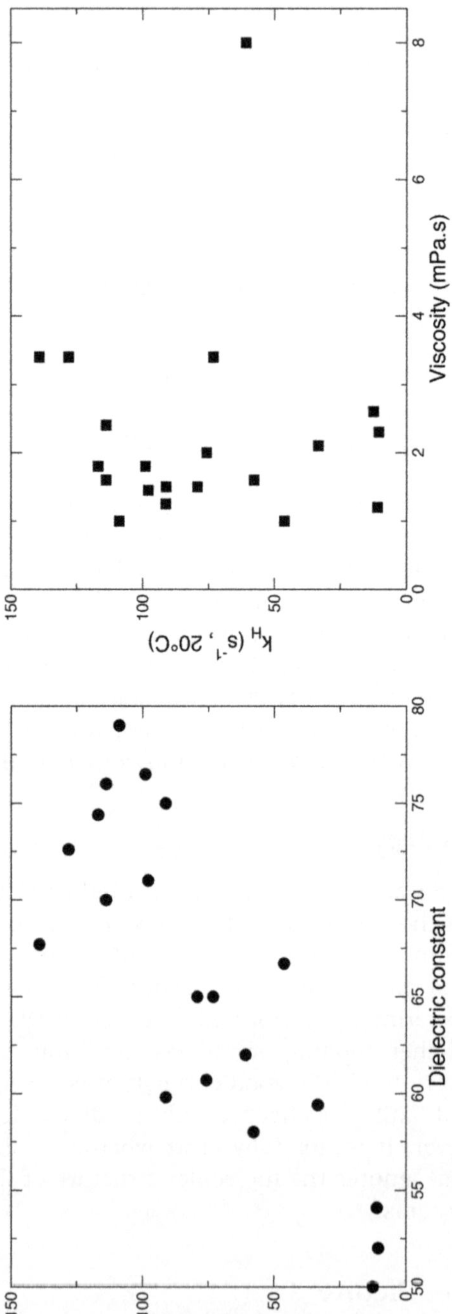

Figure 13.10 Rate constant of the hydride transfer reaction catalysed by BsDHFR in aqueous mixtures with different dielectric constants and different viscosities (data from ref. 58).

constant. The 'lubrication' terminology has been used with different meanings. It is sometimes employed in an equilibrium perspective, where the solvent would reduce the free energy barriers between different conformations (*e.g.* the open and closed M20 loop configurations in DHFR[59]) and especially the free energy cost to access enzyme conformations where the barrier for the chemical step is lower. But it also has a dynamical meaning, where the lubricating effect would imply a reduced friction on the reaction coordinate governing the chemical step (while the equilibrium free energy profile remains unchanged).

Experimental and simulation studies concluded that a decrease in the hydration level does lead to a decreased flexibility in the protein.[56] However, these studies did not specifically probe the motions involved in the reaction coordinate. While a change in the hydration level may strongly affect the flexibility of exposed protein sites, it may not have a significant effect on the active site residues directly involved in the chemical step of the catalysed reaction. For example, experimental measurements of the catalytic rate constant for a series of dihydrofolate reductase enzymes did not reveal any correlation with the viscosity of the solvent mixture[54,57,58] (see Figure 13.10b), thus suggesting that the protein flexibility may not be a key factor. In addition, other experiments have shown that adding large quantities of ions to an enzyme in an apolar organic solvent with less than a monolayer of residual water can greatly enhance the catalytic activity of a protease.[44] This was suggested to arise from the structure-breaking character of these ions, which would increase the lability of the hydrogen-bond network formed by the residual water molecules.[44] However, extensive experimental and simulation studies of the effect of ions on the dynamical properties of aqueous solutions have shown that the structure-breaker and -maker classification is often ambiguous[60] and that at the concentration levels used in these enzymatic systems, the ions actually increase the solution viscosity.[61] It thus seems difficult to explain the enhanced enzymatic activity by an enhanced lubrication, since the addition of the ions makes the solvation shell more viscous. Finally, reducing the friction on the reaction coordinate governing the chemical step would only be important if dynamical effects at the transition state played an important role in the catalytic rate constant, but recent works suggest that this effect is certainly limited.[62]

13.4.3 Competitive Inhibition

The two previous descriptions ignored the molecular structure of the solvent and focused exclusively on its relative permittivity and viscosity properties to interpret the solvent effects on catalysis. However, different solvent molecules have different abilities to penetrate into the enzymatic active site, where they can compete against the substrate for binding. It has been suggested that in some cases both the solvent and substrate molecules can be simultaneously present in the active site and the solvent molecule thus effectively changes the size of the enzymatic pocket.[63] This interplay could then lead to solvent effects that depend on the shape[64] and on the size[65] of the solvent molecule, and that would thus be very specific to each solvent molecule.

13.5 Concluding Remarks

In this chapter, we have reviewed the available experimental data and molecular models describing the effect of different solvents, including water, on the catalytic activity of enzymes. Enzymes are found to remain catalytically active in surprisingly harsh environments, including almost dry vapors and organic solvents, but the addition of water usually enhances the catalytic activity. A popular picture suggests that water acts as a lubricant of the protein conformational motions required for catalysis but this dynamical picture is not supported by recent results. Further experimental and theoretical studies will thus be needed to provide a consistent molecular interpretation of the solvent effects on the catalytic activity and specificity of enzymes.

Acknowledgements

The research leading to these results has received funding from the European Research Council under the European Union's Seventh Framework Program (FP7/2007-2013)/ERC Grant Agreement No. 279977.

References

1. L. J. Rothschild and R. L. Mancinelli, *Nature*, 2001, **409**, 1092–1101.
2. P. Ball, *Chem. Rev.*, 2008, **108**, 74–108.
3. Y. Levy and J. N. Onuchic, *Annu. Rev. Biophys. Biomol. Struct.*, 2006, **35**, 389–415.
4. M. Chaplin, *Nat. Rev. Mol. Cell Biol.*, 2006, **7**, 861–866.
5. W. Saenger, W. N. Hunter and O. Kennard, *Nature*, 1986, **324**, 385–388.
6. L. Szyc, M. Yang, E. T. J. Nibbering and T. Elsaesser, *Angew. Chem., Int. Ed. Engl.*, 2010, **49**, 3598–3610.
7. S. K. Pal and A. H. Zewail, *Chem. Rev.*, 2004, **104**, 2099–2124.
8. B. Halle, *Philos. Trans. R. Soc. London, Ser. B*, 2004, **359**, 1207–1223, discussion 1223–1224, 1323–1328.
9. C. Mattea, J. Qvist and B. Halle, *Biophys. J.*, 2008, **95**, 2951–2963.
10. F. Sterpone, G. Stirnemann and D. Laage, *J. Am. Chem. Soc.*, 2012, **134**, 4116–4119.
11. A. C. Fogarty and D. Laage, *J. Phys. Chem. B*, 2014, **118**, 7715–7729.
12. J. Qvist, E. Persson, C. Mattea and B. Halle, *Faraday Discuss.*, 2009, **141**, 131–144.
13. A. C. Fogarty, E. Duboué-Dijon, F. Sterpone, J. T. Hynes and D. Laage, *Chem. Soc. Rev.*, 2013, **42**, 5672–5683.
14. L. D. Barron, L. Hecht and G. Wilson, *Biochemistry*, 1997, **36**, 13143–13147.

15. M. R. Panman, B. H. Bakker, D. den Uyl, E. R. Kay, D. A. Leigh, W. J. Buma, A. M. Brouwer, J. A. J. Geenevasen and S. Woutersen, *Nat. Chem.*, 2013, **5**, 929–934.
16. H. Frauenfelder and D. T. Leeson, *Nat. Struct. Mol. Biol.*, 1998, **5**, 757–759.
17. K. Henzler-Wildman and D. Kern, *Nature*, 2007, **450**, 964–972.
18. J. A. Rupley and G. Careri, *Adv. Protein Chem.*, 1991, **41**, 37–172.
19. J. A. Rupley, E. Gratton and G. Careri, *Trends Biochem. Sci.*, 1983, **8**, 18–22.
20. F. Parak, E. W. Knapp and D. Kucheida, *J. Mol. Biol.*, 1982, **161**, 177–194.
21. W. Doster, S. Cusack and W. Petry, *Nature*, 1989, **337**, 754–756.
22. M. Ferrand, A. J. Dianoux, W. Petry and G. Zaccaï, *Proc. Natl. Acad. Sci. U. S. A.*, 1993, **90**, 9668–9672.
23. R. M. Daniel, R. V. Dunn, J. L. Finney and J. C. Smith, *Annu. Rev. Biophys. Biomol. Struct.*, 2003, **32**, 69–92.
24. P. A. Lind, R. M. Daniel, C. Monk and R. V. Dunn, *Biochim. Biophys. Acta*, 2004, **1702**, 103–110.
25. A. P. Brogan, K. P. Sharma, A. W. Perriman and S. Mann, *Nat. Commun.*, 2014, **5**, 5058.
26. A. W. Perriman, A. P. S. Brogan, H. Cölfen, N. Tsoureas, G. R. Owen and S. Mann, *Nat. Chem.*, 2010, **2**, 622.
27. R. M. Daniel, J. C. Smith, M. Ferrand, S. Héry, R. Dunn and J. L. Finney, *Biophys. J.*, 1998, **75**, 2504–2507.
28. J. M. Bragger, R. V. Dunn and R. M. Daniel, *BBA, Biochim. Biophys. Acta, Protein Struct. Mol. Enzymol.*, 2000, **1480**, 278–282.
29. V. Kurkal, R. M. Daniel, J. L. Finney, M. Tehei, R. V. Dunn and J. C. Smith, *Biophys. J.*, 2005, **89**, 1282–1287.
30. M. J. Liszka, M. E. Clark, E. Schneider and D. S. Clark, *Annu. Rev. Chem. Biomol. Eng.*, 2012, **3**, 77–102.
31. A. Zaks and A. M. Klibanov, *Science*, 1984, **224**, 1249–1251.
32. R. Affleck, Z. F. Xu, V. Suzawa, K. Focht, D. S. Clark and J. S. Dordick, *Proc. Natl. Acad. Sci. U. S. A.*, 1992, **89**, 1100–1104.
33. A. Koskinen and A. M. Klibanov, *Enzymatic Reactions in Organic Media*, Springer, Dordrecht, 1996.
34. A. M. Klibanov, *Nature*, 2001, **409**, 241–246.
35. G. Carrea and S. Riva, *Organic Synthesis with Enzymes in Non-Aqueous Media*, Wiley-VCH, Weinheim, 2008.
36. A. L. Serdakowski and J. S. Dordick, *Organic Synthesis with Enzymes in Non-Aqueous Media*, Wiley-VCH, Weinheim, 2008, pp. 47–71.
37. J. L. Schmitke, L. J. Stern and A. M. Klibanov, *Proc. Natl. Acad. Sci. U. S. A.*, 1998, **95**, 12918–12923.
38. D. B. Volkin, A. Staubli, R. Langer and A. M. Klibanov, *Biotechnol. Bioeng.*, 1991, **37**, 843–853.
39. L. A. S. Gorman and J. S. Dordick, *Biotechnol. Bioeng.*, 1992, **39**, 392–397.
40. Y. L. Khmelnitsky, C. Budde, J. M. Arnold, A. Usyatinsky, D. S. Clark and J. S. Dordick, *J. Am. Chem. Soc.*, 1997, **119**, 11554–11555.

41. A. Schmid, J. S. Dordick, B. Hauer, A. Kiener, M. Wubbolts and B. Witholt, *Nature*, 2001, **409**, 258–268.
42. G. Carrea and S. Riva, *Angew. Chem., Int. Ed.*, 2000, **39**, 2226–2254.
43. E. P. Hudson, R. K. Eppler, J. M. Beaudoin, J. S. Dordick, J. A. Reimer and D. S. Clark, *J. Am. Chem. Soc.*, 2009, **131**, 4294–4300.
44. R. K. Eppler, R. S. Komor, J. Huynh, J. S. Dordick, J. A. Reimer and D. S. Clark, *Proc. Natl. Acad. Sci. U. S. A.*, 2006, **103**, 5706–5710.
45. M. T. Ru, S. Y. Hirokane, A. S. Lo, J. S. Dordick, J. A. Reimer and D. S. Clark, *J. Am. Chem. Soc.*, 2000, **122**, 1565–1571.
46. C. R. Wescott and A. M. Klibanov, *BBA, Biochim. Biophys. Acta, Protein Struct. Mol. Enzymol.*, 1994, **1206**, 1–9.
47. A. Wolff, A. J. J. Straathof, J. A. Jongejan and J. J. Heijnen, *Biocatal. Biotransform.*, 1997, **15**, 175–184.
48. P. A. Fitzpatrick and A. M. Klibanov, *J. Am. Chem. Soc.*, 1991, **113**, 3166–3171.
49. Y. Kitamoto, Y. Kuruma, K. Suzuki and T. Hattori, *J. Org. Chem.*, 2015, **80**, 521–527.
50. D. Herbst, S. Peper and B. Niemeyer, *J. Biotechnol.*, 2012, **162**, 398–403.
51. B. Orsat, G. J. Drtina, M. G. Williams and A. M. Klibanov, *Biotechnol. Bioeng.*, 1994, **44**, 1265–1269.
52. G. Colombo, G. Ottolina, G. Carrea and K. M. Merz Jr, *Chem. Commun.*, 2000, 559–560.
53. M. Lopez, V. Kurkal-Siebert, R. V. Dunn, M. Tehei, J. L. Finney, J. C. Smith and R. M. Daniel, *Biophys. J.*, 2010, **99**, L62–L64.
54. E. J. Loveridge, L. H. Tey and R. K. Allemann, *J. Am. Chem. Soc.*, 2010, **132**, 1137–1143.
55. A. Zaks and A. M. Klibanov, *J. Biol. Chem.*, 1988, **263**, 3194–3201.
56. L. Yang, J. S. Dordick and S. Garde, *Biophys. J.*, 2004, **87**, 812–821.
57. E. J. Loveridge, R. M. Evans and R. K. Allemann, *Chemistry*, 2008, **14**, 10782–10788.
58. J. Guo, L. Y. P. Luk, E. J. Loveridge and R. K. Allemann, *Biochemistry*, 2014, **53**, 2855–2863.
59. D. D. Boehr, D. McElheny, H. J. Dyson and P. E. Wright, *Science*, 2006, **313**, 1638–1642.
60. R. Mancinelli, A. Botti, F. Bruni, M. A. Ricci and A. K. Soper, *J. Phys. Chem. B*, 2007, **111**, 13570–13577.
61. G. Stirnemann, E. Wernersson, P. Jungwirth and D. Laage, *J. Am. Chem. Soc.*, 2013, **135**, 11824–11831.
62. I. Tuñón, D. Laage and J. T. Hynes, *Arch. Biochem. Biophys.*, 2015, **582**, 42–55.
63. V. Léonard, L. Fransson, S. Lamare, K. Hult and M. Graber, *ChemBioChem*, 2007, **8**, 662–667.
64. K. Nakamura, M. Kinoshita and A. Ohno, *Tetrahedron*, 1995, **51**, 8799–8808.
65. J. Ottosson, L. Fransson, J. W. King and K. Hult, *BBA, Biochim. Biophys. Acta, Protein Struct. Mol. Enzymol.*, 2002, **1594**, 325–334.

CHAPTER 14

Modelling Reactivity in Metalloproteins: Hydrogen Peroxide Decomposition by Haem Enzymes

M. ALFONSO-PRIETO*[a] AND C. ROVIRA*[a,b]

[a] Departament de Química Inorgànica i Orgànica and Institut de Química Teòrica i Computacional (IQTC), Universitat de Barcelona, Carrer Martí i Franquès 1, Barcelona E-08028, Spain; [b] Institució Catalana de Recerca i Estudis Avançats (ICREA), Passeig Lluís Companys, 23, Barcelona E-08018, Spain
*Email: malfonsoprieto@gmail.com; c.rovira@ub.edu

14.1 Introduction

Although metal ions are present only in trace amounts in living organisms,[1,2] almost half of the proteome encodes for metal-binding proteins[3,4] and *ca.* 40% of the structures deposited in the Protein Data Bank contain metals.[5] Metal ions (and metal-containing cofactors, such as haem groups) play crucial structural and functional roles in biological systems. For instance, divalent metal ions (such as Ca^{2+} and Mg^{2+}) are used in signalling cascades, as well as in DNA and RNA processing. Transition metal ions (*e.g.* Fe, Co, Mn or Cu) are involved in binding and transport of oxygen and other gases, and they also act as electron transfer conduits in respiration and

RSC Theoretical and Computational Chemistry Series No. 9
Simulating Enzyme Reactivity: Computational Methods in Enzyme Catalysis
Edited by Iñaki Tuñón and Vicent Moliner

Published by the Royal Society of Chemistry, www.rsc.org

photosynthesis, because of their ability to change oxidation state. Moreover, metal-containing enzymes (hereafter metalloenzymes) are implicated in drug clearance and resistance, as well as protection against oxidative stress. Therefore, understanding the function of metalloproteins and metalloenzymes is not only a fundamental (bio)chemical question, but it can also help to understand the molecular basis of diseases and design new drugs aimed at their treatment, as well as provide hints to design biomimetic compounds to be used in biocatalysis.

Despite their biological importance, the study of metalloproteins has been hampered by the limitations of both experimental and computational methodologies when dealing with metal ions.[6–8] In recent years, significant improvements in structural, spectroscopic and theoretical techniques have resulted in an exponential growth in the number of studies in the field of metalloproteins.[9–13] In this chapter, we focus on metal-containing enzymes and their reactivity, and review some recent examples studied by us and by others in the field of computational enzymology applied to metalloenzymes.

14.2 Methodology

Quantum mechanics (QM) and quantum mechanics/molecular mechanics (QM/MM) are powerful computational tools to study the reaction mechanism of enzymes, in particular metalloenzymes.[14–20] QM and QM/MM calculations have been successfully used to characterise the possible intermediates and transition states of enzymatic reactions, which are difficult to trap using experimental structural techniques. This is particularly relevant for metalloenzymes involving redox-active metal ions, which have the additional difficulty that the metal oxidation state can change due to X-ray irradiation (photoreduction). Using QM and QM/MM methods, one can obtain the detailed molecular mechanism of the reaction, as well as the energy barriers along the reaction profile; then the highest (rate-limiting) barrier can be compared with the measured reaction rates in order to determine the most likely reaction mechanism. Moreover, when combined with molecular dynamics (MD) algorithms, QM and QM/MM calculations provide the transitions pathways connecting the stationary and metastable states along the reaction, and thus reveal further details about how protein fluctuations are coupled to the reactive events.[21,22]

Further details on the QM and QM/MM methodologies are given in Chapters 8, 9 and 11 of this book. The applications shown below were performed using density functional theory (DFT) as the QM method, because it scales efficiently with the number of atoms and shows reasonable accuracy for metal-containing molecules.[23,24] Besides the static QM and QM/MM calculations, we also employed *ab initio* MD simulations using the Car–Parrinello (CP) scheme[11,18,25] to account for thermal fluctuations. Finally, because chemical reactions are rare events that cannot be accessed within the usual timescale of *ab initio* MD, we combined MD with an enhanced sampling technique (metadynamics)[26–30] to simulate the reaction pathway and reconstruct the associated free energy surface.

Our simulations used the following protocol. An experimentally determined structure of the metalloenzyme (or a suitable model based on that structure) is considered as a starting point. This is equilibrated in a physiological-like environment (*i.e.* surrounded by water molecules and ions) using molecular mechanics (MM) MD. Details on the force field parameterisation of metal-containing sites can be found elsewhere.[13,31–33] These classical simulations allow us to sample the protein conformational space and the flexibility and hydration of the active site, from which the most likely reactive configuration can be identified. One MD snapshot – exhibiting all the protein–substrate interactions known to be important for reactivity from crystallographic, kinetic and/or mutagenesis studies – is then extracted from the classical trajectory and submitted to QM/MM calculations. First, a few picoseconds of QM/MM MD are run to re-equilibrate the system within the QM/MM Hamiltonian, and then a representative configuration is chosen to initiate the investigation of the reaction using metadynamics. A careful choice of the collective variables (or reaction coordinates) used in the metadynamics simulation allows us to obtain both the reaction mechanism and the associated free energy profile.[28–30]

We also used QM calculations on cluster models of the active site in some of the examples presented below. The aim of these gas-phase calculations is to unravel the intrinsic properties of the active site, devoid of the protein environmental effects.

14.3 Catalases and Peroxidases

14.3.1 Biological Function

Hydroperoxidases are oxidoreductase enzymes present in almost all aerobic organisms that decompose hydrogen peroxide (H_2O_2), a by-product of the aerobic metabolism.[34] These enzymes regulate the levels of H_2O_2 (and the reactive oxygen species resulting from its decomposition), and thus they have a dual biological function: (i) they protect cells against oxidative stress and (ii) they participate in redox signalling pathways. Here we focus on three of the most relevant hydroperoxidase families: catalases, peroxidases and bifunctional catalases-peroxidases (KatGs).

Monofunctional catalases are ubiquitous enzymes which dismutate H_2O_2 into water and oxygen.[35,36] They are one of the most efficient enzymes known, being able to remove up to 1 million molecules of H_2O_2 per second, and hence they constitute the main mechanism to protect cells against the toxic effects of H_2O_2. Instead, monofunctional peroxidases utilise H_2O_2 to oxidise organic substrates, such as phenols or aromatic amines.[34,37] In particular, horseradish peroxidase (HRP) is extensively used in the fields of immunochemistry, biocatalysis, bioremediation and biomedicine for its ability to oxidise a wide variety of substrates.[38] Nonetheless, when high concentrations of H_2O_2 are present, peroxidases are also able to decompose H_2O_2 into water and oxygen, although at much lower speed than catalases, and eventually they become inactivated.[39,40]

In the early 2000s, structures were determined for a new hydroperoxidase family, bifunctional catalase-peroxidases or KatGs. Surprisingly, KatGs are phylogenetically related to monofunctional (plant) peroxidases, and yet they are able to dismutate H_2O_2 at rates comparable to monofunctional catalases.[41,42] Moreover, KatGs are more catalytically versatile than peroxidases. They not only oxidise traditional peroxidatic substrates, but they are also able to catalyse other reactions, in particular the activation of the antitubercular pro-drug isoniazid.[43]

14.3.2 Reactivity

The reactions catalysed by catalases, peroxidases and KatGs are summarised in Figure 14.1. The catalytic cofactor is a haem (iron porphyrin) group. In all cases, the enzyme resting state corresponds to ferric haem (Por–Fe^{III}, where Por stands for the haem porphyrin). The catalytic cycle of all three hydroperoxidases starts with the enzyme reacting with one molecule of H_2O_2, yielding a water molecule and a high-valent iron-oxo intermediate known as Compound I (Cpd I). Cpd I formation is formally a two-electron process: one electron is removed from the iron (which changes oxidation state from Fe^{III} to Fe^{IV}) and the other from the porphyrin (which acquires a cation radical character,

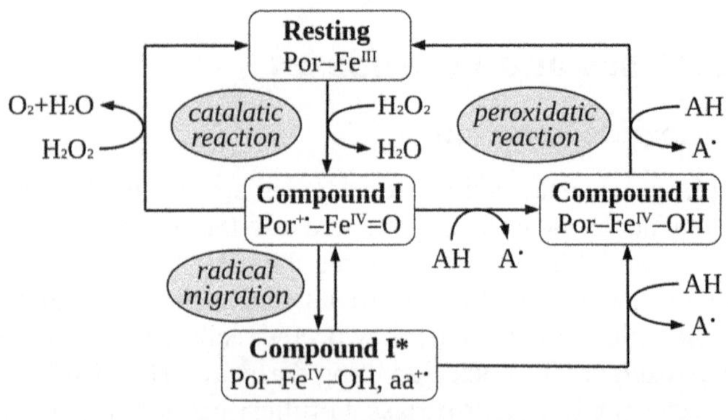

Figure 14.1 Reactions catalysed by monofunctional catalases, monofunctional peroxidases and bifunctional catalases-peroxidases (KatGs) using hydrogen peroxide (H_2O_2). The abbreviations used are: Por, haem porphyrin; AH, one-electron organic donor, *e.g.* phenol; aa, amino acid, typically Tyr or Trp.
Adapted from M. Alfonso-Prieto *et al.*, The reaction mechanisms of heme catalases: An atomistic view by *ab initio* molecular dynamics, *Arch. Biochem. Biophys.*, **525**, 121–130, Copyright 2012 with permission from Elsevier[51] and adapted from P. Vidossich *et al.*, The dynamic role of distal side residues in heme hydroperoxidase catalysis. Interplay between X-ray crystallography and *ab initio* MD simulations, *Arch. Biochem. Biophys.*, **500**, 37–44, Copyright 2010 with permission from Elsevier.[82]

$Por^{+\bullet}$); these two electrons are used to reduce H_2O_2 to water. Therefore, Cpd I contains an oxoferryl porphyrin cation radical ($Por^{+\bullet}$–Fe^{IV}=O).[44–48] All three hydroperoxidases form Cpd I; however, the fate of this highly reactive Cpd I intermediate differs for catalases, peroxidases and catalase-peroxidases.

Catalases use a second molecule of H_2O_2 to reduce Cpd I back to the resting state, releasing molecular oxygen (the catalatic pathway in Figure 14.1).[35] Cpd I reduction by H_2O_2 involves the transfer of two electrons (to reduce the ferryl ion and the porphyrin cation radical to the resting ferric porphyrin) and two protons (to protonate the oxo group, forming water). The mechanism whereby these transfers take place, as well as the spin state of the molecular oxygen released, has been debated until recently.[49–51]

In contrast, peroxidases preferentially reduce Cpd I back to the native state using one-electron donating substrates (AH), passing through another intermediate named Compound II (Cpd II).[34] The peroxidatic pathway (Figure 14.1) thus consists of two consecutive one-electron reactions, each requiring one molecule of AH. First, Cpd I reduction by AH forms Cpd II, which is generally described as an oxo-/hydroxo-ferryl porphyrin intermediate (Por–Fe^{IV}=O/Por–Fe^{IV}–OH).[52–54] Second, Cpd II is reduced by another molecule of AH to recover the ferric resting state.[54]

As explained above, monofunctional catalases and peroxidases react preferentially through the catalatic and peroxidatic pathways, respectively. However, under certain conditions, they can also undergo other reactions (Figure 14.1). For instance, when no reducing agent is present, the highly reactive Cpd I species can oxidise a residue of the protein (aa, typically Tyr or Trp). This intramolecular electron transfer process is often referred to as radical migration, and results in the formation of another intermediate named Compound I* (Cpd I*). Cpd I* is isoelectronic with Cpd II in the haem active site (Por–Fe^{IV}=O/Por–Fe^{IV}–OH), but also contains a protein radical ($aa^{+\bullet}$). In catalases, Cpd I* formation is an undesired side reaction that consumes Cpd I, decreasing the efficiency of H_2O_2 dismutation.[45,55–57] In contrast, in peroxidases, Cpd I* has been proposed to be an alternative intermediate to oxidise bulky organic substrates that cannot access the haem active site through long-range electron transfer.[58–60] For instance, cytochrome c peroxidase (CCP) forms a tryptophanyl radical to oxidise cytochrome c.[61,62] Interestingly, in KatGs, formation of a peculiar Cpd I* has been proposed to be essential for their unusual catalytic properties.[46] KatGs possess a unique triad of covalently linked protein residues stacked above the haem, the MYW adduct (Trp111, Tyr238 and Met264, using *Burkholderia pseudomallei* (*Bp*)KatG numbering).[63] Formation of a Cpd I* containing a radical adduct was first predicted using QM and QM/MM calculations[46] and was later supported by UV-visible spectroscopy,[64,65] although not all spectroscopists agree with this interpretation (A. Ivancich, private communication). Nevertheless, the mechanism by which this (radical) adduct enables the catalatic reaction in KatG has only recently been put forward by a combination of experimental data and QM/MM calculations.[66–68]

For a long time, two aspects have been puzzling the hydroperoxidase community. First, monofunctional catalases and peroxidases form the same reaction intermediate, Cpd I, and yet the reactivity of Cpd I is different for these two enzyme families. Cpd I is also used by other oxidoreductases, such as cytochrome P450, for other oxidative transformations, and changes in reactivity among haem enzymes have been explained in terms of the protein scaffold in which the haem cofactor is embedded. Nevertheless, the precise structural factors determining the distinct Cpd I reactivity of catalases and peroxidases are not well understood. The discovery of bifunctional KatGs raised a second question. Despite having a fold similar to plant peroxidases, KatGs are able to decompose H_2O_2 into water and oxygen at much higher speed than the (phylogenetically related) classical peroxidases. This high catalatic activity has been attributed to the presence of a crosslinked protein adduct,[63] but the molecular details whereby this unique protein modification imparts catalase activity are still controversial. The QM and QM/MM studies reviewed here were aimed at answering these two questions, and show how simulations can be a valuable tool in understanding the structure and reactivity of metalloenzymes.

14.3.3 Monofunctional Catalases and Peroxidases

As mentioned above, monofunctional catalases and peroxidases form the same primary reaction intermediate, Cpd I. In both cases, Cpd I can react with H_2O_2, releasing water and molecular oxygen (Figure 14.1). However, the efficiency of the catalatic reaction differs between the two enzyme families. Catalases are able to decompose up to one million molecules of H_2O_2 per second,[69] whereas peroxidases only remove one to two molecules.[39,70] The causes behind this distinct Cpd I reactivity were attributed for a long time to the structural differences between the two enzymes, but the main factors responsible for their different efficiency at dismutating H_2O_2 were not known.

Catalases are homotetramers, with each of the four subunits containing a haem group deeply buried inside the protein, whereas peroxidases are monomeric and their haem group is more solvent exposed. Hence, the H_2O_2 access channel is very long and narrow in catalases, while it is wider (funnel-like) in peroxidases. In addition, the active site residues surrounding the haem cofactor are different (Figure 14.2). In catalases, the residue coordinated to the haem iron is a negatively charged tyrosinate coupled with an arginine (Y339 and R335 in *Helicobacter pylori* catalase, HPC). Instead, in peroxidases the proximal ligand is a histidine, hydrogen bonded to an aspartate (H170 and D247 in Horseradish peroxidase, HRP). Moreover, although both enzymes have a histidine residue on the distal side of the haem that can act as an acid–base (H56 in HPC and H42 in HRP), its location with respect to the haem varies, being either parallel (catalases) or perpendicular (peroxidases). Furthermore, the distal His is accompanied by a different neighbouring residue, either Asn (N129 in HPC) or Arg (R38 in HRP)

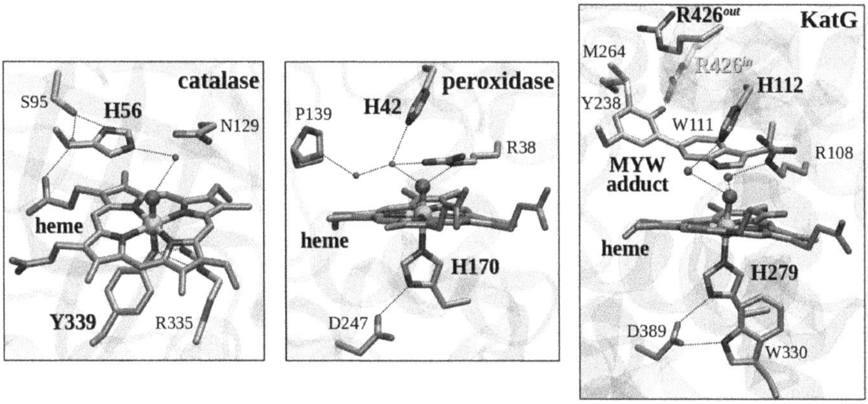

Figure 14.2 Active sites of monofunctional catalase, monofunctional peroxidase, and bifunctional catalase-peroxidase, based on the crystal structures of *Helicobacter pylori* catalase Cpd I (PDB entry 2IQF[45]), horseradish peroxidase Cpd I (PDB entry 1HCH[86]), and *Burkholderia pseudomallei* KatG Cpd I (PDB entry 2B2R[93]). For KatG, the two possible conformations of the mobile Arg are shown: the out conformation, which is the only one observed in Cpd I, and the in conformation, which is observed in native KatG at high pH (PDB entry 1MWV[93]).
Adapted from P. Vidossich *et al.*, The dynamic role of distal side residues in heme hydroperoxidase catalysis. Interplay between X-ray crystallography and *ab initio* MD simulations, *Arch. Biochem. Biophys.*, **500**, 37–44, Copyright 2010 with permission from Elsevier[82] and adapted with permission from P. Campomanes, U. Rothlisberger, M. Alfonso-Prieto and C. Rovira, *J. Am. Chem. Soc.*, 2015, **137**, 11170.[85] Copyright (2015) American Chemical Society.

(Figure 14.2). In this chapter, we show how QM and QM/MM simulations can provide molecular insight into the reaction mechanisms of catalases and peroxidases and unravel the factors causing their disparate Cpd I reactivity.

14.3.3.1 The Catalatic Reaction in Catalases

First, we investigated the mechanism of Cpd I reduction by H_2O_2 in catalase,[50] which is the most efficient enzyme at releasing molecular oxygen. The following protocol was used:

- (i) selection of an appropriate experimental structure as starting point (in this case, the X-ray structure of *Helicobacter pylori* catalase Cpd I[45]);
- (ii) equilibration in physiological-like conditions using MM simulations;
- (iii) selection of a suitable snapshot for the reactive configuration, with a water molecule bridging the oxo group of Cpd I and the distal His, which complies with the available experimental data for catalase Cpd I (see ref. 50 and references therein);

(iv) replacement of the water molecule present in the distal pocket by the H_2O_2 substrate in order to build a model of the Michaelis complex of the catalatic reaction;
(v) equilibration of the model of the Cpd I–H_2O_2 complex using QM/MM MD simulations;
(vi) modelling of the chemical reaction using an enhanced sampling technique.

The catalase QM region (see the Reactants panel in Figure 14.3) comprised the H_2O_2 substrate, the haem Cpd I, the Tyr/Arg proximal pair (Y139/R135 in HPC), and the His/Asn distal pair (H59/N129), as well as a Ser residue (S95) and a water molecule that connect the distal His with one of the haem propionate groups through a hydrogen bond network. As pointed out by us and others,[14–20] the selection of the residues to be included in the QM region is one of the crucial points when setting up a QM/MM calculation. The QM region was treated with DFT (BP86 functional, plane wave basis set with 70 Ry cut-off, Martins–Troullier pseudopotentials). This set-up was previously tested for the study of iron–porphyrin complexes.[71] The rest of the protein and the solvent was described using a force field (Amber). The monovalent carbon pseudopotential[72] and the Hamiltonian electrostatic coupling scheme developed by Laio *et al.*[73,74] were chosen to describe the QM/MM boundary and the QM–MM interactions, respectively. The Car–Parrinello (CP) method[25] was used for the *ab initio* MD simulations,

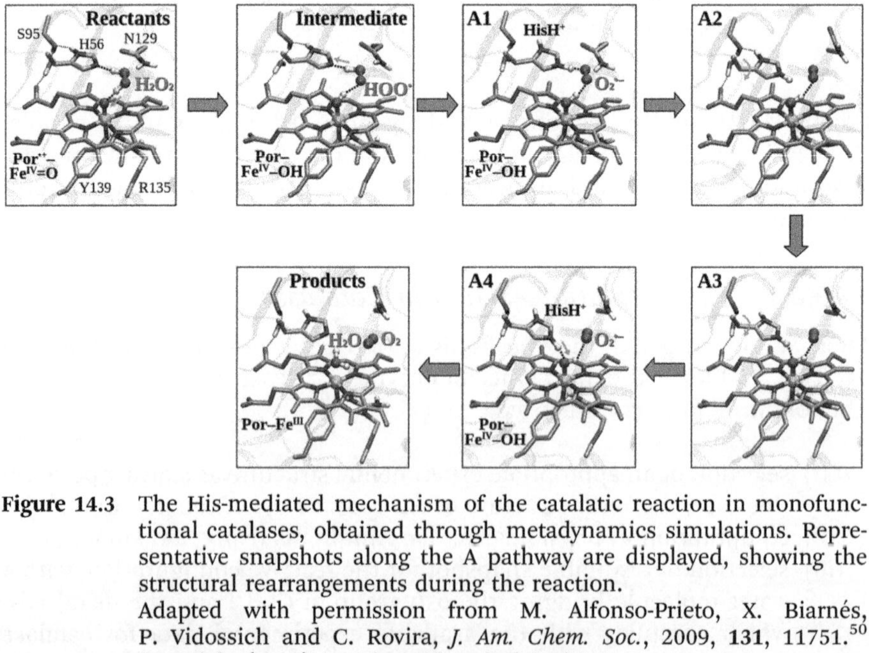

Figure 14.3 The His-mediated mechanism of the catalatic reaction in monofunctional catalases, obtained through metadynamics simulations. Representative snapshots along the A pathway are displayed, showing the structural rearrangements during the reaction.
Adapted with permission from M. Alfonso-Prieto, X. Biarnés, P. Vidossich and C. Rovira, *J. Am. Chem. Soc.*, 2009, **131**, 11751.[50] Copyright (2009) American Chemical Society.

whereas activation of the chemical reaction and reconstruction of the associated free energy surface was accomplished with metadynamics.[26–30]

Using the aforementioned QM/MM computational set-up, we found that the catalatic reaction in catalases starts with a formal hydrogen atom transfer (HAT) from H_2O_2 to Cpd I (Reactants → Intermediate in Figure 14.3). One proton is transferred to the oxoferryl and one electron quenches the haem cation radical, forming a Cpd II-like intermediate with a hydroxoferryl haem and a peroxyl radical:

$$\text{Cpd I (Por}^{+\bullet}\text{–Fe}^{\text{IV}}\text{=O)} + \text{H}_2\text{O}_2 \rightarrow \text{Cpd II (Por–Fe}^{\text{IV}}\text{–OH)} + \text{HOO}^{\bullet}$$

This first HAT step has a very low energy barrier (*i.e.* in the order of k_bT), because it occurs spontaneously during the room temperature QM/MM MD simulation. Additional calculations on a cluster model of the catalase active site, using different exchange-correlation functionals, reproduced this result.[75] In order to complete the reaction, a second hydrogen atom needs to be transferred from the peroxyl radical to the hydroxoferryl:

$$\text{Cpd II (Por–Fe}^{\text{IV}}\text{–OH)} + \text{HOO}^{\bullet} \rightarrow \text{resting (Por–Fe}^{\text{III}}\text{)} + \text{H}_2\text{O} + \text{O}_2$$

This second HAT step is expected to have a significant energy barrier and thus this process was modelled using metadynamics simulations. Two collective variables (or reaction coordinates) were employed: one describing the formation of molecular oxygen and the other the formation of a water molecule (*i.e.* the other product of the reaction). Two possible reaction pathways were found, connecting the Cpd II-like intermediate with the products of the reaction (Figures 14.3 and 14.4). The His-mediated pathway (or A) involves the participation of the distal His as an acid–base catalyst, as proposed by Fita and Rossmann.[76] Instead, the direct pathway (or B) does not require the distal His. Interestingly, regardless of the mechanism ($H^{\bullet} + H^{+}/e^{-}$ or $H^{\bullet} + H^{\bullet}$), the spin state of the molecular oxygen released is a triplet.[49,50] Although some studies claimed to have detected production of singlet oxygen by catalase,[77,78] and used this argument to rationalise a possible reaction mechanism,[79] Jakopitsch and coworkers already pointed out that it would not make biological sense for catalase to release singlet oxygen in large amounts,[79] as it would cause serious oxidative damage. In fact, careful electronic analysis of the spin state of all the species involved in the reaction (*i.e.* not only the H_2O_2 substrate, but also the haem cofactor) showed that there is no reason to assume that singlet oxygen is produced, regardless of the reaction mechanism.[49]

Analysis of the geometries and spin densities of the intermediate and transition states explored during the metadynamics simulations provide the structural and electronic details of the two possible reaction mechanisms. In pathway A, we observed that the peroxyl radical transfers first a hydrogen

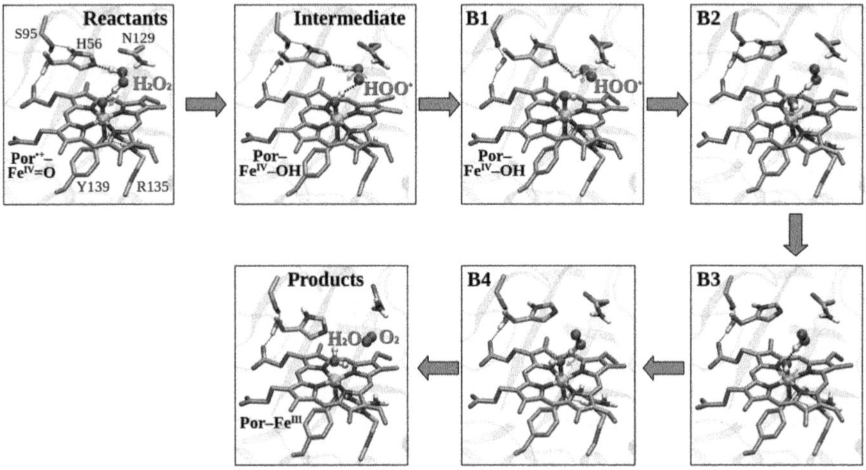

Figure 14.4 The direct mechanism of the catalatic reaction in monofunctional catalases, obtained through metadynamics simulations. Representative snapshots along the B pathway are displayed, showing the structural rearrangements during the reaction.
Adapted with permission from M. Alfonso-Prieto, X. Biarnés, P. Vidossich and C. Rovira, *J. Am. Chem. Soc.*, 2009, **131**, 11751.[50] Copyright (2009) American Chemical Society.

atom (as a proton) to the distal histidine, forming a superoxide anion (Intermediate → A1 in Figure 14.3):

$$HOO^{\bullet} + His \rightarrow O_2^{\bullet -} + HisH^+$$

In a second step, the protonated His breaks the hydrogen bond with superoxide (A1 → A2) and moves downwards, forming a hydrogen bond with the hydroxoferryl instead (A3 → A4). This new interaction allows the transfer of the proton from the distal His to the hydroxoferryl, forming water, and concomitantly an electron transfers from superoxide to the ferryl ion, yielding an oxygen molecule (A4 → Products):

$$\text{Cpd II (Por–Fe}^{IV}\text{–OH)} + O_2^{\bullet -} + HisH^+ \rightarrow \text{resting (Por–Fe}^{III}\text{)} + H_2O + O_2$$

Analysis of pathway B reveals a different mechanistic picture. Here the peroxyl radical breaks the hydrogen bond with the distal His (Intermediate → B1) and rotates, establishing a new hydrogen bond with the hydroxoferryl group (B1 → B2). Then, the Fe–OH rotates (B2 → B3 → B4), facilitating the transfer of the second hydrogen atom and resulting in formation of the products (B4 → Products):

$$\text{Cpd II (Por–Fe}^{IV}\text{–OH)} + HOO^{\bullet} \rightarrow \text{resting (Por–Fe}^{III}\text{)} + H_2O + O_2$$

Altogether, our metadynamics simulations reveal two possible routes for oxygen release in catalases, the His-mediated and the direct mechanisms.[50] Regardless of the pathway, the reaction takes place in two one-electron steps, instead of a single two-electron step, as had often been assumed.[35,80,81] Moreover, the inclusion of thermal fluctuations in our MD simulations reveals how the distal residues move along the reaction and contribute to catalysis.[51,82] On the one hand, the downward movement of the distal His (see Figure 14.3) is essential for this residue to act as a proton shuttle between the peroxyl radical and the hydroxoferryl unit. On the other, rotation of the distal Asn (see Figures 14.3 and 14.4) facilitates the departure of the produced oxygen molecule from the active site.

Both mechanisms (Figure 14.5) have similar free energy barriers, around 12 kcal mol^{-1}, which are in good agreement with the experimentally determined value (10–11 kcal mol^{-1}).[83] Therefore the His-mediated and the direct mechanisms can be competitive. Our results also rationalise previous kinetic isotope effect (KIE) data on catalase and myoglobin (Mb) mutants.[83,84] Mb is a haemprotein that can also form Cpd I when residue H64 is mutated. Kato and coworkers engineered a second mutation on Mb to investigate how Cpd I reduction by H_2O_2 is affected by the changes in the distal side of haem.

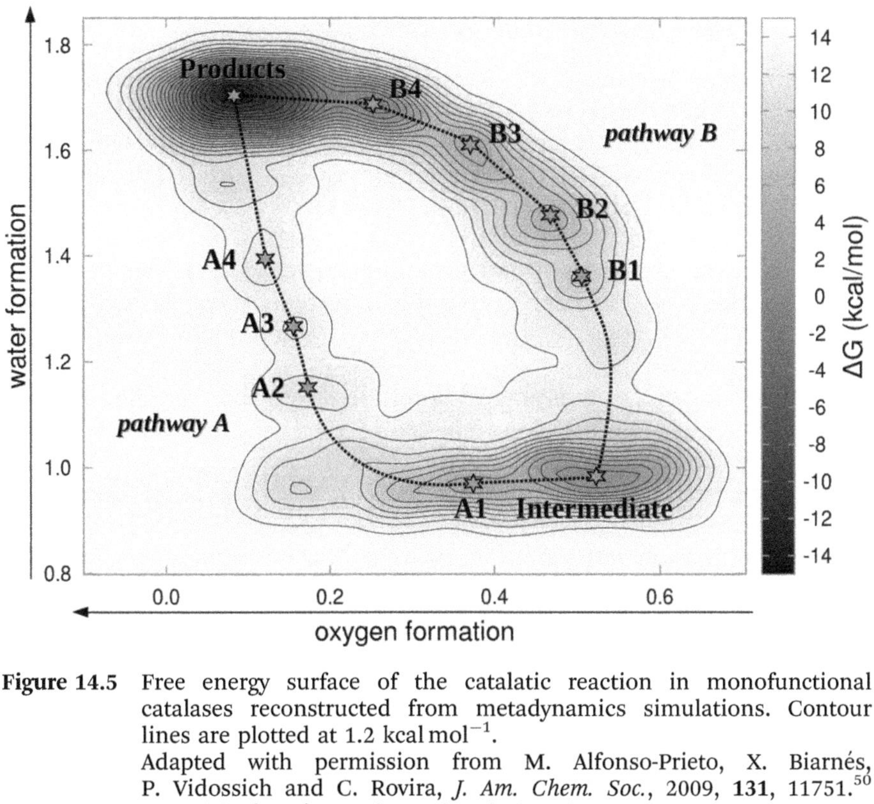

Figure 14.5 Free energy surface of the catalatic reaction in monofunctional catalases reconstructed from metadynamics simulations. Contour lines are plotted at 1.2 kcal mol^{-1}.
Adapted with permission from M. Alfonso-Prieto, X. Biarnés, P. Vidossich and C. Rovira, *J. Am. Chem. Soc.*, 2009, **131**, 11751.[50] Copyright (2009) American Chemical Society.

Catalase and the F43H/H64L Mb mutant (which displays a histidine residue on a position similar to catalase) exhibited small KIE values (2–4). Our metadynamics simulations show that the rate-limiting step in catalase does not correspond to any hydrogen transfer (in either pathway A or B), but to changes in the hydrogen bond pattern. Hence, H/D exchange is not expected to affect the reaction rate, as it is found experimentally. In contrast, the double Mb mutant L29H/H64L (in which the distal His is located further away from the haem compared to catalase) and the single Mb mutants H64D, H64A and H64S (which lack a distal His altogether) displayed large KIE values (10–29).[83,84] To compare with these Mb mutants, we created an *in silico* catalase mutant missing the distal His.[50] In the *H. pylori* catalase mutant H56G, the H_2O_2 substrate cannot establish short interactions with the oxoferryl group, and, as a result, the barrier of the initial HAT increases compared to wild-type catalase. As the first HAT step becomes rate-limiting, a large KIE is expected for this catalase mutant. Similarly to the catalase H56G mutant, the Mb mutants L29H/H64L, H64D, H64A and H64S also lack a distal His properly located to position H_2O_2 close to the oxoferryl, and thus they would have an analogous reaction energy profile, explaining the observed large KIE values.

14.3.3.2 The Catalatic Reaction in Peroxidases

As mentioned in Section 14.3.2, monofunctional peroxidases, such as HRP, can also decompose H_2O_2 into water and oxygen, although at a much lower rate than catalases.[39,40] To understand this change in reactivity, we investigated the mechanism of Cpd I reduction by H_2O_2 in HRP.[85] The following protocol, which is similar to the one for catalase, was used:

(i) selection of an appropriate experimental structure as starting point (in this case, the X-ray structure of horseradish peroxidase Cpd I[86]);
(ii) equilibration in physiological-like conditions using MM simulations;
(iii) selection of a suitable snapshot for the reactive configuration, with a water molecule hydrogen bonded to the distal His, in agreement with the crystal structure of HRP Cpd I;[86]
(iv) replacement of the water molecule by H_2O_2, in order to build a model of the Michaelis complex of the catalatic reaction;
(v) equilibration of the model of the Cpd I–H_2O_2 complex using MM MD simulations;
(vi) selection of a suitable snapshot to initiate the reaction, with H_2O_2 bridging the distal His and the oxoferryl, similar to the Cpd I–H_2O_2 complex in catalase;
(vii) equilibration of the model of the Cpd I–H_2O_2 complex using QM/MM MD simulations;
(viii) modelling of the chemical reaction using an enhanced sampling technique.

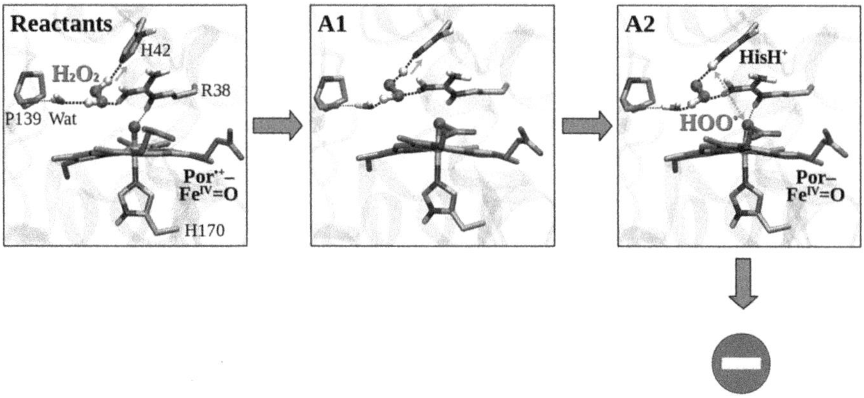

Figure 14.6 The (non-productive) His-mediated mechanism of the catalatic reaction in monofunctional peroxidases, obtained through metadynamics simulations. Representative snapshots along the (dead-end) A pathway are displayed, showing the structural rearrangements during the reaction.
Adapted with permission from P. Campomanes, U. Rothlisberger, M. Alfonso-Prieto and C. Rovira, *J. Am. Chem. Soc.*, 2015, **137**, 11170.[85] Copyright (2015) American Chemical Society.

The peroxidase QM region (see the Reactants panel in Figure 14.6) comprised the H_2O_2 substrate, the haem Cpd I, the His proximal ligand (H170 in HRP), the His/Arg distal pair (H42/R38) and an additional water molecule hydrogen bonded to H_2O_2. This is equivalent to the QM region chosen for catalase, except for the distal water molecule, which was not present in the case of the more buried active site of catalase. The QM region was treated with DFT (BLYP functional, dual basis set with a DZVP Gaussian basis set for the wavefunction and a plane wave basis set with a 320 Ry cut-off for the density, Goedecker–Tetter–Hutter pseudopotentials), while the rest of the protein and the solvent was described using a force field (Amber). The QM/MM boundary was treated with the capping hydrogen approach, whereas the QM–MM interactions were described using a real space multigrid technique.[87] The Born–Oppenheimer (BO) method was applied for the *ab initio* MD simulations, and the chemical reaction was activated using metadynamics.[26–30] The collective variables used were very similar to those applied to study the same process in catalase.[50]

Analysis of the metadynamics simulations of HRP[85] revealed significant mechanistic differences compared to catalase. The pathway equivalent to the His-mediated (or A) mechanism in catalase is not productive for HRP (Figure 14.6). As a result, only one pathway is found connecting the reactants and the products of the reaction (Figure 14.7), equivalent to the direct (or B) mechanism in catalase.

Pathway A (Figure 14.6) is characterised by a proton transfer to the distal His (Reactants→A1), accompanied by an electron transfer to the

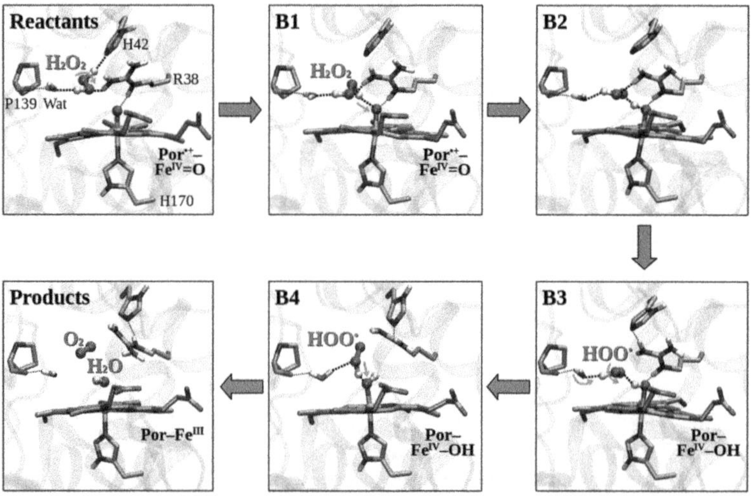

Figure 14.7 The direct mechanism of the catalatic reaction in monofunctional peroxidases, obtained through metadynamics simulations. Representative snapshots along the B pathway are displayed, showing the structural rearrangements during the reaction.
Adapted with permission from P. Campomanes, U. Rothlisberger, M. Alfonso-Prieto and C. Rovira, *J. Am. Chem. Soc.*, 2015, **137**, 11170.[85] Copyright (2015) American Chemical Society.

haem, forming a peroxyl radical and an oxoferryl Cpd II-like intermediate (A1 → A2):

$$\text{Cpd I } (\text{Por}^{+\bullet}\text{–Fe}^{\text{IV}}{=}\text{O}) + \text{H}_2\text{O}_2 + \text{His} \rightarrow \text{Cpd II } (\text{Por–Fe}^{\text{IV}}{=}\text{O}) + \text{HOO}^{\bullet} + \text{HisH}^{+}$$

In order to complete the reaction, the protonated histidine needs to transfer the proton to the Fe–O unit. In catalase, such proton transfer was possible because the distal His is located at only 3 Å from the oxoferryl. However, this distance is much longer in peroxidase, almost 5 Å, and the distal water molecule is not properly positioned to mediate the proton transfer (*i.e.* it is not bridging the distal His and the oxoferryl). Hence, the energy barrier of the proton-coupled electron transfer step increases significantly, and pathway A is no longer accessible in peroxidase.

In contrast, pathway B is still operative in peroxidase (Figure 14.7). It starts with the rotation of the H_2O_2 molecule, breaking the hydrogen bond with the distal His and establishing a new hydrogen bond with the oxoferryl (Reactants → B1). This permits the first HAT to take place (B1 → B2 → B3), forming a peroxyl radical and a hydroxoferryl Cpd II-like intermediate, similar to catalase. Afterwards, the peroxyl radical and the hydroxoferryl group rotate, such that the peroxyl hydrogen atom becomes hydrogen bonded to the oxygen atom of the hydroxoferryl (B3 → B4). This enables the second HAT (B4 → Products) to occur, yielding the products of the reaction, oxygen and water.

Altogether, only one out of the two pathways found in catalase, the direct mechanism, survives in peroxidase. The His-mediated mechanism corresponds to a dead-end pathway, due to the longer distance between the distal His and the haem compared to catalase.[85] Analysis of the conformations sampled during the reaction also show how the distal residues move to facilitate the reaction.[82] The distal His swings and the distal Arg moves upwards (see Figure 14.7) to guide the reorientation of the peroxyl radical and facilitate the departure of the oxygen molecule produced.[85]

The rate-limiting step in peroxidase turns out to be the second HAT, as in catalase. The calculated free energy barrier for the second HAT is 19 kcal mol^{-1} (see Figure 14.8), in close agreement with the experimental value (17 kcal mol^{-1}).[70,88] Moreover, the existence of the non-productive intermediate A2 (Figure 14.6) may explain another experimental observation: the inactivation of HRP after ~700 catalytic cycles.[70] We hypothesise that A2 may release the peroxyl radical (as $HOO^•$ or $O_2^{•-}$) into the solution, inactivating the enzyme. This could explain why superoxide dismutase (SOD) and other superoxide scavengers affect the HRP-catalysed production of oxygen.[88] Presumably, the resulting 'bare' Cpd II reacts with another molecule of H_2O_2 and forms Cpd III, whose spectroscopic signal has indeed been detected when HRP is incubated in the presence of excess H_2O_2.[70,88]

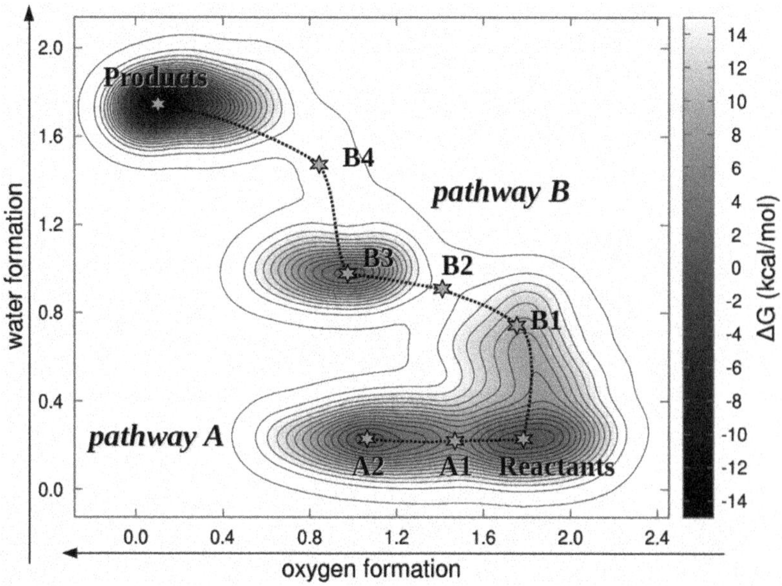

Figure 14.8 Free energy surface of the catalatic reaction in monofunctional peroxidases reconstructed from metadynamics simulations. Contour lines are plotted at 1.2 kcal mol^{-1}.
Adapted with permission from P. Campomanes, U. Rothlisberger, M. Alfonso-Prieto and C. Rovira, *J. Am. Chem. Soc.*, 2015, **137**, 11170.[85] Copyright (2015) American Chemical Society.

14.3.3.3 *Causes of the Different Reactivity of Catalases* Versus *Peroxidases*

Our QM/MM metadynamics simulations have shown that the direct mechanism is operative in both catalases and peroxidases. However, the free energy barrier in peroxidases (~19 kcal mol^{-1}) is higher than in catalases (~12 kcal mol^{-1}), in line with the lower efficiency of peroxidases at releasing molecular oxygen. Complementing these mechanistic studies, we also investigated the structural causes underlying this different Cpd I reactivity by simulations. This knowledge can provide hints to designing hydroperoxidase biomimetics to be used in biocatalysis applications.

One of the reasons put forward to explain the disparate reactivity of catalases and peroxidases was their different proximal haem ligand (a tyrosinate in catalases and a histidine in peroxidases, see Figure 14.2). The nature of the residue coordinated to the iron ion is considered a signature of the haem enzyme family and has often been used to rationalise their different reactivity.[89] To assess the effect of the proximal ligand, we used QM calculations on a cluster model of the Cpd I active site.[75] The cluster model included only the H_2O_2 substrate, the bare iron porphyrin (*i.e.* the haem without substituents), and the proximal residues (Tyr/Arg in catalases and His in peroxidases). Hence, these gas phase calculations allow us to isolate the contribution of the proximal ligand from the rest of the protein environment. In particular, the distal His is not included and thus only the direct mechanism can be explored, which is the pathway common to peroxidases and catalases.

Our QM calculations show that Cpd I reduction by H_2O_2 is not affected by the nature of the proximal ligand, and thus the tyrosinate/histidine dichotomy between catalases and peroxidases is not responsible for their different catalatic activity.[75] Instead, the distance between the substrate (H_2O_2 or $HOO^\bullet$) and the haem appears to be crucial in determining the energy barrier of the process. The barrier for HAT is almost negligible for distances shorter than 2.5 Å, and it increases as the two moieties are positioned further away. Because the approach of the substrate is regulated by the residues distal to the haem, our results suggest that the distal residues would have an impact on the efficiency of the catalatic reaction. This is in line with the kinetic experiments of Kato and coworkers discussed above,[83,84] showing that an increase in the size of the distal cavity in Mb (and thus in the haem–substrate distance) resulted in higher KIE values (*i.e.* a higher energy barrier for the HAT).

To verify whether the distal residues affect the reaction energy barrier of the catalatic reaction, we relied again on QM calculations. We focused on the second step of the reaction, that is the rate-limiting step in both catalases and peroxidases.[50,85] Therefore, we used cluster models comprising the peroxyl radical ($HOO^\bullet$), the bare iron porphyrin (*i.e.* the haem without substitutents), the proximal residues (Tyr/Arg in catalases and His in peroxidases) and the distal residues capable of interacting with $HOO^\bullet$

(His/Asn in catalases and His/Arg/Pro in peroxidases). For HRP, two water molecules present in the active site were also included; in contrast, the more buried active site of catalase does not hold any water molecules. The use of simplified cluster models allows us to disentangle the contribution of the distal residues from the rest of the protein.

As mentioned above, the barrier of the rate-limiting step of the catalatic reaction does not correspond to the actual hydrogen transfer,[50,85] but to the reorganisation of the hydrogen bond network in the distal side. On going from the Cpd II-like species (Intermediate panel in Figure 14.4 for catalase or B3 panel in Figure 14.7 for peroxidase) to the transition state (B1 or B4, respectively), the peroxyl radical breaks some hydrogen bonds and forms new ones. Therefore, we used a fragment analysis approach to evaluate the energy cost of this hydrogen bond rearrangement (see ref. 85 for details).

For catalase, the cost of breaking the hydrogen bonds of the peroxyl radical with the distal His and Asn ($+18.0$ kcal mol^{-1}) is partially compensated by the formation of a new hydrogen bond with the Fe–O unit (-8.3 kcal mol^{-1}). In contrast, in peroxidase the reorientation of the peroxyl radical is more hindered. The cost of breaking the hydrogen bonds with the distal residues (Pro and active site water molecules) is higher ($+21.3$ kcal mol^{-1}) than in catalase. In addition, the energy compensation of forming a new hydrogen bond with the oxoferryl is smaller (-3.1 kcal mol^{-1}) than in catalase. As a consequence, the net energy cost of the hydrogen bond reorganisation ($\Delta E^{\ddagger}_{\mathrm{TOTAL}}$) is higher in peroxidase ($+18.2$ kcal mol^{-1}) than in catalase ($+10.5$ kcal mol^{-1}). Interestingly, these potential energy barrier estimates are in very good agreement with the free energy barriers obtained in the metadynamics simulations ($+18.7$ and $+12.0$ kcal mol^{-1}, respectively). This gives further validation to the fragment analysis used and supports the notion that the distal side interactions are the main determinants of the different Cpd I reactivity of catalases and peroxidases.[85]

In summary, peroxidases are less efficient at releasing oxygen than in catalase because of their different distal side cavity, which was particularly designed to catalyse the oxidation of other (bulkier) substrates rather than H_2O_2. Therefore, protein engineering aimed at modifying the enzyme reactivity should focus on acting on the distal residues and the interactions formed with the substrate.

14.3.4 The Catalatic Reaction in KatGs

As noted in Section 14.3.2, KatGs are bifunctional enzymes that are phylogenetically related to plant peroxidases, such as HRP, and yet can decompose H_2O_2 into oxygen at a rate comparable to monofunctional catalases.[90] The active site of KatG exhibits the same residues as a typical peroxidase, that is a haem group with a proximal His/Asp pair (H279 and D389 using BpKatG numbering) and a distal His/Arg pair (H112 and R108) (Figure 14.2). However, there are also some particularities, present only in KatG. First, there is a unique triad of covalently linked protein residues stacked above the haem, the MYW adduct (M264, Y238 and W111), which has been found to be essential for the catalatic reaction.[63] Second, there is an Arg (R426) whose mutation has also been found to decrease the catalatic activity,[63,91] despite being 20 Å away from the haem iron. This Arg has been seen to oscillate between two conformations: closely interacting with the adduct Tyr (in conformation) or shifted away from the adduct (out conformation). The former conformation is observed in the resting state when the pH is high,[92] whereas the latter is present in the resting state at low pH[92] or in the Cpd I state at both low and high pH values.[93] According to the mutagenesis data,[63,91] these two distinctive features are responsible for the high catalatic activity of KatG; however the precise role of the adduct and the mobile Arg was not known. We used QM and QM/MM calculations to complement the structural, kinetic and spectroscopic data and give further insight into the catalatic activity of KatG.

14.3.4.1 The KatG Adduct

The covalently linked MYW adduct is an unusual modification unique to KatG. Therefore, we started by studying the intrinsic properties of this adduct using QM calculations on a cluster model containing the adduct and the mobile Arg. Our calculations showed that the acid–base and redox properties of this adduct are very peculiar.[46] First, the calculated pK_a of the adduct Tyr is 3 pK_a units lower than a normal Tyr, and it is modulated by interaction with the mobile Arg. Second, the unprotonated form of the adduct Tyr has a particularly low ionisation potential compared to a normal Tyr, and thus it is a good electron donor. As mentioned above, Cpd I is a highly oxidant species that can even oxidise nearby protein residues. Hence, it is reasonable to think that the adduct Tyr could donate one electron to the porphyrin cation radical of Cpd I, so that a Cpd I* with a radical adduct ($MYW^{+\bullet}$) would be formed. Moreover, the calculations showed that formation of a radical adduct destabilises the salt bridge between the adduct Tyr and the mobile Arg, and thus Cpd I* formation would drive the Arg away from the adduct. This is in line with the KatG Cpd I crystal structures,[93] which show the mobile Arg exclusively in an out conformation, regardless of the pH. Therefore, the mobile Arg has a dual role: in the in conformation, it decreases the pK_a of the adduct Tyr so it becomes deprotonated, whereas in the out conformation, it releases the negative charge on the oxygen of the adduct Tyr and makes it a good electron donor that delivers an electron to the haem.[46]

14.3.4.2 The KatG Compound I

How can the catalase activity be introduced in a peroxidase core?[94] Cpd I is the common reaction intermediate of catalases and peroxidases, which differ in the way Cpd I is reduced back to the enzyme resting state (see Section 14.3.2 or Figure 14.1). Therefore, the first step to understanding the reactivity of KatG is to analyse the properties of its Cpd I, which is predicted to hold a radical adduct according to gas-phase calculations in Section 14.3.4.1.[46] To elucidate the electronic structure of KatG Cpd I, we performed QM/MM calculations. Inclusion of the protein environment is crucial, because it has been shown to regulate the radical location in Cpd I of other haemproteins.[45,57,95–97] The QM region included the haem group, the distal His/Arg pair and the proximal His/Asp pair (similar to the HRP QM/MM calculations), as well as the MYW adduct (unique to KatG). In addition, we included a proximal Trp (W330), which is equivalent to the Trp of CCP that harbours a cation radical.[61,62] Our QM/MM calculations showed that the electronic structure of KatG Cpd I is different from both HRP and catalases, and does not contain a porphyrin cation radical. Instead, two different types of Cpd I* were found. When the adduct Tyr is unprotonated, the radical localises mainly on the adduct ($MYW^{+\bullet}$). In contrast, when the adduct Tyr is protonated, the unpaired electron localises mainly on the proximal Trp ($W330^{+\bullet}$), like in CCP. This is in line with the UV-visible spectra of KatG obtained upon mixing with H_2O_2, which showed two different species depending on the pH.[98] Interestingly, the catalatic activity of KatG occurs in the pH range 6–8,[90] when the adduct Tyr ($pK_a \sim 6.5$) is unprotonated. Therefore, most likely formation of this unique form of Cpd I* with a radical adduct ($MYW^{+\bullet}$) is what imparts catalatic activity to KatGs. Indeed, our prediction that a radical adduct is formed during the catalatic reaction of KatG[46] was later suggested by EPR spectroscopy.[65]

14.3.4.3 The Catalatic Reaction in KatG

Recalling the previous sections, the catalatic reaction in the monofunctional enzymes proceeds in two steps, which are formally two HATs.[50,85] The main difference between monofunctional catalases and peroxidases was the existence of the His-mediated mechanism, in which the distal His shuttles a proton between the peroxyl radical and the hydroxoferryl (see Section 14.3.3.2). This mechanism is very efficient in catalases, whereas it is a dead-end pathway that does not lead to products in peroxidases. The distal His in KatG lies perpendicular to the haem (Figure 14.2), as in HRP, and therefore it is not well positioned to act as an acid–base catalyst either. Then, how can KatG, with its peroxidase-like active site, emulate the efficiency of monofunctional catalases? We proposed that the adduct Trp could be a good alternative,[67] because its indole group occupies a position similar to the distal His in catalases (see left and right panels in Figure 14.2). In principle, the high pK_a (~16) of the indole proton makes it a very unlikely proton source. However, the radical character of the adduct in the KatG Cpd I* is expected to lower the pK_a of Trp considerably. For instance, a decrease of up to 4 pK_a units has been observed for other radical tryptophans.[99,100] Therefore, the possibility of a mechanism in which the adduct Trp plays a similar acid–base role as the distal His in catalases is conceivable.

We studied the mechanism of Cpd I reduction by H_2O_2 in KatG and the possible participation of the adduct Trp using QM/MM calculations.[46,66,67] The following protocol was used:

(i) selection of an appropriate experimental structure as starting point (in this case, the X-ray structure of KatG Cpd I,[93] with the mobile Arg in an out conformation and the adduct Tyr unprotonated);
(ii) equilibration in physiological-like conditions using MM simulations;
(iii) selection of a suitable snapshot (which conforms with the experimental data available for KatG Cpd I[66,67,93]) for the reactive configuration;
(iv) replacement of one of the active site water molecules by H_2O_2, in order to build a model of the Michaelis complex of the catalatic reaction, similar to catalase[50] and HRP;[85]
(v) equilibration of the model of the Michaelis complex using QM/MM MD simulations.

Because structural, spectroscopic and kinetic data were available for different stages of the reaction (resting state, Cpd I and Cpd III-like species), we did not attempt to compute the energetics of the whole catalatic mechanism in KatG, as we did for the simpler monofunctional enzymes. Instead we combined selected QM/MM calculations with the available structural, spectroscopic and kinetic data to elucidate the catalatic mechanism in KatG.

The QM/MM MD simulations of the Michaelis complex (Reactants panel in Figure 14.9) showed that the incoming H_2O_2 molecule bridges the distal His and the oxoferryl group, while accepting a hydrogen bond from the adduct Trp.[67] Interestingly, the adduct remains in its radical state upon H_2O_2 binding, further supporting that the intermediate of the catalatic reaction in KatG is a Cpd I* (Por–Fe^{IV}=O, $MYW^{+\bullet}$). Moreover, the distance between the peroxide oxygen atom and the oxoferryl is 2.65 Å, indicating that a HAT can occur easily (*i.e.* with a barrier less than 5 kcal mol^{-1}),[75] forming a

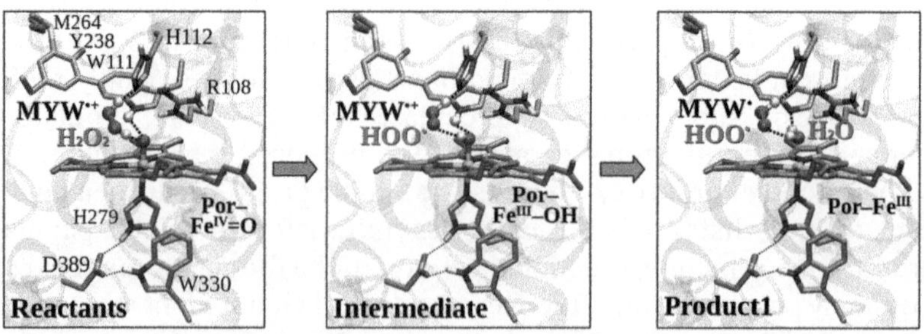

Figure 14.9 Formation of the product water molecule during the catalatic reaction in bifunctional KatG.
Adapted with permission from P. C. Loewen, X. Carpena, P. Vidossich, I. Fita and C. Rovira, *J. Am. Chem. Soc.*, 2014, **136**, 7249.[67] Copyright (2014) American Chemical Society.

peroxyl radical (Reactants → Intermediate in Figure 14.9). In monofunctional catalases or peroxidases, HAT from H_2O_2 to the haem quenches the porphyrin radical of Cpd I, yielding a Cpd II-like intermediate (Por–Fe^{IV}–OH). In contrast, in KatG, such HAT reduces the ferryl ion (to Por–Fe^{III}–OH), because the porphyrin is already in a neutral state and the radical is stored on the adduct instead:

$$\text{Cpd I* (Por–Fe}^{\text{IV}}\text{=O, MYW}^{+\bullet}) + \text{H}_2\text{O}_2 \\ \rightarrow \text{Cpd II* (Por–Fe}^{\text{III}}\text{–OH, MYW}^{+\bullet}) + \text{HOO}^{\bullet}$$

Could the adduct Trp act as the acid–base catalyst in the catalatic reaction? The indole proton is hydrogen bonded to the oxygen atom of the hydroxoferryl group, and its transfer (Figure 14.9) would seamlessly yield a water molecule, one of the two products of the catalatic reaction:

$$\text{Cpd II* (Por–Fe}^{\text{III}}\text{–OH, MYW}^{+\bullet}) + \text{HOO}^{\bullet} \\ \rightarrow \text{resting (Por–Fe}^{\text{III}}) + \text{H}_2\text{O} + \text{MYW}^{\bullet} + \text{HOO}^{\bullet}$$

Our QM/MM calculations showed that the energy barrier of this proton transfer is only ~12.5 kcal mol^{-1},[67] very similar to the rate-limiting barrier of the catalatic reaction in monofunctional catalases.[50] Therefore, it is very likely that the adduct Trp may act as the proton source, and we expect this to be the rate-determining step of the catalatic reaction in KatG.

At this stage (water product in Figure 14.10), there are two radical species in the active site, the radical adduct and the peroxyl radical, at a close distance. Given the high reactivity of radicals, it is reasonable to think that they may couple, forming a perhydroxy-modified Trp (Figure 14.10):

$$\text{MYW}^{\bullet} + \text{HOO}^{\bullet} \rightarrow \text{MYW–OOH}$$

Indeed, such modification has been observed in catalytically active crystals of KatG soaked with H_2O_2,[67] supporting the feasibility of this radical coupling step. Moreover, the disappearance of the radical on the adduct would allow the mobile Arg to swing back to the in conformation, as also seen in the X-ray structures of KatG containing Trp–OOH.[67]

The perhydroxy-modified Trp has one oxygen atom interacting with the haem iron, while the other maintains the hydrogen bond with the distal His (see Figure 14.10). Therefore, one can envisage the next step to be the cleavage of the N–OOH bond, concomitant to the transfer of the proton to the distal His, forming a ferric-superoxide species and an adduct radical, *i.e.* Compound III (Cpd III)-like (Figure 14.10):

$$\text{resting (Por–Fe}^{\text{III}}) + \text{MYW–OOH} + \text{His} \\ \rightarrow \text{Cpd III* (Por–Fe}^{\text{III}}\text{–O}_2^{\bullet -}\text{, MYW}^{+\bullet}) + \text{HisH}^{+}$$

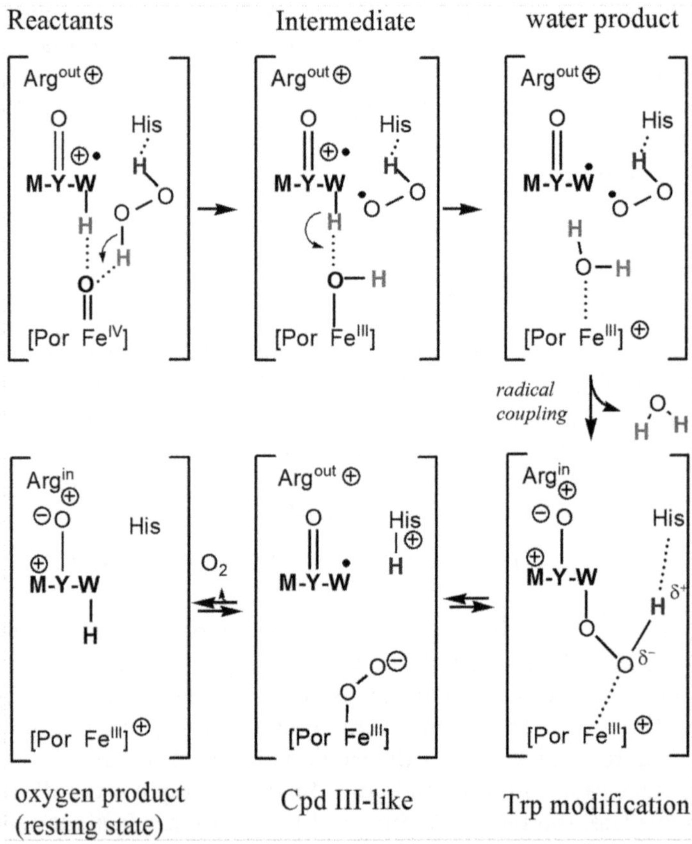

Figure 14.10 Proposed mechanism of the catalatic reaction in bifunctional KatG. All the species involved have been characterised by QM/MM calculations, X-ray crystallography or spectroscopic techniques. Adapted with permission from P. C. Loewen, X. Carpena, P. Vidossich, I. Fita and C. Rovira, *J. Am. Chem. Soc.*, 2014, **136**, 7249.[67] Copyright (2014) American Chemical Society.

This is consistent with the spectroscopic detection of a Cpd III-like species in active KatGs, with formula $Fe^{III}–O_2^{\bullet -}$ or its isoelectronic $Fe^{II}–O_2$ form.[64,65,98,101] Moreover, the reverse process (*i.e.* formation of a perhydroxy-modifed Trp) has also been observed to happen when native KatG reacts with oxygen.[65,92,102]

The last step of the catalatic reaction of KatG would be the release of molecular oxygen from the Cpd III-like species (Figure 14.10):

$$\text{Cpd III* } (\text{Por–Fe}^{\text{III}}\text{–O}_2^{\,-},\ \text{MYW}^{+\bullet}) \rightarrow \text{resting } (\text{Por–Fe}^{\text{III}}) + \text{MYW} + \text{O}_2$$

Electron transfer from superoxide to the radical adduct yields the resting state of the enzyme and generates an oxygen molecule (oxygen product in Figure 14.10). Interestingly, spectroscopic data suggest that native KatG binds oxygen (*i.e.* the reverse process of this step).[64,65] Using QM/MM calculations, we demonstrated that KatG can indeed bind oxygen, unlike monofunctional peroxidases or other haemproteins with a ferric state, because of the presence of the adduct, which transfers an electron to the oxygen molecule.[66] Applying microreversibility, oxygen release from Cpd III* is exothermic (by ~11 kcal mol^{-1}, according to our QM/MM calculations[66]), and thus a favourable process. Taken together, the experimental and computational data support the feasibility of this step. A later DFT investigation using active site models found a relatively low barrier of 5.6 kcal mol^{-1} for this step, further backing our proposal.[68]

Altogether, our QM and QM/MM results, in combination with the experimental structural, spectroscopic and kinetic data available, allowed us to formulate a novel molecular mechanism for the catalatic reaction in KatG (Figure 14.10). This mechanism relies on the MYW adduct and the mobile Arg and thus it is completely unique to KatG. The adduct changes both its oxidation and protonation state during the catalatic reaction, allowing the process to occur at a rate similar to monofunctional catalases. However, in KatG the radical is stored in the MYW adduct and the proton source is the adduct Trp, instead of the porphyrin and the distal His, respectively, in monofunctional catalases. Moreover, the changes in the adduct are accompanied by the conformational switch of the mobile Arg.

14.4 Conclusions

The number of theoretical studies in the field of metalloproteins has grown exponentially in the last years, thanks to the increase in computer power and the progress of computational methods.[9–13] In particular, QM and QM/MM calculations have been shown to be very valuable to study the reactivity of metalloenzymes. They can provide the structure of reaction intermediates not easily accessible by experiments and decipher the catalytic mechanism of (metal-containing) enzymes by computing the reaction energy profile.[14–20] It should be noted that the agreement of the calculated rate-limiting energy barrier with the kinetic measurements is only one piece of the puzzle, and the proposed mechanism should also be consistent with other experimental data, such as isotope labelling, spectroscopic measurements or mutagenesis data. Therefore, the interplay between experiments and theory is essential to obtain a more complete understanding of metalloenzymes.

Most computational studies dealing with metalloenzymes use DFT, as it provides good accuracy while maintaining a reasonable scaling.[23,24,103] Nevertheless, improvements in the description of the metal centre and the spin state energetics are still needed in certain cases.[104,105] Two different computational approaches are most commonly used: either QM calculations on a cluster model of the metal-containing active site or QM/MM calculations on the complete metalloenzyme.[14–20] The former helps to disentangle the effect of the different active site residues on catalysis, whereas the latter allows us to study the reaction (with QM accuracy), while including the effect of the whole protein environment (with an MM description). Both approaches are limited by the computational cost of the QM calculations, and therefore new, more efficient methodologies are needed. One possible solution is the use of enhanced sampling techniques,[26–30,106,107] which allow sampling of rare events (such as chemical reactions) or longer timescales. Alternatively, MM simulations can provide snapshots to be probed by QM or QM/MM simulations. Other efforts addressing this short timescale limitation include the improvement of the QM algorithms or the development of adaptive QM/MM techniques.[108,109]

In this chapter we have presented several examples of QM and QM/MM simulations applied to study the reactivity of three metalloenzymes: monofunctional catalases, monofunctional peroxidases and bifunctional KatGs. We have shown that simulations can unravel the catalytic mechanism of H_2O_2 dismutation in these three haemproteins and reconcile all available experimental information. Moreover, modelling of the same chemical reaction in three different (albeit closely related) enzymes highlight the distal side residues as the structural determinants of their different catalatic efficiency.

Acknowledgements

We are indebted to our former and current collaborators, Pietro Vidossich, Pablo Campomanes, Xevi Biarnés, Ursula Röthlisberger, Ignacio Fita and Peter C. Loewen, who contributed to some of the applications presented in this chapter. Our research is supported by the Generalitat de Catalunya (grant 2014SGR-987) and the Ministerio de Economía y Competitividad (MINECO) (grant CTQ2014-55174). M. A.-P. is also grateful for the financial support from the Government of Catalonia and the European Union through a Beatriu de Pinós fellowship (BP-B 2013). We also acknowledge the support from the COST Action CM1305 (ECOSTBio), as well as the computer support, technical expertise and assistance provided by the Barcelona Supercomputing Center–Centro Nacional de Supercomputación (BSC–CNS).

References

1. *The Biological Chemistry of the Elements: The Inorganic Chemistry of Life*, ed. J. J. R. Silva and R. J. P. Williams, Oxford University Press, New York NY, 2nd edn, 2001.
2. *Metallomics and the Cell*, ed. L. Banci, Springer, New York NY, 2013.
3. A. J. Thomson and H. B. Gray, *Curr. Opin. Chem. Biol.*, 1998, **2**, 155.
4. W. Shi and M. R. Chance, *Curr. Opin. Chem. Biol.*, 2011, **15**, 144.
5. M. M. Harding, M. W. Nowicki and M. D. Walkinshaw, *Crystallogr. Rev.*, 2010, **16**, 247.
6. P. Carloni, U. Rothlisberger and M. Parrinello, *Acc. Chem. Res.*, 2002, **35**, 455.
7. L. Banci, *Curr. Opin. Chem. Biol.*, 2003, **7**, 143.

8. T. Beitlich, K. Kühnel, C. Schulze-Briese, R. L. Shoeman and I. Schlichting, *J. Synchrotron Radiat.*, 2007, **14**, 11.
9. M. Karplus and J. A. McCammon, *Nat. Struct. Mol. Biol.*, 2002, **9**, 46.
10. L. Banci, *Curr. Opin. Chem. Biol.*, 2003, **7**, 143.
11. M. Dal Peraro, P. Ruggerone, S. Raugei, F. L. Gervasio and P. Carloni, *Curr. Opin. Struct. Biol.*, 2007, **17**, 149.
12. P. Vidossich and A. Magistrato, *Biomolecules*, 2014, **4**, 616.
13. M. Alfonso-Prieto and M. L. Klein, *Metalloproteins: Theory, Calculations and Experiments*, ed. A. E. Cho and W. A. Goddard III, CRC Press, Taylor and Francis Group, Boca Raton FL, 2015, 3, p. 95.
14. P. E. Siegbahn and T. Borowski, *Acc. Chem. Res.*, 2006, **39**, 729.
15. H. M. Senn and W. Thiel, *Angew. Chem., Int. Ed.*, 2009, **48**, 1198.
16. S. F. Sousa, P. A. Fernandes and M. J. Ramos, *Phys. Chem. Chem. Phys.*, 2012, **14**, 12431.
17. M. W. van der Kamp and A. J. Mulholland, *Biochemistry*, 2013, **52**, 2708.
18. C. Rovira, *WIREs: Comput. Mol. Sci.*, 2013, **3**, 393.
19. E. Brunk and U. Rothlisberger, *Chem. Rev.*, 2015, **115**, 6217.
20. M. G. Quesne, T. Borowski and S. P. deVisser, *Chem. - Eur. J.*, 2016, **22**, 2562.
21. K. Henzler-Wildman and D. Kern, *Nature*, 2007, **450**, 964.
22. K. A. Henzler-Wildman, M. Lei, V. Thai, S. J. Kerns, M. Karplus and D. Kern, *Nature*, 2007, **450**, 913.
23. C. J. Cramer and D. G. Truhlar, *Phys. Chem. Chem. Phys.*, 2009, **11**, 10757.
24. F. Neese, *Coord. Chem. Rev.*, 2009, **253**, 526.
25. R. Car and M. Parrinello, *Phys. Rev. Lett.*, 1985, **55**, 2471.
26. A. Laio and M. Parrinello, *Proc. Natl. Acad. Sci. U. S. A.*, 2002, **99**, 12562.
27. M. Iannuzzi, A. Laio and M. Parrinello, *Phys. Rev. Lett.*, 2003, **90**, 238302.
28. B. Ensing, M. De Vivo, Z. Liu, P. Moore and M. L. Klein, *Acc. Chem. Res.*, 2006, **39**, 73.
29. A. Laio and F. L. Gervasio, *Rep. Prog. Phys.*, 2008, **71**, 126601.
30. A. Barducci, M. Bonomi and M. Parrinello, *WIREs Comput. Mol. Sci.*, 2011, **1**, 826.
31. M. Dal Peraro, K. Spiegel, G. Lamoureux, M. De Vivo, W. F. DeGrado and M. L. Klein, *J. Struct. Biol.*, 2007, **157**, 444.
32. M. B. Peters, Y. Yang, B. Wang, L. Füsti-Molnár, M. N. Weaver and K. M. Merz Jr., *J. Chem. Theory. Comput.*, 2010, **6**, 2935.
33. A. T. Carvalho, A. F. Teixeira and M. J. Ramos, *J. Comput. Chem.*, 2013, **34**, 1540.
34. *Peroxidases and Catalases: Biochemistry, Biophysics, Biotechnology, and Physiology*, ed. H. B. Dunford, John Wiley & Sons, Hoboken NJ, 2nd edn, 2010.
35. P. Nicholls, I. Fita and P. C. Loewen, *Adv. Inorg. Chem.*, ed. A. G. Sykes and G. Mauk, Academic Press, San Diego CA, 2000, vol. 51, p. 51.
36. P. Chelikani, I. Fita and P. C. Loewen, *Cell. Mol. Life Sci.*, 2004, **61**, 192.

37. *Heme Peroxidases*, ed. H. B. Dunford, John Wiley & Sons, New York NY, 1999.
38. F. W. Krainer and A. Glieder, *Appl. Microbiol. Biotechnol.*, 2015, **99**, 1611.
39. J. Hernández-Ruiz, M. Arnao, A. Hiner, F. García-Cánovas and M. Acosta, *Biochem. J.*, 2001, **354**, 107.
40. J. Vlasits, C. Jakopitsch, M. Bernroitner, M. Zámocký, P. G. Furtmüller and C. Obinger, *Arch. Biochem. Biophys.*, 2010, **500**, 74.
41. A. Díaz, P. C. Loewen, I. Fita and X. Carpena, *Arch. Biochem. Biophys.*, 2012, **525**, 102.
42. I. Fita, P. C. Loewen and X. Carpena, *Encyclopedia of Inorganic and Bioinorganic Chemistry*, ed. John Wiley, New York NY, 2015, pp. 1–14.
43. B. Wiseman, X. Carpena, M. Feliz, L. J. Donald, M. Pons, I. Fita and P. C. Loewen, *J. Biol. Chem.*, 2010, **285**, 26662.
44. A. Ivancich, H. M. Jouve, B. Sartor and J. Gaillard, *Biochemistry*, 1997, **36**, 9356.
45. M. Alfonso-Prieto, A. Borovik, X. Carpena, G. Murshudov, W. Melik-Adamyan, I. Fita, C. Rovira and P. C. Loewen, *J. Am. Chem. Soc.*, 2007, **129**, 4193.
46. P. Vidossich, M. Alfonso-Prieto, X. Carpena, P. C. Loewen, I. Fita and C. Rovira, *J. Am. Chem. Soc.*, 2007, **129**, 13436.
47. W. J. Chuang and H. E. Van Wart, *J. Biol. Chem.*, 1992, **267**, 13293.
48. E. Derat, S. Cohen, S. Shaik, A. Altun and W. Thiel, *J. Am. Chem. Soc.*, 2005, **127**, 13611.
49. M. Alfonso-Prieto, P. Vidossich, A. Rodríguez-Fortea, X. Carpena, I. Fita, P. C. Loewen and C. Rovira, *J. Phys. Chem. A*, 2008, **112**, 12842.
50. M. Alfonso-Prieto, X. Biarnés, P. Vidossich and C. Rovira, *J. Am. Chem. Soc.*, 2009, **131**, 11751.
51. M. Alfonso-Prieto, P. Vidossich and C. Rovira, *Arch. Biochem. Biophys.*, 2012, **525**, 121.
52. C. E. Schulz, R. Rutter, J. T. Sage, P. G. Debrunner and L. P. Hager, *Biochemistry*, 1984, **23**, 4743.
53. E. Derat and S. Shaik, *J. Am. Chem. Soc.*, 2006, **128**, 8185.
54. E. Derat and S. Shaik, *J. Am. Chem. Soc.*, 2006, **128**, 13940.
55. A. Ivancich, H. M. Jouve and J. Gaillard, *J. Am. Chem. Soc.*, 1996, **118**, 12852.
56. C. D. Putnam, A. S. Arvai, Y. Bourne and J. A. Tainer, *J. Mol. Biol.*, 2000, **296**, 295.
57. M. Alfonso-Prieto, H. Oberhofer, M. L. Klein, C. Rovira and J. Blumberger, *J. Am. Chem. Soc.*, 2011, **133**, 4285.
58. A. Ivancich, C. Jakopitsch, M. Auer, S. Un and C. Obinger, *J. Am. Chem. Soc.*, 2003, **125**, 14093.
59. T. Spolitak, J. H. Dawson and D. P. Ballou, *J. Biol. Inorg. Chem.*, 2008, **13**, 599.
60. A. T. Smith, W. A. Doyle, P. Dorlet and A. Ivancich, *Proc. Natl. Acad. Sci. U. S. A.*, 2009, **106**, 16084.

61. M. Sivaraja, D. B. Goodin, M. Smith and B. M. Hoffman, *Science*, 1989, **245**, 738.
62. C. A. Bonagura, B. Bhaskar, H. Shimizu, H. Li, M. Sundaramoorthy, D. E. McRee, D. B. Goodin and T. L. Poulos, *Biochemistry*, 2003, **42**, 5600.
63. G. Smulevich, C. Jakopitsch, E. Droghetti and C. Obinger, *J. Inorg. Biochem.*, 2006, **100**, 568.
64. J. Suarez, K. Ranguelova, A. A. Jarzecki, J. Manzerova, V. Krymov, X. Zhao, S. Yu, L. Metlitsky, G. J. Gerfen and R. S. Magliozzo, *J. Biol. Chem.*, 2009, **284**, 7017.
65. X. Zhao, J. Suarez, A. Khajo, S. Yu, L. Metlitsky and R. S. Magliozzo, *J. Am. Chem. Soc.*, 2010, **132**, 8268.
66. P. Vidossich, X. Carpena, P. C. Loewen, I. Fita and C. Rovira, *J. Phys. Chem. Lett.*, 2011, **2**, 196.
67. P. C. Loewen, X. Carpena, P. Vidossich, I. Fita and C. Rovira, *J. Am. Chem. Soc.*, 2014, **136**, 7249.
68. B. I. Kruft, R. S. Magliozzo and A. A. Jarzecki, *J. Phys. Chem. A*, 2015, **119**, 6850.
69. J. Switala and P. C. Loewen, *Arch. Biochem. Biophys.*, 2002, **401**, 145.
70. A. N. Hiner, J. Hernández-Ruiz, G. A. Williams, M. B. Arnao, F. García and M. Acosta, *Arch. Biochem. Biophys.*, 2001, **392**, 295.
71. C. Rovira, K. Kunc, J. Hutter, P. Ballone and M. Parrinello, *J. Phys. Chem. A*, 1997, **101**, 8914.
72. O. A. von Lilienfeld, I. Tavernelli, U. Röthlisberger and D. Sebastiani, *Phys. Rev. B*, 2005, **71**, 195119.
73. A. Laio, J. VandeVondele and U. Röthlisberger, *J. Chem. Phys.*, 2002, **116**, 6941.
74. A. Laio, J. VandeVondele and U. Röthlisberger, *J. Phys. Chem. B*, 2002, **106**, 7300.
75. P. Vidossich, M. Alfonso-Prieto and C. Rovira, *J. Inorg. Biochem.*, 2012, **117**, 292.
76. I. Fita and M. G. Rossmann, *J. Mol. Biol.*, 1985, **185**, 21.
77. F. Lledías, P. Rangel and W. Hansberg, *J. Biol. Chem.*, 1998, **273**, 10630.
78. S. Michán, F. Lledías, J. D. Baldwin, D. O. Natvig and W. Hansberg, *Free Radical Biol. Med.*, 2002, **33**, 521.
79. C. Jakopitsch, M. Auer, A. Ivancich, F. Rüker, P. G. Furtmüller and C. Obinger, *J. Biol. Chem.*, 2003, **278**, 20185.
80. M. Zámocký and F. Koller, *Prog. Biophys. Mol. Biol.*, 1999, **72**, 19.
81. H. N. Kirkman and G. F. Gaetani, *Trends Biochem. Sci.*, 2007, **32**, 44.
82. P. Vidossich, M. Alfonso-Prieto, X. Carpena, I. Fita, P. C. Loewen and C. Rovira, *Arch. Biochem. Biophys.*, 2010, **500**, 37.
83. S. Kato, T. Ueno, S. Fukuzumi and Y. Watanabe, *J. Biol. Chem.*, 2004, **279**, 52376.
84. Y. Watanabe, H. Nakajima and T. Ueno, *Acc. Chem. Res.*, 2007, **40**, 554.
85. P. Campomanes, U. Rothlisberger, M. Alfonso-Prieto and C. Rovira, *J. Am. Chem. Soc.*, 2015, **137**, 11170.

86. G. I. Berglund, G. H. Carlsson, A. T. Smith, H. Szöke, A. Henriksen and J. Hajdu, *Nature*, 2002, **417**, 463.
87. T. Laino, F. Mohamed, A. Laio and M. Parrinello, *J. Chem. Theory Comput.*, 2005, **1**, 1176.
88. C. J. Baker, K. Deahl, J. Domek and E. W. Orlandi, *Arch. Biochem. Biophys.*, 2000, **382**, 232.
89. M. T. Green, J. H. Dawson and H. B. Gray, *Science*, 2004, **304**, 1653.
90. R. Singh, B. Wiseman, T. Deemagarn, L. J. Donald, H. W. Duckworth, X. Carpena, I. Fita and P. C. Loewen, *J. Biol. Chem.*, 2004, **279**, 43098.
91. X. Carpena, B. Wiseman, T. Deemagarn, R. Singh, J. Switala, A. Ivancich, I. Fita and P. C. Loewen, *EMBO Rep.*, 2005, **6**, 1156.
92. X. Carpena, B. Wiseman, T. Deemagarn, B. Herguedas, A. Ivancich, R. Singh, P. C. Loewen and I. Fita, *Biochemistry*, 2006, **45**, 5171.
93. X. Carpena, S. Loprasert, S. Mongkolsuk, J. Switala, P. C. Loewen and I. Fita, *J. Mol. Biol.*, 2003, **327**, 475.
94. O. J. Njuma, E. N. Ndontsa and D. C. Goodwin, *Arch. Biochem. Biophys.*, 2014, **544**, 27.
95. G. M. Jensen, S. W. Bunte, A. Warshel and D. B. Goodin, *J. Phys. Chem. B*, 1998, **102**, 8221.
96. S. P. deVisser, S. Shaik, P. K. Sharma, D. Kumar and W. Thiel, *J. Am. Chem. Soc.*, 2003, **125**, 15779.
97. J. N. Harvey, C. M. Bathelt and A. J. Mulholland, *J. Comput. Chem.*, 2006, **27**, 1352.
98. C. Jakopitsch, J. Vlasits, B. Wiseman, P. C. Loewen and C. Obinger, *Biochemistry*, 2007, **46**, 1183.
99. C. Tommos, J. J. Skalicky, D. L. Pilloud, A. J. Wand and P. L. Dutton, *Biochemistry*, 1999, **38**, 9495.
100. M. T. Zhang, T. Irebo, O. Johansson and L. Hammarström, *J. Am. Chem. Soc.*, 2011, **133**, 13224.
101. X. Zhao, A. Khajo, S. Jarrett, J. Suarez, Y. Levitsky, R. M. Burger, A. A. Jarzecki and R. S. Magliozzo, *J. Biol. Chem.*, 2012, **287**, 37057.
102. T. Deemagarn, X. Carpena, R. Singh, B. Wiseman, I. Fita and P. C. Loewen, *J. Mol. Biol.*, 2005, **345**, 21.
103. H. Chen, M. Ikeda-Saito and S. Shaik, *J. Am. Chem. Soc.*, 2008, **130**, 14778.
104. P. H. Sit, A. Migliore, M. H. Ho and M. L. Klein, *J. Chem. Theory Comput.*, 2010, **6**, 2896.
105. *Spin States in Biochemistry and Inorganic Chemistry: Influence on Structure and Reactivity*, ed. M. Swart and M. Costas, John Wiley & Sons, New York NY, 2015.
106. M. Sprik and G. Ciccotti, *J. Chem. Phys.*, 1998, **109**, 7737.
107. J. Hénin and C. Chipot, *J. Chem. Phys.*, 2004, **121**, 2904.
108. R. E. Bulo, B. Ensing, J. Sikkema and L. Visscher, *J. Chem. Theory. Comput.*, 2009, **5**, 2212.
109. M. Böckmann, N. L. Doltsinis and D. Marx, *J. Chem. Theory Comput.*, 2015, **11**, 2429.

CHAPTER 15

Enzyme Design

LUR ALONSO-COTCHICO, JAIME RODRÍGUEZ-GUERRA, AGUSTÍ LLEDÓS AND JEAN-DIDIER MARÉCHAL*

Departament de Química, Universitat Autònoma de Barcelona, 08193 Cerdanyola del Vallès, Spain
*Email: jeandidier.marechal@uab.cat

15.1 Introduction

Enzymes are one of the cornerstones of living organisms. They are catalysts that provide the catabolic and metabolic needs of the cells and the number of physiological processes they are involved in is legion. From their primary sequence to their expression, enzymes are one of the most important markers of evolution as they are modulated according to the necessities of cells. Amongst the examples that crystallise the importance of enzymes for living organisms is their involvement in frontline mechanisms of cellular defence. Cells would not be able to promptly respond to harmful stimuli generated by their surrounding (*i.e.* the absorption of pathogenic molecules) without a battery of efficient enzymes.[1]

Millions of years of evolution have led enzymes to display properties that are of the most complex to achieve by man-made catalysts. They have great substrate and catalytic selectivity and specificities. Enzymes tend to catalyse the reactions they are specifically designed for on a narrow scope of substrates and at a very well-defined position. They carry out their function with a minimum of by-products so that wastage of vital resources is reduced to a minimum and the formation of potentially harmful compounds is avoided to protect the organism. The particularity of enzymes in term of chemical specificity added to their capacities to work under mild conditions and water

RSC Theoretical and Computational Chemistry Series No. 9
Simulating Enzyme Reactivity: Computational Methods in Enzyme Catalysis
Edited by Iñaki Tuñón and Vicent Moliner

Published by the Royal Society of Chemistry, www.rsc.org

solvent have greatly interested chemical industries for years, especially in relation to their transition towards greener alternatives.[2–5]

Using biological catalysts to perform chemical transformations on organic compounds is known as biocatalysis. The list of biotechnological approaches to sustain successful biocatalysts is long and includes isolated enzymes, immobilised enzymes or living cells in culture.[3–5] The initial applications of biocatalysis could be considered more an art and ancient knowledge than a technology. Since prehistory, mankind has used biocatalytic processes to craft many different alimentary products, like beer, wine and cheese. It was not until the end of the 19th century and the early 20th centuries that biocatalysis was first used as a new technology based on scientific facts and rigour as illustrated by the hydrolysis of starch using the diastase enzyme.[6] Since then, natural enzymes have emerged as a potential alternative in many chemical processes. The most successful applications to date are likely but not uniquely acyl transferases (with the most famous examples being lipases and liases), carbohydrate processing enzymes (*i.e.* glycosydases), and enzymes able to perform C–O, C–N reduction and oxidations. Biocatalysis represents a very useful tool for controlling chemical reactions in terms of substrate and catalytic specificity with particular relevance for the production of enantiopure molecules.[2–5] However, limiting the field to enzymes that appear in Nature is inconceivable when considering biocatalysts as a key foundation in the transition towards green chemistry. Despite their wide spectrum of activities, natural enzymes only provide a narrow scope of the need of chemical industries. Expanding the scope of enzymology has become a hot topic at the interface between chemistry, biology and physics. Generating new enzymes both increases our technological knowledge as well as allowing us to learn from Nature's mechanisms and evolution.

Many molecular variables intervene in the design of a new biocatalyst (*i.e.* the efficient geometric disposition of a putative binding site for the reaction to occur, the interplay between enzyme candidates and their substrate, *etc.*). Their control by trial and error represents a tremendous challenge on experimental basis and it is undeniable that computation has a lot to offer to this field. From the prediction of single point mutations to the identification of the dynamical effects that affect transition state energies, computational tools exist. Nevertheless, when looking at the tremendous amount of work produced by computational means in the last decades at the interface between chemistry, biology and physics, *in silico* inputs for novel biocatalysts are still rather marginal. Despite the recent efforts that have been made, there is still a long way to go to reach standardised computational methodologies to lead to catalytically efficient *de novo* enzymes.

15.2 Scope and Objectives

This chapter aims to present how molecular modelling is involved in the design and modification of non-natural enzymes. The first part focuses on the main families of non-natural enzymes that are generated today, independently of whether computation is involved in their design or not. The second part focuses on general concepts of molecular modelling and how they could be specifically applied to the overall field of enzyme design. In order to avoid repetition of concepts covered in detail in previous chapters, only general ideas will be presented. The third part focuses on reviewing the most relevant applications of modelling in computer-aided enzyme design.

This chapter is aimed to be didactic and will not enter into the details of technicalities of the computational approaches carried out. The reader will be provided with a series of references that we considered illustrative for the aim of this book. Finally, aspects regarding works performed on enzymes to adapt them to industrial scales and non-conventional engineering, like enzymes supported on material and nanomaterials are not overviewed here but the reader could refer to several recent reviews.[7–10]

15.3 Man-made Enzymes

15.3.1 An Overview of Novel Enzymes

The concept of novel enzymes is broad and encompasses a vast panel of strategies. In function of the step of the design at which non-natural activity is reached, novel enzymes can be classified into two different families. To introduce them, a parallel between enzyme and drug design (far more widespread in the chemical biology community) could be drawn.

In drug design projects two phases exist: discovery and optimisation. Drug discovery corresponds to the identification of a hit compound, rationally or randomly, that displays some – and generally weak – affinity with the target. Then a series of successive hit-to-lead optimisations take place during which the initial hit compound is altered to improve its properties (like affinity to the target, solubility, bioavailability, *etc.*). This ultimately leads to the stage of clinical trials and characterisation of efficient and safe drugs. Enzyme designs could be classified following another dichotomy, which mainly depends on how the man-made enzymes diverge from those previously available at the beginning of the work. On one side, when designs are performed from scratch and lead to three dimensional folds absent from Nature or give naturally occurring folds completely different uses, the resulting enzymes are generally referred as *de novo* or artificial. On the other, when novel enzymes are obtained by tuning a catalytically competent scaffold, either natural or synthetic, towards more interesting chemical uses (modification of substrate, change in regio- or enantioselectivity profiles, stability, yield, *etc.*), they are generally referred as optimised or redesigned enzymes.

15.3.2 Tricking Nature's Enzymes

Given the wide range of advantages that enzymes offer, many fields soon set their sights on them including organic and fine chemistry.[3–5] Natural enzymes offer a very wide pool of activities with interesting applicability.[2] Only a small amount of enzymes display naturally promiscuous recognition or catalytic profiles and most have evolved to work efficiently under particular conditions and a narrow range of substrates.[11] This precisely represents a limitation when it comes to using them for chemical ends. What appeared beneficial for life results restrictive for chemical applications. As a consequence, enzyme designers intend to alter or develop new biocatalysts that diverge from their primary function.

Re-orienting the activity of natural enzymes is one of the most common approach to identify novel biocatalysts and has been the focus of attention of chemists and biotechnologists for several decades. Strategies have been already standardised and tremendous amounts of *savoir faire* have been spread. The two major ways for tailoring enzymes are rational redesign[12] and directed evolution (Figure 15.1).[13–15] The former is not far from any other molecular design strategy and basically introduces changes on the system based on structural criteria. Genetic alterations (mutations) are the most usual approach but chemical alterations can also be considered like the use of chemical crafting or non-canonical amino acids.[16] The latter is a two-step procedure that starts with a random, and generally massive, mutagenesis approach to obtain altered copies of the targeted enzyme. In the second step, those mutants pass through a selective pressure (in this case the desired reaction) and only the most efficient ones are selected. These mutants can be used as the starting point of a new iteration in the directed evolution scheme (Figure 15.1). The number of cycles will condition the quality of the final candidate.[15]

Both strategies have strengths and weaknesses. Modifications of the system under rational design are, generally, less costly than directed evolution. However, the condition *sine qua non* for these developments is to previously own structural knowledge of the system; a situation not necessarily accessible at the early stages of a design. Additionally, the exact impact of mutations on a biomolecular scaffold in terms of stability and/or reactivity is not necessarily easy to foresee even with state-of-the-art modelling techniques. On its side, directed evolution is a framework that eludes molecular information and intrinsically considers the viability and stability of the enzyme at the molecular and cellular level. This procedure is strongly dependent on the efforts invested on the mutagenesis and screening stages which could be extremely time and resource consuming.

With the advent structural biology, strategies that combine directed evolution and rational design are frequent. It generally consists of performing directed evolution on the amino acids that are at short distance from the reactive residues (*i.e.* 10 Å). When 3D structures are available on these species, random mutagenesis could be performed in a much more focused manner (Figure 15.1). This kind of procedure drastically reduces the size of the mutant libraries and the corresponding screening process with respect to directed evolution.[18] Increasing the probability to alter key amino acids in the active site regions also increases the chances of identifying cumulative

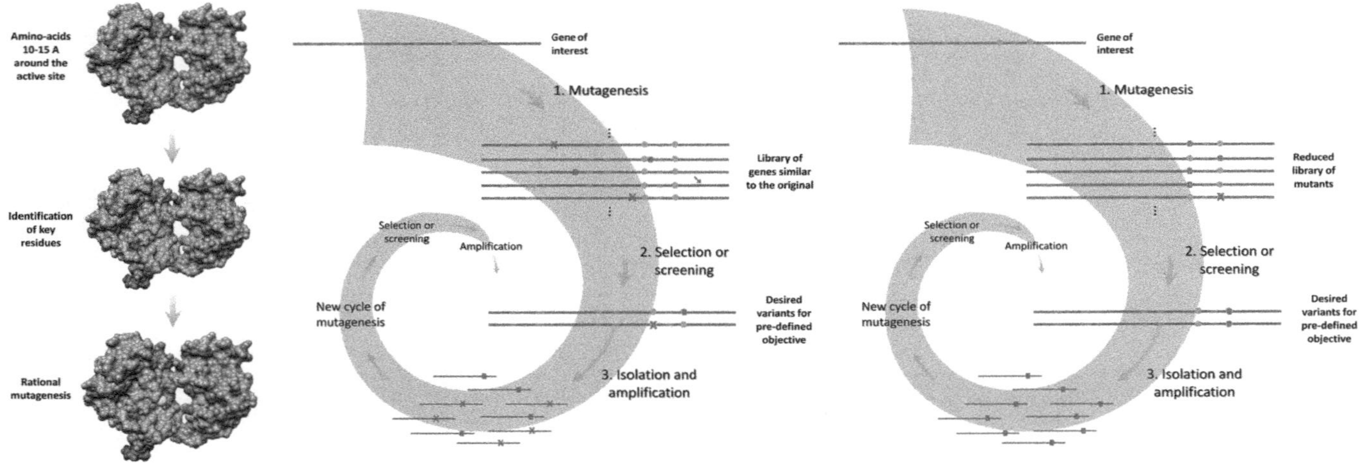

Figure 15.1 Processes for rational redesigning (left), directed evolution (middle) and semi-rational (right). Cre recombinase enzyme (PDB code: 3MGV) has been selected as an illustrative example.[17]

effects (*i.e.* cooperative effects between two different amino acids) that enhance the enzyme toward the desired reactivity. Despite summing advantages of both previous strategies, semi-rational approaches do not lack of drawbacks. Even if the size of the mutational space is reduced with respect to directed evolution, deciding what position to alter on the active site is not necessarily straight forward because the molecular factors that influence the catalytic efficiency are not limited to the amino acids in the very close vicinity of the reactive site; long-range interactions and cascade effects could be responsible of positive and negative effects.

15.3.3 New Folds for New Activities

Manipulating enzymes from the natural realm and tailoring them so that their activity meets chemical needs is a major challenge for structural biology and protein engineering. However, engineered enzymes generally remain with the activity that Nature evolved to and human modifications lead to changes on substrate specificity or eventually the regio- and enantiospecificity. Playing with enzyme promiscuity in terms of catalytic and substrate selectivities and specificities could be an interesting way forward (*i.e.* replacing a metal ion in a binding site of a metalloenzyme, discovering way to enhance a secondary catalytic route, *etc.*). However, such features still involve a reduced number of enzymes.[11] A second-order expansion of biocatalysis depends on our ability to identify non-natural biocatalytic scaffolds. The holy grail in this sense is the identification from scratch of a primary sequence, the subsequent expression of the biomolecule and the final characterisation of a stable fold desired catalytic activity and substrate, regio- and enantiospecificity and selectivity.

These ambitious *de novo* designs not only challenge our understanding of proteins and chemistry in aqueous medium but also oblige us to find strategies able to work with wide combinatorial spaces. Indeed, these designs need explorations of mutational, chemical and conformational spaces; something unreachable from scratch even under a computational framework built around the next generation of architectures and cutting edge software. Because enzyme designs from scratch are still inaccessible, the most common strategies rely on reducing the combinatorial space by gathering molecular rules from structural knowledge.

Some designers take advantage of structural rules observed in Nature and intend to reproduce them on systems of small dimension.[19,20] Most of these systems are built on lessons learned on peptidic scaffolds and/or by seizing on interactions between secondary structure elements. These systems are generally below a hundred residues and their final fold frequently takes advantage of well identified stabilising hydrophobic contacts (*i.e.* between α-helices) to generate auto-assembling folds (Figure 15.2). It is to mention that most of these systems tend to work in association with metallic cofactors.[21] The main interest is these designs is to be easily handled both in chemical and biological ways. On the counterpart, their structure is rather

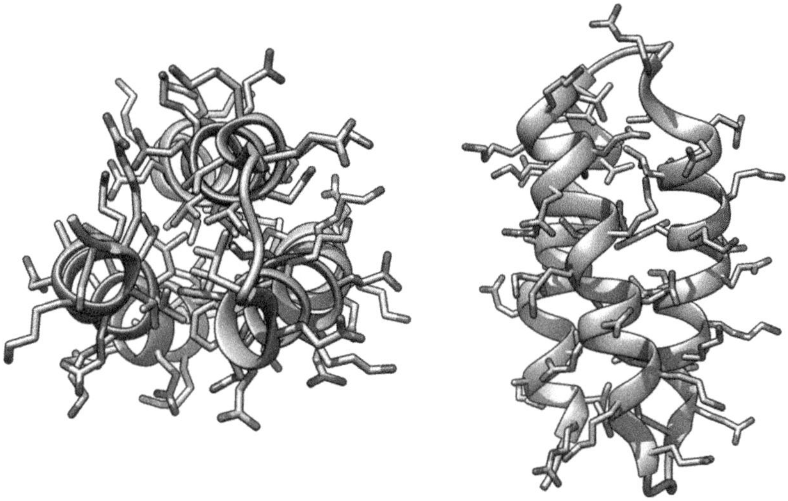

Figure 15.2 Side (left) and down (right) views of a *de novo* protein maquette (pdb code: 1lq7).[22]

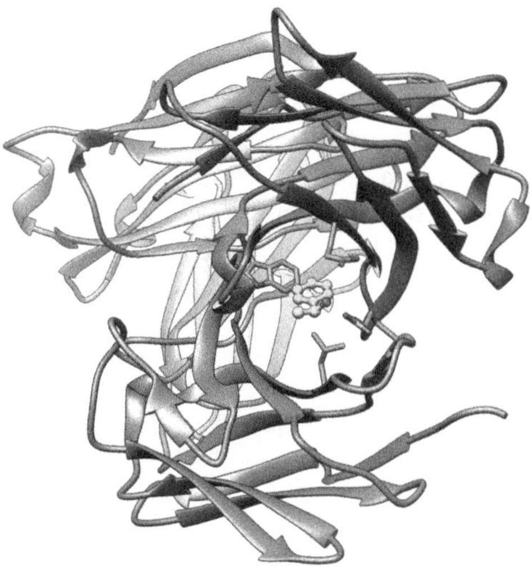

Figure 15.3 Overview of an abzyme (PDB code: 1UM5). Amino acids that directly interact with the substrate at the enzyme active site are represented as orange sticks and the substrate is represented as yellow balls and sticks.

compact and allow a reduced number of changes of substrate specificity with respect to scaffold of higher dimensions. In general, these systems have been built in absence of molecular modelling inputs.

Another pioneer strategy to generate artificial enzymes consists in generating antibodies against a chemical compound corresponding to a transition state analogue of a given reaction (Figure 15.3).[23] If successful, the resulting antibodies present a hapten site with efficient complementarity with the transition state structure and therefore own the main structural and physicochemical properties to assist in the catalysis. These so-called abzymes have been the object of intensive research for years but have shown a series of limitations with respect to other strategies, such as cost and time; obtaining monoclonal antibodies is a lengthy and delicate process. By gaining knowledge on abzymes, it has been demonstrated that the active site based on pure binding of TS analogue structures does not necessarily endorse the antibody with the sufficient electronic effect necessary for a reaction to take place.

Somehow reminiscent of the previous strategy, is the one that consists in spotting proteic scaffolds that under appropriate modification of their primary sequence could stabilise a transition state structure of a given reaction.[24] Protocols based on this idea are two-step processes with the first one being computationally based and the second experimental in nature. Regarding the computational part, first quantum chemistry calculations serve to identify transition state structures that can be reliable in a biological medium. These structures then serve to guide a wide screening of large datasets of macromolecules to find those that better satisfy the molecular requirements for the TS structures. The best candidates provided by this screening process are produced and tested experimentally (Figure 15.4). Iterative computational and/or experimental refinements can be performed at this stage in hopes of improving the final enzyme. These 'inside out' approaches have been mainly initiated and expanded by Kiss *et al.*[25] and will be further detailed in Section 15.5 of this chapter.

De novo enzymes present numerous advantages to respect to abzymes, the most important being that they are not limited to antibodies. Moreover, they are built by identifying complementarity with TS structures and not

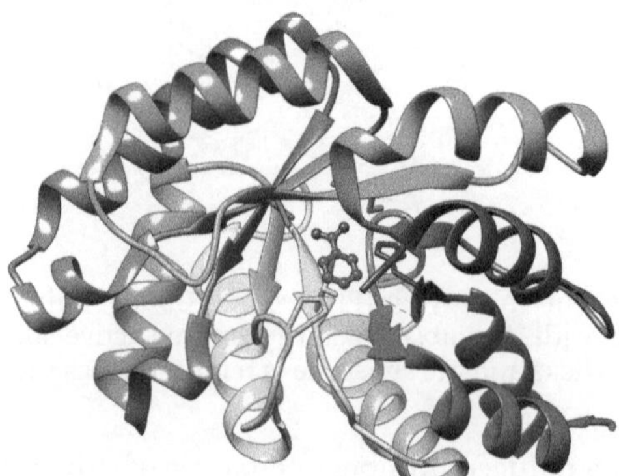

Figure 15.4 Kemp eliminase (KE) is one of the cornerstones of *de novo* enzyme design. Depicted in this figure together with substrate benzamidine, KE70 (PDB code: 3NQV) underwent nine rounds of directed evolution, resulting in 400-fold improvement in catalytic efficiency over the original design.[26]

analogues. This means that the selection of the best candidates considers explicitly the structural but also the physicochemical requisites for the reaction to take place under a biological media, not only the recognition parameters for the binding of a TS-like structure.

With about a third of all the genomes corresponding to metal-containing proteins, designers soon looked at generating non-natural enzymes that take advantage of the rich chemistry provided by transition metal ions. Several strategies exist but one in particular stands at the crossroads between *de novo* and redesign approaches. It consists in the alteration of the first and/or second coordination sphere of natural metalloenzymes so that the electronic properties of metal could be tuned, the same metal exchanged by another or the binding site modelled to access novel substrate or regiospecificities. Those strategies are interesting because the overall fold of the macromolecule is not drastically affected and the overall stability is not in danger. Moreover, the use of non-canonical amino acids or the involvement of supporting metal coordinating ligands can help to modulate the geometry and behaviour of the novel enzyme.[27] Of the most recent and impressive successes obtained with this strategy are the redesign of a blue copper azurin for superoxide scavenging[28] and the increase of oxidase activity by incorporating a tyrosine–histidine crosslink in a myoglobin model of haem-copper oxidase.[29]

15.3.4 Bringing Homogenous Catalysts into the Game

Another approach has been gaining interest during the last decades and consists in the insertion of coordination complexes into macromolecular cavities.[30] The resulting hybrids also known as artificial metalloenzymes (ArM) merge the best of both homogeneous and enzymatic catalysis (Figure 15.5).[31]

ArM are biohybrids conceptually reminiscent of naturally occurring hemoenzymes. While the coordination complex brings most of the catalytic features to the biohybrid, the proteic scaffold constitutes an asymmetric environment that provides a protective embedding of the complex and also dictates substrate selectivity and specificities.[33] Moreover, its role as a second coordination sphere is key for stabilising the transition state structure and dictates enantioselectivities (Figure 15.6). As homogeneous catalysis is such a vivid field of research and is so widely used in organic synthesis, the number of these biohybrids is rapidly growing. The activities obtained with this strategy already include from Diels–Alder reactions to oxidation processes and, in some cases, impressive enantioselective profiles have been reached (Table 15.1).[30,34] Only in the latter months, three prototypical organometallic reactions have been reached under this framework: Suzuki–Miyaura cross-couplings,[35] olefin metathesis[36,37] and olefin cyclopropanation.[38]

The main challenge in this field is to identify the right partners so that the biomolecule, substrate and artificial cofactor can behave in synergy. Somehow, one must realise that this approach depends on an unnatural triad for

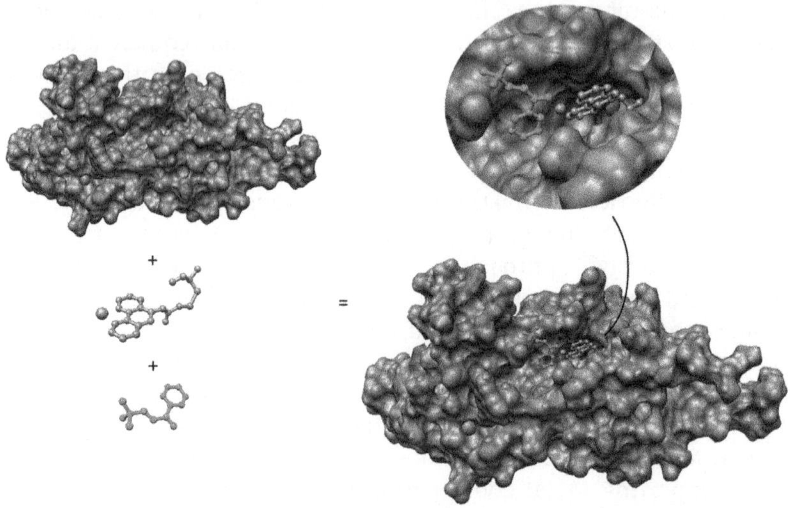

Figure 15.5 Key players for the synthesis of an artificial enzyme. A proper selection of the biomolecular scaffold (in blue), the cofactor (in orange) and the substrate (in green) must be carried out in order to obtain an effective triad fitting (at right side) which allows the performance of a predefined reaction. Drawing inspired by the work of Roelfes and coworkers.[32]

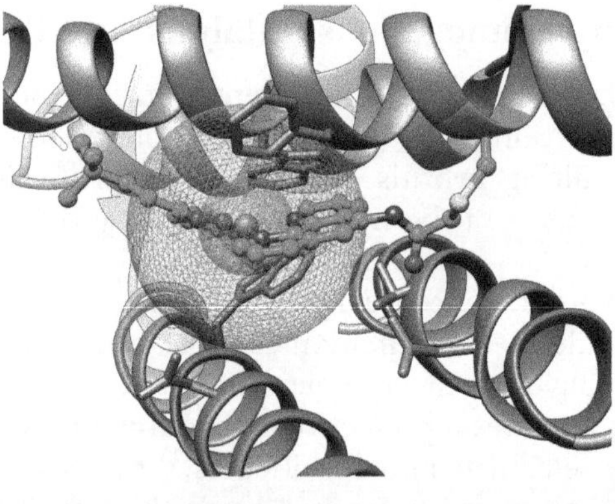

Figure 15.6 First (in orange) and second (in green) coordination spheres around a copper-containing organometallic cofactor coordinated to its substrate. The example illustrates the LmrR-Phenanthroline-Cu(II) artificial metalloenzymes designed by Roelfes' group.[32]

which none of the components have been naturally optimised by evolution. To overcome this difficulty most designers tend to use highly stable proteins and DNA strands[44] that are easy to express. The stability of the host is fundamental because its fold must resist the insertion of a coordination complex whose geometry could be far from the natural compounds it normally binds to. The design starts generally on molecular grounds using a previously resolved 3D structure for which accessible cavities are very well

Table 15.1 Non-extensive list of ArMs developed on the insertion of homogenous catalysts into biomolecular hosts.

Biomolecular scaffold	Organic cofactor	Transition metal	Substrate	Catalytic activity	ee	Enantiomer	Ref.
NikA transport protein	Organic ligand L1	Iron(III)	Sulfides	Oxidation	10		39
LmrR	Phenanthroline	Copper(II)	Ketones	Hydration	87	?	
			Azachalcone	Diels–Alder reaction	>97	+	32
β-Helical bio-nanotube	Bipyridine, Ser, Thr	Scandium(III)	*cis*-Stilbene oxide	Epoxide ring-opening reaction	17	*R*	40
Streptavidin mutants	Quinidine or quinine derivatives	Osmium(VIII)	Olefins	Asymmetric dihydroxylation	95	*R*	41
	Biotinylated complex	Iridium(III)	Cyclic imines	Hydrogenation	96, 78	*RS*	42
DNA	DNA intercalating moiety		Dienophiles	Diels–Alder reaction	90	*exo*	43
Prolyl oligopeptidase	Alkyne-substituted	Rhodium(II)	Olefins	Cyclopropanation	92	*exo*	38
b-Barrel protein nitrobidin	Grubbs–Hoveyda type catalyst	Ruthenium	Olefins	Metathesis	70, 30	*cis*/*trans*	37
Carbonic anhydrase II	Grubbs–Hoveyda type catalyst arylsulfonamide	Ruthenium	Olefins	Ring-closing metathesis	—	—	36
Streptavidin	Biotinylated cofactor	Palladium	2-Methoxy-binaphthyl	Suzuki–Miyaura cross-coupling	90	*R*	35

identified. Even though, the first generation of these systems requires several runs of post-optimisation that could involve deep changes in both the chemical and biological nature of some parts of the receptor–cofactor–substrate triad (*i.e.* covalent binding between the homogenous catalyst and the host, mutation of the receptor to satisfy strong polar interaction with the artificial cofactor, *etc.*).

To date, computational chemistry has been little applied on these systems in part because finding the recipe for the best host–cofactor–substrate is highly dependent on the nature of the coordination compound, but also because these non-natural complementarities push the limits of our bioinorganic knowledge.[34] Both chemical and biochemical intuitions are challenged and it is frequent to observe that the resulting ArM present molecular behaviours different to what designers could expect.[45] In a sense, a lot of fundamental work on these designed enzymes needs to be done to develop newer design strategies and improve the existing ones.

15.4 Computational Tools and Designed Enzymes

With the expansion of computational chemistry and bioinformatics, modellers can now make use of a wide spectrum of methods in their toolbox. Several chapters of this book presented the grounds of molecular modelling. For details, in-depth explanation, examples and references, the reader should refer to these chapters. Here we summarise the most important concepts of the methodologies that molecular modelling encompasses and aim to answer questions like: How does computation fit into the general needs of enzyme design? What would be the molecular variables we should take into account to generate a computational framework that covers the necessity in studying, designing or optimising novel enzymes?

15.4.1 Accuracy *vs.* Sampling

At this stage, either if the reader already has previous knowledge of computational chemistry or is discovering this field by reading this book, he or she should have grasp one of its most important aspects: accuracy is intrinsically linked to computational power and both condition the dimensionality of the system one could study. In a way, quantum mechanical and molecular mechanics approaches crystallise this idea very well.

The physicochemical models used to describe atoms in the modelling of biochemical systems are basically divided in two families: quantum mechanics (QM) and molecular mechanics (MM). A way to summarise their differences could be that QM methods deal explicitly with both electrons and nuclei while MM stands on a description of atoms as unified objects with simplified (but not necessarily simple) sets of parameters for both covalent and non-covalent interactions between them – the force field. At least two major differences of QM and MM have to be mentioned to understand how they apply to the context of enzyme design. On one side, the classical representation of atoms and bonds in standard MM methods do not allow to deal with chemical reactivity. The force field approximation also entails a series of technical simplifications that allow MM methods to solve the energy of a given system using several orders of magnitude fewer resources than the QM ones. On the other, QM methods and more particularly DFT and *ab initio* ones, are fundamental when accurate electronic representation is needed like in the case of simulating catalytic processes. Despite increasing efforts from the community in generating quantum-biased reactive force fields so that wider sampling could be reached, only a few are available.[46]

When it comes to identifying *in silico* reactive novel enzymes it is vital to consider which level of accuracy for energies is needed, what amount of sampling is required to study the molecular phenomenon and what could be the computational cost in term of resources and/or methodological development.

15.4.2 Reactivity

Speaking of enzymes, the key element that computation should address is the enzymatic reaction itself. The most common idea about why biocatalysts are so efficient is that the proteic scaffold decreases the energy of the transition state that leads from reactant to product. Several molecular reasons could be key in dictating this effect and include electrostatics, better pre-organisation and/or entropic effects.[47–49] Despite years of intensive study, the evidence of one predominant effect has not yet reached unanimity.

To study enzymatic reactions, calculations are carried out using quantum-based methods and two different approaches can be adopted. The first corresponds to QM calculations on a reduced model of the enzyme (Figure 15.7). By reducing the size of the system, it becomes feasible to calculate reaction paths in a reasonable amount of time and with a reduced amount of resources still using high level of theory like functional density theory (DFT) or post-Hartree Fock methods (post-HF). These so-called cluster models focus on the reactive part of the enzyme and thanks to the raise of computational power, their size have been constantly growing.[50,51] Removing the full protein environment around the active site has to be taken with caution. Modellers have to avoid that their calculations (typically optimisations) explore regions of the potential energy surface (PES) that would not be accessible in the real size system and maintain the electronic and steric coherence of the clusters with the crystal structure of the enzyme.

The other main approach considers the full size enzyme and solves the calculation of the energy under a hybrid QM/MM scheme (Figure 15.7). In these frameworks, the reactive region of the enzyme is modelled using QM methods while the rest of the system is studied using a MM approximation. The way the interaction between both parts is described is the main difference between the different QM/MM methods, as well as the force field and

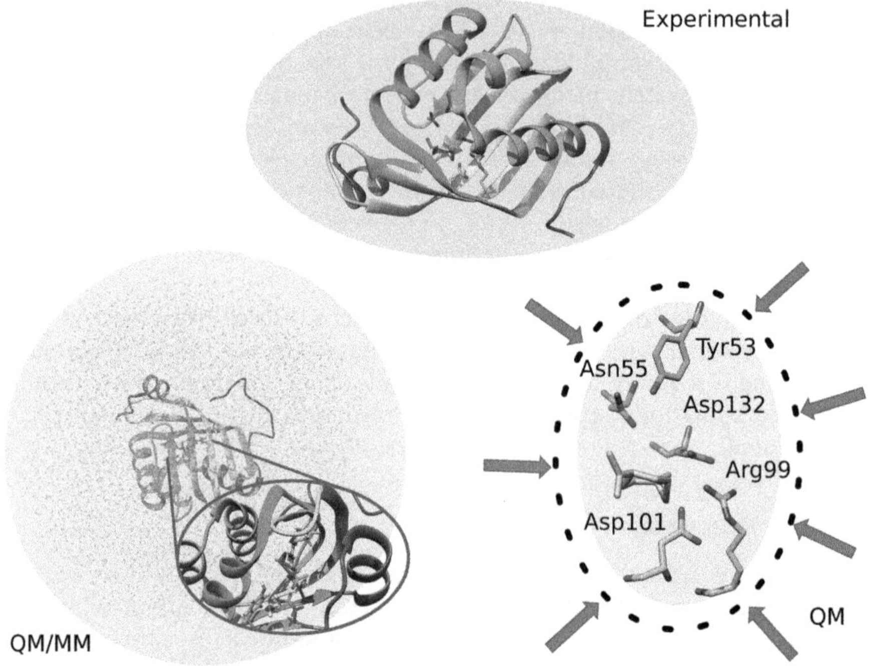

Figure 15.7 A general description of QM *versus* QM/MM view on enzyme modelling. The example is taken on the limonene-1,2-hydratase and inspired by the work of Lind and coworkers for the QM52 model and Hou and coworkers for the QM/MM description.[53]

electronic methods available for each part.[54] A frequent misconception regarding QM/MM calculation is to believe that introducing MM terms into the calculation reduces the computation cost. In reality, what defines the cost of exploration of the potential or free energy landscapes is the quality of the calculation involved in the QM region of the molecule. The cheaper the QM methods or the smaller the QM part, the most extensive could be the QM/MM sampling along the chemical reaction.

However, QM/MM methods present an interesting advantage with respect to cluster models as they enable evaluating how the protein participates along the reaction in terms of organisation and electrostatic effects. When selecting computationally demanding methods of the QM part, typically DFT, sampling along the reaction is reduced to the minimum and the simulation tends to be reduced to the exploration of the PES. This approach generally gives answers with chemical sense but could be too limited when it comes to reach quantitative effects related to the sampling of the system during the simulation of the chemical reactions and evaluating entropic contributions. When aiming at the calculation of free energy values, high level description by QM calculations have to be avoided. Reducing the cost of the calculation of the energy without substantial loss of quality in the electronic accuracy has been the focus of many works and clever solutions are available that include EVB theory,[55] semi-empirical like DFT methods as SCC-DFTB[56] and Classical Dynamics biased QM techniques like Carr–Parrinello.[57] The sampling reachable by these QM based approaches could be fundamental to access several effects but still remains quite limited when compared with many biochemical events of wider scale and important for enzyme design.

15.4.3 Substrate Binding

One of the most important variables in modelling enzymes is the identification of catalytically competent enzyme–substrate complexes (or protein–cofactor–substrate in the case of ArM). The complexity to reach such structures depends on the structural knowledge available at the starting point of the study. Experimental structures obtained with NMR or X-rays rarely correspond to geometries with a bound substrate. In the best case scenario, an inhibitor or a transition state analogue (TSA) forming a Michaelis-like complex is bound into the binding site. In those cases, simulations with QM and/or QM/MM methods could be initiated rather easily by locating the substrate 'by hand' to reach an orientation that satisfies the catalytic requirements (*i.e.* distance between reactive atoms of the substrate and known reactive residues, orientation consistency with enantiospecificity, *etc.*).

When information on the location and orientation of the substrate is less clear, like it happens in *in silico* predictions for synthetic enzymes, protein–ligand dockings represent an interesting way forward. Protein–ligand dockings are techniques that identify low energy geometries in the conformational space formed by the chemical compound(s) and the binding site cavity of a protein. In most protein–ligand docking techniques, global translation and rotation of the ligand, some of its internal deformations as well as a limited amount of the reorganisation of the protein upon binding are explored. In general, those calculations are performed on a well located region of the protein (the active site).

Each protein–ligand docking program allows to deal with the problem of the high dimensionality thanks to two key elements: an explorative algorithm optimised for wide conformational search spaces and a simplified evaluation of the energy. The first is generally based on random optimisers, the most frequent being Monte Carlo and Genetic Algorithms.[58] The second is a higher level of simplification in force field description of molecules and correspond to so-called scoring functions for which the energy of the system is lowered down to functions that only account for non-covalent interactions.[58] To conveniently sample the search space, protein–ligand dockings needs several runs of millions of explorations. In the case of enzymes, docking rarely provides with a unique low energy orientation of the substrate inside the binding site and the eye of the modeller is fundamental to verify the chemical sense of the results.

Protein–ligand docking algorithms have been primarily designed to identify low energy complexes in a short amount of time. One of their main objectives is the screening of massive libraries to detect drug candidates. As such, their main application to the enzymatic world is the identification of inhibitors. But the design of enzyme requires considering substrates not inhibitors and in this case dockings require to elucidate other aspects than pure binding affinity. A common mistake is to consider that the complex with the lowest energy should be the reactive one. In fact, it is frequent that this complex present orientations with extremely good affinity but with no chemical sense in terms of reactivity. Executing and analysing docking essays with focus towards catalytic mechanism needs the formulation of the correct molecular assumptions, chemical sense and technical *savoir faire*.

One of the most useful ways to adapt docking techniques to the field of enzymology consists in imposing structural features related to the reaction during the explorative runs. Those constraints are generally distances between reactive atoms of the substrate and those of the protein. Another way consists in performing very extensive docking runs so that a large number of low energy complexes are generated and finally analysing them to reveal where the substrate is oriented in the correct manner for the reaction to proceed (Figure 15.8). This kind of hands-off procedures has been particularly useful for decoding the mechanism of structural and/or catalytic promiscuous enzymes like metabolising cytochromes P450[59] or lipooxygenases.[60]

Another important aspect regarding standard protein–ligand dockings is that they have sense when only a reduced amount of sampling of the protein is involved. That means that their effectiveness is maximum when the

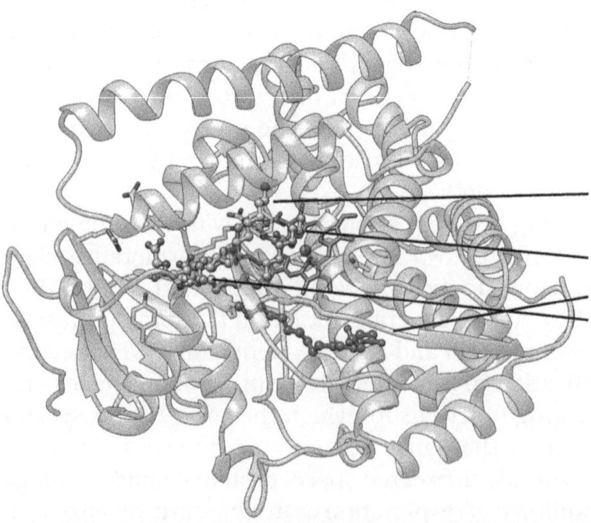

#	Score	Catalytically consistent
1	30.64	NO
2	30.99	YES
3	31.89	NO
4	32.16	NO
5	32.20	YES
6	32.37	NO
7	32.56	NO
8	32.57	YES
9	32.58	YES
10	32.69	NO

Figure 15.8 A schematic representation of how several binding modes of low energy could be classified under the limit of precision of protein-ligand docking scoring functions but have different catalytic implication.

structure of the protein is pre-organised enough for catalysis and only local changes of the binding site may occur. The farther is the structure of the enzyme from a geometry convenient for the catalytic process, the harder it will become to identify the substrate well posed and oriented. In general, docking approaches only consider a limited flexibility of the enzyme with some backbone configuration and local changes of a selected number of side chain (generally under a rotameric approach). Other conformational samplings may be needed when it comes to work on designing novel enzymes.

In most cases, sampling of large scale motions is performed and aimed at guiding the systems towards convenient structures. One way to move forward consists in allying MD with protein–ligand dockings. Those calculations could be run in two ways. One kind of calculations consists in running long MD runs and subsequently performing the docking on a series of discrete representative structures.[61] This approach is particularly interesting if the experimental structure of the enzyme requires major conformational changes to reach a substrate bound one. Another kind of approaches consists in accelerating the exploration of the protein–ligand landscape and somehow modelling the entrance of the substrate into the protein. A family of these techniques considered steered MD dynamics with guiding restraints to fly over the energetic landscape.[62,63] Another kind of alternatives consider normal mode displacements on the backbone to insert collective motions into an advanced docking protocol.[64–66]

15.4.4 Folds

All the techniques mentioned in the previous sections are applicable when a 3D structure of the enzyme on which stands the modelling presents its native fold and only local or collective changes are required for the reaction to occur (*i.e.* those corresponding to the binding of the substrate). However, one of the cornerstones in enzyme design is the identification of new folds or mutants that are absent from Nature. To date, computational approaches to simulate the entire folding path of an enzyme from its primary sequence cannot be achieved under a routine protocol. Despite the increase in explorative algorithms and their implementation on new generations of processor architectures, *ab initio* folding are still limited to small proteins and peptides.[67]

When dealing with folds of higher dimension, one possible solution is to use homology modelling (HM).[68] These methods are useful when the enzyme of interest (the target) has one or various family member(s) whose structure(s) have been reported previously in the Protein Data Bank (the templates).[69] A common protocol for HM, like the one used by the popular package Modeller[68] stands on a sequential strategy that includes (1) a fold identification based on sequence similarity between the target and structures that have already been resolved, (2) an alignment between the sequence of the target and the sequence of the template(s), (3) the structural building of 3D models of the target based on the structure of the template(s) beginning by the overall backbone, the side chains and the modelling of the loops and finally (4) the identification of possible errors in the structure of the model and their subsequent elimination by filtering and/or refinements (*i.e.* energy minimisation or molecular dynamics).[68] Despite being wide spread for modelling protein structures, HM methodology shows its limitations when sequence identity between the template and the target is lower than 25–30%. Below this threshold, the sequence of the target may not find any correspondence with the template(s). Generally centred in regions of the loops, those regions need to be modelled by *ab initio* prediction. All HM platforms tend to include loop optimisers whose accuracy reaches it limit for loops of 7–8 residues length.[70]

Another strategy consists in generating the entire fold of the target by assembling building blocks that correspond to regions of the primary sequence. These blocks can be of different lengths and pre-established from structural statistics on the Protein Data Bank. The way to assemble the different fragments to generate the final full size model (*i.e.* Monte Carlo, Genetic algorithms, *etc.*) and how to evaluate its energy is what differentiates these folding protocols. The Rosetta package and its particular server version for folding prediction, Robetta, is probably of the most successful strategies to date in *de novo* fold generation.[71,72] The use of structurally meaningful fragments allows to drastically reduce the conformational space exploration with respect to *ab initio* techniques and it is not as limitative as homology modelling approaches.

15.4.5 Chemogenetic Spaces

The chemogenetic space comprises all those modifications and/or insertions that alter the enzyme amino acid sequence and those that add, replace or remove substituents on small molecules, such as cofactors or substrates of such enzyme. Since each atomic modification has an effect on both the chemical and structural spaces – replacing a glycine residue with a lysine not only affects the polarity of its environment, but also adds a considerable steric bulk to that region – considering all possible changes at once can be overwhelming (Figure 15.9). The accurate prediction of the effect of mutations or chemical changes of an enzyme in terms of binding and catalytic processes needs high level of theory as it should account with fine electrostatic and sampling modelling. However, wide chemogenetic space explorations are not accessible with these methods and are better achieved with empirical approaches.

In this sense, the enzyme designer can resort to some intelligent approaches. Instead of relying on systematic torsions on the rotatable bonds of each residue, conformational variability of side chains can be studied with rotameric libraries, which offer a curated selection of torsion angles for each residue after studying the most common occurrences in experimentally characterised structures.[73,74] The designer can also take advantage of these libraries to mutate the selected residues to other amino acids, whose side

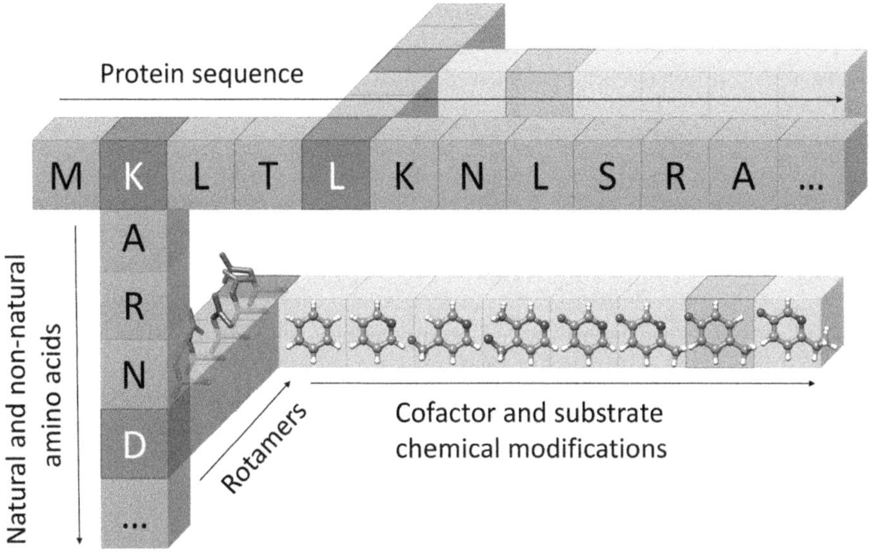

Figure 15.9 The chemogenetic space is highly dimensional and considers combinatorial and conformational spaces for substrate, residues (canonical or not) and cofactors. For example, for each residue position, a huge number of conformational configurations need to be explored. Reducing the search space can be reached by discrete rotameric states.

chains will be optimised given the original backbone dihedral angles. This strategy could be expanded to non-canonical residues[75] for which interest are constantly growing including for specialised force fields such as NCAA or PTM implementations of the AMBER force field.[76,77] When it comes to the insertion additional residues in the protein, loop optimisation protocols as those provided by homology modelling programs (*i.e.* Modeller or Rosetta) can be useful.

Regarding the chemical modification of small molecules, an usual approach consists in recombining essential building blocks to create novel compounds, usually with the help of evolutionary algorithms to guide the process[78] or by mimicking chemical synthesis pathways.[79,80]

15.4.6 Multi-scale

Let's imagine a computational framework dedicated to the design of *de novo* enzymes. It could account with (1) identification of new folds consistent with the binding of substrate, (2) identification of substrate (and/or cofactor) bound structures in a catalytically competent manner, (3) identification of the energy barrier of the transition state, and (4) prediction of the alteration of the energy barrier by changes in the protein environment, the nature of the substrate or an eventual cofactor. Each of the methodologies we mentioned in the previous part of the chapter could be extremely relevant for one or several aspects but not for all. The key point is that wide conformational or biochemical searches are incompatible with accurate, and consequently expensive, methods. Depending on which stage of the design computation is involved, the precision of the calculation of the energy and the amount of combinational and conformational space to explore could differ.

For this reason, many strategies to study or help at the design of non-natural enzymes stand on the combination of several computational approaches (Figure 15.10). Based on the same ideas than those that lead Warshel, Karplus and Levitt to their Nobel Prize in chemistry in 2013, multi-scale (also known as multilevel, integrative or hierarchical) protocols offer very interesting alternatives to work on enzyme design. In fact, the most successful *in silico* strategies like those designed by Houk, Mayo and Backer are already based on this ground, which starts with quantum mechanical calculations, followed by massive screening of folds in Nature, looking for those that, conveniently modified, could host the desired transition state geometry.

One of the major problems with multi-scale strategies is to reach and efficient flow of inputs/outputs exchange of information between programs. In this sense commercial suites are likely the best candidates when heading toward multi-scale strategies as they are sustained by a series of developers

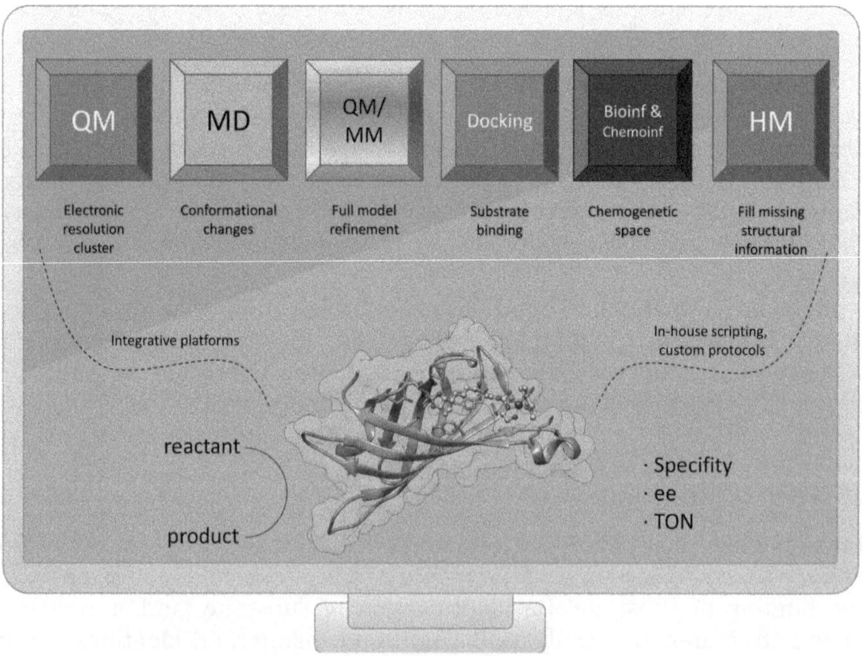

Figure 15.10 The computational enzyme designer has a number of utilities and techniques at hand. Individually, each one serves a very specific purpose, but together they can act as an integrative toolkit that can transform mere hypothesis in actual designs with catalytic properties.

whose efforts are dedicated to optimise the program and the workflow between the different units of the package. However, the coupling between non-commercial software and more particularly those with academic licenses is becoming increasingly easier thanks to the growing presence of glue code written in languages like Python or Ruby, or the embedding of older projects in Python interfaces. We can only highly reckon for those interested in studying artificial enzymes to dedicate time in their formation to scripting procedures. While enzyme design demands expertise on individual methodologies, solid grounds on complementary areas are much welcome. It comes to the creativity of the researcher to appreciate what combination of tools is the most convenient for the task at hand and how to build an adequate platform to execute it.

15.5 Applications

The design of novel enzymes is a boiling field of research and computation constantly demonstrate its suitability to help and guide enzyme designers. Here we overview cases that we believe are clear examples of this statement. Some may be classified as proof-of-concepts while others are running strategies that have already spread over the community.

15.5.1 *De novo* Enzymes

15.5.1.1 Background

Actual rational design of enzymes from scratch is far from reachable nowadays. One cannot simply enumerate a series of desired features and catalytic properties and expect to obtain a novel protein sequence ready for synthesis. The current state-of-the-art techniques requires previous knowledge of the chemical reaction to be catalysed: the transition state and the chemostructural conditions that support its stabilisation.[24] This information is then gathered in a reduced model of the active site called 'theozyme',[81] consisting of amino acids located at their optimal geometry around the transition state compound. QM calculations are pivotal at this initial stage.

Once the theozyme is modelled, it must be accommodated in a protein scaffold. This is done with a search algorithm that tries to match the active site coordinates to an already characterised structure, such as an entry of the Protein Data Bank database.[82] A handful of programs have been developed for this task, such as Dezymer,[83] ORBIT,[84] RosettaMatch,[85] OptGraft,[86] Scaffold-Selection,[87] PRODA_MATCH,[88] or SABER.[89]

Amongst all of them, RossetaMatch is probably the most popular. Part of the Rosetta Modelling suite, this module finds suitable scaffolds in a sequential process. First, a set of potentially binding scaffolds is searched for backbone positions that can hold the transition model and the disembodied residues in an adequate orientation. For this step, two methods are

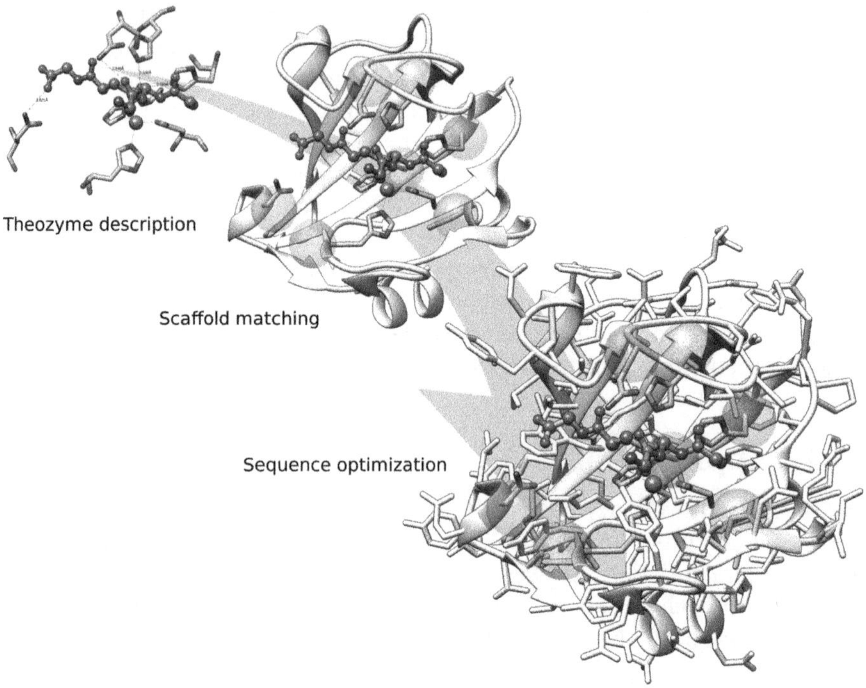

Figure 15.11 Rosetta design protocols usually start with a model of the transition state compound and its surroundings, called 'theozyme'. This is later accommodated in a suitable scaffold protein by an inverse rotamer lookup (inside-out strategy) or the RosettaMatch algorithm (outside-in strategy). A number of consensus matches is returned.[91]

provided: an inverse rotamer lookup (à la Mayo's ORBIT) and an extension of the MetalSearch algorithm.[90] Next step takes the previously obtained results and refine them to minimise steric clashes and optimise the catalytic geometry (Figure 15.11).

15.5.1.2 Remarkable Events of Success

The first big step in *de novo* enzyme design was taken in 2001, when Mayo *et al.* transformed a catalytically inert *E. coli* thioredoxin into an esterase 'protozyme' (enzyme-like protein) by computational design.[92] They identified two promising catalytic positions and a set of stabilising mutations that allowed 180-fold rate acceleration for a *p*-nitrophenyl acetate hydrolysis reaction. This big achievement was later followed by three main victories in *de novo* enzyme design, accomplished by Baker and coworkers,[93–95] whose common feature is being able to catalyse a reaction with low or non-existing representation in the enzymatic world.[96]

Kemp eliminase. Being a not very complex and well-studied reaction, with no known enzymatic catalyst, the Kemp elimination is usually considered one of the benchmarks for enzyme design. The first *de novo* design of a Kemp eliminase was obtained with Rosetta,[93] which returned 59 models ready for experimental characterisation. Of those, eight models were active and one of them was resolved by X-ray diffraction analysis, whose structure superimposed well with the computational model. Interestingly enough, these computational designs did not get close to the efficiency of previously obtained abzymes for Kemp elimination, but some of them did serve as excellent starting points for directed evolution experiments. In another study, one of the results, KE70, worked great for computational refinement and further directed evolution.[26]

Retro-Aldolase. This reaction involves breaking a C–C bond in several steps, which results in higher complexity of design, due to the active site having to support different transition states and intermediates. In 2008, Baker *et al.* presented a design that resulted in 72 significative models for experimental validation, 32 of which showed catalytic activity.[94] The design started considering four different models of the active site, instead of a single one. While the number of catalytically active models was remarkably high, the activities were two or three orders of magnitude lower than the abzymatic equivalents. Detailed inspection on the models revealed the limitations of the design: the protein itself had an effect on the structure of the intermediates, highlighting the drawbacks of rigid backbone assumptions. When the *de novo* designs were submitted to directed evolution, the specific activity increased 4400-fold, getting close to natural enzyme efficiencies.[97] Later on, another set of directed evolution essays gave a boost of over 1000-fold on the previous stage.[98] However, when the structures were assessed crystalographically, the binding sites revealed dramatic changes with respect to the original design.

Diels–Alderase. After studying the QM details of a bimolecular Diels–Alder cycloaddition, Siegel *et al.* were able to report a high stereoselectivity and substrate specificity *de novo* design in 2010.[95] This work produced two designs with catalytic activity, which were both refined with several rounds of directed evolution resulting in one particular macromolecule with 100-fold activity improvement over the initial version.

15.5.1.3 *Difficulties and Perspectives*

While Baker and coworkers have been very successful using Rosetta for *de novo* enzyme design, there have been a lot of unsuccessful attempts. One of the most known failures is the design of a triosephosphate isomerase, retractions included.[99] Though there is no clear feedback as why these have not succeeded, one may wonder why the others did happen. Kemp elimination can be promoted easily by any base, and aldol reactions can be accelerated by common prolines.[100] That may mean that obtaining designs for these reactions might be as easy as applying some rational redesign to rudimentary constructions. As a matter of fact, the idea has been explored by several authors, and some results indicated that the activities and yields were in the same order of magnitude as the first generations of Rosetta-generated models.[101–104]

Even though, one cannot deny that computation algorithms do improve the chances of successful designs. Thus the main issue may have to do with the robustness of protocols. This is evidenced when shotgun approaches are needed due to the time-accuracy trade-offs designers are forced to apply (fixed backbone, lack of long-range interactions, simplified force fields...) and several rounds of directed evolution refinements are almost compulsory. The situation is expected to improve in the years to come, as computational resources become more and more powerful. In fact, some of those expected updates may be already here: RosettaRemodel supports flexible backbones,[105] and new possibilities of sequence optimisation are allowed with non-canonical residue insertion.[106,107] Furthermore, multi-scale design protocols may be the next big thing if applied successfully, since they will consider an ensemble of possible conformational and chemical states instead of a single, fixed scaffold template.[108]

In the meantime, the designer is expected to apply ulterior refinements to the initial results of the in silico design. Even failed designs can be rescued and converted into high performant models if the failure is carefully inspected and rationalised. Mayo *et al.* propose an iterative protocol that insists on paying attention to initially inactive designs in hopes of not discarding useful data implicitly contained in such models.[109] It has been reported very useful in rescuing a previously inactive Kemp eliminase, HG-1. After understanding the root cause of such inactivity with MD simulations and X-ray crystallography, they introduced changes in the interior of the protein, obtaining an already active model, HG-2. Further iterations on this scheme led to HG-3, which produced a 3-fold improvement in activity.[110]

Another approach insists on the computational side of the optimisation, but using a very different strategy. Instead of programming complex algorithms to mimic the atomic interactions of molecular design, they harness the full potential of our human problem solving and intuition skills in a multiplayer online game called FoldIt, where the players are expected to solve folding problems presented as a puzzles.[111] Applied to enzyme design, it has been successful in improving the *in silico* design of the Diels–Alderase DA_20_10 with an outstanding 18-fold activity increase.[112]

15.5.2 Redesigning, Optimising and Filtering Enzymes

The rational redesign of naturally occurring enzymes or the improvement of *de novo* ones stand on decoding crucial aspects of their molecular mechanism. At this stage, computation can be descriptive in nature or intend to filter convenient candidates from ones with low activity. Generally, more than one method is involved. Still, we decided to separate our illustrative cases in discrete sections identifying that some methods are more relevant than others in specific cases. This may be sometime an empiric decision but without doubt it eases the flow of reading.

15.5.2.1 Protein–Ligand Dockings and HM

Understanding how the substrate orients in the active pocket of a putative enzyme is an indispensable step in rational (re-)designs. Protein–ligand dockings have been applied in several interesting cases, generally at the early stage of a project and in combination with experimental works.

A particularly successful example is the work of Savile and coworkers that performed several docking experiments on a homology model of the R-selective transaminase ATA-117 (a homologous of a transaminase from *Arthrobacter* sp.). Based on the *in silico* insights, the authors were able to alter the recognition properties of the enzyme so that it accepts wider substrates but perform the desired transamination. Combined with directed evolution, the final biocatalyst was able to perform at manufacturing scale the transmination of prositagliptin ketone to form with a 99.95% ee the antidiabetic stagliptin drug. Interestingly, this biocatalyst is able to compete with a recently implemented at industrial case rhodium-catalysed asymmetric enamine hydrogenation strategy.[113]

Another successful work where docking is a determinant element is the strategy reported by Lippow *et al.* that consists in a computational/experimental approach using a library-based enzyme screening. After docking of D-glucose into a glucose 6-oxidase and exploring convenient mutations with Rosetta, the subsequent experimental screening identified mutants approximately 400-fold more active than the wild-type enzyme in the transformation of D-glucose to D-glutaric acid.[114]

Protein–ligand dockings are generally part of major multidisciplinary efforts and can have a discrete but pivotal role. Amongst studies along this way are the engineering of Cytochrome P450 (CYP) BM3 from *Bacillus megaterium* to alter its substrate and catalytic specificity. Cytochromes P450 are monooxygenases with biotechnological potential since they are able to perform a wide range of oxidative mechanisms, primarily hydroxylation, on a wide range of substrates. CYPBM3 is probably the most promising member of this family because its reactive heme binding subdomain is naturally linked to its electron transfer partner. Huang and coworkers could redesign the binding site of CYPBM3 thanks to a preliminary dockings followed by full characterisation by X-ray of A82F and A82W mutants; two excellent catalysts for indole hydroxylation.[115] The redesign of the binding site cavity obtained in these mutant leads to an increase of affinity for the indole substrates of about ∼800-fold and similarly increased catalytic efficiency. Hu *et al.* manipulated the same species to reach a D168W mutation that lead to an improved regioselectivity for indole hydroxylation and optimised the production of indirubin. They rationalise the effect of the D168W by combining HM with protein–ligand dockings.[116]

Homology modelling and protein–ligand dockings are excellent companions when the starting material for design is low or fast molecular insight is sufficient or even necessary. But docking does not allow the accurate description of the catalytic mechanisms. They represent interesting explorative works and at the most a relevant starting point of a more accurate modelling.[117] Recent examples show the pivotal role of protein–ligand docking in a multilevel protocol augmented by MD and QM calculations for the design of new Diels–Alderases.[118,119] This is also the case of the application of last generation docking algorithm like PELE merged with MD and QM/MM and applied on laccase engineering.[120]

15.5.2.2 Molecular Dynamics

Molecular Dynamics is frequently included in redesigning protocols due, in between others, to its ability to identify the quality of the (pre-)organisation of the binding site and the stability of an entire scaffold.

One of the most illustrative examples is the framework developed by Wijma *et al.* FRESCO (Framework for Rapid Enzyme Stabilisation by Computational libraries).[121] At initial stages, structure predicting techniques are executed to generate a library of mutants with promising designs. Next, this library is reduced through successive screening steps and the resulting pool of systems is experimentally assessed to find out stable mutants. Further combinations of successful mutations are implemented to obtain highly stabilised specimens.[121] Amongst promising results, FRESCO led to a series of stabilising mutations and functional variants of the mesostable enzyme limonene epoxide hydrolase (Figure 15.12).

The Houk's group now intensively involve MD calculations in their quest for enzyme designs.[122] Recently, they investigated ketoreductases (KRED), a family of enzymes commonly used to synthetise alcohols from ketones to generate enantiopure chiral alcohols. The authors compared a mixed protocol with QM theozyme and long MD with experimental results obtained on a series of KRED mutants. Their most significant results demonstrated that certain mutations affect the conformational space of the substrate-binding loop that consequently influence on the stereochemical features of the substrates involved in the reaction. Other simulations focused on MD, with or without QM or docking inputs, have also been applied for asymmetric biocatalysis.[123–125]

Although MD implementations have been useful in the rationalisation of enzyme design in particular when it comes to identify failures in active site organisation, information related to transition state structures remains qualitative. QM and QM/MM methodologies provide with the identification of transition energies under an accurate framework.

15.5.2.3 QM and QM/MM

Despite the high number of QM studies on enzyme cluster models, their applications on system with biocatalytic application are rather limited. On the basis of what has been described in the previous sections and chapters, one could argue that long-range effects of mutations on the (pre-)organisation of the enzyme makes these models too limitative. However, the

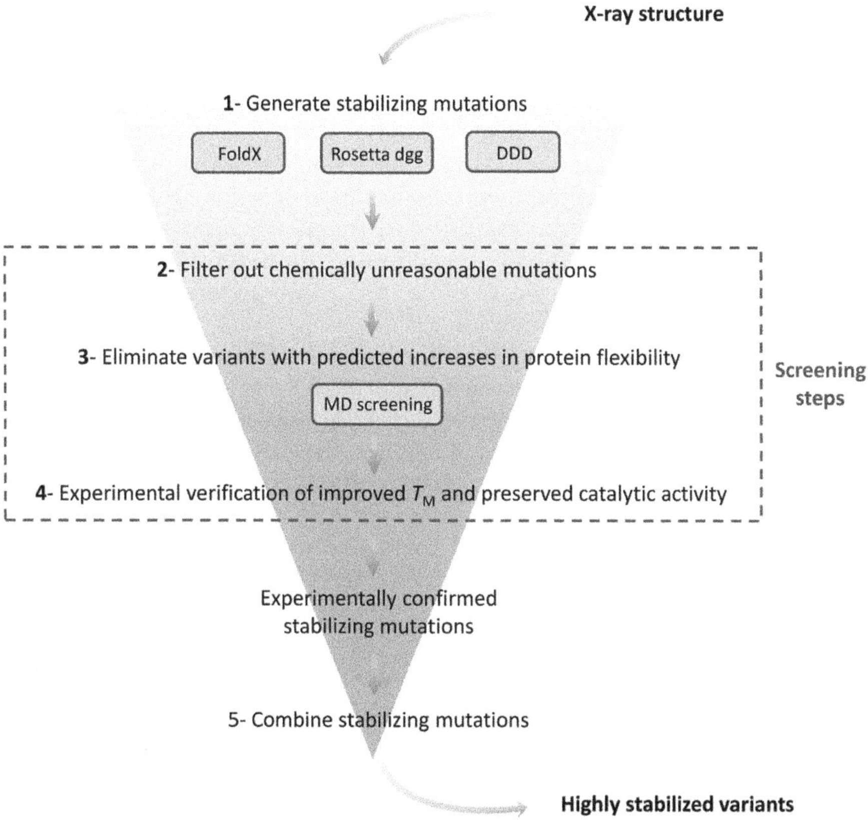

Figure 15.12 The FRESCO workflow.

proof-of-concept that QM clusters could also be valuable in biocatalytic designs have been brought by Himo *et al.* on the limonene-1,2-epoxide hydrolase (Figure 15.7). Their DFT calculations were able to reproduce the experimental enantioselectivity for both the wild-type and mutated versions of the enzyme.[52] Another successful example was reported by Valdez *et al.* who used DFT calculations on a cluster representing the active site of a β-lactamase and showed that this enzyme could be improved by the replacement of zinc by nickel.[126]

QM/MM approaches are more frequently applied than QM clusters in the field of biocatalysis and enzyme design. A substantial number of applications have focused on Kemp eliminases. Houk and coworkers compared hybrid QM/MM approaches, QM clusters and MD simulations to analyse the reason of inactive scaffold positively identified as interesting candidates by RosettaDesign.[91] On the basis of the cost of the calculation of quantum based approaches and the importance of correct physicochemical preorganisation of the enzyme, MD protocol was established as the more adequate for the prediction on active scaffolds. Warshel *et al.* worked on the same systems as a benchmark to their Empirical Valence Bond approach. With an excellent ratio of success, their calculations highlight the interest of involving wide sampling at reduced computational cost to efficient account with fine electrostatic effects and so screen *de novo* candidates from less energetically accurate procedures.[127] Finally, Tuñon and coworkers work on one particularly efficient design, HG3.17 by a QM/MM MD approach. Of the most important results of their work is the demonstration that the catalytic properties of the enzyme are maximised by conformations that are better electrostatically prepared for the reaction and that identifying those conformations is not necessarily accessible in the interpretation of X-ray data but could be revealed by computation.[128]

With the apparent relevance of binding processes and enzyme flexibility in enzyme design, QM based approaches are likely to find their niche when gathering with docking and large scale dynamics. Works that combine the simulation of substrate binding and catalytic mechanisms are likely to become increasingly spread in our community. Recent studies by Acebes *et al.*[129] on a manganese peroxidase, Ni *et al.*[124] on Lipase A and Muñoz Robles *et al.*[130] on different kinds of enzymes illustrate this aspect.

15.5.2.4 Others

Rational redesign of proteins as well as making reasonable predictions on how these changes may affect the final protein structure, activity or substrate specificity and selectivity continue being an enormous challenge. Aiming to somehow avoid drawbacks, several computational tools and integrative procedures have emerged.

One of them is OptZyme, which attempts to lower the TS barriers occurring in an enzyme throughout its rational redesign. This tool was used to improve the enzymatic activity of GUS, the *Escherichia coli* b-glucuronidase, for two similar substrates.[131] The procedure was able to specifically detect different enzyme–ligand contacts for both cases and efficiently predicted mutations which enhanced activity, stabilisation and packing of the active site.

Obviously, Rosetta represents a major strategy for redesign prospect. Further than the pioneer cases of Baker's lab, another success to mention is the redesign of the substrate specificity of kumamolisin-As from *Alicyclobacillus sendaiensis* by Gordon *et al.*[132] The resulting enzyme exhibited a 116-fold greater proteolitic activity and 800-fold increase in substrate specificity for immunogenic portions of gluten peptides, becoming a potential system for therapeutic for celiac disease.

15.5.3 Artificial Metalloenzymes

15.5.3.1 Particularities

Part of the research in designing non-natural metalloenzymes have been centred on reconfiguring the metal binding site or identifying new spots in a protein scaffold that could host a metal ion.[54–56] The development of ArM based on the insertion of homogenous catalysts inside a macromolecular host is still highly limited and has not been extensively studied by computational means so far. Remarkably, no *in silico de novo* design has been successfully performed yet. In fact, *de novo* ArM have always been designed on experimental trials so far. This is mainly due to the complexity to predict the complementarity between three different molecular entities -cofactor, host and substrate- while 'standard' *de novo* enzymes only account with protein–substrate recognition and activation. Moreover, dealing with homogenous catalysts implies possible changes in the coordination of the metal during the binding process or during la reaction with the substrate, something far more complicated to model than for pure organic systems. Finally, the parametrisation of metal ions in simplified energetic calculations like force fields and scoring functions is not a trivial task.

We can separate the way that molecular modelling has been involved with ArM in three different types of events: the prediction of the binding of the cofactor to its receptor, the prediction of the binding of the substrate to the resulting biohybrid and the activation and/or process that takes place in a catalytically competent ArM. Interestingly, calculations have also allowed to shed light on several bioinorganic aspects that were not considered so far because absent from natural systems.

15.5.3.2 Cofactor Binding and Protein–Ligand Dockings

A fundamental step in the design of ArM is the identification of a convenient resting state: a geometry in which the metal has a stable configuration but needs proper activation so that the reaction starts (entrance of the substrate in the active site, movement of a given residue, electronic switch, *etc.*). Those resting states are efficient when at least the first coordination sphere of the metal is well adapted to the host in which the homogenous catalyst binds. Failure at this stage may lead to unsuccessful designs of ArM and is a direct consequence of the lack of evolution pressure in the formation of the host–cofactor–substrate triad.[45,133] For this purpose, strategies based on protein–ligand dockings are of the most interesting.

Despite their relevance in drug design projects, protein–ligand dockings have barely been introduced in the prediction of the interaction of co-ordination compounds with proteins. One of reasons is the complexity in modelling dynamical effects of the first coordination sphere of the metal during the formation of the complex. However, several workarounds could be found by breaking down the way a homogenous catalyst could interact with a biological host.

One scenario is to consider the first coordination sphere of the homogenous compound insensitive to ligand exchanges or changes in the electronic state of the metal occur upon binding. From a protein–ligand docking perspective, these 'inert' interactions, as defined by Meggers,[134] could be modelled rather easily using a simple 'dummy atom' that maintains the scaffold of the organometallic first coordination sphere but has no electrostatic or coordination interaction. This smart tuning of protein–ligand dockings programs apparently suffices and we showed recently they are able to produce the same accuracy with organometallic systems than with standard drug like molecules.[133] A second scenario consists in considering that (1) the metal accounts with labile ligand(s) in one or several coordination sites (*i.e.* providing from the elimination of weak ligand like water and chloride ions), (2) these sites could interact with adequate protein residues (typically histidine, glutamate, aspartate) but (3) the main electronic and geometric features of the artificial cofactor is not deteriorated during the binding. As protein–ligand dockings only account with one coordination state at the time, one need to predict (*i.e.* by experimental knowledge and/or QM based calculations) what sites of the isolated cofactor could exchange with proteic ligands. Then calculations can be carried out with a model of a vacant coordination site (*i.e.* hydrogen bond like functions) so to detect interaction with Lewis basis of the binding site during the docking run.[135] Despite their simplicity, both approximations actually reproduce most of the few ArM and metallodrugs systems that have been reported in the Protein Data Bank and mainly reside on solving technical problems. Their main advantages is to allow to reach crucial information on molecular features like shape, hydrophobic and hydrogen bonding complementarity. Docking essays performed under these schemes have been fundamental in the investigation of the mechanism of a series of ArM like artificial oxidases and asymmetric transfer hydrogenases.[136,137]

On the former, our group predict the 3D structure of complexes resulting from the binding of iron and manganese porphyrinic complexes in the binding cavities of Xylanase A (Xln10A) from *Streptomyces lividans* and that lead to a biohybrid with promising peroxidase activities.[138] These studies demonstrate that the decrease of catalytic power of the different candidates in function of the substitution of the porphyrin rings was due to different interactions and orientations of the artificial cofactor in the Xln10A cleft (Figure 15.13). We also showed that Xln10A was not a good receptor for metallic Schiff bases and that a particular Mn-Porphyrin derivative (Mn(TpCPP)) afforded extremely good cavities to accommodate a substrate for the epoxidation of the series of aromatic styrenes. Interestingly, one of the residues of the receptor (Arg139) is identified to control the access of the substrates.[139] This docking strategy was expanded to other scaffolds like an engineered Neocarzinostatine (NCS), a 113 amino acid chromoprotein secreted by *Streptomyces* and an peroxidase based on the abzyme concept.

The 'worst' case scenario in the binding of artificial cofactor to protein happens when changes of the first coordination sphere of the metal, its electronic state and the structure of the cofactor and/or the protein occur during binding. We have recently showed that an integrative procedure that combines docking, structural statistics and quantum mechanical based calculations could be highly relevant. The benchmark study of this approach

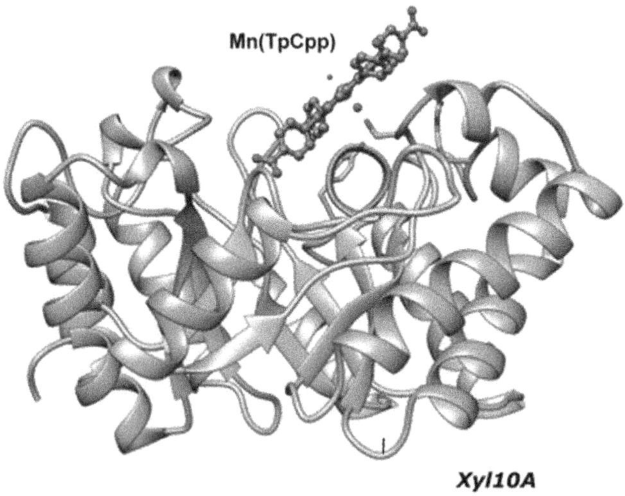

Figure 15.13 General view of the lowest energy predicted model of the artificial metalloenzymes host–cofactor complexes of the supramolecular anchoring of a manganese porphyrin (concretely Mn(II)TCPP) to the Xylanase10A.[138]

was performed on an artificial metalloenzyme obtained by the substitution of the heme by a Fe(Schiff base) salophen in *Corynebacterium diphtheria* heme oxygenase (cdHO).[45] Probably aimed at mimicking peroxidases, this novel Fe(Schiff base)-cdHO enzyme is in fact a superoxidase and is not able to oxidise any substrate. The calculations showed that the flexibility of the helices that embed the cofactor binding site and the distribution the polar patches next to the position of the metal are responsible of an unreported and surprising highly distorted cofactor observed in the crystal structure. There the metal presents a non-conventional octahedral geometries were two residues of the receptor (a histidine and a glutamate) bind in one axial and equatorial position respectively. We also showed that this geometry represents the true resting state of the system. Modelling gives here important insight on molecular processes of ArMs that differ tremendously from the natural hemoenzymes there are aimed at mimicking and condusing to novel consideration in enzyme design.[133,135]

The same scheme was more recently applied on the artificial imine reductases designed by Ward and coworkers and resulting from the incorporation of a biotinylated Cp*Ir-Noyori's like catalyst (Cp* = C5Me5) within different mutants of the homotetrameric streptavidin (Sav) (referred to as Cp*Ir(Biot-*p*-L)Cl] ⊂ Sav). The computation filled the gap between the X-ray structures regarding the position and orientation of the homogenous catalyst in the cavity of the host and sustains that the two more reactive mutants S112A and S112K Sav mutants prefer binding the SIr and RIr enantiomeric forms of the cofactor, respectively; a phenomenon not observed on natural hemoenzymes.[137]

15.5.3.3 Mechanistic Study and Tuning of Artificial Metalloenzymes

The prediction of the binding of an artificial cofactor is an important aspect in understanding the activity of ArM. Nonetheless, only the knowledge of how the substrate interacts with the binding site of the enzymes, and the reaction finally proceeds, has impact on catalytic questions relevant for biocatalytic prospects.

What range of substrates can be metabolised by an artificial metalloenzyme is a natural question when an efficient scaffold has been characterised. The work of Ménage and coworkers based on Nika loaded with a series of iron complexes is a perfect example on how protein–ligand docking strategies allow identifying these active molecules. Taking advantage of the stability of the Nika scaffold to bind iron complexes, the macromolecule was enriched with Fe(II)N2Py2 systems. After resolving the structure of one of their candidates by X-ray analysis, a massive computational screening was performed to identify potential substrates. The binding site of the substrates and those with higher catalytic potential were identified and tested experimentally with good reproduction of the substrate preferences displayed by the computation.[39,140]

Substrate specificity is only a part of the interest of ArM; enantioselectivity is another. We tested until which level docking could be useful to this end on one of the best composites mentioned earlier and resulting from the insertion of Mn(III)-*meso*-tetrakis(*p*-carboxyphenyl)porphyrin (Mn(TpCPP)) into Xln10A. As the structure of Mn(TpCPP)-Xln10A could not be resolved experimentally, we had to rely on structures generated by cofactor docking. The dockings of a series of substrates into this model lead to binding orientations consistent with the formation of the *S*-epoxide product over the *R*-epoxide in all cases except for the *p*-methoxystyrene for with the trend is inverted. Those observations are in good agreement with the *ee* observed experimentally. Such inversion is associated with an additional H-bond between tyrosine 172 and the oxygen atom of the *p*-methoxy substituent (Figure 15.14).[139]

Screening for chemogenetic changes of the biohybrid is the natural step forward in order to control enantioselective profiles. Recently, Ward and Baker tuned the [(η5-Cp*)Ir(pico)Cl] ⊂ WT human carbonic anhydrase II (Cp* = C5Me5), for the asymmetric reduction of a cyclic imine, based on a crystal structure of the biohybrid and combined with the explorative power of RosettaDesign. The resulting best candidates presented a notable improvement of the artificial metalloenzyme both in term of enantioselectivy, with a change from 70% (*R*) to 96% *ee* (*S*) and an increase from 9 to 59 TON.[141]

Obviously, docking approaches or screening chemogenetic strategies are not aimed at identifying true transition state structures.[117] Their application only enlightens us with orientations of substrates that could be catalytically relevant or mutational modifications that could be worthy to explore for

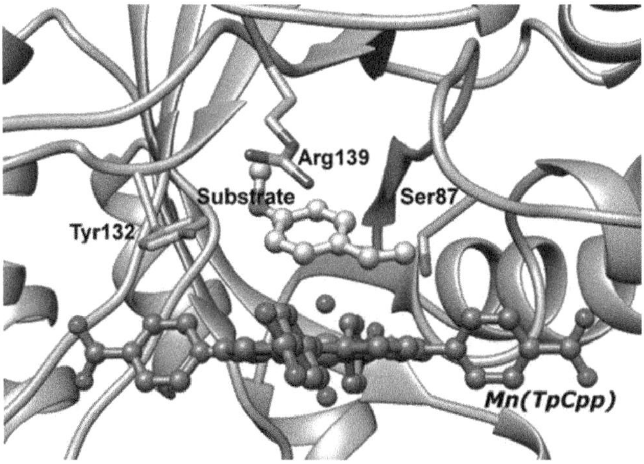

Figure 15.14 Predicted low energy complexes of the *p*-methoxystyrene inside the cavity of the Mn(TpCPP))-Xyl10A model that is catalytically consistent with the formation of *R* epoxide product.[139]

catalytic purposes. True pathways and transition state structures have to be performed by QM means. The first study of ArM under a QM/MM protocol was proposed by Morokuma and coworkers. They analysed with the ONIOM scheme the mechanism of the polymerisation of phenylacetylene by a [Rh(norbornadiene)Cl]2 compound embedded in a horse L-chain apo-Ferritin. After investigating different reaction mechanisms, the authors were able to characterise the most likely cavity in the apo-Ferritin structure that could shelter the inorganic cofactor and favor the polymerisation reaction. Amongst the interesting facts was the necessity of a substantial change of the location of the reactive sphere and the migration of the cofactor and the substrate in different pockets of the host from the starting structure. Luckily this could happen naturally along the optimisation process but this shed lights on the importance to well explore how the homogenous catalyst and the substrate are located inside the host.[142]

But yet calculations under the QM/MM umbrella require a clear idea on where and how stands the substrate. With systems like ArM, where none or very little information is available on the geometry of the reactive host–cofactor–substrate triad, novel protocols are needed so that energetically relevant pathways could be identified. Muñoz and coworkers established a first procedure that bridges QM and QM/MM with protein–ligand dockings to this end.[34] Reminiscent in concepts in *de novo* designs developed by Houk, Mayo and Baker, the pipeline accounts with three differentiated steps: (1) The identification of the energetic paths and transition state structures of the homogenous catalysis in a cluster model of the enzyme by DFT calculations (2) a docking of the cofactor-substrate geometries of the transition state structures obtained in 1 in the active site of the host and (3) the pseudo-transition states generated in 2 are treated with QM/MM calculations first to identify real transition state structure in the PES landscape of the entire biohybrid and the entire reaction paths are then determine throughout IRC calculations (Figure 15.15).

Our first study with this methodology was performed on the catalytic mechanism of the artificial [Cp*Ir(Biot-p-L)Cl] ⊂ S112A transfer hydrogenase mutants and aimed at understanding the molecular grounds of the change in enantioselective profile between them in the reduction of imine.[137] The study implied several pathways in isolated and enzymatic media that account from different sources of protons (a hydronium from the medium or a positively charged lysine that active site could afford with), stepwise and concerted mechanisms and both production of *R* and *S* reduced imine. The final lowest energy paths for the formation of *R* and *S* products clearly indicate preference towards the formation of the former. The final exercise, somehow optimistic, was to calculate the *ee* based on the difference in energy of the transition states between both paths and compare it with the experimental one. Interestingly, the calculated value reaches 80%; a magnitude in good agreement with the experimental 98%. Importantly, the calculation showed that the predicted transition states structures of *R* and *S* mechanisms display a large drift of the cofactor and the substrate into the SAV cavity. So far, this mechanism has never been observed in natural hemoenzymes. At the most, in Nature, the cofactor shows little conformational changes in the overall reaction pathway, only implying subtle arrangements induced by electronic modifications along the activation and catalytic steps of the reaction.

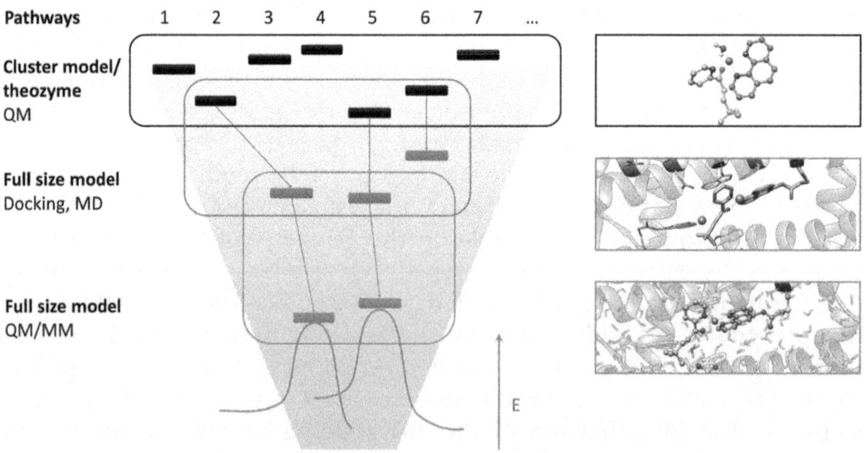

Figure 15.15 Schematic procedure for the identification of catalytic mechanism inside a proteic scaffold of artificial metalloenzymes considering the uncertainty of the location of the cofactor and the formation of a catalytically competent host–cofactor–substrate triad.[130]

15.6 Conclusion and Perspectives

Looking at Nature's molecular interplay is one of the most exciting visions that could reach (bio)chemical eyes. But as we learn more and more on how Nature manages to transform matter, our anxiety to use it for our own technological purposes increases. The most established way to reach non-natural enzymes consists in modifying those that the biological realm already accounts for. Many bioengineered enzymes have already reached industries. However, it is not easy to surpass millions of years of evolution and redesigning natural enzymes does not cover all the needs of chemical industries. Designing pure *de novo* enzymes with reactivity that are absent from biology is one clear objective of biocatalysis. Although our computational skills in this field are still to improve, proofs-of-concept are there and the first successful design of enzymes from 'almost' zero has been reached. From small scaffolds to ArM, in the last years, designers have based most of their work on (bio)chemical intuitions and structural information. Not accidentally, designers have increasingly involved computers in their work.

Designing novel enzymes can highly benefit from molecular modelling but it is undeniable that accounting with all the energetic and structural aspects that need to be considered for these designs represents a major tour de force. Stand-alone methodologies can shed light on some molecular aspects and trigger discovery and/or optimisation of non-natural enzymes. However, the future of this field seems clearer when considering multi-scale approaches. Computational chemists and biophysicists are likely to enter in a new dimension in which creativity will be as important as accuracy and finding convenient combinations of methods will likely be driven by ingenious and easy flowing programming focused on understanding and predicting novel reactivity. In this chapter, we intended to describe all the general approaches that are used experimentally or in combination with computational methods for novel enzymes. We also intended to focus on the most important aspects that computational chemistry bring into the game and describe some of the most illustrative *in silico* work on designed enzymes so far.

A lot is still to do and it is with expectation that we should look at the role of computation in this exciting field. The route is paved with uncertainties but enthusiasm and stubbornness in overcoming methodological limits are likely to guide us toward a fantastic experience: having fun at challenging Nature.

Acknowledgements

We are thankful for the support given by the Spanish grant CTQ2014-54071-P, the Generalitat de Catalunya grant 2014SGR989. Support of COST Action CM1306 is kindly acknowledged. LAC and JRGP thank the Generalitat de Catalunya for its support to their PhD grants.

References

1. N. Furnham, N. L. Dawson, S. A. Rahman, J. M. Thornton and C. A. Orengo, *J. Mol. Biol.*, 2015, **428**, 253.
2. B. G. Davis and V. Boyer, *Nat. Prod. Rep.*, 2001, **18**, 618.
3. B. M. Nestl, S. C. Hammer, B. A. Nebel and B. Hauer, *Angew. Chem., Int. Ed.*, 2014, **53**, 3070.
4. M. T. Reetz, *J. Am. Chem. Soc.*, 2013, **135**, 12480.
5. U. T. Bornscheuer, G. W. Huisman, R. J. Kazlauskas, S. Lutz, J. C. Moore and K. Robins, *Nature*, 2012, **485**, 185.
6. H. T. Brown and T. A. Glendinning, *J. Chem. Soc., Trans.*, 1902, **81**, 388.
7. S. A. Ansari and Q. Husain, *Biotechnol. Adv.*, 2012, **30**, 512.
8. J. N. Talbert and J. M. Goddard, *Colloids Surf., B*, 2012, **93**, 8.
9. R. A. Sheldon and S. Van Pelt, *Chem. Soc. Rev.*, 2013, **42**, 6223.
10. H. Wei and E. Wang, *Chem. Soc. Rev.*, 2013, **42**, 6060.
11. U. T. Bornscheuer and R. J. Kazlauskas, *Angew. Chem., Int. Ed.*, 2004, **43**, 6032.
12. F. Cedrone, A. Ménez and E. Quéméneur, *Curr. Opin. Struct. Biol.*, 2000, 405.
13. T. W. Johannes and H. Zhao, *Curr. Opin. Microbiol.*, 2006, **9**, 261.
14. F. H. Arnold, *Trends Biochem. Sci.*, 2009, **74**, 41.
15. F. Valetti and G. Gilardi, *Nat. Prod. Rep.*, 2004, **21**, 490.
16. B. G. Davis, *Curr. Opin. Biotechnol.*, 2003, **14**, 379.
17. B. Gibb, K. Gupta, K. Ghosh, R. Sharp, J. Chen and G. D. van Duyne, *Nucleic Acids Res.*, 2010, **38**, 5817.
18. S. Lutz, *Curr. Opin. Biotechnol.*, 2010, **21**, 734.
19. A. G. Tebo and V. L. Pecoraro, *Curr. Opin. Chem. Biol.*, 2015, **25**, 65.
20. J. C. Lewis, *Curr. Opin. Chem. Biol.*, 2015, **25**, 27.
21. M. M. Sheehan, N. M. Ennist, B. A. Fry, S. E. Chobot, J. A. Mancini, C. T. Armstrong, Z. Zhao, T. V. Esipova, S. A. Vinogradov, B. M. Discher, C. C. Moser and L. P. Dutton, *Biochem. Soc. Trans.*, 2013, **40**, 561.
22. Q.-H. Dai, C. Tommos, E. Fuentes, M. Blomberg, P. Dutton and A. J. Wand, *J. Am. Chem. Soc.*, 2002, **124**, 10952.
23. P. G. Schultz, *Angew. Chem., Int. Ed. Engl.*, 1989, **28**, 1283.
24. F. Richter, A. Leaver-Fay, S. D. Khare, S. Bjelic and D. Baker, *PLoS One*, 2011, **6**, e19230.
25. G. Kiss, Celebi-Ölçüm, R. Moretti, D. Baker and K. N. Houk, *Angew. Chem., Int. Ed.*, 2013, **52**, 5700.
26. O. Khersonsky, D. Röthlisberger, A. M. Wollacott, P. Murphy, O. Dym, S. Albeck, G. Kiss, K. N. Houk, D. Baker and D. S. Tawfik, *J. Mol. Biol.*, 2011, **407**, 391.
27. A. Pordea, *Curr. Opin. Chem. Biol.*, 2015, **25**, 124.
28. J. Liu, K. K. Meier, S. Tian, J. L. Zhang, H. Guo, C. E. Schulz, H. Robinson, M. J. Nilges, E. Münck and Y. Lu, *J. Am. Chem. Soc.*, 2014, **136**, 12337.
29. X. Liu, Y. Yu, C. Hu, W. Zhang, Y. Lu and J. Wang, *Angew. Chem., Int. Ed.*, 2012, **51**, 4312.

30. M. Hoarau, C. Hureau, E. Gras and P. Faller, *Coord. Chem. Rev.*, 2015, **308**, 445.
31. M. R. Ringenberg and T. R. Ward, *Chem. Commun.*, 2011, **47**, 8470.
32. J. Bos, F. Fusetti, A. J. M. Driessen and G. Roelfes, *Angew. Chem., Int. Ed.*, 2012, **51**, 7472.
33. O. Pàmies, M. Diéguez and J. E. Bäckvall, *Adv. Synth. Catal.*, 2015, **357**, 1567.
34. V. Muñoz Robles, E. Ortega-Carrasco, L. Alonso-Cotchico, J. Rodriguez-Guerra, A. Lledós and J.-D. Maréchal, *ACS Catal.*, 2015, **5**, 2469.
35. A. Chatterjee, H. Mallin, J. Klehr, J. Vallapurackal, A. Finke, L. Vera, M. Marsh and T. R. Ward, *Chem. Sci.*, 2015, **7**, 673.
36. J. Zhao, A. Kajetanowicz and T. R. Ward, *Org. Biomol. Chem.*, 2015, **13**, 5652.
37. D. F. Sauer, T. Himiyama, K. Tachikawa, K. Fukumoto, A. Onoda, E. Mizohata, T. Inoue, M. Bocola, U. Schwaneberg, T. Hayashi and J. Okuda, *ACS Catal.*, 2015, **5**, 7519.
38. P. Srivastava, H. Yang, K. Ellis-Guardiola and J. C. Lewis, *Nat. Commun.*, 2015, **6**, 7789.
39. C. Esmieu, M. V. Cherrier, P. Amara, E. Girgenti, C. Marchi-Delapierre, F. Oddon, M. Iannello, A. Jorge-Robin, C. Cavazza and S. Ménage, *Angew. Chem., Int. Ed.*, 2013, **52**, 3922.
40. H. Inaba, S. Kanamaru, F. Arisaka, S. Kitagawa and T. Ueno, *Dalton Trans.*, 2012, **41**, 11424.
41. V. Köhler, Y. M. Wilson, C. Lo, A. Sardo and T. R. Ward, *Curr. Opin. Biotechnol.*, 2010, **21**, 744.
42. M. Dürrenberger, T. Heinisch, Y. M. Wilson, T. Rossel, E. Nogueira, L. Knörr, A. Mutschler, K. Kersten, M. J. Zimbron, J. Pierron, T. Schirmer and T. R. Ward, *Angew. Chem.*, 2011, **123**, 3082.
43. G. Roelfes and B. L. Feringa, *Angew. Chem., Int. Ed.*, 2005, **44**, 3230.
44. A. Rioz-Martínez and G. Roelfes, *Curr. Opin. Chem. Biol.*, 2015, **25**, 80.
45. T. Ueno, N. Yokoi, M. Unno, T. Matsui, Y. Tokita, M. Yamada, M. Ikeda-Saito, H. Nakajima and Y. Watanabe, *Proc. Natl. Acad. Sci. U. S. A.*, 2006, **103**, 9416.
46. B. Hartke and S. Grimme, *Phys. Chem. Chem. Phys.*, 2015, **17**, 16715.
47. A. Warshel, *J. Biol. Chem.*, 1998, **273**, 27035.
48. I. Tuñón, D. Laage and J. T. Hynes, *Arch. Biochem. Biophys.*, 2015, **582**, 42.
49. S. Osuna, G. Jiménez-Osés, E. L. Noey and K. N. Houk, *Acc. Chem. Res.*, 2015, **48**, 1080.
50. M. R. a Blomberg, T. Borowski, F. Himo, R.-Z. Liao and P. E. M. Siegbahn, *Chem. Rev.*, 2014, **114**, 3601.
51. P. E. M. Siegbahn and F. Himo, *Wiley Interdiscip. Rev.: Comput. Mol. Sci.*, 2011, **1**, 323.
52. M. E. S. Lind and F. Himo, *Angew. Chem., Int. Ed.*, 2013, **125**, 4661.
53. Q. Q. Hou, X. Sheng, J. H. Wang, Y. J. Liu and C. B. Liu, *Biochim. Biophys. Acta, Proteins Proteomics*, 2012, **1824**, 263.

54. H. M. Senn and W. Thiel, *Angew. Chem., Int. Ed.*, 2009, **48**, 1198.
55. A. Warshel and R. M. Weiss, *J. Am. Chem. Soc.*, 1980, **102**, 6218.
56. M. Gaus, Q. Cui and M. Elstner, *J. Chem. Theory Comput.*, 2011, **7**, 931.
57. M. Sulpizi, A. Laio, J. VandeVondele, A. Cattaneo, U. Rothlisberger and P. Carloni, *Proteins: Struct., Funct., Bioinf.*, 2003, **52**, 212.
58. G. L. Warren, C. W. Andrews, A.-M. Capelli, B. Clarke, J. LaLonde, M. H. Lambert, M. Lindvall, N. Nevins, S. F. Semus, S. Senger, G. Tedesco, I. D. Wall, J. M. Woolven, C. E. Peishoff and M. S. Head, *J. Med. Chem.*, 2006, **49**, 5912.
59. J. Maréchal, J. Yu, S. Brown, I. Kapelioukh, E. M. Rankin, C. R. Wolf, G. C. K. Roberts, M. J. I. Paine and M. J. Sutcliffe, *Drug. Metab. Dispos.*, 2006, **34**, 534.
60. L. Toledo, L. Masgrau, J.-D. Maréchal, J. M. Lluch and A. González-Lafont, *J. Phys. Chem. B*, 2010, **114**, 7037.
61. R. E. Amaro, R. Baron and J. A. McCammon, *J. Comput. -Aided Mol. Des.*, 2008, **22**, 693.
62. S. K. Lüdemann, V. Lounnas and R. C. Wade, *J. Mol. Biol.*, 2000, **303**, 813.
63. X. Liu, X. Wang and H. Jiang, *J. Biochem. Biophys. Methods*, 2008, **70**, 857.
64. A. N. Lima, E. A. Philot, D. Perahia, A. S. K. Braz and L. P. B. Scott, *Appl. Math. Comput.*, 2012, **219**, 511.
65. K. W. Borrelli, A. Vitalis, R. Alcantara and V. Guallar, *J. Chem. Theory Comput.*, 2005, **1**, 1304.
66. C. N. Cavasotto, J. A. Kovacs and R. A. Abagyan, *J. Am. Chem. Soc.*, 2005, **127**, 9632.
67. J. Abbass and J.-C. Nebel, *BMC Bioinf.*, 2015, **16**, 136.
68. B. Webb and A. Sali, Comparative protein structure modeling using MODELLER, in *Current Protocols in Bioinformatics*, John Wiley & Sons, Inc., U. S. A., 2014, p. 5.6.1.
69. F. C. Bernstein, T. F. Koetzle, G. J. Williams, E. E. Meyer Jr., M. D. Brice, J. R. Rodgers, O. Kennard, T. Shimanouchi and M. Tasumi, *J. Mol. Biol.*, 1977, **112**, 535.
70. A. Fiser, R. K. Do and A. Sali, *Protein Sci.*, 2000, **9**, 1753.
71. C. A. Rohl, C. E. M. Strauss, K. M. S. Misura and D. Baker, *Methods Enzymol.*, 2004, **383**, 66.
72. K. W. Kaufmann, G. H. Lemmon, S. L. Deluca, J. H. Sheehan and J. Meiler, *Biochemistry*, 2010, **49**, 2987.
73. R. L. Dunbrack and M. Karplus, *Nat. Struct. Biol.*, 1994, **1**, 334.
74. A. D. Scouras and V. Daggett, *Protein Sci.*, 2011, **20**, 341.
75. D. Gfeller, O. Michielin and V. Zoete, *Nucleic Acids Res.*, 2013, **41**, 327.
76. G. A. Khoury, J. P. Thompson, J. Smadbeck, C. A. Kieslich and C. A. Floudas, *J. Chem. Theory Comput.*, 2013, **9**, 5653.
77. G. A. Khoury, J. Smadbeck, P. Tamamis, A. C. Vandris, C. A. Kieslich and C. A. Floudas, *ACS Synth. Biol.*, 2014, **3**, 855.

78. D. Douguet, E. Thoreau and G. Grassy, *J. Comput. -Aided Mol. Des.*, 2000, **14**, 449.
79. M. Hartenfeller, H. Zettl, M. Walter, M. Rupp, F. Reisen, E. Proschak, S. Weggen, H. Stark and G. Schneider, *PLoS Comput. Biol.*, 2012, **8**, e1002380.
80. H. C. Kolb, M. G. Finn and K. B. Sharpless, *Angew. Chem., Int. Ed.*, 2001, **40**, 2004.
81. D. J. Tantillo, J. Chen and K. N. Houk, *Curr. Opin. Chem. Biol.*, 1998, **2**, 743.
82. H. M. Berman, *Nucleic Acids Res.*, 2000, **28**, 235.
83. H. W. Hellinga and F. M. Richards, *J. Mol. Biol.*, 1991, **222**, 763.
84. B. I. Dahiyat and S. L. Mayo, *Protein Sci.*, 1996, **5**, 895.
85. A. Zanghellini, L. Jiang, A. M. Wollacott, G. Cheng, J. Meiler, E. A. Althoff, D. Röthlisberger and D. Baker, *Protein Sci.*, 2006, **15**, 2785.
86. H. Fazelinia, P. C. Cirino and C. D. Maranas, *Protein Sci.*, 2009, **18**, 180.
87. C. Malisi, O. Kohlbacher and B. Höcker, *Proteins*, 2009, **77**, 74.
88. Y. Lei, W. Luo and Y. Zhu, *Protein Sci.*, 2011, **20**, 1566.
89. G. R. Nosrati and K. N. Houk, *Protein Sci.*, 2012, **21**, 697.
90. N. D. Clarke and S. M. Yuan, *Proteins*, 1995, **23**, 256.
91. G. Kiss, D. Rothlisberger, D. Baker and K. N. Houk, *Protein Sci.*, 2010, **19**, 1760.
92. D. N. Bolon and S. L. Mayo, *Proc. Natl. Acad. Sci. U. S. A.*, 2001, **98**, 14274.
93. D. Röthlisberger, O. Khersonsky, A. M. Wollacott, L. Jiang, J. DeChancie, J. Betker, J. L. Gallaher, E. A. Althoff, A. Zanghellini, O. Dym, S. Albeck, K. N. Houk, D. S. Tawfik and D. Baker, *Nature*, 2008, **453**, 190.
94. L. Jiang, E. A. Althoff, F. R. Clemente, L. Doyle, D. Rothlisberger, A. Zanghellini, J. L. Gallaher, J. L. Betker, F. Tanaka, C. F. Barbas, D. Hilvert, K. N. Houk, B. L. Stoddard and D. Baker, *Science*, 2008, **319**, 1387.
95. J. B. Siegel, A. Zanghellini, H. M. Lovick, G. Kiss, A. R. Lambert, J. L. Clair, J. L. Gallaher, D. Hilvert, M. H. Gelb, B. L. Stoddard, K. N. Houk, F. E. Michael and D. Baker, *Science*, 2010, **329**, 309.
96. K. Świderek, I. Tuñón, V. Moliner and J. Bertran, *Arch. Biochem. Biophys.*, 2015, **582**, 68.
97. E. A. Althoff, L. Wang, L. Jiang, L. Giger, J. K. Lassila, Z. Wang, M. Smith, S. Hari, P. Kast, D. Herschlag, D. Hilvert and D. Baker, *Protein Sci.*, 2012, **21**, 717.
98. R. Obexer, S. Studer, L. Giger, D. M. Pinkas, M. G. Grütter, D. Baker and D. Hilvert, *ChemCatChem*, 2014, **6**, 1043.
99. E. Check Hayden, *Nature*, 2008, **453**, 275.
100. B. List, R. A. Lerner and C. F. Barbas, *J. Am. Chem. Soc.*, 2000, **122**, 2395.
101. M. Merski and B. K. Shoichet, *Proc. Natl. Acad. Sci.*, 2012, **109**, 16179.
102. M. M. Müller, M. A. Windsor, W. C. Pomerantz, S. H. Gellman and D. Hilvert, *Angew. Chem., Int. Ed.*, 2009, **48**, 922.

103. B. Wörsdörfer, L. M. Henning, R. Obexer and D. Hilvert, *ACS Catal.*, 2012, **2**, 982.
104. I. V. Korendovych, D. W. Kulp, Y. Wu, H. Cheng, H. Roder and W. F. DeGrado, *Proc. Natl. Acad. Sci. U. S. A.*, 2011, **108**, 6823.
105. P.-S. Huang, Y.-E. A. Ban, F. Richter, I. Andre, R. Vernon, W. R. Schief and D. Baker, *PLoS One*, 2011, **6**, e24109.
106. J. H. Mills, S. D. Khare, J. M. Bolduc, F. Forouhar, V. K. Mulligan, S. Lew, J. Seetharaman, L. Tong, B. L. Stoddard and D. Baker, *J. Am. Chem. Soc.*, 2013, **135**, 13393.
107. P. D. Renfrew, E. J. Choi, R. Bonneau and B. Kuhlman, *PLoS One*, 2012, **7**, e32637.
108. J. A. Davey and R. A. Chica, *Protein Sci.*, 2012, **21**, 1241.
109. H. K. Privett, G. Kiss, T. M. Lee, R. Blomberg, R. A. Chica, L. M. Thomas, D. Hilvert, K. N. Houk and S. L. Mayo, *Proc. Natl. Acad. Sci. U. S. A.*, 2012, **109**, 3790.
110. R. Blomberg, H. Kries, D. M. Pinkas, P. R. E. Mittl, M. G. Grütter, H. K. Privett, S. L. Mayo and D. Hilvert, *Nature*, 2013, 418–421.
111. S. Cooper, F. Khatib, A. Treuille, J. Barbero, J. Lee, M. Beenen, A. Leaver-Fay, D. Baker, Z. Popović and F. Players, *Nature*, 2010, **466**, 756.
112. C. B. Eiben, J. B. Siegel, J. B. Bale, S. Cooper, F. Khatib, B. W. Shen, F. Players, B. L. Stoddard, Z. Popovic and D. Baker, *Nat. Biotechnol.*, 2012, **30**, 190.
113. C. K. Savile, J. M. Janey, E. C. Mundorff, J. C. Moore, S. Tam, W. R. Jarvis, J. C. Colbeck, A. Krebber, F. J. Fleitz, J. Brands, P. N. Devine, G. W. Huisman and G. J. Hughes, *Science*, 2010, **329**, 305.
114. S. M. Lippow, T. S. Moon, S. Basu, S. H. Yoon, X. Li, B. A. Chapman, K. Robison, D. Lipovsek and K. L. J. Prather, *Chem. Biol.*, 2010, **17**, 1306.
115. W.-C. Huang, A. C. G. Westlake, J.-D. Maréchal, M. G. Joyce, P. C. E. Moody and G. C. K. Roberts, *J. Mol. Biol.*, 2007, **373**, 633.
116. S. Hu, J. Huang, L. Mei, Q. Yu, S. Yao and Z. Jin, *J. Mol. Catal. B: Enzym.*, 2010, **67**, 29.
117. J. Oláh, A. J. Mulholland and J. N. Harvey, *Proc. Natl. Acad. Sci.*, 2011, **108**, 6050.
118. M. Linder, A. J. Johansson, T. S. G. Olsson, J. Liebeschuetz and T. Brinck, *J. Chem. Inf. Model.*, 2011, **51**, 1906.
119. M. Linder, A. J. Johansson, T. S. G. Olsson, J. Liebeschuetz and T. Brinck, *J. Comput. -Aided Mol. Des.*, 2012, **26**, 1079.
120. E. Monza, M. F. Lucas, S. Camarero, L. C. Alejaldre, A. T. Martínez and V. Guallar, *J. Phys. Chem. Lett.*, 2015, **6**, 1447.
121. H. J. Wijma, R. J. Floor, P. A. Jekel, D. Baker, S. J. Marrink and D. B. Janssen, *Protein Eng., Des. Sel.*, 2014, **27**, 49.
122. E. L. Noey, N. Tibrewal, G. Jiménez-Osés, S. Osuna, J. Park, C. M. Bond, D. Cascio, J. Liang, X. Zhang, G. W. Huisman, Y. Tang and K. N. Houk, *Proc. Natl. Acad. Sci. U. S. A.*, 2015, **112**, E7065.
123. P. Braiuca, C. Ebert, A. Basso, P. Linda and L. Gardossi, *Trends Biotechnol.*, 2006, **24**, 419.

124. Z. Ni, P. Zhou, X. Jin and X. F. Lin, *Chem. Biol. Drug Des.*, 2011, **78**, 301.
125. M. T. Reetz, M. Puls, J. D. Carballeira, A. Vogel, K. E. Jaeger, T. Eggert, W. Thiel, M. Bocola and N. Otte, *ChemBioChem*, 2007, **8**, 106.
126. C. E. Valdez and A. N. Alexandrova, *J. Phys. Chem. B*, 2012, **116**, 10649.
127. M. P. Frushicheva, J. Cao and A. Warshel, *Biochemistry*, 2011, **50**, 3849.
128. K. Swiderek, I. Tuñon, V. Moliner and J. Bertran, *ACS Catal.*, 2015, **5**, 2587.
129. S. Acebes, E. Fernandez-Fueyo, E. Monza, M. F. Lucas, D. Almendral, F. J. Ruiz-Dueñas, H. Lund, A. T. Martinez and V. Guallar, *ACS Catal.*, 2016, 1624.
130. V. Muñoz Robles, P. Vidossich, A. Lledós, T. R. Ward and J.-D. Maréchal, *ACS Catal.*, 2014, **4**, 833.
131. M. J. Grisewood, N. P. Gifford, R. J. Pantazes, Y. Li, P. C. Cirino, M. J. Janik and C. D. Maranas, *PLoS One*, 2013, 8.
132. S. R. Gordon, E. J. Stanley, S. Wolf, A. Toland, S. J. Wu, D. Hadidi, J. H. Mills, D. Baker, I. S. Pultz and J. B. Siegel, *J. Am. Chem. Soc.*, 2012, **134**, 20513.
133. E. Ortega-Carrasco, A. Lledós and J.-D. Maréchal, *J. R. Soc., Interface*, 2014, **11**, 20140090.
134. E. Meggers, *Chem. Commun.*, 2009, 1001.
135. V. Muñoz Robles, E. Ortega-Carrasco, E. G. Fuentes, A. Lledós and J.-D. Maréchal, *Faraday Discuss.*, 2011, **148**, 137.
136. J.-P. Mahy, J.-D. Maréchal and R. Ricoux, *Chem. Commun.*, 2014, **51**, 2476.
137. V. Muñoz Robles, M. Dürrenberger, T. Heinisch, A. Lledós, T. Schirmeister, T. R. Ward and J.-D. Maréchal, *J. Am. Chem. Soc.*, 2014, **136**, 15676.
138. R. Ricoux, R. Dubuc, C. Dupont, J.-D. Maréchal, A. Martin and M. Sellier, *Bioconjugate Chem.*, 2008, **19**, 899.
139. M. Allard, C. Dupont, V. Muñoz Robles, N. Doucet, A. Lledós, J.-D. Maréchal, A. Urvoas, J.-P. Mahy and R. Ricoux, *ChemBioChem*, 2012, **13**, 240.
140. C. Cavazza, C. Bochot, P. Rousselot-Pailley, P. Carpentier, M. V. Cherrier, L. Martin, C. Marchi-Delapierre, J. C. Fontecilla-Camps and S. Ménage, *Nat. Chem.*, 2010, **2**, 1069.
141. T. Heinisch, M. Pellizzoni, M. Dürrenberger, C. E. Tinberg, V. Köhler, J. Klehr, D. Häussinger, D. Baker and T. R. Ward, *J. Am. Chem. Soc.*, 2015, **137**, 10414.
142. Z. Ke, S. Abe, T. Ueno and K. Morokuma, *J. Am. Chem. Soc.*, 2012, **134**, 15418.

Subject Index